MATHEMATICS RESEARCH DEVELOPMENTS

CLASSIFICATION AND APPLICATION OF FRACTALS

MATHEMATICS RESEARCH DEVELOPMENTS

Additional books in this series can be found on Nova's website
under the Series tab.

Additional E-books in this series can be found on Nova's website
under the E-book tab.

MATHEMATICS RESEARCH DEVELOPMENTS

CLASSIFICATION AND APPLICATION OF FRACTALS

WILLIAM L. HAGEN
EDITOR

Nova Science Publishers, Inc.
New York

Copyright © 2012 by Nova Science Publishers, Inc.

All rights reserved. No part of this book may be reproduced, stored in a retrieval system or transmitted in any form or by any means: electronic, electrostatic, magnetic, tape, mechanical photocopying, recording or otherwise without the written permission of the Publisher.

For permission to use material from this book please contact us:
Telephone 631-231-7269; Fax 631-231-8175
Web Site: http://www.novapublishers.com

NOTICE TO THE READER

The Publisher has taken reasonable care in the preparation of this book, but makes no expressed or implied warranty of any kind and assumes no responsibility for any errors or omissions. No liability is assumed for incidental or consequential damages in connection with or arising out of information contained in this book. The Publisher shall not be liable for any special, consequential, or exemplary damages resulting, in whole or in part, from the readers' use of, or reliance upon, this material. Any parts of this book based on government reports are so indicated and copyright is claimed for those parts to the extent applicable to compilations of such works.

Independent verification should be sought for any data, advice or recommendations contained in this book. In addition, no responsibility is assumed by the publisher for any injury and/or damage to persons or property arising from any methods, products, instructions, ideas or otherwise contained in this publication.

This publication is designed to provide accurate and authoritative information with regard to the subject matter covered herein. It is sold with the clear understanding that the Publisher is not engaged in rendering legal or any other professional services. If legal or any other expert assistance is required, the services of a competent person should be sought. FROM A DECLARATION OF PARTICIPANTS JOINTLY ADOPTED BY A COMMITTEE OF THE AMERICAN BAR ASSOCIATION AND A COMMITTEE OF PUBLISHERS.

Additional color graphics may be available in the e-book version of this book.

Library of Congress Cataloging-in-Publication Data

Classification and application of fractals / [edited by] William L. Hagen.
p. cm.
Includes index.
ISBN 978-1-61209-967-5 (hardcover)
1. Fractals. 2. Mathematical physics. I. Hagen, William L. (William
Larry), 1969-
QC20.7.F73C53 2011
514'.742--dc22
2011006689

Published by Nova Science Publishers, Inc. † New York

CONTENTS

Preface		**vii**
Chapter 1	Modeling the Cluster Structure of Dissolved Air Nanobubbles in Liquid Media *N. F. Bunkin, S. O. Yurchenk, N. V.Suyazov, A. V. Starosvetskiy, A. V. Shkirin and V. A. Kozlov*	**1**
Chapter 2	Fractal Dynamics of Complex Systems *Oswaldo Morales-Matamoros, Teresa I. Contreras-Troya, Mauricio Flores-Cadena and Ricardo Tejeida-Padilla*	**49**
Chapter 3	Fractal Analysis of Electromagnetic Emissions in Possible Association with Earthquakes *M. Hayakawa, N. Yonaiguchi, Y. Ida, S. Masuda and Y. Hobara*	**83**
Chapter 4	Design of Frequency Selective Surface Using Fractals Geometries *Antonio Luiz Pereira de Siqueira Campos and Paulo Henrique da Fonsêca Silva*	**103**
Chapter 5	Fractal-Based Mathematical Models of Cancer in the Era of Systems Biology *Fabio Grizzi and Irene Guaraldo*	**125**
Chapter 6	Fractal Analysis of Soil Structure *S. H. Anderson*	**137**
Chapter 7	Dynamics of Miscellaneous Fractal Structures in Higher-Dimensional Evolution Model Systems *Victor K. Kuetche, Thomas B. Bouetou and Timoleon C. Kofane*	**149**
Chapter 8	Multifractals: Concepts and Applications *Ashok Razdan*	**247**
Chapter 9	Subgrid Modeling of Steady-State Flow Processes in Fractal Medium *O. N. Soboleva and E. P. Kurochkina*	**289**

vi   Contents

Chapter 10	Generating Euclidean Structures Using IFS with Memory *Michael Frame, Brenda Johnson and Kathleen Meloney*	**329**
Chapter 11	Trace Theorems on Scale Irregular Fractals *Raffaela Capitanelli and Maria Agostina Vivaldi*	**363**
Chapter 12	Physics on the Net Fractals *Zygmunt Bak*	**383**
Chapter 13	Fractal Properties of Solutions of Differential Equations *Mervan Pavsic, Darkov Zubrinic and Vesnav Zupanovic*	**405**
Index		**467**

PREFACE

This new book presents topical research in the study of the classification and application of fractals, including the fractal analysis of oil crude market volatility and supply chain volatility in the telcom industry; a stochastic analysis of fractal properties of clusters composed of stable gas nanobubbles suspended in aqueous electrolyte solutions; the fractal analysis of electromagnetic emissions in earthquake detection; applications of fractal geometries to design Frequency Selective Surfaces (FSS); fractal-based models of cancer in systems biology; soil structure fractal analysis and applications of multifractals in diverse fields like astronomy and the stock market.

Chapter 1 – A stochastic analysis of fractal properties of clusters composed of stable gas nanobubbles suspended in aqueous electrolyte solutions was carried out. The model of nucleating the stable bubbles in water at room temperature was suggested. This model is completely based on the property of the affinity of water at the nanometer scale; it was shown that under certain conditions the extent of disorder in the liquid starts growing, which results in a spontaneous decrease of the local density of a liquid and in the formation of nanometer-sized voids. These voids can capture the molecules of dissolved gas preserving the void from collapse. After adsorption of likely charged ions at the "void – liquid" interface, the latter is being kept from collapse by the Coulomb repulsion forces, acting along the interface. This way the steady bubbles at the nanometer size are generated; these bubbles serve as nuclei for the following generation of the so-termed bubstons (the abbreviation for *bubbles, stabilized by ions*). The model of charging the bubstons by the ions, which are capable of adsorption, and the screening by a cloud of counter-ions, which are incapable of adsorption, was analyzed. It was shown that, subject to the charge of bubston, two regimes of such screening can be realized. At low charge of bubston the screening is described in the framework of the known linearized Debye – Huckel approach, for which the sign of the counter-ion cloud preserves its sign everywhere in the liquid surrounding the bubston, whereas at large charge this sign is changed at some distance from the bubston surface. This effect provides the emergence mechanism for two types of compound particles having the opposite polarity, which leads to the aggregation of such compound particles by a ballistic kinetics. The fractal properties of clusters, generated by consecutive attachments of separate particles under the condition of central falling (i.e. ballistic "particle-cluster" aggregation), were investigated. This analysis was essentially based on experimental data measured by laser polarimetric scatterometry of NaCl aqueous solution. The fractal dimension for such clusters was built up as a function of monomer number. General properties of the light scattering matrix of bubston-cluster ensembles were found. The calculations of the light scattering matrix were performed for random ballistic-type clusters, comprising the bubstons of various sizes, on the assumption that the bubstons have lognormal size distribution with an effective radius 90 nm and effective variance 0.02; to fit the experimental scattering indicatrix of aqueous salt solutions, containing dissolved air, the mean number of bubstons in a cluster was set to 100.

Chapter 2 - One of the most important properties of systems is complexity. In a simple way, the authors can define the complexity of a system in terms of the number of elements that it contains, the nature and number of interrelations, and the number of levels of embeddedness. When a high level of complexity exists in a system, it is considered a complex system. Although there is no single agreed-on definition of complex systems, they share some themes: (i) they are inherently complicated or intricate, so that they have factors such as the number of parameters affecting the system or the rules governing interactions of components of the system; (ii) they are rarely completely deterministic, and state parameters or measurement data may only be known in terms of probabilities; (iii) mathematical models of the system, are usually complex and involve non-linear, ill-posed, or chaotic behavior; and (iv) the systems are predisposed to unexpected outcomes (so-called emergent behavior). To try to understand the dynamics of these systems diverse mathematical tools have been developed. A new scientific discipline with great impact in the analysis of the complex systems has been developed in recent years, called fractal analysis.

The study of the complex systems in the framework of fractal theory has been recognized as a new scientific discipline, being sustained by advances that have been made in diverse fields ranging from physics to economics. In this chapter the basic concepts of fractal analysis of complex systems are briefly explained and three examples of fractal analysis are provided: epilepsy, oil crude price market volatility, and supply chain volatility in the telecom industry.

Chapter 3 - An earthquake (EQ) is known to be a large-scale fracture phenomenon in the Earth's crust and a vital problem in the short-term EQ prediction is the identification of precursors of EQs. When a heterogeneous crust is strained, its nonlinear evolution toward the final rupture is characterized by self-organization toward the critical point including the local nucleation and coalescence of microcracks (i.e., self-organized criticality (SOC)). Both acoustic as well as electromagnetic emissions in a wide frequency range from DC, ULF (Ultra-low-frequency) up to VHF are produced by those microcracks during the preparatory phase of EQs. This nonlinear dynamics in the lithosphere can be extensively investigated with the use of fractal analysis. This paper deals with the reviews on those fractal analyses especially on the two types of seismogenic emissions; one is ULF electromagnetic electromagnetic emissions and the second is VHF electromagnetic noises. Significant changes in the fractal properties of those electromagnetic emissions are found mainly prior to an EQ, which provides a rather promising candidate for predicting EQs.

Chapter 4 - This chapter describes applications of fractal geometries to design Frequency Selective Surfaces (FSSs), where the authors emphasize the use of fractal geometries to improve their frequency responses. Initially, they do a review about fractals and fractal geometries. After this, they describe the L-system and the Iterated Function Systems – IFS to generate fractal geometries. Thereafter, the authors show a review of literature about the use of fractal geometries in FSS design. Then, they present some numerical results of FSS designing using fractal geometries. The obtained numerical results are compared with the experimental ones. Finally, conclusions are listed.

Chapter 5 - Cancer research has undergone radical changes in the past few years. Amount of information both at the basic and clinical levels is no longer the issue. Rather, how to handle this information has become the major obstacle to progress. System biology is the latest fashion in cancer biology, driven by advances in technology that have provided us with a suite of "omics" techniques. It can be seen as a conceptual approach to biological research that combines "reductionist" (parts) and "integrationist" (interactions) research, to understand the nature and maintenance of entities. In geometrical terms, cancerous lesions can be depicted as fractal entities mainly characterized by their irregular shape, self-similar structure, scaling relationship and non-integer or fractal dimension. It is indubitable that The Fractal Geometry of Nature has provided an innovative paradigm, a novel epistemological approach for interpreting the anatomical world. It is also known that mathematical methods and their derivatives have proved to be possible and practical in oncology. Viewing cancer as a system that is dynamically complex in time and space will probably reveal more about its underlying behavioral characteristics. It is encouraging that mathematicians, biologists and clinicians contribute together towards a common quantitative understanding of cancer complexity.

Chapter 6 - Fractal analysis has been a very useful characterization tool to assist in understanding physical properties and processes of earth systems. This analysis has been applied to soil systems to help in assessing management and landscape effects on soil properties. Applications include fractal analysis of soil pores, hydraulic properties, solute transport properties, and cracking processes. Fractal dimensions of X-ray computed tomography (CT)–measured porosity vary depending upon land management and landscape position. Investigations have found that fractal dimension of soil macropores increases 19% under vegetative buffers as compared to row crop management. Similar increases (26%, 9% and 18%) in fractal dimension were found for agroforestry (tree/grass) buffers compared to row crop management, agroforestry buffers compared to pasture management, and native prairie compared to row crop management, respectively. Fractal dimension of macropores was found to be highly correlated with saturated hydraulic conductivity ($r = 0.87$). CT-measured solute pore-water velocity and dispersivity were found to be fractal, and fractal dimension increased with average grain size in soil cores. Soil cracking patterns have also been characterized by both mass fractal dimension and crack edge fractal dimension; mass fractal dimension was found to be a function of soil landscape position, while crack edge fractal dimension values did not vary with landscape. Fractal analysis has been shown to be a useful characterization tool to differentiate management influences on critical soil physical properties and processes.

Chapter 7 - Throughout the present chapter, based upon the viewpoint of Weiss–Tabor–Carnevale formalism, the authors study the integrability properties of a set of higherdimensional evolution model systems, namely, the coupled nonlinear extension of the reaction-diffusion equation modeling the development of highly complex organisms based upon nonlinear interactions between common genes, the two-coupled nonlinear Schr"odinger equation arising in the description of dynamics of miscellaneous Bose–Einstein condensate mixtures confined within a time-independent anisotropic parabolic trap potential mapped onto the higher-dimensional time-gated Manakov system up to a first-order of accuracy, the dynamics of bulk polaritons in ferromagnetic slab through the single-oscillation two-dimensional soliton system, and the three-coupled Gross–Pitaevskii type nonlinear equations arising in the context of spinor Bose–Einstein condensates of atomic hyperfine spin $f = 1$ species. As a result, due to the arbitrariness of some functions stemming from the Laurent expansion up to a suitable truncation, the authors unearth an interesting family of fractal structures of miscellaneous patterns and dynamics worthy to the understanding of many physical phenomena occurring in the nature.

Chapter 8 - In the present chapter on multifractals, attempt has been made to present concepts and application of multifractals in simplified manner. Applications from diverse fields like astronomy to stock markets have been discussed. Discussion on lacunarity and wavelets and their relationship with multifractals have been discussed.

Chapter 9 - The effective coefficients in the equations of steady-state flow processes are calculated for a multiscale medium by using a subgrid modeling approach. The physical parameters in the medium are mathematically represented by a Kolmogorov multiplicative continuous cascades with a log-normal or log-stable probability distributions. The scale of the solution domain is assumed to be large as compared with the scale of heterogeneities of the medium. The theoretical results obtained in the chapter are compared with the results of a direct 3D numerical simulation and the results of the conventional perturbation theory.

Chapter 10 – The authors find conditions under which the attractor of an n-IFS, an IFS determined by a prescribed set of allowed compositions of length $n+1$, consists of a finite family of parallel lines, a modest step in the problem of characterizing the topological types of the attractors of n-IFS. A step in this analysis is showing that the attractor of the n-IFS with allowed compositions $R_1 \cup R_2$ is the union of the attractor of the n-IFS with allowed

com! positions R_1 and the attractor of the n-IFS with allowed compositions R_2, if the edge transition graphs of R_1 and R_2 are disjoint. Another is that the attractor is nonempty if and only if the edge transition graph contains a cycle. For attractors consisting of lines, the endpoints of the lines must constitute unions of cycles, and images of cycle points, on opposite edges of the unit square. These lines can be generated by n-IFS if and only if each endpoint lies in a distinct address length $(n+1)$-square along that edge of the unit square.

Chapter 11 - Scale irregular fractals" are irregular objects that exhibit some fractal properties but do not satisfy any exact scaling relation. In this chapter, the authors state some trace results on "scale irregular fractals" and on prefractal structures approximating the limit objects. These results are crucial tools in the study of the asymptotic convergence of energy forms defined on prefractal structures approximating the "scale irregular fractals".

Chapter 12 - The idea of a fractal has become an effective tool in the analysis of common features of many complex processes observed in physics, biology, chemistry or earth sciences. Any fractality of a physical system can be generated two-ways, it can arise due to the fractality of underlying medium (material) or due to the fractality of the process. This means that physical quantities can have fractal (power-law) characteristics even if the material itself does not need to have fractal microstructure. The hallmark of fractality of a geometrical set is a hierarchical organization of its elements, described by discrete scaling laws, which makes the fractal self-similar or self-affine. This effect can be observed at both classical and quantum levels. For example in quantum systems under some conditions, e.g. at the critical energy separating localized and extended states, the wave functions are shown to have fractal structure. Evidence for that comes from the measurements of the participation numbers N_q in some systems.

Chapter 13 - The authors give a survey of recent results by the authors and their collaborators concerning fractal analysis of trajectories of dynamical systems, oscillatory solutions of ODE's and singular sets of elliptic PDE's.

The idea of fractal dimension, i.e. noninteger dimension, has a long history, going back to the very beginning of the 20th century. The notion of box dimension is related to Minkowski and Bouligand. There are many other names for box dimension appearing in the literature, usually meaning the upper box dimension. One can encounter other equivalent names such as box counting dimension, Minkowski-Bouligand dimension, the Cantor-Minkowski order, Minkowski dimension, Bouligand dimension, Borel logarithmic rarefaction, Besicovitch-Taylor index, entropy dimension, Kolmogorov dimension, fractal dimension, capacity dimension, and limit capacity.

In: Classification and Application of Fractals
Editor: William L. Hagen

ISBN 978-1-61209-967-5
© 2012 Nova Science Publishers, Inc.

Chapter 1

MODELING THE CLUSTER STRUCTURE
OF DISSOLVED AIR NANOBUBBLES IN LIQUID MEDIA

N. F. Bunkin[1], S. O. Yurchenk[2], N. V. Suyazov[1],
A. V. Starosvetskiy[3], A. V. Shkirin[1] and V. A. Kozlov[2]

[1]A.M.Prokhorov General Physics Institute of Russian Academy of Sciences,
Vavilova, Moscow, Russia
[2]Bauman Moscow State Technical University, Baumanskaya, Moscow, Russia
[3]Moscow Engineering Physics Institute, Kashirskoe,
Moscow, Russia

ABSTRACT

A stochastic analysis of fractal properties of clusters composed of stable gas nanobubbles suspended in aqueous electrolyte solutions was carried out. The model of nucleating the stable bubbles in water at room temperature was suggested. This model is completely based on the property of the affinity of water at the nanometer scale; it was shown that under certain conditions the extent of disorder in the liquid starts growing, which results in a spontaneous decrease of the local density of a liquid and in the formation of nanometer-sized voids. These voids can capture the molecules of dissolved gas preserving the void from collapse. After adsorption of likely charged ions at the "void – liquid" interface, the latter is being kept from collapse by the Coulomb repulsion forces, acting along the interface. This way the steady bubbles at the nanometer size are generated; these bubbles serve as nuclei for the following generation of the so-termed bubstons (the abbreviation for *bubbles, stabilized by ions*). The model of charging the bubstons by the ions, which are capable of adsorption, and the screening by a cloud of counter-ions, which are incapable of adsorption, was analyzed. It was shown that, subject to the charge of bubston, two regimes of such screening can be realized. At low charge of bubston the screening is described in the framework of the known linearized Debye – Huckel approach, for which the sign of the counter-ion cloud preserves its sign everywhere in the liquid surrounding the bubston, whereas at large charge this sign is changed at some distance from the bubston surface. This effect provides the emergence mechanism for two types of compound particles having the opposite polarity, which leads to the aggregation of such compound particles by a ballistic kinetics. The fractal properties of clusters, generated by consecutive attachments of separate particles under the condition of central falling (i.e. ballistic "particle-cluster" aggregation), were investigated. This analysis was essentially based on experimental data measured by laser polarimetric scatterometry of NaCl aqueous solution. The fractal dimension for such clusters was built up as a function of monomer number. General properties of the light scattering matrix of bubston-cluster ensembles were found. The calculations of the light scattering matrix were performed for random ballistic-type clusters, comprising the bubstons of various sizes, on the assumption that the bubstons have lognormal size distribution with an effective radius 90 nm and effective variance 0.02; to fit the experimental scattering indicatrix of aqueous salt solutions, containing dissolved air, the mean number of bubstons in a cluster was set to 100.

1. STATEMENT OF A PROBLEM. INTRODUCTION

The present study is devoted to the problem of spontaneous nucleation of the quasi-stable phase of dissolved gas nanobubbles in an equilibrium liquid, which is kept under normal conditions, i.e. at room temperature and atmospheric pressure. It will be shown that these nanobubbles are capable of coagulation with one another with the formation of clusters composed of such nanobubbles. The given problem arises, for example, at the interpretation of the ultrasonic cavitation phenomenon. For example, if we focus an ultrasonic wave of high enough intensity in a liquid, we can see in the focal volume of an ultrasonic lens a track of vapor-gas bubbles. This phenomenon is termed as a cavitation and is widely explored by the specialists in the field of underwater acoustics, as the cavity effect plays one of key roles in the destruction process of ship screw propellers.

It is well known that the rupture strength of a liquid can be expressed by the formula $\sigma n^{1/3} \sim 10^4$ atmospheres, where σ is the coefficient of the surface tension of a liquid, n is the bulk density of molecules of a liquid (all numerical estimates will be hereinafter made for water). At the same time, experimental data on a cavitation indicate that it can be induced already at the amplitudes of a sound wave of about 1 atmosphere. It follows from here that long-lived (quasi-stable) centers of the cavitation must be present in the liquid for the cavitation effect to occur. The steady gas bubbles at the micron scale are evidently related to such centers. Let's mark that we will restrict our analysis by considering the micron-sized bubbles, since the bubbles of smaller size should be squeezed by the surface tension forces and disappear, whereas larger bubbles (whose size achieves 1 mm) promptly float up.

The mechanical equilibrium condition for the micron-sized bubble is given by the known Euler equation:

$$P_{in} = P_0 + \frac{2\sigma}{R} > P_{atm},\tag{1}$$

where R is radius of a bubble, P_{in} is the pressure of gas inside a bubble, P_{atm} is the pressure of atmospheric gas above a surface of a liquid, i.e. the atmospheric pressure. Let us note that in the formula (1) we did not take into account the hydrostatic pressure associated with the weight of the liquid volume above the bubble: we can obviously do it, as all experiments on a cavitation are carried out, as a rule, under the laboratory conditions, when the hydrostatic pressure is much less than the atmospheric pressure. The formula (1) implicates that the pressure of gas inside the micron-sized bubble is always higher, than the pressure of the same gas above the liquid surface. Thus, the solution of the gas in the liquid, whose content, according to the Henry law, is dictated by the pressure P_{atm}, appears to be unsaturated with respect to the pressure P_{in} of the same gas inside the bubble. It follows from here that such bubble is diffusively non-stable: the gas escapes from such bubble by the diffusion kinetics, and such bubble eventually disappears. This fact was analyzed (probably for the first time) in Ref. [1]. As was shown in this study, if we deal with a bubble with the initial radius of 10^{-3} cm, located in water saturated with dissolved air, the lifetime of that bubble does not exceed 10 sec. This time drastically falls at decreasing the radius of the bubble. For example, if the bubble radius is about 1 nm, the lifetime of such bubble in a wide range of temperatures does not exceed 10 ps, see, for example, [2]. Hence, to observe regularly the cavitation effect, we should require the forces of the surface tension to be somehow neutralized.

One of the mechanisms of such neutralization can be associated with externally introduced solid impurities, i.e., bubbles arise at solid (possibly hydrophobic) micro-particles suspended in the liquid. The wetting angle changes along the rough and non-planar (in general) surface of a hydrophobic particle, and the formation of a dissolved gas bubble at the solid–liquid interface becomes energetically favorable. This is why for the gas bubbles, attached to a solid interface, the surface tension forces appear to be balanced and such bubbles become stabilized both mechanically and diffusively. The model where the stable heterogeneous centers of a cavitation arise due to the presence of solid impurities, is widely accepted, see, e.g., [3 - 6]. It is necessary to note, however, that even a fine filtration of a liquid from solid impurities cannot completely suppress the cavitation effect. At the same time, it is well known (see, e.g., [7]) that the cavitation ability in aqueous media increases by adding various salts, albeit new solid particles are definitely not introduced into the water together with a salt. Thus, it should be recognized that the above considered mechanism of micro-bubbles stabilization is inconsistent.

The study presented below is devoted to theoretical description of an alternative (not associated with the solid impurities) model of the origin and stabilization of micro-bubbles in water and aqueous solutions of salts. The liquids are considered to be in the equilibrium state under normal conditions and saturated by dissolved gas (for example, atmospheric air). The suggested mechanism of stabilization is stipulated by a selective adsorption of ions of the same sign on a bubble interface. These ions give rise to occurrence of the Coulomb repulsion forces along the bubble interface; these forces are added vectorially to the surface tension forces compressing the bubble. It is necessary at once to note that the adsorbed ions in no way influence the magnitude of the surface tension coefficient. Such bubble was termed as *bubston* (abbreviation of *bubble, stabilized by ions*). The bubston model was first put forward in the study [8], and then was developed in subsequent works [9, 10]. Refs. [11 - 17] contain a direct experimental proof of the bubston existence in an equilibrium aqueous solution of salts, saturated by dissolved air. However, key evidence supporting this hypothesis was inferred from observations of low-threshold laser-induced breakdown in water and aqueous electrolytic solutions that were transparent to the laser beams employed [18 – 23]. Laser-induced breakdown results from the development of electron avalanches inside separate bubstons followed by cluster coalescence into a large bubble (usually called *cavitation bubble*).

The structure of this work is the following. In the second chapter we analyze the mechanisms of the bubston nucleation in the equilibrium water, the third chapter is devoted to the mechanisms of bubston coagulation, leading to the formation of bubston clusters, and in the fourth chapter we describe the fractal properties of the bubston clusters.

Prior to finishing this chapter, it is worth saying that for a long time the question of the adsorption of ions at the gas – liquid interface was not discussed in literature. It was coupled to the fact that, as was demonstrated in the studies [24, 25], such adsorption should result in growth of the surface tension coefficient σ, which obviously means an increase of a surface energy, i.e. such adsorption is unfavorable. Note that these works have been carried out on the basis of the mean field approach, and the liquid was considered as a continuum characterized only by its macroscopic parameters (the surface tension coefficient σ, the dielectric permittivity ε etc.), i.e. the specific local properties of the liquid were by no means taken into account; here we speak about the dipole moments and polarizabilities of molecules, and their sizes. The subsequent numerous theoretical works allowed for the geometrical characteristics of water molecule [26], numerical MD simulation in the view of the effect of polarization of water molecules [27 - 37], and also series of experimental results including vibrational sum frequency generation spectroscopy [38 - 43], second harmonic generation spectroscopy [44, 45], and high-pressure VUV photoelectron spectroscopy [46, 47], and X-ray photoelectron spectroscopy combined with scanning electron microscopy [48, 49] allow us to make a conclusion about an opportunity of adsorption of some anions, for example, Cl^-, I^-, Br^- on the water - gas interface. The fact that basically negative ions are capable of adsorption at this interface is also supported by the data obtained in Ref. [50], where the negative density of the charge on the surface of the gas bubble in pure water was directly measured: at pH = 6.9 its value is $\alpha = -18 \cdot 10^{-6}$ C/m^2.

2. THE MECHANISM OF NUCLEATION OF THE BUBSTONS

The occurrence of mesoscopic nanoscale structures in water basically has a statistical nature. Therefore, hereinafter we will use statistical methods for the description of nucleation of such structures. This approach is essentially based on the model of nanoscale water as a quasi-crystal ice-like medium caused by the presence of the hydrogen bonds. Such a model was apparently first put forward in the literature in the known study [51] and then developed in the works [52 - 60]. Within the framework of this approach it is possible to explain many anomalies of liquid water (it is possible to find details on the Website [61]). The considered model is based on the assumption that water is generally a disorder crystal. Let's note, however, that the statistical description of the disordered structures still remains poorly developed.

For the case of an ideal crystal the pair distribution function of distances between two atoms can be represented by a linear combination of the kind

$$n(r) = \sum_k n_k \delta(r - r_k),$$

where r_k is radius of k-th coordinate sphere, n_k is the number of atoms at the surface of k-th coordinate sphere, $\delta(r)$ is the Dirac delta-function, and the summation is carried out over all coordinate spheres.

The peaks of the function $n(r)$ appear to be "diffuse" in the amorphous state, in comparison with the ordered state. The essential disadvantage of the function $n(r)$ is its dependence on the number of atoms in the structure in question, and the quadratic divergence $n(r) \sim r^2$ at a large r, which means the impossibility of normalizing. However, the basic properties of the disordered structures are defined by energy of the lattice, and the remote diverging "tail" of the function $n(r)$ practically does not contribute to the lattice energy, as the interaction forces quickly decrease with the distance r. This is why the tail part of the function $n(r)$ can be ignored, and we should take into account the part of this function, which mainly contributes to the lattice energy. In the case of ideal crystal such contribution is provided by the atoms of the nearest coordination spheres. For the amorphous structures we cannot outline any coordination spheres; it is possible only to specify the coordination peaks in the pair distribution function. Thus, it is necessary to develop a new technique for describing a structure, which is crystal in a particular case, and is amorphous in others cases. The ordered structure can be preset by the geometry of bonds in an elementary cell; the crystal structure formation can be modelled by the series of translations and spatial turns of the elementary cell. In our case the icosahedral structure of the water matrix is generated by the combinations of translations and spatial turns of the elementary tetrahedrons of water molecules [62].

Let's study the transition from the ordered state to the disordered state by considering an example of a one-dimensional chain of atoms, and then generalize the results to a more common case. To construct a one-dimensional chain of atoms, it is sufficient to choose the initial lattice point and to set the transfer vector between two nearest neighbors. In case of the disordered lattice it is necessary to set the density of probability for the distance between two lattice points, which will be termed as s-function. According to the definition, the s-function is normalized to unity

$$\int_0^\infty s(r)dr = 1, \tag{2}$$

is positive $s(r) \geq 0$, and in the extreme case of an ideal crystal it should transform into the Dirac delta - function $\delta(r-a)$, where a is the distance between two lattice points in the ordered structure. Besides, the condition

$$s(r) = 0, \qquad r \leq 0$$

should be met. Thus the complete density of probability for localization of an atom in the given point is expressed by the sum:

$$p(r) = \sum_n p_n(r),$$

where $p_n(r)$ is the density of probability for the atom related to n-th coordination peak to be located in a point with the coordinate r. The density of probability of finding an atom in the first coordination peak is s-function, i.e.

$$p_1(r) = s(r).$$

The considered structure has the property of affinity: the ratio of the second atom coordinate to the first atom coordinate is the same as for the third and the second atoms and so on. The one-dimensional chain of atoms is constructed by the consecutive shift of an initial cell in the positive direction of r. Therefore, all $p_n(r)$ - functions can be calculated in the recurring way, i.e. by using only the s-function. For example, the probability for an atom to be located in the second peak is given as

$$p_2(r) = \int s(r_1)s(r - r_1)dr_1,$$

where the integrand is the density of probability for the interatomic bond to occur in the point r through all intermediate points r_1. For the third peak we can write

$$p_3(r) = \iint s(r_1)s(r_2 - r_1)s(r - r_2)dr_1 dr_2 = \int p_2(r_2)s(r - r_2)dr_2.$$

It follows from here that $p_{n+1}(r)$ function can be expressed as

$$p_{n+1}(r) = \int p_n(z)s(r - z)dz.$$

It is straightforward to verify that all functions of $p_n(r)$ are normalized.

Let us illustrate this approach by using an example of s-function of the Gaussian type:

$$s(r) = \frac{1}{\sigma\sqrt{2\pi}}\exp\left[-\frac{(r-a)^2}{2\sigma^2}\right], \tag{3}$$

where a is the mean value of distance between the nearest neighboring atoms and has the sense of the lattice constant, σ is the dispersion, characterizing the scatter of distances between the neighboring atoms. The Gaussian function, generally speaking, does not obey the normalizing condition (2) at integrating for $r \in (0, \infty)$, and also $s(0) \neq 0$. However, in the case of small values of σ it is possible to assume that $s(r) = 0$ at $r = a \pm 3\sigma$, and therefore the errors of the model cannot be significant. After integrating over the interval $r \in (-\infty, +\infty)$ we arrive at

$$p_n(r) = \frac{1}{\sigma\sqrt{2n\pi}} \exp\left[-\frac{(r-na)^2}{2n\sigma^2}\right].$$

Figure 1 illustrates the functions $p(r)$ calculated for the cases of $\sigma/a = 0.2, 0.3$. At $\sigma \to 0$ the function $p(r)$ transforms to a series of the Dirac delta - functions in the points $r = na$.

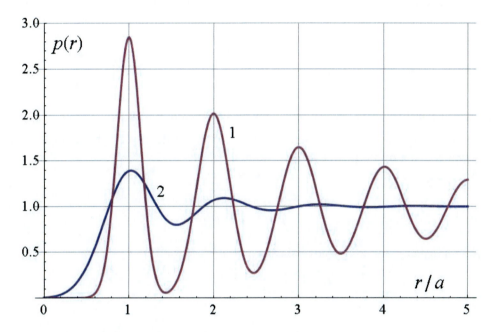

Figure 1. The probability density function of the distances between two atoms at $\sigma/a = 0.15$ (1); $\sigma/a = 0,3$ (2).

In case the degree of disorder is low it is possible to distinguish the coordination peaks, but at high degrees of disorder we can see only a pronounced but essentially diffuse first peak. As expected, the transition to the amorphous state is accompanied by washing out the distant coordination peaks, i.e. the long range ordering disappears.

If one knows the function of pair distribution $p(r)$ of the distances and potential $\varphi(r)$ of interaction between two particles, it is possible to find the average energy of pair interaction:

$$\Phi = \int p(r)\varphi(r)dr.$$

The lattice constant a relaxes to its equilibrium magnitude with the sound velocity, as it is connected with the change of specific volume.

Therefore, the dependence $a(\sigma)$ is defined from the condition $\Phi \rightarrow min$ at the given value of σ.

Thus, the effective lattice constant a, the function of pair distribution of the distances $p(r)$, average energy of interaction Φ (and, accordingly, the elasticity modulus) appear to be dependent upon the parameter σ. This is why it is possible to consider σ as the dimensional disorder parameter, on which the properties of the amorphous structure depend.

The described way of using s-function under the assumption of affinity of the structure can be generalized in the case of a three-dimensional spatial amorphous structure. Let us introduce three-dimensional s-function, with the help of the above pair distribution function $p(r)$, which can be represented as

$$p(\mathbf{r}) = \sum_{\alpha} p_{\alpha}(\mathbf{r}), \tag{4}$$

where $p_{\alpha}(\mathbf{r})$ is the probability density to find the particle of the lattice node α at the point with the radius-vector $\mathbf{r}$. By definition, s-function is the density of probability for some chosen lattice point of the first coordination peak to be located in the point with the radius – vector $\mathbf{r}$, i.e.

$$p_{1}(\mathbf{r}) = s(\mathbf{r}).$$

The normalizing condition here is the following

$$\int s(\mathbf{r})d\mathbf{r} = 1,$$

where $d\mathbf{r} = dr_{x}dr_{y}dr_{z}$ is an element of volume. It is necessary to note that the spatial s-function is positively defined ($s(\mathbf{r}) \geq 0$), is equal to zero in the point of origin ($s(0) = 0$), and the ordered structure should reduce to the Dirac delta - function with a vector argument:

$$s(\mathbf{r}) = \delta(\mathbf{r} - \mathbf{a}),$$

where $\mathbf{a}$ is the radius - vector of the location of the lattice point in the ordered structure. As an example, let us consider a quadratic lattice illustrated in Figure 2. Let's choose the lattice point A and assume that s-function for this point has the form $s_{A}(\mathbf{r}) = s(\mathbf{r})$. Then the two-dimension s-function for the atom D looks like

$$s_{D}(\mathbf{r}) = s[T\mathbf{r}], \tag{5}$$

where T is operator of turning the system of coordinates Oxy at $\frac{\pi}{2}$, where the point D coincides with the point A. By using the property of affinity of the spatial ratio between the points O, A and B, it is easy to find the density of probability for the point B to locate in the point with the radius – vector $\mathbf{r}$:

$$p_B(\mathbf{r}) = \int s(\mathbf{r}_1)s(\mathbf{r}-\mathbf{r}_1)d\mathbf{r}_1.$$

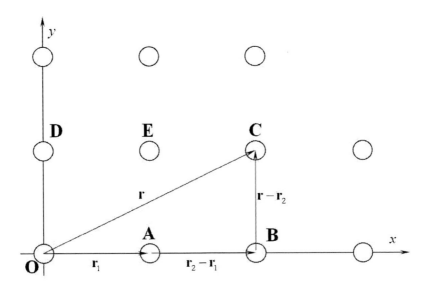

Figure 2. The quadratic lattice.

For the point E we find accordingly

$$p_E(\mathbf{r}) = \int s(\mathbf{r}_1)s[T(\mathbf{r}-\mathbf{r}_1)]d\mathbf{r}_1 = \int s[T\mathbf{r}_1]s(\mathbf{r}-\mathbf{r}_1)d\mathbf{r}_1.$$

The two definitions of $p_E(\mathbf{r})$ presented here are connected with the two possible shortest graphs of the movement to area of the point E: we can reach the point E by the two-step graphs OAE or ODE.

For the point C we have three possible shortest graphs: $OABC$, $OAEC$ and $ODEC$ (see Figure 2). It follows from here that we have three equivalent expressions for $p_C(\mathbf{r})$:

$$\begin{aligned}p_C(\mathbf{r}) &= \iint s(\mathbf{r}_1)s(\mathbf{r}_2-\mathbf{r}_1)s[T(\mathbf{r}-\mathbf{r}_2)]d\mathbf{r}_1 d\mathbf{r}_2 = \\ &= \iint s(\mathbf{r}_1)s[T(\mathbf{r}_2-\mathbf{r}_1)]s(\mathbf{r}-\mathbf{r}_1)d\mathbf{r}_1 d\mathbf{r}_2 = \\ &= \iint s[T(\mathbf{r}_1)]s(\mathbf{r}_2-\mathbf{r}_1)s(\mathbf{r}-\mathbf{r}_2)d\mathbf{r}_1 d\mathbf{r}_2.\end{aligned} \qquad (6)$$

It is clear that the integrands can be written explicitly for the movement along each graph:

$$p_C(\mathbf{r}) = \iint s_{OA}(\mathbf{r}_1)s_{AB}(\mathbf{r}_2 - \mathbf{r}_1)s_{BE}(\mathbf{r} - \mathbf{r}_2)d\mathbf{r}_1 d\mathbf{r}_2 =$$
$$= \iint s_{OA}(\mathbf{r}_1)s_{AD}(\mathbf{r}_2 - \mathbf{r}_1)s_{DE}(\mathbf{r} - \mathbf{r}_2)d\mathbf{r}_1 d\mathbf{r}_2 = \qquad (7)$$
$$= \iint s_{OC}(\mathbf{r}_1)s_{CD}(\mathbf{r}_2 - \mathbf{r}_1)s_{DE}(\mathbf{r} - \mathbf{r}_2)d\mathbf{r}_1 d\mathbf{r}_2.$$

A simple routine for calculating any function for n-th coordination peak follows from here: it is necessary to choose a node α in the ordered lattice, then to construct a graph with the minimum number of steps $1, 2, \ldots n$ that leads to the lattice point n. After that the function $p_\alpha(\mathbf{r})$ can be found as a $n-1$-multiple integral

$$p_\alpha(\mathbf{r}) = \int \ldots \int s_{0,1}(\mathbf{r}_1)s_{1,2}(\mathbf{r}_2 - \mathbf{r}_1) \ldots s_{n-1,n}(\mathbf{r} - \mathbf{r}_{n-1})d\mathbf{r}_1 \ldots d\mathbf{r}_{n-1}. \qquad (8)$$

Thus the function $s_{k-1,k}(\mathbf{r})$ describes the density of probability for a k-th lattice point to locate in the point $\mathbf{r}$ at the movement from the $k-1$-th lattice, and is connected with the function $s_{01}(\mathbf{r})$ as

$$s_{k-1,k}(\mathbf{r}) = s(T\mathbf{r}),$$

where T is the operator of turn, which transforms the bond-vector $(k-1,k)$ (connecting vector from node $k-1$ to node k) to the bond-vector, for which the s-function was introduced. For example, if the s-function is set for the lattice point A, then

$$s_{OA}(\mathbf{r}) = s_{AB}(\mathbf{r}) = s_{DE}(\mathbf{r}) = s_{EC}(\mathbf{r}) = s(\mathbf{r}). \qquad (9)$$

For other pairs of bonds between the points illustrated in Figure 2 we can write

$$s_{OD}(\mathbf{r}) = s_{AE}(\mathbf{r}) = s_{BC}(\mathbf{r}) = s(T\mathbf{r}), \qquad (10)$$

where T is the operator of the clockwise turn to the angle of $\dfrac{\pi}{2}$. At the same time, it is possible to set the function for the point D instead of A. Then in the formulas (9) and (10) it is necessary to replace $s(\mathbf{r})$ by $s(T\mathbf{r})$, and the T- operator will now mean the counter-clockwise turn at an angle of $\dfrac{\pi}{2}$. For the spatial structures the process of calculating the

$p_\alpha(\mathbf{r})$ function by using the algorythm (8) can be presented in more direct kind (see Figure 3).

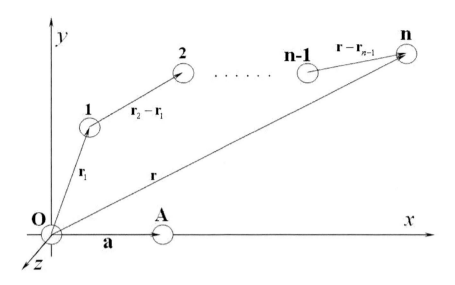

Figure 3. The schematic picture explaining the procedure of the $p_\alpha(\mathbf{r})$ function calculating.

The $p_\alpha(\mathbf{r})$ function is calculated through integrating the products of $s_{k-1,k}$-functions over all intermediate coordinates $\mathbf{r}_1...\mathbf{r}_{n-1}$ for subsequent lattice points $1,2,...n-1$. According to Figure 3, the initial s-function is preset for the point A, and therefore, the operator of turn for each $(n-1,n)$ transition transforms the direction of the bond-vector $(n-1,n)$ to the radius – vector $\mathbf{a}$. The function of radial distribution of the lattice points can be found by integrating $p(\mathbf{r})$ function in spherical coordinates over all directions:

$$p(r) = \iint p(\mathbf{r})r^2 \sin\varphi \, d\varphi \, d\theta. \qquad (11)$$

This function does not depend any more on a direction in space. Thus, the calculation of density of pair distribution of probability, according to the formulas (8) and (11), is reduced to setting of some function, whose form should be chosen from the criterion of the best fit to experimental data. If one accepts that the number bonds, on which the statistics of the amorphous cells is designed, is approximately equal to $\sim 10^3$, then the scale, for which the s-function is determined, appears to be of the order of 1 nm^3. As a matter of fact, it is the size of area, where the statistical averaging is carried out, and the structural properties are defined. The way of using the function of the probability density for the distance between two nearest lattice points (the s-function) represents parametrization of the disordered structure: each small area of that structure is associated with the corresponding statistics of the given kind, which should agree with the experimental data and should depend on certain parameters. The parameters of s-function are not local (these are determined not in a point). On the contrary, these are determined within a small area (where the s-function is constructed) and can be considered as the parameters of disorder (or amorphism) of the structure. The field of the disorder parameter appears to be connected with the properties of the structure under study; this field influences the kinetics and spatial features of evolution of the system.

When choosing a concrete form of the disorder parameter, let us use the s-function given by Eqn. (3). Namely, denoting such a dimensionless parameter as $\eta \equiv \sigma / a,$ we put that in the amorphous state $\eta \neq 0$, while $\eta \equiv 0$ in the ordered state. Let us note that we consider slow (in comparison with the velocity of sound) transformations of isotropic amorphous structures. We suppose that the properties of each small area of the liquid can be completely described with the help of some parameter (or a set of parameters) related to the degree of disorder of the system. Thus, the "slow" transformations are such ones, at which the velocity of spatial relaxation of the field η is much less than the sound velocity. The phenomenological theory reported below does not allow us to speak generally about the peculiarities of evolution of strongly disordered structures, for which η is not small any more. The initial symmetry of the ordered structure is broken in the amorphous state, and therefore, the corresponding considerations should be based on the statistical description, since we deal with an average statistical cell instead of a crystal cell of the amorphous lattice. In this particular case the statistics of the structure corresponds to a crystal, whereas in all other cases the structures are disordered. Thus, the parameter η should be connected to the statistics of distribution of interatomic bonds. It is then possible to consider that the evolution of the disordered structures is the quasi-equilibrium process, whose kinetics is defined only by evolution of the field η. Let's consider that at small deformations the expansion of the bulk density of the free energy of isotropic amorphous structure is given by the formula [62, 63]

$$F = A(\eta) - B(\eta)u_{ll} + \frac{1}{2}C(\eta)u_{ll}^2 + D(\eta)(u_{jk} - \frac{1}{3}u_{ll}\delta_{jk})^2, \tag{12}$$

where $A(\eta), B(\eta), C(\eta), D(\eta)$ are some coefficients, u_{jk} is the deformation tensor, which is defined through the displacements u_j of the points of the structure as

$$u_{jk} = \frac{1}{2}(\nabla_k u_j + \nabla_j u_k), \qquad \nabla_k = \partial/\partial x_k;$$

here the summation over the repeating indices is implied.

In the absence of deformations we have $u_{jj} = 0$, and the free energy is expressed as $F = A(\eta)$, i.e. $A(\eta)$ is the bulk density of energy, which is required to form the structure having the degree of disorder η. In the absence of the amorphous transitions, i.e. when $\eta = 0$ the Eqn. (12) should coincide with the known expression for the bulk density of the free energy related to the elastically deformed isotropic homogeneous body. It follows from here that $C(\eta) = K(\eta)$ is the compression modulus, and $D(\eta) = \mu(\eta)$ is the shear modulus.

The strain tensor has the form (see Eqn. (12))

$$\sigma_{jk} = \frac{\partial F}{\partial u_{jk}} = ((K - \frac{2}{3}\mu)u_{ll} - B(\eta))\delta_{jk} + 2\mu u_{jk}. \tag{13}$$

As is seen from here, even in the case where the strain $\sigma_{ik} = 0$, we nonetheless may have that $u_{jk} \neq 0$, i.e. the coefficient $B(\eta)$ is associated with expansion or compression accompanied the transition to the amorphous state:

$$B(\eta) = \alpha\eta K, \qquad \alpha(\eta) = \frac{1}{V}\frac{\partial V}{\partial \eta},$$

where α is the amorphous expansion coefficient.

In the quasi-static mechanical equilibrium we have $\nabla_k \sigma_{jk} = 0$. In view of (13), we arrive at

$$\nabla_j \left[\left(K - \frac{2}{3}\mu \right) \nabla_l u_l \right] + \nabla_k [\mu(\nabla_k u_j + \nabla_j u_k)] = \beta \nabla_j \eta, \tag{14}$$

where we introduce the function $\beta(\eta) = \dfrac{dB}{d\eta}$.

Eqn. (14) should be supplemented with the equation of the evolution of the field η. Let us consider that the kinetics of changing the disorder parameter η obeys the Landau – Khalatnikov equation [64]:

$$\frac{\partial \eta}{\partial t} = -\gamma \frac{\partial F}{\partial \eta}, \tag{15}$$

where $\gamma > 0$ is the generalized "viscosity" as related to the parameter η, while the sign "-" in the right-hand side of this equation reflects the tendency of the system to achieve the minimum of the free energy.

Substituting (12) in Eqn. (15) and making allowance for the connection of the coefficients A, B, C, D with the properties of the amorphous structure gives the equation of the field η evolution:

$$\frac{\partial \eta}{\partial t} = -\gamma \left[\lambda - \beta u_{ll} + \frac{1}{2}K_\eta u_{ll}^2 + \mu_\eta \left(u_{jk} - \frac{1}{3}u_{ll}\delta_{jk} \right)^2 \right], \tag{16}$$

where $\lambda(\eta) = \dfrac{dA}{d\eta}$, $K_\eta = \dfrac{dK}{d\eta}$, $\mu_\eta = \dfrac{d\mu}{d\eta}$. Formulas (14), (16) together with the initial conditions define the model of evolution of the isotropic amorphous structure. In the case where the shear deformations can be ignored (this is the case for liquids, for which $\mu \ll K$), the system of the equations (14), (16) can be expressed as

$$\begin{cases} \nabla_j (K\nabla_k u_k) = \beta\nabla_j \eta, \\ \dfrac{\partial \eta}{\partial t} = -\gamma[\lambda - \beta\nabla_k u_k + \dfrac{1}{2}K_\eta(\nabla_k u_k)^2]. \end{cases} \tag{17}$$

The first equation of this system has the solution in the form

$$\nabla_k u_k = \frac{1}{K}(B + p(t)), \tag{18}$$

where $p(t)$ is the function of time, which stands for the pressure in a liquid. With regards to (18), the system of the equations (17) can be reduced only to the equation of evolution of the field η:

$$\frac{\partial \eta}{\partial t} = -\gamma\left[\lambda - \frac{\beta(B + p(t))}{K} + \frac{1}{2}\frac{K_\eta}{K^2}(B + p(t))^2\right]. \tag{19}$$

To solve this equation it is necessary to know the explicit dependences of $B(\eta), K(\eta), \lambda(\eta)$. However, in some cases we can establish the form of such dependences from the general considerations. For instance, close to the ordered state (i.e., at $\eta \ll 1$) the dependences $\lambda(\eta), K(\eta), \alpha(\eta)$ can be expanded in the series of the the parameter η:

$$\lambda(\eta) = \lambda_0 + \lambda_1\eta + \lambda_2\eta^2 + ..., \qquad K(\eta) = K_0 + K_1\eta + K_2\eta^2 + ... \tag{20}$$

$$\alpha(\eta) = \alpha_0 + \alpha_1\eta + \alpha_2\eta^2 + ...$$

The expansion for the function $B(\eta)$ can be found as

$$B(\eta) = \alpha\eta K = \alpha_0 K_0 \eta + (\alpha_0 K_1 + \alpha_1 K_0)\eta^2 + (\alpha_0 K_2 + \alpha_1 K_1 + \alpha_2 K_0)\eta^3 ...$$

Substituting these expansions to the right-hand side of Eqn. (19) and implying $p(t) \equiv 0$, we arrive at

$$\dot{\eta} = -\gamma(\lambda_0 + a_1\eta + a_2\eta^2 + a_3\eta^3),$$ (21)

$$a_1 = \lambda_1 - \alpha_0^2 K_0, \qquad a_2 = \lambda_2 - 3K_0\alpha_0\alpha_1 - \frac{3}{2}\alpha_0^2 K_1$$

$$a_3 = \lambda_3 - 4\alpha_0\alpha_1 K_1 - 4\alpha_0\alpha_2 K_0 - 2\alpha_1^2 K_0 - 2\alpha_0^2 K_2$$

Let us assume that the ordered state is steady (then $\lambda_0 = 0$), and the transition to the amorphous state is always accompanied by the expansion of the structure, i.e. $\alpha_0 = 0$. In this situation Eqn. (21) has the form

$$\dot{\eta} = -\gamma\eta(\lambda_1 + \lambda_2\eta - (2\alpha_1^2 K_0 - \lambda_3)\eta^2).$$ (22)

Factorization of the right-hand side of this formula gives

$$\frac{d\eta}{d\tau} = \eta(\eta - \eta_1)(\eta - \eta_2),$$ (23)

where $\tau = \gamma(2\alpha_1^2 K_0 - \lambda_3)t$ is the dimensionless time, and

$$\eta_{1,2} = \frac{\lambda_2 \pm \sqrt{\lambda_2^2 + 4\lambda_1(2\alpha_1^2 K_0 - \lambda_3)}}{2(2\alpha_1^2 K_0 - \lambda_3)}.$$

Integrating Eqn. (23) we obtain an inverse dependence $\tau(\eta)$

$$\tau(\eta) = \frac{1}{\eta_1 - \eta_2}\left[\frac{1}{\eta_2}\ln\left(\frac{\eta(\eta_0 - \eta_2)}{\eta_0(\eta - \eta_2)}\right) - \frac{1}{\eta_1}\ln\left(\frac{\eta(\eta_0 - \eta_1)}{\eta_0(\eta - \eta_1)}\right)\right],$$ (24)

where η_0 is the value of the disorder parameter in the initial moment of time.

Figure 4 illustrates the curves $\eta(\tau)$ for the case where $\lambda_2 = 0$ (i.e., $\eta_1 = -\eta_2$) at the various initial conditions preset by the parameter η_0. The parameter η_1 in this case is defined as

$$\eta_1 = \sqrt{\frac{\lambda_1}{2\alpha_1^2 K_0 - \lambda_3}}$$

The found solutions enable us to analyze the process of transition between the ordered and amorphous states. First, providing that $\eta_1 \ll 1$, the limit equilibrium states, for which $\dot{\eta} = 0$, appear to be possible. The parameters $\eta_{1,2}$ are defined by the ratio between the heat release at the ordering of the amorphous structure (the amorphous condensation) and the energy consumptions, which are necessary for the structure deformation. Generally, various scenarios of evolution of the amorphous structure become possible, which depends on the initial conditions for the parameter $\eta(t)$ and the external forces $p(t)$.

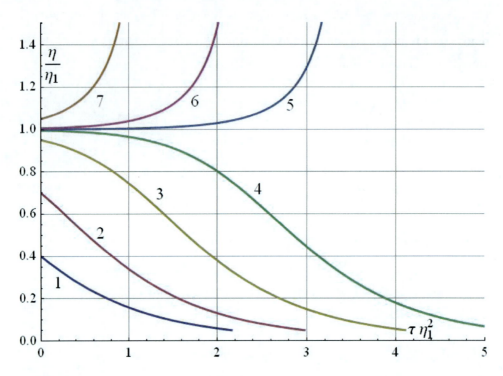

Figure 4. Dependences $\eta(\tau)$ at the various initial conditions η_0/η_1: (1) $\eta_0/\eta_1 = 0{,}4$; (2) $\eta_0/\eta_1 = 0{,}7$; (3) $\eta_0/\eta_1 = 0{,}95$; (4) $\eta_0/\eta_1 = 0{,}99$; (5) $\eta_0/\eta_1 = 1{,}0005$; (6) $\eta_0/\eta_1 = 1{,}005$; (7) $\eta_0/\eta_1 = 1{,}05$.

As is seen in Figure 4, at $|\eta_0/\eta_1| < 1$ the developing deformations cannot stop the amorphous condensation (or the ordering). At $\eta = \eta_{1,2}$ the unsteady equilibrium state of the amorphous structure becomes possible. However, even the small perturbations can result in either the amorphous condensation, or growing the disorder. The last can be treated as formation of less dense structures, which develop, eventually turning in a cavity. When the spatial gradients of density become significant, the interface between the high and low density media arise. Let us remark that the condition of the growth disorder does not depend on the symmetry of a field, i.e. it supposes the existence of cavities of spherical, cylindrical and other forms, which was obtained in the method of molecular dynamics [65].

The critical characteristics of the disordered structure evolution are the amorphous condensation heat and the compression modulus. In the case where the heat release is higher than the increase of the deformation energy, the transition is directed towards the ordering, while in the opposite situation the disordering grows. In the terms of the field η the processes of destruction and disordering are quite similar and connected with the growth of parameter η. The perturbation of the parameter η can be also be associated with the shear deformations. Let's consider the case, where at the initial moment of time the bulk deformation is absent, i.e. $\nabla_k u_k = 0$. Then in the equations (14) and (16) it is impossible to neglect the terms, connected with μ, even if the condition $\mu \ll K$ is met. It follows from here that

$$
\begin{cases}
\nabla_k[\mu(\nabla_k u_j + \nabla_j u_k)] = \beta \nabla_j \eta \\
\dfrac{\partial \eta}{\partial t} = -\gamma[\lambda + \mu_\eta (u_{jk})^2]
\end{cases}
\tag{25}
$$

Provided that at the initial moment of time we have $\eta \equiv 0$, the first equation of the system (25) can be integrated, and the decision looks like

$$
u_{jk} = \frac{1}{2}(\nabla_k u_j + \nabla_j u_k) = \frac{p_{jk}(t)}{2\mu_0}, \qquad \mu_0 = \mu(\eta)\big|_{\eta=0},
$$

where $p_{jk}(t)$ is the tensor of external shear stress, causing the deformation. For the known deformations we find the equation for the disorder parameter

$$
\dot{\eta} = -\gamma\left[\lambda + \mu_\eta \left(\frac{p_{jk}}{2\mu_0}\right)^2\right].
\tag{26}
$$

Let's consider that close to the value of $\eta = 0$ the parameter μ can be expanded in the series

$$
\mu = \mu_0 + \mu_1 \eta + \mu_2 \eta^2 + \ldots
$$

By using the expansion for the parameter $\lambda(\eta)$ (see (20)), we find

$$
\dot{\eta} = -\gamma\left(\lambda_1 + \frac{\mu_2 p_{jk}^2}{2\mu_0^2}\right)\eta.
\tag{27}
$$

In deducing Eqn. (27) we assume that the ordered state is steady in relation to small shears, so we shall put $\mu_1 = 0$. Let us imagine that $\mu_2 < 0$, i.e. the shear modulus decreases with the disordering. Then, as is seen from Eqn. (27), provided that $p_{jk}^2 = 2\lambda_1 \mu_0 / |\mu_2|$, the state where $\eta = 0$ becomes unsteady and the disorder parameter η starts growing exponentially. The following evolution of the field η depends on the initial state and can result in formation of cavities, which in our opinion is one possible scenario for nucleation of the bubston. For example, at flowing an isotropic liquid a certain molecular layer slides on another one; in the area of relative slippage of such layers with respect to one another the tangent component of a stress exceed the critical magnitude, and the conditions for the disorder occurrence can be fulfilled. Thus, formed disorder structures together with the field of disorder η start to develop. As a result one of the two scenarios, described above, can be realized. Obviously, the exponential growth of the parameter η means a drastic decrease of the material density within the area of such growth; hereinafter we will term this process as the early stage of nucleation of the bubstons. This mechanism of local reduction of the liquid density should be taken into consideration as explaining the discrepancy between the volume, which should be occupied by the densely packed molecules of a liquid, and its real volume. Indeed, the volume of the water gram-molecule under normal conditions is 18 cm^3. At the same time, it is generally accepted that the radius of the water molecule is equal to 1.38 Å (it is just half of the distance between the two nearest neighbors in ice, see, for example, [61]). Thus, with allowing for the packing factor of spheres we can obtain that the volume, occupied by the Avogadro number of spheres having such radius, is approximately equal to 12.6 cm^3.

Summarizing, the early stage is associated with relative shear movements of the liquid molecular layers on the nano-scale level. It should be noted that a specific type of potential of the interaction between the molecules of the liquid was not taken into account here. This is why the basic results obtained within the framework of such model are of general character, and free of specific properties of a liquid and its molecular structure. They should remain fair in case of polar liquids, i.e. to the water and aqueous electrolyte solutions, which is the subject of the subsequent sections of the present work.

In the sequel, we will study the spherical voids formed as a result of the processes reported above. So far we did not consider the formation of surface of such a void. At high enough gradients of the material density the interface "the void – liquid" is formed. It is clear that when the surface is formed, the void is subject to collapse due to the surface tension forces. During the time of squeezing the void the ions of the same sign are being adsorbed at the void interface, which results in effectively charging that interface, keeping the void from the irreversible collapse; the processes of the interface formation and adsorption of the ions, which will be termed as the final stage of nucleation of the bubstons. Below, we give the thermodynamic description of this final stage. Let's consider that the bubston nucleus (a tiny bubble) is already formed, i.e. the kinetic processes are completed.

The model of this nucleation process in an aqueous solution described here is essentially based on several assumptions. We consider ions with different adsorptivity to the surface of a liquid (ions adsorbed at the surface of a void will be hereinafter referred to as basic ions, while the ions, not capable of adsorption, will be termed as counter-ions). This assumption is confirmed by the data on the selective adsorption of ions at the water – gas interface, see [26 – 50]; these results are briefly reported in the Introduction. In accordance with those results we suppose that the basic ions are essentially anions. The probability of the formation of a nucleus of this kind is proportional to $\exp\left(-\Delta\Phi_{\min}/kT\right)$, where $\Delta\Phi_{\min}$ is the minimum increase in the Gibbs free energy of the system caused by the formation of a void of radius r

. An expression for $\Delta\Phi_{min}$ is derived here by assuming that $r \ll a_D$ (where $a_D = \sqrt{\dfrac{\varepsilon kT}{8\pi Z^2 e^2 n_{i0}}}$ is the so-called Debye length, Z is the valence of dissolved ions, e is the elementary charge, T is the temperature, and ε is the permittivity, n_{i0} is the bulk density of ions) and using CGSE units. The electrostatic free energy of a sphere of radius a having a charge q (this charge is associated with the presence of basic ions at the void interface) is

$$W = \frac{\varepsilon}{8\pi} \int_V E^2(r)\, dV = \frac{\varepsilon}{8\pi} \int_a^\infty 4\pi r^2 E^2(r)\, dr = \frac{\varepsilon}{2} \int_a^\infty r^2 E^2(r)\, dr.$$

Since the field strength is $E(r) = \dfrac{q}{\varepsilon r^2}$, we have $W = \dfrac{q^2}{2\varepsilon} \int_a^\infty \dfrac{dr}{r^2} = \dfrac{q^2}{2\varepsilon a}$. Prior to the void formation, the free energy of N_i dissolved ions with charge e (treated as spherical particles of radius δ_i) is $\dfrac{N_i e^2}{2\varepsilon \delta_i}$; it is the so-called hydration energy. The charge of a void whose surface has N_i adsorbed ions is $q = N_i e$ (after the counter-ions have diffused into the solution). Then, the electrostatic free energy of the void is $W = 4\pi r^2 \sigma + \dfrac{(N_i e)^2}{2\varepsilon r}$, where the first and second terms represent the free energy of surface tension and the free energy of the adsorbed ions (σ is surface tension). Note that this expression for the free energy does not include the contribution of the electrical double layer that screens the adsorbed ion charge $q = N_i e$, because it cannot develop at the stage of void formation. The resulting Gibbs free energy is

$$\Phi = W + PV = 4\pi r^2 \sigma + \frac{(N_i e)^2}{2\varepsilon r} + \frac{4\pi}{3} r^3 P_{atm}, \tag{28}$$

where P_{atm} is the pressure of ambient gas squeezing the void (see Eqn. (1)); note that the last term in this equation is extremely small as it is proportional to r^3, where r is extremely small in size, see below. Therefore, the increase in the Gibbs free energy is

$$\Delta\Phi(r) = 4\pi r^2 \sigma + N_i^2 e^2 / 2\varepsilon r + \frac{4\pi}{3} r^3 P_{atm} - N_i e^2 / 2\varepsilon \delta_i. \tag{29}$$

The probability of formation of a void of radius r is proportional to $\exp\left(-\dfrac{\Delta\Phi(r)}{kT}\right)$. When N_i is held constant, its minimum corresponds to $\Delta\Phi_{min}(r) \equiv \Delta\Phi(r)\big|_{r=r_0} = \Delta\Phi(r_0)$. The radius r_0, where $\Delta\Phi$ reaches the minimum, is determined by two conditions:

$$\frac{\partial}{\partial r}\Delta\Phi(r)\bigg|_{r=r_0} = 0, \tag{30}$$

$$\frac{\partial^2}{\partial r^2}\Delta\Phi(r)\bigg|_{r=r_0} > 0. \tag{31}$$

Using expression (30), (31) we arrive at

$$\frac{\partial}{\partial r}\Delta\Phi(r)\bigg|_{r=r_0} = 8\pi r_0\sigma - \frac{(N_i e)^2}{2\varepsilon r_0^2} + 4\pi r_0^2 P_{atm} = 0, \tag{32}$$

$$\frac{\partial^2}{\partial r^2}\Delta\Phi(r)\bigg|_{r=r_0} = 8\pi\sigma + \frac{(N_i e)^2}{\varepsilon r_0^3} + 8\pi r_0 P_{atm} > 0. \tag{33}$$

It is easy to assure that the inequality (33) is satisfied at any positive r, and the root of the equation (32) corresponds to a minimum of $\Delta\Phi(r)$. We represent this equation as

$$r_0^3\left(1 + \frac{P_{atm}r_0}{2\sigma}\right) = \frac{(N_i e)^2}{16\pi\varepsilon\sigma}. \tag{34}$$

Estimations show that for the realistic values of N_i the following inequality should be met

$$\frac{P_{atm}}{2\sigma}\left(\frac{(N_i e)^2}{16\pi\varepsilon\sigma}\right)^{1/3} \ll 1, \tag{35}$$

and the root of Eqn. (34) is

$$r_0 = \left(\frac{(N_i e)^2}{16\pi\varepsilon\sigma}\right)^{1/3}. \tag{36}$$

The values of N_i satisfying (35) are estimated by rewriting this condition as follows:

$$N_i^{2/3} \ll \frac{2}{P_{atm}}\left(\frac{16\pi\varepsilon\sigma^4}{e^2}\right)^{1/3} = \frac{2\cdot10^{-6}}{P_{atm}[atm]}\left(\frac{16\pi\cdot80\cdot(73)^4}{2.3\cdot10^{-19}}\right)^{1/3} \sim \frac{10^5}{P_{atm}[atm]}.$$

Here, we use the values of the material constants of water measured in CGSE units under normal conditions, and P_{atm} [atm] is the ambient gas pressure measured in atmospheres, P_{atm} = 1 atm by definition. We see that expression (36) for the void radius r_0 holds even for very large N_i. Since the increase $\Delta\Phi$ should be minimized as a function of N_i as well as of r, we substitute $r = r_0 = \left(\dfrac{(N_i e)^2}{16\pi\varepsilon\sigma}\right)^{1/3}$ into expression (29) and find that its derivative with respect to N_i is

$$\frac{\partial \Delta\Phi}{\partial N_i} = \frac{N_i e^2}{\varepsilon r_0} - \frac{e^2}{2\varepsilon\delta_i}.$$

The derivative is positive if N_i satisfies the inequality

$$N_i^{1/3} > \left(\frac{e^2}{8 \cdot 16\pi\varepsilon\sigma\delta_i^3}\right)^{1/3} \approx 1/2,$$

which is obviously true. Thus, $\Delta\Phi(r_0)$ is an increasing function of N_i. We presume that $N_i = 6$. An estimate of N_i can be obtained on the basis of other considerations. Indeed, two rectangular pyramids with a common base inscribed in a sphere have exactly six vertices. The Coulomb repulsion between the ions occupying the vertices prevents a spherical void from collapsing. While any of a variety of different symmetric polyhedrons can obviously be inscribed in a sphere, the minimal number of vertices corresponds to the regular tetrahedron (four), two regular tetrahedrons with a common base (five), the two pyramids with a common base (six), and a cube (eight). Since the void radius is $r_0 \sim N_i^{2/3}$, any of the numbers 4, 5, 6 and 8 can be used to obtain a qualitative estimate of r_0.

Under these assumptions (and having taken for definiteness $N_i = 6$), the void radius given by expression (36) is

$$r_0 = \left(\frac{9e^2}{4\pi\varepsilon\sigma}\right)^{1/3} \approx 3 \text{ Å;} \tag{37}$$

this is the equilibrium radius of the bubston nucleus. The estimation obtained is in conformity with our initial assumption that the size of spatial domain of the amorphous structure, where the disorder parameter η grows exponentially (the early stage of nucleation, see Figure 4 and Eqn. (26), is of the order of ~ 1 nm^3. Indeed, the bubble with the radius r_0 occurs at the final stage of nucleation as a result of balance of the surface tension forces and the Coulomb repulsion forces generated by the adsorbed ions. At the same time the surface tension forces start acting immediately after the interface "the void – liquid" was formed, while the ions are adsorbed yet in the process of collapsing the void. It is obvious that the surface tension should result in reduction of the initial size of the void.

Further, it is necessary to estimate (not involving the thermodynamic reasonings) the time of duration of the final stage of nucleating the bubston nucleus, whose size is approximately equal to 3 Å. Let's assume that at the early stage of nucleation (remember that this stage is related to the exponential growth of the disorder parameter η, see Figure 4 and Eqn. (27)) the molecules of the dissolved gas penetrate into the growing cavities. After the interface formation (the final stage of the nucleation) the void starts to be compressed by the surface tension forces and external pressure P_{atm}. If the void manages to be collapsed before the adsorption of ions at its interface (the number N_i of such ions should be necessary for keeping the void from collapse and in accordance with our estimates is equal to six, see above), such void cannot turn into a thermodynamically steady bubston nucleus with radius $r_0 \approx 3$ Å. Apparently inside the steady nucleus of that size the molecules of dissolved gas cannot be present, i.e. such void should be empty. Indeed, by estimations, if at least one molecule of dissolved gas were to exist inside the sphere of such radius, this molecule would create the pressure of about 10^3 atmosphere at room temperature, i.e. such gas bubble would not be steady with respect to diffusion processes, see our reasonings in the Introduction[1]. It is clear however that the pressure of such magnitude is capable of keeping the void from collapsing before the ionic adsorption processes are complete. It follows that the kinetics of the void collapse is controlled by diffusion of gas molecules across "the void – liquid" interface, as this process is rather slow. Having designated the duration time of the final stage as $\tau \sim \dfrac{S}{D}$, where S is the interfacial area, and D is diffusivity of gas molecules in the liquid, let us estimate the time τ for room temperature and $r_0 \sim 3$ Å, assuming the atoms of He (its diffusivity $D_{He} \sim 7 \cdot 10^{-5}$ cm^2/s, see [66 - 68]) or the molecules of nitrogen ($D_{N_2} \sim 2 \cdot 10^{-5}$ sm^2/s, see [69]) are initially situated inside the void. Therefore, contrary to the case of nitrogen, where $\tau \approx 10^{-9}$ s, in the case of helium we have $\tau \approx 2.8 \cdot 10^{-10}$ sec. Thus, in the first case the ionic adsorption process can be completed, while in the last case the ions probably will not manage to be adsorbed at "the void – liquid" interface. Note that the process of ionic adsorption is essentially determined by the concentration of ions: the higher this concentration, the more quickly the ions can adsorb at the bubble interface. This is in agreement with the results of our experimental study [23], where we explored the specific effect of different gases on the bubston formation in water, As was demonstrated in this work, the replacement of dissolved atmospheric gas by helium (under the condition that the concentration of ions is low) results in suppressing the bubston nucleation. This analysis also shows that the bubston nucleation should develop more effectively at growing the concentration of ions; this also conforms to our experimental results, see, e.g., [13-15, 22].

Let us now estimate the value of $\Delta\Phi_{min}$: after substituting (37) into (29) we arrive at

$$\Delta\Phi_{min} = 30\left(\frac{\sigma e^4}{\varepsilon^2}\right)^{1/3}\left[1 - \frac{1}{10}\left(\frac{e^2}{6\varepsilon\delta_i^3}\right)^{1/3}\right] + \frac{3e^2 P_{atm}}{6\varepsilon}. \tag{38}$$

---

[1]For this reason we have neglected the term, corresponding to the contribution of pressure of the gas inside the void in Eqn. (28) for the thermodynamic potential.

Here the second factor in a right-hand part is ~ 1, and the ratio of last term to the first factor is equal to $\dfrac{1}{10}\left(e^2 P_{atm}^3 / \sigma^4 \varepsilon\right)^{1/3}$ and is small in comparison with unity. For I$^-$ ions of radius δ_i = 2.2 Å (in accordance with the data of the review [37], these ions are capable of being adsorbed), we find that the minimal work required for a void of radius $r_0 = 3.0$ Å to occur is $\Delta\Phi_{min} \sim 1$ eV. It is important to note that bubstons can nucleate in a liquid without superheat, negative pressure or dissolved gas supersaturation.

The bubston nucleation rate $\delta n_b / \delta t$ depends on the probability $\exp\left(-\Delta\Phi_{min} / kT\right)$, dissolved gas concentration n_g, dissolved ion concentration n_i, and diffusion rate of basic ions adsorbed to the surface of a bubston nucleus:

$$\delta n_b / \delta t = 4\pi r_0 \cdot D_i n_g \cdot n_i \exp\left(-\Delta\Phi_{min} / kT\right), \tag{39}$$

where D_i is the ion diffusivity. At equilibrium with atmospheric air, this expression yields $\delta n_b / \delta t \sim 3 \cdot 10^{-12} \cdot n_i$ [cm^{-3}] 1/cm^3·s. For purified water with resistivity $\rho = 1.0 \cdot 10^7$ Ω·cm, where the concentration of ions is $4.5 \cdot 10^{14}$ cm^{-3}, we obtain $\delta n_b / \delta t \sim 10^3$ 1/cm^3·s. For solutions having the ion concentration about 10^{-2} M, the corresponding estimate is 10^7 1/cm^3·s.

After a nucleus of radius r_0 has been formed, it starts to grow up to reaching the bubston state. Such growth is the result of the alternate adsorption of the basic ions and counter-ions on the bubston nucleus surface, and the repulsive Coulomb forces effectively extend this surface. The bubble stops growing when at a certain radius of the bubble the diffusive flow of basic ions, directed toward the surface (it is the adsorption flow) becomes to be balanced by the flow of ions of the same sign, directed outward the surface (the desorption flow). The corresponding bubble size is the critical bubston radius. Under these conditions the bubble reaches the state of mechanical and diffusive equilibrium. This way the bubston is being formed. In this process the bubstons are being filled with the molecules of dissolved gas, so such bubbles are no longer empty. The estimation of the equilibrium bubston radius should be made by using the condition of a minimum of the thermodynamic potential of the whole system including the liquid, being in the contact with ambient gas and containing the dissolved ionic component. The solution to this problem is the subject of separate work; the corresponding results will be published elsewhere. It is obvious that the mobility of ions bound to the bubston is lower than that of free ions. Therefore, the conductivity of the liquid must increase when it is completely degassed (bubstons are removed), since the ion concentration does not change. The experiments reported in [70, 71] have shown that degassing does lead to a higher conductivity, which provides indirect evidence for the existence of the bubstonic structure. In the next chapter we analyze the screening of bubstons by mixing basic ions and counter-ions. As will be shown, such screening together with the viscous friction forces applied to the moving charged bubbles provide the mechanism for coagulation of such bubbles with the formation of the bubston clusters.

3. Screening of Strongly Charged Gas Bubbles in Aqueous Salt Solution as Applied to the Mechanism of Their Stabilization

A gas bubble in a liquid solution of electrolyte acquires an electric charge Q_0 owing to certain physicochemical processes on its surface. Under equilibrium conditions, this charge is screened by the ionic shell with electric charge density $\rho(\mathbf{r})$. Henceforth we will consider only spherical bubston of radius R (this radius remains yet uncertain) and, accordingly, assume the charge density $\rho(\mathbf{r})$ is spherically symmetric. The screening (quasineutrality) condition can be written in the form

$$Q_0 + 4\pi \int_R^\infty r^2 \rho(r) dr = 0 \tag{40}$$

(r is the distance from the particle center). Here and below we will consider (for the sake of definiteness) that the sign of Q_0 is negative, i.e. $Q_0 < 0$. Indeed, in accordance with the results of study [37], the ions capable of adsorption at the water interface are essentially negative ions. Generally speaking, the bubbles (not necessarily the bubstons) can be divided into weakly charged when the parameter $B \equiv |ZeQ_0 / \varepsilon RkT| << 1$ and strongly charged when $B >> 1$. We will consider a binary $Z : Z$ solution; the CGSE system of units is used. In the case when $B << 1$, using the Poisson – Boltzmann equation in the Debye – Huckel approximation, we obtain the well-known result [72] for the charge density distribution $\rho(r)$, satisfying condition (40):

$$\rho(r) = -\frac{Q_0 \kappa^2}{4\pi (1 + \kappa R)} \frac{e^{-\kappa(r-R)}}{r}, \ r \geq R, \tag{41}$$

$$\kappa = a_D^{-1} = \sqrt{8\pi Z^2 e^2 n_{i0} / \varepsilon kT} = Z\sqrt{8\pi l_B n_{i0}}. \tag{42}$$

Here, $l_B = e^2 / \varepsilon kT$ is the Bjerrum length ($l_B = 7$ Å for water at room temperature), a_D is the Debye radius, and n_{i0} is the equilibrium density of ions (for $r = \infty$). A distinguishing feature of weakly charged bubbles is that, in accordance with Eqn. (41), the sign of density $\rho(r)$ is opposite to the sign of the surface charge Q_0 for all values of $r \geq R$; in other words, counter-ions always prevail in the screening ionic shell. As a result, ionic screening of charged bubbles is impossible for finite values of radius r, in accordance with Eqn. (40). For a long time, this circumstance was indisputable for specialists in colloidal systems irrespective of values of parameter B. However, some experimental data obtained in recent years obviously have contradicted the concept of screening the ionic shell with absolute prevalence of counter-ions. Above all, this concerns the experimental proof of attraction of likely charged macro-particles in aqueous solutions of salts under certain conditions [73 – 86], which contradicts the Deryaguin – Landau – Verwey – Overbeek (DLVO) theory [87, 88] based on the Debye screening (41). Such contradictory facts also include recent experimental results on electrophoresis of macro-particles in solutions of various salts, indicating an ambiguous dependence of the electrophoretic mobility of particles on the polarity of their surface charge [89 – 91].

These contradictions, as well as other facts, stimulated a number of theoretical publications [92 - 106], in which the so-called inversion (or superscreening) effect of the charge of macro-particles in aqueous solutions of electrolytes is predicted. The effect is that for strongly charged particles the charge density distribution $\rho(r)$ in the ionic shell can reverse its sign upon an increase in r. In this case, the sign of the total charge within the sphere with a certain finite value of radius $r > R$,

$$Q(r) = Q_0 + 4\pi \int_R^r x^2 \rho(x)\, dx, \tag{43}$$

can be inverted, i.e. sgn $Q(r)$ = - sgn Q_0. The existence of this effect opens wide possibilities in interpreting the experimental facts mentioned above. In the cited publications, theoretical concepts are developed, according to which condensation of counter-ions can occur on the surface of a bubble with the formation of a two-dimensional strongly correlated liquid for large values of B. Under certain conditions, the charge density (per unit area) of such a liquid may exceed in absolute value the unscreened (initial) surface charge density $Q_0/4\pi R^2$ that is the main reason for charge inversion. According to the results obtained by these authors, the screening of the inverted charge can be successfully described on the basis of the Poisson – Boltzmann equation.

An important step in explaining the details of the charge inversion effect were publications [107 - 109], which described the results of numerical calculation aimed at the study of the structure of the screening ionic shell of a strongly charged macro-particle in an aqueous salt solution. For instance, in the study [107] a strongly charged macro-particle (a macro-ion) with the charge $|Q_0|$ = 20e and that the ionic salt solution is binary 2 : 2 or 1 : 1 solution was considered. In this case, parameter $B = 20Z(l_B/R)$; consequently, parameter B for 2 : 2 and 1 : 1 solutions was equal to 280/R(Å) and 140/R(Å), respectively. In calculations, only the Coulomb interactions between dissolved ions, as well as between the ions and the charged macroparticle was taken into account for T = 300 K. For various concentrations of both types of solutions, only three values of parameter B were taken into account: B = 14 (2 : 2 solution, R = 20 Å, and 1 : 1 solution, R = 10 Å), B = 22.4 (2 : 2 solution, R = 12.5 Å), and B = 28 (2 : 2 solution, R = 10 Å). The main result obtained in [107] is as follows: inversion is absent for B = 14 and inversion is present for B = 22.4 and 28. Note that qualitatively similar results are summarized in Refs. [107, 108], but in these works the authors considered the planar negatively charged surface. It is important that the sign reversal of charge $Q(r)$ always occurs not jump-wise, as follows from the concepts of the formation of a two-dimensional liquid of counter-ions, but as a result of a smooth tendency of $Q(r)$ to zero for $r \to R + \Delta r$, where Δr is the scale of considerable decrease in the volume density of counter-ions near the particle surface (in the immediate vicinity of the surface, their density is maximal, but finite). Thus, the results obtained in [107 - 109] indicate that, first; the inversion effect is indeed of the threshold type in parameter B. Second, charge inversion occurs not at the particle surface proper, but as a result of a gradual mixing of counter-ions with basic ions upon an increase in $(r - R)$. For large values of B, such mixing leads to the prevalence of basic ions, i.e. to the sign reversal of the charge density $\rho(r)$ and to the emergence of the inversion of charge $Q(r)$. These conclusions make it possible to refute the model of the inversion effect, which stems from the ideas of surface condensation of counter-ions (formation of two-dimensional liquid) and consider another model, which is free from such concepts. This model is proposed here as applied to dilute solutions of electrolytes, for which the parameter $\zeta \equiv \kappa R \ll 1$. We proceed from the fact that the charge Q_0 of a bubble emerges on its surface as a result of adsorption of one of two types of dissolved ions ($Z : Z$ solution). Here, we consider the possibilities of the ion-adsorption mechanism for spherical gas bubbles. Despite the fact that still don't know the exact value of the bubston radius R, we will assume that this size is $R \gg 3\delta_l \sim 0.3 - 1.0$ nm (δ_l is the radius of a liquid molecule). This concept will, of course, be refined in our following publications, but in all cases we will speak of nanometer-size and coarser bubbles. Hereinafter we will consider that we deal with strongly charged particles, i.e. the condition $B \gg 1$ is met in our case.

Quasineutrality condition (40), which should be valid for arbitrary values of B, implies the ionic charge $Q(r)$ contained within a sphere of radius $r > R$ tends to zero with increasing $(r - R)$. This leads to the situation when the value of $Ze|Q(r)|/\varepsilon r k T$ becomes smaller than unity for large values of $(r - R)$ and, hence, the density $\rho(r)$ of the screening ionic charge for such values of $(r - R)$ must have the Debye form, i.e. $\sim (1/r)\exp\left[-\kappa(r - R)\right]$, where κ is defined by (42). On the other hand, as was shown above, in accordance with numerical modeling [107 - 109], the decrease in $\rho(r)$ from the maximal value $\rho(R)$ for small values of $(r - R)$ must be much more rapid, but smooth. On the basis of these two features of the behavior of density $\rho(r)$, our model of screening stems from the representation of $\rho(r)$ in the entire range of $(r - R)$ in the form

$$\rho(r) = C_1 \frac{e^{-b(r-R)}}{r} + C_2 \frac{e^{-\kappa(r-R)}}{r}, \ r \geq R, \tag{44}$$

where C_1 and C_2 are the sought parameters independent of r, having the opposite signs. Here C_1 has the sign opposite to Q_0, whereas C_2 has the same sign as Q_0. We will consider only dilute solutions for which $\zeta \equiv \kappa R \ll 1$. As will be clear from the subsequent analysis, parameter $\xi \equiv bR$ is on the order of or greater than unity, where $b \gg \kappa$ is a yet unknown factor; consequently, $\zeta/\xi \ll 1$ in all cases.

The substitution of Eqn. (44) into (40) and the inclusion of the boundary condition $C_1 + C_2 = R\rho(R)$ gives a system of two equations in C_1 and C_2. Its solution, accurate to terms on the order of $\sim \zeta^2$, has the form

$$C_1 = -\frac{Q_0 F}{4\pi R^2}\left(1 + \frac{f}{F}\zeta^2\right), \ C_2 = \frac{Q_0 f}{4\pi R^2}\zeta^2, \tag{45}$$

where

$$F = -\frac{4\pi R^3 \rho(R)}{Q_0} > 0, \quad f = F\frac{\xi+1}{\xi^2} - 1. \tag{46}$$

For strongly charged particles, the density of charge $\rho(R)$ on the surface of a particle is completely determined by the density $n_i(R)$ of counter-ions having the charge Ze; sgn $\rho(R)$ = - sgn Q_0. Consequently, $\rho(R) = Zen_i(R)$ (as Q_0 is negative), and parameter $F > 0$ is defined by the formula

$$F = \frac{4\pi R^3 Zen_i(R)}{|Q_0|}. \tag{47}$$

Eqs. (46), (47) contain two as yet unknown parameters, viz., the number density $n_i(R)$ of counter-ions at the surface, and $\xi = bR$. In order to determine $n_i(R)$, we use the Boltzmann equation

$$n_i(R) = n_{i0} \exp\left[\frac{Ze\varphi(R)}{kT}\right]. \tag{48}$$

Here, $\varphi(R)$ is the potential of the electric field produced by the charge density distribution $\rho(r)$ in (44). The number density $n_i(R)$ of counter-ions defined in this way increases indefinitely with growing B. It is this circumstance that mainly stimulated other authors to develop the concept of counter-ion condensation at the surface of a bubble, leading to the formation of a two-dimensional liquid layer. Remaining in the framework of our model, we proceed from the fact that the counter-ion density $n_i(R)$ for $B \gg 1$ attains its maximal but finite value $n_i^{\max}(R)$ in accordance with (48). Thus, parameter ξ can be defined in terms of $n_i^{\max}(R)$.

First of all, we determine the dependence of $\varphi(R)$ on ξ and $n_i^{\max}(R)$. The solution of the Poisson equation $\nabla^2 \varphi = -(4\pi/\varepsilon)\rho(r)$ for an arbitrary distribution $\rho(r)$ satisfying quasineutrality condition (40) with the boundary conditions $\varphi(\infty) = 0$, $\varphi'(R) = -Q_0/\varepsilon R^2$ can be represented in the form

$$\varphi(r) = -\frac{4\pi}{\varepsilon r} \int_r^\infty (x^2 - xr)\rho(x)\,dx. \tag{49}$$

Substituting expression (44) into this formula, integrating, setting $r = R$, and taking into account (45), we obtain

$$\varphi(R) = \frac{Q_0}{\varepsilon R}\left(1 - \frac{F_0}{\xi}\right), \tag{50}$$

where, in accordance with Eqn. (47),

$$F_0 = \frac{4\pi R^3 Z e n_i^{\max}(R)}{|Q_0|} = \frac{R n_i^{\max}(R)}{\left|\gamma_i^{AD}\right|}, \tag{51}$$

where γ_i^{AD} is the surface density of adsorbed ions. Substituting now $n_i(R) = n_i^{\max}(R)$ into the left-hand side of Eqn. (48) and expression (50) into the right-hand side, we obtain the following relation between ξ and $n_i^{\max}(R)$:

$$\xi = \frac{F_0}{1-\alpha} = \frac{R}{1-\alpha}\frac{n_i^{\max}(R)}{\left|\gamma_i^{AD}\right|}, \tag{52}$$

where $\alpha = L/B$, $L \equiv \ln\left[n_i^{\max}(R)/n_{i0}\right]$. In this case, quantity f (see the second Eqn. of (46)) has the form

$$f = \frac{1-\alpha}{\xi} - \alpha. \tag{53}$$

It remains for us to determine the value of $n_i^{\max}(R)$. Counter-ions are not capable of adsorption, but being near the surface of a bubble, they are attracted to it by the Coulomb forces exerted by adsorbed ions. At modeling the adsorption of basic ions we make two assumptions. First, we assume that the water molecules form a regular quasi-crystalline lattice of the Wigner-Seitz type at the "water – gas" interface. Second, we consider that the adsorbed basic ions form the sub-lattice, adjacent to the interfacial water molecules; this sub-lattice settles totally inside the bulk of the liquid. This model differs from the ionic adsorption model described in [37], where a spatial arrangement of ions and the water molecules is non-correlated. Thus, the basic ions are located in the centers of their Wigner-Seitz cells, covering the entire surface of the bubble. The maximal surface density $\Gamma_i^{\max}$ of counter-ions obviously corresponds to their arrangement at the same distance $2\delta_i$ from three adsorbed ions; each adsorbed ion mainly interacts with only one counter-ion (see Figure 5). It is this arrangement that corresponds to a stable state. Since each counter-ion in this case corresponds to three individual Wigner-Seitz cells, we have $\Gamma_i^{\max} = 1/3S_{WS} = 1/8\sqrt{3}\delta_i^2$. Assuming $\Gamma_i^{\max} = \left[n_i^{\max}(R)\right]^{2/3}$, we obtain

$$n_i^{\max}(R) = \frac{\delta_l^{-3}}{162(4/3)^{1/4}}, \quad \frac{n_i^{\max}(R)}{\left|\gamma_i^{AD}\right|} = \frac{\delta_l^{-1}}{9(12)^{1/4}}. \tag{54}$$

For aqueous solutions, the number density $n_i^{\max}(R) = 3.1 \cdot 10^{21}$ cm^{-3}. It can be seen from Eqn. (44) that distribution $\rho(r)$ must vanish at a certain point $r = r_0$ if quantities C_1 and C_2 have opposite signs. In accordance with (45), this is valid for $f > 0$ or, due to Eqn. (53), for $\xi < (1 - \alpha)/\alpha$. Introducing a new variable,

$$\lambda = \frac{R/\delta_l}{18(12)^{1/4}} \equiv \frac{F_0}{2}, \tag{55}$$

we have

$$\xi = \frac{2\lambda}{1-\alpha}; \tag{56}$$

consequently, condition $f > 0$ assumes the form of inequality

$$\frac{\alpha}{(1-\alpha)^2} < \frac{1}{2\lambda}. \tag{57}$$

When this inequality holds, we can easily obtain the expression for the position of point r_0:

$$r_0 = R\left[1 + \frac{1}{\xi}\ln\frac{F_0}{f\zeta^2}\right] = R\left\{1 + \frac{1}{\xi}\ln\frac{(\xi/\zeta)^2}{\left[1 - 2\lambda\dfrac{\alpha}{(1-\alpha)^2}\right]}\right\}. \tag{58}$$

At this point, the charge density $\rho(r)$ of the ionic shell changes its sign: sgn $\rho(r < r_0) = -$ sgn $\rho(r > r_0)$, see Figure 6. This leads to sign inversion of the total charge $Q(r)$, occurring at a certain value of radius $r_1 \neq r_0$. Let us determine r_1. On the basis of Eqns. (40), (42), (47), and (48), we have

$$Q(r) = -4\pi\int_r^\infty x^2\rho(x)dx = Q_0\left\{\frac{F_0}{\xi^2}(1+br)e^{-b(r-R)} - f(1+\kappa r)e^{-\kappa(r-R)}\right\}, \tag{59}$$

$$\frac{dQ}{dr} = 4\pi r^2\rho(r). \tag{60}$$

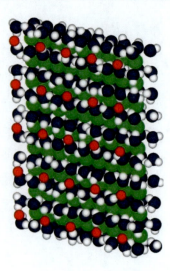

Figure 5. Bubston surface as is seen from the side of water. Molecules of water form the regular Wigner – Seitz cells at the surface. The basic ions are located in the vertices of the corresponding Wigner – Seitz cells, whereas the counter-ions are located farther from the interface.

It follows from (59) that the necessary condition for $Q(r)$ vanishing is the inequality $f > 0$, i.e. inequality (57). In this case, the equation for $y_1 \equiv r_1/R$ has the form

$$y_1 = 1 + \frac{1}{\xi} \ln \frac{1+\xi y_1}{1 - \frac{\alpha \xi}{1-\alpha}} \equiv \frac{r_0}{R} - \frac{1}{\xi} \ln \frac{(\xi/\zeta)^2}{1+\xi y_1}. \tag{61}$$

Since $(\xi/\zeta)^2 \gg 1$, it follows from (59) that radius $r_1 < r_0$; i.e., the sign inversion of the total charge $Q(r)$ (sgn $Q(r < r_1)$ = sgn Q_0 = -sgn $Q(r > r_1)$) occurs at such distances from the bubble surface where the sign of the charge density remains unchanged (sgn $\rho(r_1)$ = sgn $\rho(R)$ = -sgn Q_0, see Figure 6). In accordance with (60), charge $Q(r)$ attains its extreme value at point $r = r_0$. On the basis of Eqns. (58), (59) we obtain the following relation (accurate to terms on the order of $(\zeta/\xi)^2$) for the value of $Q(r_0)$:

$$Q(r_0) = -Q_0 f \left(1 + \xi r_0 / R\right). \tag{62}$$

As the value of r increases further, charge $Q(r)$ tends to zero.

While considering the screening of bubbles by the ionic shell, we tacitly assumed that a bubble is at rest, and its ionic shell is not subjected to hydrodynamic action; such a shell will be referred to as an equilibrium shell. Indeed, bubbles perform thermal Brownian motion; in this case, a compound particle with total charge Q_c, which consists of the initial bubble of radius R and surface charge Q_0 and a certain layer of the liquid with the "frozen-in" ionic shell, distorted by viscous forces, is involved in motion. If the velocity of the particle is small, we can assume that the distortion of the ionic shell under steady-state conditions boils down to washing out (vanishing) of the equilibrium charge density distribution $\rho(r)$ due to viscous forces of a peripheral layer $r > a_c$, while the remaining part of this distribution ($r < a_c$) is spherically symmetric as before. Such a pattern is widely used in theoretical treatment of electrophoresis, see, e.g., [110]. In this case, the radius a_c of a compound particle is usually referred to as the hydrodynamic radius of a bubble and the spherical surface with such a radius is called the glide surface. Our task is to establish the relation between parameters Q_c and a_c of a compound particle and the initial parameters Q_0 and R of the bubble and of the solution itself. As before, we assume the solution is dilute ($\zeta \ll 1$).

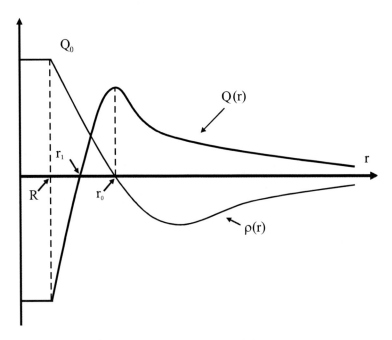

Figure 6. The total charge $Q(r)$ and the charge density $\rho(r)$ in the liquid surrounding the bubston of radius R. пузырек with radius The curve for $Q(r)$ was modeled to have the same form as the corresponding graphs for the symmetric $Z:Z$ electrolyte solution in works [107 - 109].

The physical foundation for solving the formulated problem is that the pressure of viscous forces destroying the ionic shell is always opposed by the electrostriction-induced pressure p_{str} compressing the compound particle and associated with its charge Q_c, which coincides with the charge $Q(r)$ for $r = a_c$ (see Eqn. (43)), i.e.,

$$Q_c = Q_0 + 4\pi \int_R^{a_c} x^2 \rho(x)\,dx = -4\pi \int_{a_c}^{\infty} x^2 \rho(x)\,dx, \qquad (63)$$

and $\rho(r)$ is the equilibrium charge density distribution defined in the previous section. The electric field strength at the surface of a compound particle ($r = a_c$) is $E_c = Q_c/\varepsilon a_c^2$, while the pressure is given by

$$p_{str} = \frac{\varepsilon E_c^2}{8\pi} = \frac{Q_c^2}{8\pi\varepsilon a_c^4}. \tag{64}$$

The stationary state corresponds to the equality of the spherically symmetric pressure p_{str} to the maximal absolute value of the negative pressure of viscous forces (i.e., the pressure of detachment from the rear side as related to motion), which sets in for a certain value of radius a_c of the glide surface. Under the laminar flow conditions for a particle with velocity u, the maximal pressure of detachment is given by [111]

$$p_v = \frac{3\eta u}{2a_c} \tag{65}$$

(η is the viscosity of the liquid). Consequently, for the given velocity u, parameters Q_c and a_c can be defined from the equation

$$\frac{3\eta u}{2a_c} = \frac{Q_c^2}{8\pi\varepsilon a_c^4} \tag{66}$$

and Eqn. (63). In the study of thermal motion of bubbles, we can use the approach based on stochastic differential equation of the Langevin type, see Ref. [112]

$$m_c \frac{du}{dt} = -F(u) + f(t), \tag{67}$$

where

$$F = 4\pi\eta a_c u \tag{68}$$

is the regular Stokes frictional force corresponding to the motion of a gas particle of radius a_c in a liquid of viscosity η (see, e.g., [111]), m_c is the mass of a compound particle, and $f(t)$ is a stochastic force standing for instantaneous random hitches onto the gas particles,

$$\langle f(t) \rangle = 0, \ \langle f(t) f(t+\tau) \rangle = \delta(\tau),$$

where the angular brackets mean the averaging, $\delta(\tau)$ is the Dirac delta – function. It is implied in the framework of this approach that the velocity $u = u(t)$ (here and below, $u(t)$ stands for the velocity component along a certain fixed direction s) is a stationary random process with the average value $\langle u(t) \rangle = 0$, and the stationary autocorrelation function

$\langle u(t)u(t+\tau)\rangle \sim \exp\left(-\dfrac{|\tau|}{\tau_0}\right)$. Here τ_0 is the correlation time of random process $u(t)$. The spectral intensity of this process is expressed by

$$g(\omega) = \frac{2\tau_0}{\pi} \frac{kT/m_c}{1+(\omega\tau_0)^2}, \quad \int_0^\infty g(\omega)\,d\omega = \langle u^2\rangle = \frac{kT}{m_c}, \tag{69}$$

as the Maxwellian distribution should be established in the stationary regime. In this formula the correlation time τ_0 should be close to the relaxation time, standing for appearance of the system on the stationary level; this time depends directly on the dissipative properties of the system. Thus we arrive at the following formula

$$\tau_0 \approx m_c/4\pi\eta a_c. \tag{70}$$

Obviously, it would be incorrect to substitute the quantity $\langle u^2\rangle^{1/2} = (kT/m_c)^{1/2}$ for u into Eqns. (68), (69). In fact, the velocity $u(t)$ can be represented as the sum $u(t) = u_1(t) + u_2(t)$. Indeed, the motion of a compound particle consists of fast "vibrations" with amplitudes smaller than its radius a_c; we will denote the characteristic velocity of such process as $u_1(t)$. It is easy to show that the relative amplitude of such vibrations is $\langle u_1^2\rangle^{1/2}(\tau_0/a_c) \sim 10^{-2}/a_c^{1/2}[\mathrm{nm}] \ll 1$. Besides, there exist "smooth" displacements with amplitudes comparable to a_c. The second type of motion is interesting, since only such movements of a compound particle may lead to viscous washing out of the peripheral layer of the ionic shell. The smooth velocity process will be denoted as $u_2(t)$; this process corresponds to the low-frequency part of spectrum $g(\omega)$ in the interval $(0, \tau_2^{-1})$, where $\tau_2 \gg \tau_0$, and can be represented as random process $u(t)$ averaged over the time interval τ_2:

$$u_2(t) = \frac{1}{\tau_2} \int_{-\infty}^{t} e^{-(t-t')/\tau_2} u(t')\,dt', \ \tau_2 \gg \tau_0. \tag{71}$$

The spectral intensity $g_2(\omega)$ of this process is given by

$$g_2(\omega) = \frac{g(\omega)}{1+(\omega\tau_2)^2}, \quad \int_0^\infty g_2(\omega)\,d\omega = \langle u_2^2\rangle = \langle u^2\rangle\frac{\tau_0}{\tau_2}, \tag{72}$$

and the spectral intensity for the complete random process $u(t) = u_1(t) + u_2(t)$ can be written in the form

$$g(\omega) = g_1(\omega) + g_2(\omega), \quad g_1(\omega) = g(\omega)\frac{(\omega\tau_2)^2}{1+(\omega\tau_2)^2}. \tag{73}$$

Spectrum $g_1(\omega)$ stands for a rapid component $u_1(t)$ of velocity $u(t)$; it vanishes for $\omega = 0$ and has a width $\Delta\omega_1 \cong \tau_0^{-1}$. The rapid component $u_1(t)$ is responsible only for small vibrations of the gas particles under consideration, and makes zero contribution to their diffusive movements. Indeed, as is known, for the random Wiener (diffusion) processes the diffusion coefficient is expressed as $D = \pi g(0)$; thus for the process $u_1(t)$ the diffusion coefficient $D = 0$. Smooth displacement of particles are controlled by spectrum $g_2(\omega)$ having a width of $\Delta\omega_2 \cong \tau_2^{-1}$ as τ_2 is the correlation time of random process $u_2(t)$. Washing out of the peripheral layer is most effective on intervals of unidirectional motion of a particle, i.e., in the intervals between successive extrema of process $u_2(t)$. It is well known (see, e.g., [113]) that the mean value of such intervals coincides in order of magnitude with the correlation time of a random process, i.e., with time τ_2. During this time, the particle is displaced on the average by $\left\langle \Delta S^2 \right\rangle^{1/2} = \left\langle u_2^2 \right\rangle^{1/2} \tau_2 \approx a_c$. However $\tau_2 = \left\langle u^2 \right\rangle \tau_0 / \left\langle u_2^2 \right\rangle$ in accordance with the second Eqn. in (72). This leads to the expression $\left\langle u_2^2 \right\rangle^{1/2} \approx \left\langle u^2 \right\rangle \tau_0 / a_c \approx kT/4\pi\eta a_c^2$, where we used (69), (70). After substituting $\left\langle u_2^2 \right\rangle^{1/2}$ instead of u into formula (66), we arrive at

$$\frac{Q_c^2}{2\varepsilon a_c} = \frac{3kT}{2}, \tag{74}$$

where $(3/2)kT$ is the average thermal energy of the particle; the left-hand side of this formula defines the energy of the electric field created by the compound particle. Note that the formula establishing the relation between a_c and Q_c was derived on the basis of Eqns. (65), (66), (68), and (69), i.e., universal results of macroscopic electrodynamics, hydrodynamics, and the theory of Brownian motion. Thus, it is not associated with our model (including the condition $\kappa R \ll 1$) and, being independent of viscosity η, is of the thermodynamic nature: in the steady state, the energy of the electric field of a compound particle for an arbitrary value of the adsorbed charge Q_0 is a function of the temperature of the medium only, and is exactly equal to its mean kinetic energy. It should also be noted that this result is not associated with the sign inversion effect and is also valid for weakly charged macroparticles, for which $B \ll 1$, and the Debye screening (41) takes place. The charge of a compound particle in this case is given by

$$Q_c = Q_0 \frac{1+\kappa a_c}{1+\kappa R} e^{-\kappa(a_c-R)}. \tag{75}$$

Here, for all values of a_c, charge Q_c has the same sign as charge Q_0 and tends monotonically to zero with increasing a_c. As a result, the system of Eqns. (74) and (75) has only one pair of roots (Q_c, a_c) for all possible temperatures, which corresponds to a single type of compound particle. In the case of strongly charged particles, in accordance with Eqns. (46) and (59), the quantity Q_c^2 as a function of a_c has a two-hump shape (Figure 7), and the system of Eqns. (59), (74) for $T < T^*$ (the temperature T^* will be defined below) has three pair of roots $\left(Q_c^{(i)}, a_c^{(i)}\right)$ ($i = 1, 2, 3$), the relation between Q_c and a_c in each pair being

$$\left(Q_c^{(i)}/e\right)^2 = 3\left(a_c^{(i)}/l_B\right) \ (i = 1, 2, 3). \tag{76}$$

As can be seen, only two pairs of roots $\left(Q_c^{(1)}, a_c^{(1)}\right)$ and $\left(Q_c^{(3)}, a_c^{(3)}\right)$ correspond to stable states of compound particles, i.e., only these two types of such particles can exist, while parameters $\left(Q_c^{(2)}, a_c^{(2)}\right)$ are found to be impracticable. This follows from the fact that for radii $a_c < a_c^{(2)}$ (but for $a_c > a_c^{(1)}$), pressure $p_{str} < p_v$ (see Figure 7); consequently, the particle size should continue to decrease in the course of Brownian motion (ionic shell is washed out) down to value $a_c^{(1)}$. For $a_c > a_c^{(2)}$ (but $a_c < a_c^{(3)}$), pressure $p_{str} > p_v$, and the motion does not hamper the growth of the ionic shell of a compound particle up to value $a_c^{(3)}$. At the same time, radii $a_c^{(1)}$ and $a_c^{(3)}$ are stable since the condition $\text{sgn}\,(a_c - a_c^{(1,3)}) = -\text{sgn}\,(p_{str} - p_v)$ holds for small deviations ($a_c - a_c^{(1,3)}$).

Thus, compound particles can be divided into two types: "small" particles with radii $a_c^{sm} \equiv a_c^{(1)}$ and charges $(Q_c^{sm}/e) = (3\,a_c^{sm}/l_B)^{1/2}$ ($\text{sgn}\,Q_c^{sm} = \text{sgn}\,Q_0$) and "coarse" particles with radii $a_c^{cr} \equiv a_c^{(3)}$ and charges $-Q_c^{cr}/e = 3\left(a_c^{cr}/l_B\right)^{1/2}$ ($\text{sgn}\,Q_c^{cr} = -\text{sgn}\,Q_0$). Such a division of compound particles into two types with opposite charge polarities is a peculiar manifestation of the charge inversion effect; its presence follows from independent data on parameters of compound particles.

In accordance with Eqns. (62), (74), temperature T^* is defined by the formula

$$kT^* = \frac{Q_0^2 f^2 \left(1 + \zeta r_0/R\right)^2}{3\varepsilon r_0}. \tag{77}$$

Using formulas (52), (53), (58) and taking into account the fact that $\alpha \sim 0.1$ and $\zeta \leq 0.1$, we obtain the approximate formula $\left(T^*/T\right) \approx 10^2 Z^2 \left(Rl_B \left|\gamma_i^{AD}\right|\right)$. For aqueous solutions ($Z = 1$), this gives $\left(T^*/T\right) \geq 10^3$, i.e., for any temperatures T, all three pairs of roots

$\left(Q_c^{(i)}, a_c^{(i)}\right)$, i.e., both types of compound particles should exist. It can be easily verified that $a_c^{(1,2)} = r_1\left[1 \mp O\left(T/T^*\right)^{1/2}\right]$. Since $\left(T/T^*\right)^{1/2} \ll 1$, for small compound particles we have

$$a_c^{sm} \approx r_1, \quad \frac{|Q_c^{sm}|}{e} \approx 4\sqrt{\left(\lambda + \frac{1}{2}\right)}, \tag{78}$$

where the parameter λ is given by Eqn. (55). The roots $\left(Q_c^{(i)}, a_c^{(i)}\right)$ for coarse compound particles can be deduced from Eqn. (74), into which we should substitute only the second term of formula (59) (the first term is exponentially small since $\left(a_c^{(3)} - R\right)b \gg 1$).

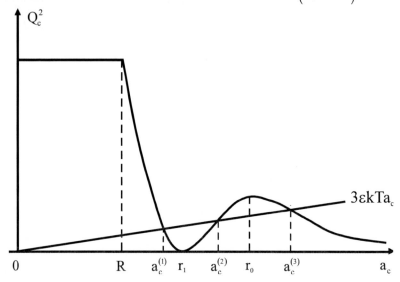

Figure 7. The graphic solution of the equations (59) and (74).

Using (77) and the above approximate formulas for (T^*/T), we can obtain the following approximate results (we assume $(r_0/R) \sim 3$, $(T^*/T) \geq 10^3$):

$$\frac{a_c^{cr}}{R} = \frac{1}{2\zeta}\ln\left[\frac{Q_0^2 f^2}{3\varepsilon r_0 kT} \frac{r_0\left(1+\zeta a_c^{cr}/R\right)^2}{a_c^{cr}/R}\right] \approx \frac{1}{2\zeta}\ln\left(\frac{3T^*}{T}\right). \tag{79}$$

For $\ln(3T^*/T) \approx 10$ and $\zeta = 0.1$, we obtain

$$a_c^{cr} \approx 50R, \quad \frac{Q_c^{cr}}{e} = -\left(\frac{42}{1+2/4\lambda}\right)^{1/2}\left(\frac{Q_c^{sm}}{e}\right). \tag{80}$$

It is clear that the absolute values of the corresponding charges obey the inequality $\left|Q_c^{cr}\right| > \left|Q_c^{sm}\right|$. It is well to say that parameters a_c^{sm} and Q_c^{sm} do not depend on the adsorbed charge Q_0 to within terms on the order of $(T/T^*)^{1/2}$, while a_c^{cr} and Q_c^{cr} depend on it only logarithmically.

The analysis of the effect of Brownian motion of charged bubbles on their screening shows that the effect of charge inversion manifests itself in the formation of two types of compound-particles differing in their radii by an order of magnitude and having opposite charges, see (78) – (80). Thus, the mutual attraction of such compound-particles can play an essential role in the cluster formation. As the mass of a coarse compound-particle and its charge (by the absolute magnitude) exceed drastically the mass and the charge of a small compound-particle, it seems possible to consider the particle-cluster aggregation model [114]. In this model, a small, light and highly mobile compound-particle coagulates with a heavy, coarse and low mobile compound particle, thus forming a charged dimeric particle. As the attraction between the oppositely charged compound-particles, which are far from one another, is the dominant factor, we can use the ballistic particle-cluster model of bubbles aggregation [114]. As a result of such process, a cluster composed of the small and the coarse compound-particles occurs. The charge of such cluster is obviously defined by a ratio between small and coarse compound-particles in the cluster. It is clear however that the cluster, consisting of limited number of compound-particles of both signs, cannot be neutral. The sign of the charge of such cluster should be determined in direct experiment. Let us note that the offered model of the coagulation of strongly charged macroparticles should be considered as an alternative in relation to the known model of coagulation put forward in the DLVO theory [87, 88]. In the next chapter we analyze the fractal properties of the bubston clusters based on the experimental data obtained earlier..

4. MODELING THE AGGREGATION OF BUBSTONS; THE CALCULATION OF THE SCATTERING MATRIX OF BUBSTON CLUSTERS

We have modeled laser light scattering characteristics of bubston clusters via the calculation of their scattering matrix, because the light scattering measurements offer an effective contactless method for studying disperse media, see our experimental studies [12 – 15]. Here we refer to Muller matrix (4x4), which completely describes the polarization state transformation of an incoming light after interaction with an optical object. In order to generate examples of possible stochastic realizations of the clusters, a numerical algorithm of their formation, resulting from a certain mechanism of aggregation of spherical air nanobubbles, was implemented. The data obtained with the help of laser interference phase microscope allow us to make the assumption that the structure of bubston clusters is not very friable and their shape does not strongly deviate from spherical. Therefore, the cluster-particle aggregation model was chosen [114, 115], where a growing cluster is formed owing to sequential attachment of separate particles (in our case, bubstons) to it. As an argument in favor of this model, there can serve the fact that, contrary to, for instance, the aggregates of solid particles in air, the viscosity of the liquid medium decelerates the relative motion of large aggregates and thereby hinders their joining. Being oriented to the opportunity of the

Coulomb attraction between a bubston and the cluster, see the previous chapter; in this work we used the ballistic aggregation model with zero aiming distance. In such model, the trajectories of all bubstons are rectilinear and directed to the center of mass of the growing cluster [114]. Meanwhile, the bubstons of variable radius were considered and the distribution of their radii r was assumed to satisfy the logarithmically normal law

$$p(r) = (2\pi B)^{-1/2} r^{-1} \exp\left[(2B)^{-1} \ln^2\left(A^{-1}r\right)\right].$$ (81)

In the light scattering theory the most convenient similarity criterions of different dispersed systems are two integral parameters, characterizing their particle size distributions, the effective radius r_{eff} and the effective variance v_{eff} that are defined as follows:

$$r_{\text{eff}} = \frac{\int\limits_0^\infty r^3 p(r)\,dr}{\int\limits_0^\infty r^2 p(r)\,dr}, \qquad v_{\text{eff}} = \frac{\int\limits_0^\infty (r - r_{\text{eff}})^2 r^2 p(r)\,dr}{r_{\text{eff}}^2 \int\limits_0^\infty r^2 p(r)\,dr}.$$ (82)

In the case of lognormal size distribution these parameters take form

$$r_{\text{eff}} = A \cdot \exp(5B/2), \qquad v_{\text{eff}} = \exp(B) - 1.$$

Alongside with the number of monomer particles within a cluster N, these parameters are independent variables in the aggregation model of the cluster that define the average components of the light scattering matrix. The results of the numerical computations of the scattering matrix for a random sampling of clusters, generated in accordance with the model described above, are used here to compare them with the experimental data on the scattering matrix of liquid media. The model parameters r_{eff}, v_{eff}, N are subject to determination, and their values are selected by the principle of the best conformity between calculated and experimentally measured scattering-matrix elements. The comparison of the experimental data with the calculations, obtained through the diffusion limited aggregation model describing the other extreme case of the absence of non-contact interaction between a bubston and a growing cluster, where separate bubstons undergo Brownian motion, and also through cluster – cluster aggregation models, will be a subject of further particulars.

The dimension of a physical cluster with a finite number N of particles - bubstons we determine from the dependence $V(R)$ of the cumulative volume of bubstons, confined to a sphere of radius R, circumscribed around the cluster center, on the magnitude of this radius [114, 115].

For simplicity we consider, that a bubble is wholly inside the sphere, if at least the bubble center belongs to it; the error of such approach is negligibly small at $r_j \ll R$. In this approximation we have

$$V(R) = \frac{4\pi}{3} \sum_{|R_j - R_0| < R} r_j^3. \quad (83)$$

Here R_j - the radius vector of the center of a bubble, r_j - the radius of this bubble, and

$$R_0 = \left(\sum_{j=1}^{N_c} r_j^3\right)^{-1} \sum_{j=1}^{N_c} r_j^3 R_j \quad (84)$$

is the vector of the coordinates of the mass center of the cluster. The dependence $V(R)$ is approximated with good precision everywhere (except for more friable area close to the border of the cluster and also close to the area of small values of the sphere radius R, where R is less or approximately equal to $(2 - 3) \cdot r_{eff}$), by the power-law function

$$V(R) = C \cdot R^{D_c}, \quad (85)$$

where C is a dimensional constant. The index D_c defines effective dimension of the real physical cluster. An example of embodying such a dependence is shown in Figure 8. This dependence corresponds to a computer model of the cluster, depicted in Figure 9; this cluster was obtained as a result of the ballistic cluster-particle aggregation of poly-disperse spheres with zero aiming distance (the model of falling towards the cluster center) with the help of a computing algorithm, in which the lognormal generators of random numbers are applied.

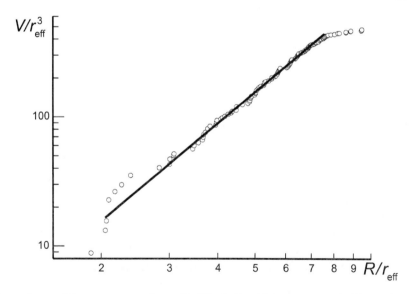

Figure 8. Dependence of the aggregate volume V of the balls which centers are inside a sphere of radius R, circumscribed around the center of a cluster, on this radius - is shown by points; power-law approximation of this dependence - solid line. The radii of the balls constituting the cluster are distributed lognormally with the parameters $r_{eff} = 90$ nm, $v_{eff} = 0.02$; the whole number of particles in the cluster $N = 120$.

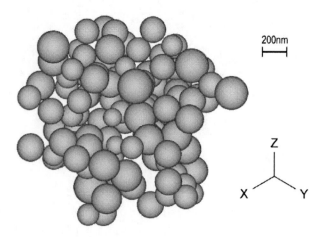

Figure 9. Stochastic model of ballistic cluster-particle aggregation cluster with the monomer size distribution parameters: r_{eff} = 90 nm, v_{eff} = 0.02, N = 120.

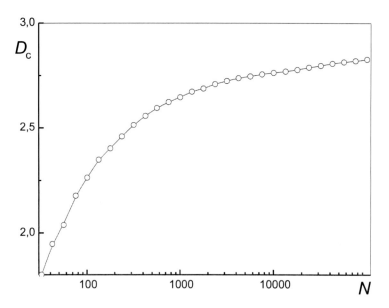

Figure 10. Dependence of the effective cluster dimension of a D_c (averaged over a sampling of 100 cluster realizations) on the number of particles N for ballistic cluster-particle aggregation model with zero aiming distance in the case of lognormal radius distribution of spherical monomer particles with the fixed parameters r_{eff} = 90 nm, v_{eff} = 0.02.

The approximating line in the plot (Figure 8) gives us the estimate of the fractal dimension of this sample of real cluster D_c (Figure 9), which in this particular case is equal to 2.6. It is necessary to note that for real physical clusters made up of a finite number of particles N, the dimension D_c, defined according to (85), generally gives a random quantity, whose ensemble average depends on the particles number N. In Figure 10 the dimension D_c as an average over an ensemble of one hundred realizations of the ballistic cluster-particle aggregation clusters, generated by a computer at the same values of r_{eff}, v_{eff} and N, is shown as a function of the quantity N.

It is known (see, for example, [114, 116]), that in the extreme case of infinite number of particles the dimensionality of such clusters D_c tends to the Euclidean dimension of the space, which in our case is three. However, as opposed to the idealized mathematical models of infinite clusters the real clusters (see Figure 10) have fractal dimension D_c, differing from three. According to preliminary estimations of the relation between the sizes of the clusters and the bubstons, basing on the interference-microscopy data [12 - 15], the number of particles in the bubston cluster N belongs to a range of 10^2 - 10^3. Hence, the estimate of the dimension D_c for the clusters of such extent (see Figure 10) varies from 2.3 to 2.6. Thus, the clusters that were used by us in calculating the characteristics of light scattering have an effective dimension, which is remarkably distinct from the dimension of space; so they should be treated as fractal objects.

In order to calculate the scattering matrix for bubston clusters we used a program code, based on development in far-field solution for the electromagnetic scattering by aggregates of spheres [117]. The computations were performed in two situations. The scattering matrix was calculated first for a set of spatial rotations of a single cluster and second for rather large samplings of same-type clusters varying in size (specifically, in the number of constituent particles). Computer-generated stochastic cluster realizations were assembled by the ballistic mechanism of aggregation and corresponded to the lognormal distribution of monomer sizes with r_{eff} = 90 nm, v_{eff} = 0.02. We compare two stochastic samplings of such clusters, that were obtained by randomizing the number of monomers in a cluster N according to the lognormal distributions (less and more wide) with the average $\langle N \rangle$ = 100 and two different standard deviations σ_N =50, 100, respectively. This magnitude of average number of bubstons in a cluster was chosen due to the experimental scattering indicatrix (the element F_{11} of the scattering matrix) of aqueous salt solutions, containing dissolved air [12, 13]. The scattering matrix elements, calculated as functions of scattering angle for light with a wavelength of 532 nm, are shown in a Figure 11, 12, where they are confronted with the experiment in water and salt aqueous solution. The theoretical curves are related to light scattering by ballistic aggregation clusters as averaged over spatial rotations of the cluster depicted at Figure 11 and over the sampling of clusters with the above indicated ensemble parameters. From the examination of the angular variation of matrix elements we put forward an assertion about the general form of the ensemble scattering matrix:

1. A system of randomly distributed clusters with respect to size and orientation composed of spherical particles that are small relative to lightwavelength (Rayleigh particles) is described by light scattering matrix, which has a block diagonal structure: "zero" matrix elements are at least an order of magnitude smaller than the minimal nonzero matrix element.

2. The scattering indicatrix (element F_{11}) is characterized by an angular profile that is typical for some nonspherical Mie particles (which size is about lightwavelength) and depends mainly on the monomer size and the number of monomers in clusters. While the other nonzero matrix elements demonstrate an angular behavior that is similar to formulas for elements of the Rayleigh scattering matrix.

3. The small differences from the Rayleigh matrix consist of the appearance of local periodic extremums, coupled to the mean characteristic size of clusters as wholes,

and in some descent of the elements f_{12} and f_{22} nearby the region of transverse scattering. The extremums decay quickly with increase of the scattering angle.

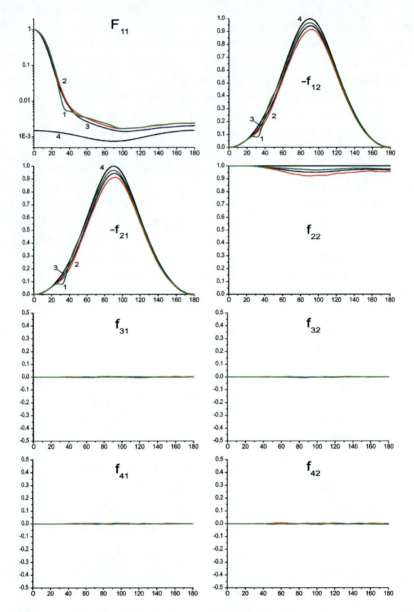

Figure 11. Normalized scattering matrix elements $f_{ij} = F_{ij}/F_{11}$ (F_{11} is normalized by its value at $0°$) F_{11}, f_{12}, f_{21}, f_{22}, f_{31}, f_{32}, f_{41}, f_{42} for the bubston clusters. The curve (1) corresponds to the average over a set of equidistant rotations over three Euler angles with a step of $20°$ for one ballistic cluster with parameters of the monomer size distribution $r_{\text{eff}} = 90$ nm, $v_{\text{eff}} = 0.02$, $N = 120$, which is illustrated in Figure 9; the curves (2), (3) correspond to the averages over two systems of clusters of the same type with varying monomer number, having distribution parameters $\langle N \rangle = 100$ and $\sigma_N = 50, 100$.

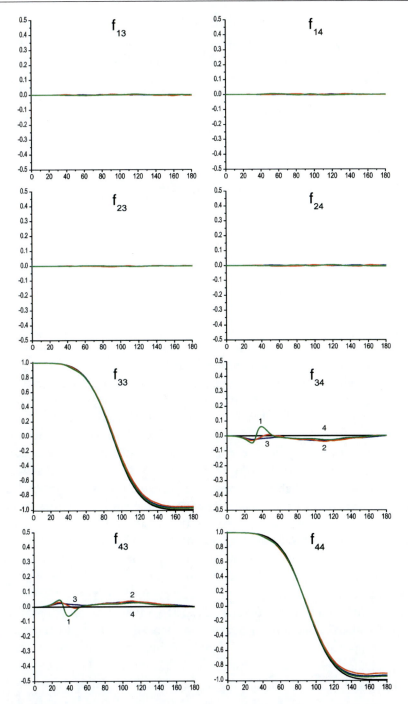

Figure 12. Normalized scattering matrix elements $f_{ij} = F_{ij}/F_{11}$ (F_{11} is normalized by its value at $0°$) f_{13}, f_{14}, f_{23}, f_{24}, f_{33}, f_{34}, f_{43}, f_{44} for the bubston clusters. The curve (1) corresponds to the average over a set of equidistant rotations over three Euler angles with a step of $20°$ for one ballistic cluster with parameters of the monomer size distribution $r_{\text{eff}} = 90$ nm, $v_{\text{eff}} = 0.02$, $N = 120$, which is illustrated in Figure 9; the curves (2), (3) correspond to the averages over two systems of clusters of the same type with a varying monomer number, having distribution parameters $\langle N \rangle = 100$ and $\sigma_N = 50, 100$.

As is seen from the graph, the oscillations in the angular dependencies of a single cluster realization get smoothed when we proceed to the ensembles. According to the analytic estimation made within Born approximation, the position of the first angular with corresponds to the scale about 0.5 µm. In the case of very wide distribution over cluster sizes the extremums almost disappear. The best conformity with the experiments that we reported in [12 - 15] takes place exactly for the wide distribution, as the experimental angular profiles of scattering-matrix elements reveal a rather monotonic character.

CONCLUSION

We have shown that nanometer-sized bubbles in an aqueous ionic solution at room temperature can be nucleated in the two-stage process. The first (early) stage includes the spontaneous growth of the disorder parameter of a liquid system at the nanometer scale, which is accompanied by the density decrease and capturing the molecules of dissolved gas inside the voids being formed. The second (final) stage includes the formation of "the void – liquid" interface and the adsorption of the basic ions at this interface. The adsorbed ions keep the void from collapse, and the void appears in the thermodynamically equilibrium state. The critical size of such nucleus was estimated from the condition of minimum of the thermodynamic potential and appears to be approximately 3 Å.

Besides, the model of the screening of strongly charged particles shows that the charge inversion effect can be interpreted in the framework of a modified Poisson–Boltzmann model. The quantitative basis of this model is the assumption that liquid molecules are arranged on the surface of bubbles in the form of a densely packed two-dimensional hexagonal lattice. This assumption makes it possible to connect the surface density of the basic ions and the bulk density of counter-ions on the surface of a bubble. Our analysis of the effect of the Brownian motion of charged bubbles shows that the charge inversion effect is manifested in the formation of two types of compound particles of the opposite polarities, which strongly differ in their radii and in the absolute values of their charges. In this case, the equipartition law (Eqn. (74)) is observed for the mean energy of the electric field, created by both types of compound particles. This result transfers our concepts of the interaction of the charged bubbles in aqueous solutions of electrolytes to a new qualitative level. For example, according to these conceptions, coagulation of identical nanobubbles in a solution should be attributed not to the interaction of likely charged particles (as was done in numerous theoretical publications), but to the interaction of oppositely charged compound particles. Naturally, the proposed model requires further refinement, primarily in respect of the requirement that the ionic solution must be dilute ($\zeta = \kappa R \ll 1$). Of course, the charge inversion effect should also exist in the case where this requirement is not satisfied, i.e. when the solution is not dilute. It can obviously be stated that the effect of Brownian movement on screening must also survive, but the quantitative parameters of the compound particles, formed in this case, may change. Thereby, we may figuratively conclude that in all cases "like likes unlike".

Finally, we have studied numerically the fractal characteristics of ensembles of random clusters, comprising polydisperse spherical particles, whose parameters comply with the theoretical characteristics of bubstons, which have been outlined here. The numerical simulation of stochastic growth based on the ballistic particle-cluster aggregation model with the lognormal law of size distribution of monomers in such clusters have been carried out together with computation of the elements of the light scattering matrix with the use of appropriate T-matrix program codes for optical scattering by clustered spheres, reported in [117]. We have introduced the concept of effective fractal dimension of cluster as the average fractal dimension over the ensemble of random clusters with the same statistical parameters (a fixed number of monomers and monomer size distribution). The evolution of the effective cluster fractal dimension with the growing number of monomers (i.e. the bubstons), composing each realization of the bubston cluster within a stochastic sampling, is investigated. We established principal features of the scattering matrix for ensembles of bubston clusters. It was obtained that the distribution of monomer number of clusters in ensemble must be sufficiently wide so as to meet experimental data on the light scattering matrix of aqueous salt solutions results.

REFERENCES

[1]   Epstein, P.S.; Plesset, M.S., *J. Chem. Phys.* 1950, 18, 1505-1509.
[2]   Lugli, F.; Hofinger, S.; Zerbetto, F., *J. Am. Chem. Soc.* 2005, 127, 8020-8021.
[3]   Crum, L.A. *Nature* (London) 1979, 278, 148-149.
[4]   Crum, L.A. In Cavitation and Inhomogeneities in Underwater Acoustics; Lauterborn, W.; Ed.; Springer-Verlag: New York, NY, 1980, pp 3-12.
[5]   Crum, L.A., *J. Appl. Sci. Res.* 1982, 38, 101-115.
[6]   Crum, L.A. In Mechanics and Physics of Bubbles in Liquids; van Wijngaarden L.; Ed.; Martinus Nijhoff Publishers: The Hague, 1982, pp 101 – 115.
[7]   Sirotyuk, M.G. In High Intensity Ultrasonic Fields; Rozenberg, L.D.; Ed.; Plenum Press: New York, NY, 1971, pp 319-337.
[8]   Bunkin, N.F.; Bunkin, F.V., *J. Exp. Theor. Phys.* 1992, 74, 271-278.
[9]   Bunkin, N.F.; Bunkin, F.V., *J. Exp. Theor. Phys.* 2003, 96, 730–746.
[10] Bunkin, N.F.; Bunkin, F.V., Z. *Phys. Chem.*, 2001, 215, 111-132.
[11] Bunkin, N.F.; Lobeyev, A.V.; Lyakhov, G.A.; Ninham, B.W., *J. Phys. Rev. E,* 1999, 60, 1681 - 1690.
[12] Bunkin, N.F.; Suyazov, N.V.; Shkirin, A.V.; Ignatiev, P.S.; Indukaev, K.V., *J. Chem. Phys.* 2009, 130, 134308.
[13] Bunkin, N.F.; Suyazov, N.V.; Shkirin, A.V.; Ignatiev, P.S.; Indukaev, K.V., JETP, 2009, 108, 800-816.
[14] Bunkin, N.F.; Shkirin, A.V.; Kozlov, V.A.; Starosvetskiy, A.V., Proc SPIE, 2010, 7376, 73761D.
[15] Bunkin, N.F.; Ninham, B.W.; Shkirin, A.V.; Ignatiev, P.S.; Kozlov, V.A.; Starosvetskij, A.V., J Biophotonics, 2011, (in press), DOI 10.1002/jbio.201000093 (http://onlinelibrary.wiley.com/doi/10.1002/jbio.201000093/pdf).
[16] Bunkin, N.F.; Bunkin, F.V., *Laser Phys.*, 1993, 3, 63 - 78.
[17] Bunkin, N.F.; Lobeyev, A.V., *Quantum Electron*, 1994, 24, 297 - 301.
[18] Vinogradova, O.I.; Bunkin, N.F.; Churaev, N.V.; Kiseleva, O.A.; Lobeyev, A.V.; Ninham, B.W., *J. Col. Int. Sci.*, 1995, 173, 443 – 447.

[19] Bunkin, N.F.; Lyakhov, G.A., *Phys of Wave Phen*, 2005, 13, 61 - 80.

[20] Bunkin, N.F.; Bakum, S.I., *Quantum Electron*, 2006, 36, 117 - 124.

[21] Bunkin, N.F.; Kochergin, A.V.; Lobeyev, A.V.; Ninham, B.W.; Vinogradova, O.I., *Col. and Surf A*, 1996, 110, 207 - 212.

[22] Bunkin, N.F.; Kiseleva, O.A.; Lobeyev, A.V.; Movchan, T.G.; Ninham, B.W.; Vinogradova, O.I., *Langmuir,* 1997, 13, 3024 – 3028.

[23] Bunkin, N.F.; Ninham, B.W.; Babenko, V.A.; Suyazov, N.V.; Sychev, A.A., J. Phys. Chem. B, 2010, 114, 7743-7752.

[24] Wagner, C., *Phys. Z*, 1924, 25, 474-477.

[25] Onsager, L.; Samaras, N.N.T., *J. Chem. Phys.*, 1934, 2, 528-536.

[26] Chattoraj, D.K.; Birdi, K.S. Adsorption and the Gibbs Surface Excess; Plenum Press: New York, NY, 1984, pp 1 – 471.

[27] Wilson, M.A.; Pohorille, A.; Pratt, L.R., *J. Phys. Chem.*, 1987, 91, 4873-4878.

[28] Wilson, M.A.; Pohorille, A., *J. Chem. Phys.*, 1991, 95, 6005-6013.

[29] Benjamin, I., *J. Chem. Phys.*, 1991, 95, 3698-3709.

[30] Jungwirth, P.; Tobias, D.J., *J. Phys. Chem. B*, 2000, 104, 7702-7706.

[31] Jungwirth, P.; Tobias, D.J., *J. Phys. Chem. B,* 2005, 105, 10468-10472.

[32] Mucha, M.; Frigato, T.; Levering, L.M.; Allen, H.C.; Tobias, D.J.; Dang, L.X.; Yungwirth, P., *J. Phys. Chem. B*, 2005, 109, 7617-7623.

[33] Kniping, E.M.; Lakin, M.J.; Foster, K.L.; Yungwirth, P.; Tobias, D.J.; Dabdub, R.B.; Finlayson-Pitts, B.J., *Science,* 2000, 288, 301-306.

[34] Stuart, S.J.; Berne, B.J. *J. Phys Chem A,* 1999, 103, 10300-10307.

[35] Herce, D.H.; Perera, L.; Darden, T.A.; Sagui, C., *J. Chem. Phys.*, 2005, 122, 024513.

[36] Vrbka, L.; Mucha, M.; Minofar, B.; Yungwirth, P.; Brown, E.C.; Tobias, D.J., *Curr. Opin. Colloid Interface Sci.,* 2004, 9, 67-73.

[37] Jungwirth, P.; Tobias, D.J., *Chem. Rev.*, 2006, 106, 1259-1281.

[38] Baldelli, S.; Schnitzer, C.; Shultz, M.J., *Chem. Phys. Lett.,* 1999, 302, 157-163.

[39] Schnitzer, C.; Baldelli, S.; Shultz, M.J., *J. Phys. Chem. B*, 2000, 104, 585-590.

[40] Shultz, M.J.; Schnitzer, C.; Simonelli, D.; Baldelli, S., *Int. Rev. Phys. Chem.*, 2000, 19, 123-153.

[41] Shultz, M.J.; Baldelli, S.; Schnitzer, C.; Simonelli, D., *J. Phys. Chem. B,* 2002, 106, 5313-5324.

[42] Liu, D.F.; Ma, G.; Levering, L.M.; Allen, H.C., *J. Phys. Chem. B*, 2004, 108, 2252-2260.

[43] Scatena L.F.; Richmond, G.L., *J. Phys. Chem. B*, 2004, 108, 12518-12528.

[44] Petersen, P.B.; Saykally, R.J., *Chem. Phys. Lett.*, 2004, 397, 51-55.

[45] Petersen, P.B.; Johnson, J.C.; Knutsen, K.P.; Saykally, R.J., *Chem. Phys. Lett.*, 2004, 397, 46-50.

[46] Weber, R.; Winter, B.; Schmidt, P.M.; Widdra, W.; Hertel, I.V.; Dittmar, M.; Faubel, M., *J. Phys. Chem. B*, 2004, 108, 4729-4736.

[47] Winter, B.; Weber, R.; Schmidt, P.M.; Hertel, I.V.; Faubel, M.; Vrbka, M.; Yungwirth, P., *J. Phys. Chem. B*, 2004, 108, 14558-14564.

[48] Ghosal, S.; Shbeeb, A.; Hemminger, J.C., *Geophys Res. Lett.*, 2000, 27, 1879-1882.

[49] Finlayson-Pitts, B.J.; Hemminger, J.C., *J. Phys. Chem. A,* 2000, 104, 11463-11477.

[50] Kelsall, G.H.; Tang, S.; Yurdakul, S.; Smith, A., *J. Chem. Soc. Faraday Trans.*, 1996, 92, 3887-3893.

[51] Bernal, J.D.; Fowler, R.H., *J. Chem. Phys.*, 1933, 1, 515-548.

[52] Eisenberg, D.; Kauzmann, W., The structure and properties of water; Oxford University Press: London, 1969, pp 1 – 296.

[53] Pauling, L., In Hydrogen bonding; Hadzi D.; Thompson, H.W.; Ed.; Pergamon Press Ltd: London, 1959, pp 1- 6.

[54] Kamb, B., In Structural Chemistry and Molecular Biology; Rich, A.; Davidson, N.; Ed.; W.H. Freeman: San Francisco, CA, 1968, pp 507-542.

[55] Dore, J.C., *J. Mol. Struct.*, 1991, 250, 193-211.

[56] Boutron, P.; Alben, A., *J. Chem. Phys*, 1975, 62, 4848-4853.

[57] Svishchev, I.M.; Kusalik, P.G., *J. Chem. Phys*, 1993, 99, 3049-3058.

[58] Kusalik, P.G.; Svishchev, I.M., *Science,* 1994, 265, 1219-1221.

[59] Narten, A.H.; Thiessen W.; Blum, L., *Science*, 1982, 217, 1033-1034.

[60] Chialvo, A.A.; Cummings, P.T.; Simonson, J.M.; Mesmer, J.M.; Cochran, H.D., *Ind. Eng. Chem. Res.*, 1998, 37, 3021-3025.

[61] http://www.lsbu.ac.uk/water.

[62] Landau, L.D.; Lifshitz, E.M., Theory of Elasticity; Butterworth-Heinemann: Oxford, UK, 1986, pp 1-187.

[63] Landau, L.D.; Lifshitz, E.M., Statistical Physics, Part 1; Butterworth-Heinemann: Oxford, UK, 1980, pp 1 – 542.

[64] Pitaevskii, L.P.; Lifshitz, E.M., Physical Kinetics; Pergamon Press Ltd: London, 1981, pp 1-452.

[65] Archer, A.J.; Wilding, N.B., *Phys. Rev. E*, 2007, 76, 031501.

[66] Houghton, G.; Ritchi, P.D.; Thomson, J.A., *Chem. Eng. Science*, 1962, 17, 221 – 227.

[67] Roetzel, W.; Blomker, D.; Czarnetzki, W., *Chem. Eng. Tech*, 1997, 69, 674 – 678.

[68] Wise, D.L.; Houghton, G., *Chem. Eng. Science*, 1966, 21, 999 – 1010.

[69] Pfeiffer, W.F.; Krieger, I.M., *J. Phys. Chem.*, 1974, 78, 2516-2521.

[70] Pashley, R.M.; Rzechowicz, M.; Pashley, L.R.; Francis, M.J., *J. Phys. Chem. B*, 2005, 109, 1231-1238.

[71] Francis, M.J.; Gulati, N.; Pashley, R.M., *J. Col. Int. Sci.*, 2006, 299, 673-677.

[72] Debye, P.W.; Huckel, E., *Phys. Z*, 1923, 24, 185-206.

[73] Ise, N.; Okubo, T., *Acc. Chem. Res.*, 1980, 13, 303-309.

[74] Kepler, G.M.; Fraden, S., *Phys. Rev Lett.*, 1994, 73, 356-359.

[75] Dosho, S.; Ise, N.; Ito, K.; Iwai, S., et al, *Langmuir*, 1993, 9, 394-411.

[76] Matsuoka, Y.; Harada, T.; Yamaoka, H., *Langmuir*, 1994, 10, 4423-4425.

[77] Crocker, J.C.; Grier, D.G., *Phys. Rev. Lett.*, 1994, 73, 352-355.

[78] Crocker, J.C.; Grier, D.G., *Phys. Rev. Lett.*, 1996, 77, 1897-1900.

[79] Larsen, A.E.; Grier, D.G., *Nature,* 1997, 385, 230-233.

[80] Ise, N., *Proc. Jpn Acad. B*, 2002, 78, 129-137.

[81] Zheng, J.; Pollack, G.H., *Phys. Rev. E*, 2003, 68, 031408.

[82] Gomez-Guzman, O.; Ruiz-Garcia, J., *J. Col. Int. Sci.*, 2005, 291, 1-6.

[83] Zheng, J.; Chin, W.; Khijniak, E.; Khijniak Jr., E.; Pollack, G.H., *Adv. Col. Int. Sci.*, 2006, 127, 19-27.

[84] Liang, Y.; Hilal, N.; Langston, P.; Starov, V., *Adv Col Int Sci*, 2007, 134 – 35, 151-166.

[85] Zheng, J.; Wexler, A.; Pollack, G.H., *J. Col. Int. Sci.*, 2009, 332, 511-514.

[86] Nagornyak, E.; Yoo, H.; Pollack, G. H., *Soft Matter,* 2009, 5, 3850-3857.

[87] Derjaguin, B.V.; Landau, L.D., *Acta Physicochimica* (USSR), 1941, 14, 633-645.

[88] Verwey, E.J.; Overbeek, J.T.G., Theory of the Stabilization of Lyophobic Colloids; Elsevier: Amsterdam, 1948, pp 1 – 321.

[89] Yoon, R.; Yordan, J.L., *J. Col. Int. Sci.*, 1986, 113, 430-438.

[90] Li, C.; Somasundaran, P., *J. Col. Int. Sci.*, 1991, 146, 215-218.

[91] Li, C.; Somasundaran, P., *Col. Surf. A*, 1993, 81, 13-15.

[92] Mateescu, E.M.; Jeppesen, C.; Pincus, P., *Europhys Lett.*, 1999, 46, 493-498.

[93] Park, S.Y.; Bruinsma, R.F.; Gelbart, W.M., *Europhys Lett.*, 1999, 46, 454-460.

[94] Joanny, J.F., *Eur. J. Phys. B*, 1999, 9, 117-122.

[95] Perel, V.I.; Shklovskii, B.I., Physica A (Amsterdam), 1999, 274, 446-453.

[96] Shklovskii, B.I., *Phys. Rev. E*, 1999, 60, 5802-5811.

[97] Nguyen, T.T.; Grosberg, F.Yu.; Shklovskii, B.I., *J. Chem. Phys.*, 2000, 113, 1110-1125.

[98] Nguyen, T.T.; Grosberg, F.Yu.; Shklovskii, B.I., *Phys. Rev. Lett.*, 2000, 85, 1568-1571.

[99] Y. Levin, *Rep. Prog. Phys.*, 65, 1577-1632 (2002).

[100] Quesada-Perez, M.; Gonzalez-Tovar, E.; Martin-Molina, A.; Lozada-Cassou, M.; Hidalgo-Alvarez, R., *Chem. Phys. Chem.*, 2003, 4, 235-248.

[101] Grosberg, F.Yu.; Nguyen, T.T.; Shklovskii, B.I., *Rev. Mod. Phys*, 2002, 74, 329-345.

[102] Zhang, R.; Shklovskii, B.I., *Phys. Rev. E*, 2005, 72, 021405.

[103] Pittler, J.; Bu, W.; Vakhnin, D.; Travesset, A.; McGillivray, D.J.; Losche, M., *Phys. Rev. Lett.*, 2006, 97, 046102.

[104] Gracheva, M.E.; Leburton, J.P., *Nanotechnology*, 2007, 18, 145704.

[105] J. Faraudo, and A. Travesset, *J. Phys. Chem. C*, 111, 987-994 (2007).

[106] C. Calero, and J. Faraudo, *Phys. Rev. E* 80, 042601-1 – 042601-4 (2009).

[107] T. Terao, T. Nakayama, *Phys. Rev. E* 2001, 63, 041401.

[108] Z.Y. Wang, and Y.Q. Ma, *J. Chem. Phys.*, 2009, 131, 244715.

[109] Z.Y. Wang, and Y.Q. Ma, *J. Chem. Phys.*, 2010, 133, 064704.

[110] Dukhin, S.S.; Deryaguin, B.V., Electrophoresis; Nauka: Moscow, 1976, pp 1 - 168 (in Russian).

[111] Landau, L.D.; Lifshitz, E.M., Fluid Mechanics; Pergamon Press: Oxford, 1987, pp 1 – 533.

[112] Langevin, P., *Coptes Rendus* (Paris), 1908, 146, 530-533.

[113] Tikhonov, V.I., Outliers in Random Processes; Nauka: Moscow, 1970, pp 1 - 325 (in Russian).

[114] Jullien, R., *Cont Phys*, 1987, 28, 477-493.

[115] Smirnov, B.M., *Sov. Phys. Usp*, 1986, 29, 481-505.

[116] Ball, R.; Nauenberg, M.; Witten, T.A., *Phys. Rev. A*, 1984, 29, 2017-2020.

[117] Xu, Y.-l.; Wang, R.T., *Phys. Rev. E*, 1998, 58, 3931-3948.

In: Classification and Application of Fractals
Editor: William L. Hagen

ISBN 978-1-61209-967-5
© 2012 Nova Science Publishers, Inc.

Chapter 2

FRACTAL DYNAMICS OF COMPLEX SYSTEMS

Oswaldo Morales-Matamoros, Teresa I. Contreras-Troya, Mauricio Flores-Cadena and Ricardo Tejeida-Padilla*
Instituto Politécnico Nacional, Distrito Federal, Mexico

ABSTRACT

One of the most important properties of systems is complexity. In a simple way, we can define the complexity of a system in terms of the number of elements that it contains, the nature and number of interrelations, and the number of levels of embeddedness. When a high level of complexity exists in a system, it is considered a complex system. Although there is no single agreed-on definition of complex systems, they share some themes: (i) they are inherently complicated or intricate, so that they have factors such as the number of parameters affecting the system or the rules governing interactions of components of the system; (ii) they are rarely completely deterministic, and state parameters or measurement data may only be known in terms of probabilities; (iii) mathematical models of the system, are usually complex and involve non-linear, ill-posed, or chaotic behavior; and (iv) the systems are predisposed to unexpected outcomes (so-called emergent behavior). To try to understand the dynamics of these systems diverse mathematical tools have been developed. A new scientific discipline with greatimpact in the analysis of the complex systems has been developed in recent years, called fractal analysis.

The study of the complex systems in the framework of fractal theory has been recognized as a new scientific discipline, being sustained by advances that have been made in diverse fields ranging from physics to economics. In this chapter the basic concepts of fractal analysis of complex systems are briefly explained and three examples of fractal analysis are provided: epilepsy, oil crude price market volatility, and supply chain volatility in the telecom industry.

---

*E-mail: omoralesm@ipn.mx

INTRODUCTION

One of the most exciting and rapidly developing areas of modern research is the quantitative study of complexity (Balankin, 1991). Complexity has had an important impact in the fields of physics, mathematics, information science, biology, medicine, sociology, and economics. The main goal of the science of complexity is to develop mathematical methods able to discriminate among the fundamental constituents of a complex system and to describe their interrelations in a concise way (Scafetta, 2002). Although there is no single agreed-on definition of complex systems, here we pointed out some common themes they share as stated by Foote (2007): (i) complicated or intricate; (ii) they are rarely completely deterministic, and state parameters may only be known in terms of probabilities; (iii) mathematical models of the system, are usually complex and involve non-linear, ill-posed, or chaotic behavior; and (iv) the systems exhibit "emergent behavior", i.e. *"P is a global [or collective or non-distributive] property of a system of kind K, none of whose components possesses P"* (Bunge, 2003). Complex systems contain many constituents interacting non-linearly in such a way that they are capable of emergent behavior that usually is responsible for power laws which are universal and independent of the microscopic details of the phenomenon.

The emergence behavior is originated in the interrelations among the elements of the concerned system. To account for the emergence of a system the corresponding combination or assembly process must be discovered, in particular, the bonds resulting in the formation in the whole. In this chapter fractal analysis was used to gain insight about the emergence of fluctuations in three different complex systems, in order to getthe simplest and most parsimonious description of the phenomena under study.

FRACTALS

The term *fractal* (from Latin *fractus* –irregular, fragmented) applies to objects in space or fluctuations in time that possess a form of self-similarity and cannot be described within a single absolute scale of measurement. Fractals are recurrently irregular in space or time, with themes repeated like the layers of an onion at different levels or scales. Fragments of a fractal object or sequence are exact or statistical copies of the whole by shifting and stretching. Fractal geometry has evoked a fundamentally new view of how both nonliving and living systems result from the coalescence of spontaneous self-similar fluctuations over many orders of time and how systems are organized into complex recursively nested patterns over multiple levels of space (Klonowsk, 2000).

A fractal object has a property similar to morphological complexity, which means that more fine structure (increased resolution and detail) is revealed with increasing magnification (Klonowsk, 2000).

In both natural and engineering systems, it is often inappropriate to discuss objects in terms of one, two, or three dimensions. These Euclidean dimensions do not adequately describe the morphology and behavior of the complex objects and relationships which are found in practice. Moreover, just a small group of fractals have one certain dimension, which is scale invariant. These fractals are (statistically self-similar) mono-fractals (characterized by the unique fractional metric "Hausdorff" dimension). Nevertheless, the most natural fractals have different fractal dimensions depending on the scale. They are composed of many fractals with different fractal dimensions (Zmescal, 2001). They are called "multifractals" and are characterized by the infinite spectrum of generalized dimension (also called Rény dimension spectrum (Ivanova*et al.*, 1994)).

Self-similar fractals are invariant under the similarity transformation (Figure 1). Therefore they are essentially isotropic, whereas many natural fractals and time series often have anisotropic asperity distributions. For description of such patterns, the concept of self-similarity has been extended to account for anisotropy through the notion of self-affinity. The fundamental difference between self-similar and self-affine fractals is the way in that scaling will produce statistical equivalence.

Self-similar fractals should be scaled equally in all directions to produce statistically equivalent patterns, whereas self-affine fractals (Figure 2) must be scaled by different amounts in different directions to produce statistical equivalence.

Let us assume that we have a function, $Y(t)$, of one variable only. Here t (usually time) is the horizontal variable, while Y is the vertical variable. Self-affinity is defined through statistical invariance under the transformation

$$t \to \lambda t \qquad (1)$$

$$Y \to \lambda^H Y \qquad (2)$$

where H is called the Hurst exponent. By combining such transformations, one can construct the affine group (Balankin, 1997). An alternative way of expressing this invariance is by the standard definition of self-affine that says that a process of continuous time $Y = Y(t); t > 0$ is self-affine if the distribution probability of $Y(t)$ has the same distribution probability of $\lambda^H Y(\lambda t)$ for $\lambda > 0$ (Gao*et al.*, 2007), that is

$$Y(\lambda t) \triangleq \lambda^H Y(t) \qquad (3)$$

where the parameter H measures the correlation persistence of data, and $\triangleq$ denotes equality in distribution.

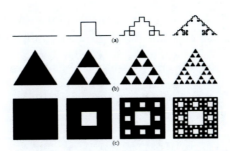

Figure 1. Iterative constructions of self-similar fractals: (a) Koch square room, (b) Sierpinski's triangle, and (c) Sierpinski's carpet.

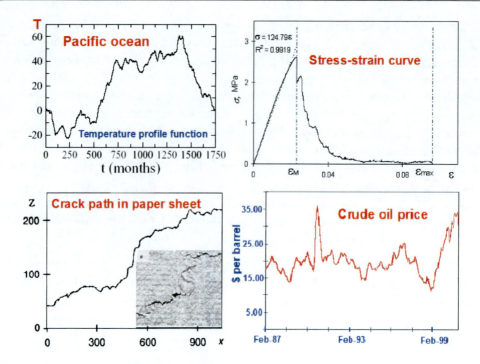

Figure 2. Self-affine fractals: (a) daily measured temperature of Pacific Ocean (Morales, 2004), (b) crack path in paper sheet (Chave and Levin, 2002), (c) Stress-strain-curve (Balankin, 1997), and (d) the daily record of WTI crude oil price (Balankin *et al.*, 2004).

One can see that, if the exponent H is less than 0.5, the time series displays "anti-persistence" (pink noise). This means that positive excess return is more likely to be reversed and therefore the next period's performance is likely to be below average. If the exponent H is greater than 0.5, the process displays "persistence" (black noise). This means that the positive excess return is more likely to remain above average. However, if the exponent H is equal to 0.5 (random white noise), the returns do not display any memory and thus positive returns is equally to be followed by above or below average performance (Morales, 2003).

Analysis of Self-Affine Curves

Various methods can be used to provide estimates of the Hurst exponent. Below we show some most popular methods to calculate the Hurts exponent for self-affine curves (Morales, 2003).

Rescaled-range method: $\left(\dfrac{R}{\sigma} \propto t^H\right)$ (4)

Power-spectrum method: $\left(P \propto \tau^{-2H-1}\right)$ (5)

Roughness-length method: $\left(SD \propto \tau^{\wedge}(H)\right)$ (6)

Variogram method:	$\left(\left(V \propto \tau^{2H}\right)\right)$	(7)

Wavelets method:	$(W[X](a) = ((\|W(a, b)\|)\_b \propto a^{\wedge}(H + 1/2) \,)$	(8)

Fractal analysis has been used for studies of DNA sequences, proteins' structure, metabolism, cardiovascular and pulmonary systems, and other areas of biology and physiology. The use of fractal geometry in microscopic anatomy is now well established and it will be increasingly useful in establishing links between structure and function. (Klonowsk, 2000).

The fractal theory has also been increasingly applied in the field of materials science and engineering (Balankin, 1991,1997; Ivanova*et al.*, 1994;Rodriguez andPandolfelli,1998). Models of fractal lines and surfaces have been generated to describe the microstructural features of materials. Special interest is placed upon a description of the fracture surface based on fractal geometry in order to understand the crack path in materials.

Human society is, by far, one of the most complex extended systems. Its self-organization is evident: every single characteristic (historical, cultural, etc.) of a member of a society influences all the others members around him (her) and through them all the others, in a non-trivial way. However, despite this complexity, it is universally accepted that cultural details play an interesting but somewhat 'marginal' role in a human history. Economic development has always been considered the driving (or relevant) force in determining the relationship inside a society. During the last years there has been greater interest in applications of statistical physics to financial market dynamic (Cuniberti, 2001).

In the following three sections of this chapter there is shown the application of the fractal analysis into three different complex systems: epilepsy, crude oil price market volatility and supply chain volatility in the telecom industry.

FRACTAL DYNAMICS OF EPILEPSY

The human brain is a complex system made up of billions of nerve cells, called neurons, which transmit signals inside the brain, and between the brain and the rest of the body. Thehuman brain controls a wide range of tasks, such as consciousness, awareness, movement and posture. The brain sends and receives messages to make these tasks happen, and if there is a mistake in this communication a seizure can occur.

A seizure is a sudden surge of electrical activity in the brain that usually affects how a person feels or acts for a short time. *Epilepsy* is the tendency to have repeated seizures that begin in the brain, due to clusters of neurons communicating abnormally.

Biomedical science tries to predict the clinical onset of epileptic seizures, characterized by sudden changes in the growthof wave rhythmicity; electroencephalograms (EEG) analysis provides a window through epilepsy dynamics can be studied.

An electrographic seizure is the beginning of an observed crisis in the EEG, it has four stages:

- Pre-ictal: the period before the seizure.
- Ictal: it is the moment when the seizure takes place.

- Post-ictal: after the seizure.
- Inter-ictal: period between crises.

By applying fractal analysis it is possible to detect transitions and non-stationarities in signals, in order to explore the beginning of seizures in EEG, and to provide important help to the surgical process (Contreras, 2007).

In this section of the chapter, we show a fractal analysis to time series, generated from EEG from epileptic patients and non-epileptic ones (but they had had a neurological problem as headache, stress or depression), in order to decide which method of self-affine trace methods can characterize statistical patterns from epilepsy.

Data Analysis

The collected data was from EEG´s on the scalp. The results that are shown are from data that refers to five people with a neurological problem like depression, headache, and stress but no epilepsy, and data from three epileptic patients. Each data represents voltage from neuronal activity, expressed in micro-volts per second ($\mu \frac{v}{s}$). This data was given by the "20 de Noviembre Hospital", ISSSTE, Mexico).

Data from five non-epileptic patients was analyzed.16 electrodes (channels) were placed on the scalp of each patient, and each channel gave different numeric data).

First, numeric data from each channel was transformed into time series (16 per patient), each one with 32,000 rows; after that, from each channel standard deviation (fluctuations) for 100 time horizons (from $n = 2$ to $n = 102$) were obtained, then time series fluctuations were obtained; in total there were generated 1,616 time series (16 form original data and 1,600 with fluctuation data) per each patient; so 8,080 temporal series for non-epileptic patients were obtained.

In the *epilepticpatients* a total of 18 time series of 450,000 data for each one were obtained. There were windows of $2^{15}=32,768$ data per each channel, so 417 temporal series per channel were generated. Thus, for each patient there were 435 time series (18 original series and 417 for each window of size $2^{15}=32,768$). So, in total there were 1,035 temporal series of epileptic patients.

Fractal Analysis

Fractal analysis assisted indetermination of the Hurst exponent (H) to each time series under study. All temporal series were run in Benoit software (Scion Corporation) in order to obtain the value of H . They used four self-affine trace methods from Benoit: R/S analysis, Roughness-Length, Variogram, Wavelets: Truncate Data and Zoom. And finally, an average of these methods was applied to each time horizon or window, respectively. There was used data from epileptic and non-epileptic people to determine which method could be used in both cases. These results are presented on graphs of H values per method and average of each one of the study signals.

In studies of the EEG of five non-epileptic patients, just one graph of H values for an original signal (without statistical treatment) of one patient (Figure 3) is shown. H values are different for each method, and it is necessary to determine which of them produce similar values and they are the nearest to the average. In this case, the R/S (curve 1) and Zoom-Wavelets (curve 4) present consistent values. The same way, the curves generated by these methods have a similar behavior to the average (curve 6). Then, for this case, the elected methods to analyze complex signals are the R/S and the Zoom-Wavelets.

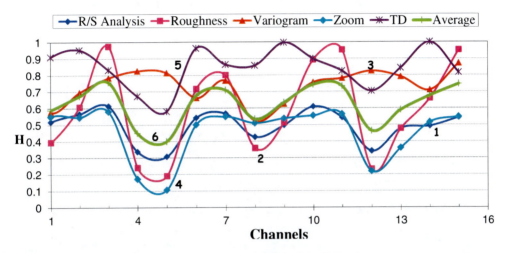

Figure 3. Graph of Hurst values of a non-epileptic patient (original signal).

In Figures 4a-4c, note the obtained curves of H values versus the time horizons ($n = 2, 3, 4, ..., 100, 101$). It should be mentioned that each H value represents a time series of standard deviation (or fluctuations); so, there are 100 H values (100 hundred temporal series of standard deviation). Similar H values with Wavelets were obtained: Truncate data (curve 4) and Zoom (curve 5); also H values are similar to Variogram (curve 3) and Roughness-Length (curve 2).

To establish what pair of methods should be used in the analysis of EEG, it is necessary to select those whose H values are closer to the average of both (curve 6); however, the resultant curves of each pair of methods are the same distance of the average, so, the pair of selected methods will depend on the chosen methods of epileptic patients, so it is important to compare the EGG behavior of patients with neurological problems and epileptic people.

The generated signal from Variogram method (curve 3) shows a special behavior: although H values are not similar to the others, it has a nearly stationary behavior: from $n = 2$ to $n = 11$, with a jump between $n = 12$ and $n = 1$ 5, to be stational from $n = 20$, so this method must be considered for analysis of the EEG.

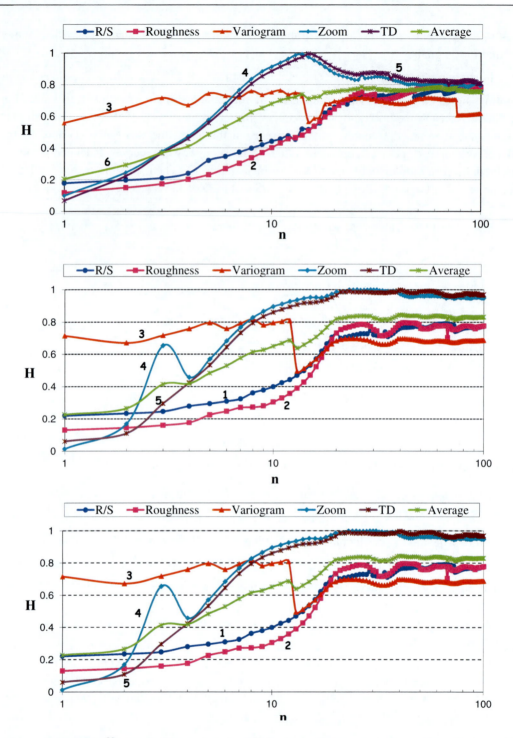

Figure 4. Curves with H of different time horizons (n), generated from the 100 time series of standard deviation: (a) channel 2, (b) channel 13, and (c) channel 16.

The EEG of three epileptic patients were studied. In Figure 5 there are H values from the original signal of an epileptic patient. As it can be seen the methods that have consistent values are R/S (curve 1) and Variogram (curve 3), these curves have a similar behavior to the one from the average (curve 6). It is concluded that for this case, the chosen methods are the R/S and the Variogram.

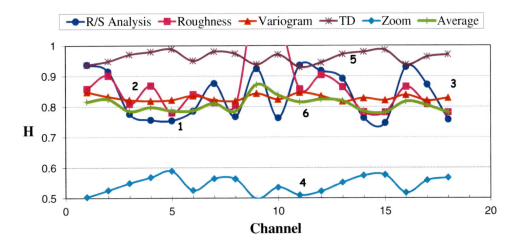

Figure 5. Graph of H values of an epileptic patient (original signal).

Figures 6a-6c show the obtained curves of H values of the windows of size $2^{15}=32,768$ (there are shown three out of 18 graphs). Each H value represents a time series generated from each window, thus, there are 417 H values.

In this case, similar H values were obtained with Wavelets-Zoom (curve 4), Variogram (curve 3) and Roughness-Length (curve 1). As the last two methods were selected to analyze the complex signal of epileptic patients in the original series, then, the chosen methods are R/S and Variogram.

It is important to mention that the curve generated from Variogram method presents a stationary behavior in every single channel, but in channel five (Figure 10), it shows stationarity from $n = 1$ to $n = 12$ ($H \approx 0.8$), but it decreases in $n = 13$ ($H = 0.45$) and has similar values up to $n = 44$ ($H = 0.47$), since $n = 45$ ($H = 0.81$) it goes back to be stational in that rank of values. There is not another signal with this behavior.

It was found that the most consistent self-affine trace methods to analyze and characterize the EEG of patients with neurological problems and epileptic patients are R/S and Variogram.

In Figure 10, the generated curve with the Variogram method (curve 3) presents a stationary behavior "persistent" from $n = 2$ to $n = 12$ ($H = 0.80$); after this point it presents a sudden jump $n = 13$ ($H = 0.45$) keeping "antipersistent" up to $n = 44$ ($H = 0.47$); and since $n = 45$ its behavior "persistent" is constant ($H = 0.80$) until $n = 101$. The behavior of the generated Variogram curve indicates that fluctuations of channel 5 have an abrupt transition from a persistent state (positive correlation) to an antipersistent one (negative correlation but nearly random) to go in an abrupt way from an antipersistent state to another persistent. Thus, it can be said that fluctuations of electric discharges among neurons that cause epilepsy, have a multifractal behavior because they have, at least, three different H values (0.80, 0.45 and, again, 0.80).

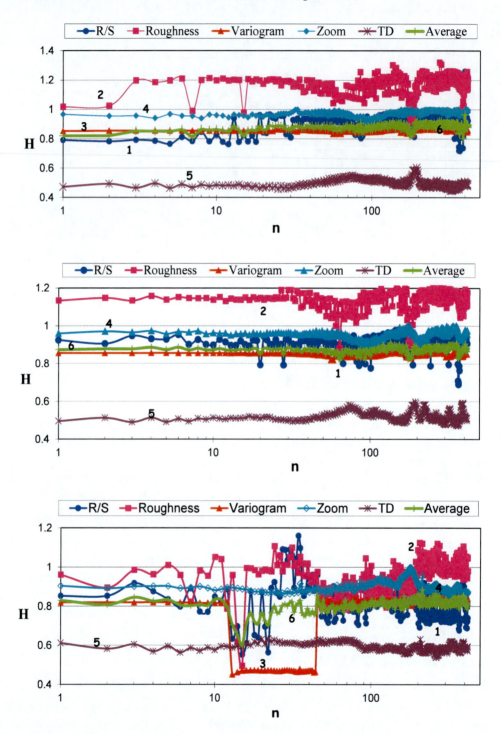

Figure 6. Curves of H values of different windows of size 2^{15}: (a) channel 1, (b) channel 3, and (c) channel 5.

FRACTAL DYNAMICS OF OIL CRUDE PRICE MARKET VOLATILITY

Dynamics of all realistic complex systems commonly exhibit scaling invariance, i.e., their behavior does not change under rescaling of variables (for example, space and time). This allows us to use the dynamic scaling approach to study the kinetic roughening of growing interfaces and the dynamics of financial markets. Many statistical properties of financial markets have already been explored, and have revealed striking similarities between price volatility dynamics and the kinetic roughening of growing interfaces. Based on this fact, the general dynamic scaling approach was used to study the scaling properties of the crude oil market (Balankin *et al.*, 2004).

The world oil is a capital-intensive environment characterized by complex interactions deriving from the wide variety of products, transportation-storage issues, and stringent environmental regulation. Crude oil is the world's most actively traded commodity, accounting for about 10% of total world trade. The crude oil market is characterized by extremely high levels of price volatility. Fluctuations in crude oil price are caused by supply and demand imbalances arising from events such as wars, changes in political regimes, economic crises, formation breakdown of trade agreements, unexpected weather patterns, etc. At the same time, many of the price forecasting models are based on the belief that historical price series exhibit some statistical properties that allow prediction of future price movements.

Data Analysis

To quantify the scaling dynamics of the crude oil market, a study of the daily records of the spot prices $P(t)$ (Figure 7a) and the price volatilities $V_n(t)$ (Figures 7b-7d) was done from the West Texas Intermediate (WTI) crude oil price listings. Specifically analyzed werethe WTI crude oil price in constant 1983 US dollars over the period from December 30, 1984–June 23, 2003 representing 5,181 observations (weekends and business holidays are excluded). Then they construct 699 time series of realized volatility

$$V_n(\tau) = (n-1)^{-1}\sqrt{\sum_{i=1}^{n}(P^2(\tau+i) - \langle P^2(\tau)\rangle)}$$

(9)

of length $T = 5,4096$ business days (about 16 business years) for different time horizons $n = 2,3,...,700$ (from 2 business days to about 3 business years), where t is the business time and $\langle \cdots \rangle_n$ denotes the business time average within a window of size n. In this study, all records of volatility (Figures 7b-7d) correspond to the period from October 15, 1986 to December 23, 2002. One can see that the price volatility changes from day to day in such a way that the time series of volatilities realized at different time intervals n look similar.

Fractal Analysis

To detect and quantify the dynamic scaling behavior of price volatility within a framework of interface roughening dynamics, Balankin *et al.* (2004) treated the volatility horizon n as an analog of time variable (t), while the business time t was treated as an analog of lateral extent (X) of growing interface.

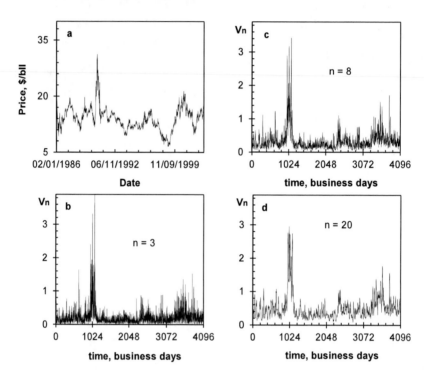

Figure 7. (a) Time record of West Texas Intermediate crude oil spot price in 1983 constant dollars per barrel $/bbl); and (b)–(d) realized price volatilities for the period of 4,096 business days for different horizons: (b) $n = 3$, (c) $n = 8$, and (d) $n = 18$ business days. All time series correspond to the period from October 15, 1986 to December 23, 2002 Balankin *et al.* (2004).

Accordingly, the price volatility fluctuations (σ) were characterized by the analog of interface height fluctuations defined as

$$\sigma(\Delta, \tau) = \langle [V_n(t) - \langle V_n(t) \rangle_\tau]^2 \rangle_R^{\frac{1}{2}} \propto \Phi(\tau^{H_n}, n^{\beta(\tau)}), \qquad (10)$$

where $\langle \ldots \rangle_\tau$ denotes the business time average within a window of size τ and $\langle \ldots \rangle_R$ denotes the average over different realizations, Φ is the scaling function, and β is the volatility growth exponent.

To characterize the scaling properties of time series, within a framework of the general dynamic scaling concept (10), the volatility growth exponent β can be determined from the scaling behavior

$$\sigma(n\Delta t) \propto n^{\beta(\tau)}, \qquad (11)$$

for different intervals of business time.

It was found that the realized volatilities (Figures 7b–7d) possess a statistical self-affine invariance within wide ranges of business time scale [$3 < \xi < \tau\_c(n)$] characterized by the Hurst exponent H_n for each horizon n (Figures 8a–8d):

$$H_n = 0.0621n, \text{ when } n < 12 \qquad (12)$$

and

$$H_n = 0.83 + 0.04, \text{ when } n > 18 \qquad (13)$$

This finding means that the long-horizon realized volatilities ($n > 8$) are persistent, i.e. volatility increments are positively correlated in business time, whereas the short-horizon volatilities ($n < 8$) are anti-persistent, i.e. $V_n < 8(t)$ displays negative autocorrelations in business time.

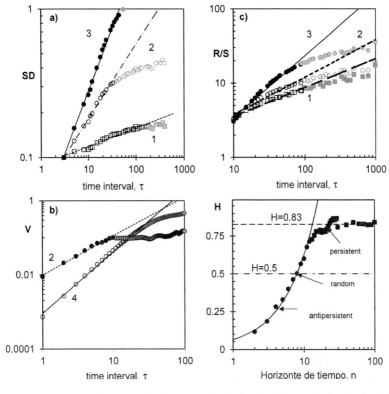

Figure 8. (a)-(c) Fractal graphs of price volatility records shown in Figures 7a-7c obtained by (a) the roughness-length, (b) the variogram, and (c) the rescaled-range methods [numbers correspond to different horizons: (1) corresponds to $n = 3$, (2) corresponds to $n = 8$, (3) corresponds to $n = 20$, and (4) corresponds to $n = 60$ business days]. (d) Horizon dependence of the Hurst exponent (values of H_n are averaged through five methods) in the semilog coordinates [time scale in business days, circles and squares: experimental data, solid line: data fitting by Eq. (13), $R^2 = 0.9882$].

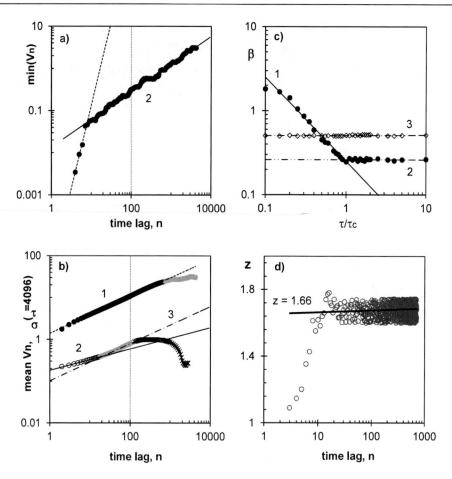

Figure 9. Results of the scaling analysis of crude oil price volatility. (a) The minimum of realized volatility versus time horizon [points: experimental data, lines: power-law fits (1) $\min(V_n) = 10^{-5} n^{3.95}, R^2 = 0.97$ and (2) $\min(V_n) = 0.014 n^{0.66}, R^2 = 0.993$]. (b) The mean (1) and the standard deviation (2, 3) of realized volatility ($t = 4,096$) versus time lag, n [points: experimental data, lines: power-law fittings. (1) $\langle V_n(\tau = 4096) \rangle_\tau = 1.5 n^{0.5}, R^2 = 0.999$, (2) $\sigma\{V_n(\tau = 4096)\} = 0.18 n^{0.26}, R^2 = 0.995$, and (3) $\sigma\{V_n(\tau = 4096)\} = 0.1 n^{0.45}, R^2 = 0.998$; the y axis labels indicate the quantity]. (c) The volatility growth exponent β versus normalized time interval for volatility horizons $n = 3$ (1,2) and $n = 60$ (3); points: results of analysis, lines: fitting when $n < 18$ and $\beta = 0.5$ for any τ, when $n > 18$. (d) Horizon dependence of dynamic exponent for price volatility (time scale in business days).

Furthermore, it was found that the transition from anti-persistent to persistent volatility at $n = 8$ is accompanied by an abrupt change in the behavior of $\min_n \{V_n(T = 4096)\}$ versus n (Figure 9), nevertheless the time-average and the standard deviations of $V_n(t)$ have no anomaly at $n = 8$ (Figure 9b). Specifically, the time averaged volatility behaves as $\langle V_n(\tau) \rangle_{\tau=4096} \propto n^{0.5}$ up to $n = 700$, while the standard deviation of realized volatility

Behaves as $\sigma(\tau = T = 4096) \propto n^{0.25}$ up to $n = 18$, but it scales as $\sigma \propto n^{0.5}$, when the realized volatility is characterized by the constant Hurst exponent $H = 0.83 + 0.04$ (constant).

Moreover, the growth exponent behaves (Figures 9a-9c) as $\beta = 0.25\tau_C / \tau$ if $\tau < \tau_c(n)$ and $\beta = 0.25$ if $\tau > \tau_c(n)$, when $n < 8$, while $(= 0.5$ for any τ, when $n > 18$.

It was also found that the interval of correlations τ_C in the business time scale increases with the horizon of volatility as $\tau_C \propto n^{z(n,\tau)}$, where the dynamic exponent is a function of n and t if $n < 18$, while $z = H / (= 1.66$ for long horizons, $n > 18$ (Figure 9d).

Accordingly, it was found that the long-horizon volatility ($n > 18$) satisfies the Family-Viscek dynamic scaling ansatz

$$\sigma(\Delta t) \propto t^b f\left[\frac{\Delta}{\xi(t)}\right],\tag{14}$$

Where $((t) (t^{\wedge}(1 / z)$ is the correlation length of the "space" scale and the scaling function behaves as $[((y) (y]^{\wedge}H$, if $y \ll 1$, or $((y) (1$, if $y \gg 1$; here H is the so-called local randomness (or Hurst) exponent; z is the dynamic exponent, and $(= H/z$ is the growth exponent. The Hurst exponent gives an indication of whether the system behavior is random ($H = 0.5$) or displays persistence ($0.5 < H < 1$) or anti-persistence ($0 < H < 0.5$).

Whereas, for horizons smaller than 18 business days, the crude oil price volatility satisfies the generalized dynamic scaling law

$$\sigma(n,\tau) \alpha F_{\ast}\left[(\tau)^{H(n)}, n^{\beta(\tau)}\right]\tag{15}$$

with continuously varying scaling exponents (13) and $n < 18$.

The crossover from anti-persistent to persistent behavior indicates the existence of intrinsic horizon scale of price volatility $n_c \approx 8$.

It also was found that for short time horizons $n < 8$, the conditional probability of realized volatility is best fitted by the light-tailed Pearson distribution (Figure. 10a)

$$f(V_n) = \frac{(V_n - \emptyset)^{(\gamma+1)}}{\rho^{-\gamma}\Gamma(\gamma)} exp\left[\frac{V_n - \phi}{\rho}\right]\tag{16}$$

with horizon dependent parameters $\gamma = 3.6 \pm 0.1$, $\rho = 0.48n^{0.34}(R^2 = 0.95)$, and $\phi = 0.13n^{-0.69}(R^2 = 0.97)$, where $\Gamma(\gamma)$ is the gamma function.

At the same time, for the larger time horizons, $n > 8$, the conditional probability of realized volatility is the log-logistic distribution (Figure 10b):

$$f(V_n) = \frac{\gamma \left[\frac{V_n - \emptyset}{\rho}\right]^{\gamma-1}}{\rho \left\{1 + \left[V_n - \frac{\phi}{\rho}\right]^\gamma\right\}^2} \tag{17}$$

where $\gamma = 2.55 \pm 0.07$, $\rho = 0.104 n^{0.385} (R^2 = 0.998)$, and $\phi = 0.0082 n^{0.744} (R^2 = 0.996)$. i.e., the long-horizon volatility distribution is fat-tailed, but it is well outside the stable Levy range $0 < (< 2$.

Furthermore, it was found that the statistical distribution of avalanches is also best fitted by the log-logistic distribution with $p = 0.48$, $((= 0$, $(= 0.75 n^{\wedge}(-0.68)$ and $(= 1.46 + 0.02$, obeying a power-law tail (Figure 10c):

$$F \alpha 1 - \left\{\frac{Q}{mediana[Q(\tau)]}\right\} - 1.46 \tag{18}$$

with the scaling exponent within the stable Levy range.

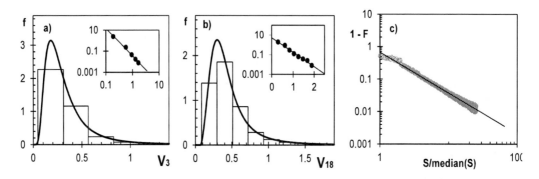

Figure 10. (a) Conditional probability distributions of price volatility for horizon $n = 3$ [bins: experimental data, solid lines fitting by the Pearson distribution (16) (p value is 0.41), inset: the distribution tail in the semi-log coordinates]. (b) Conditional probability distributions of price volatility for horizon $n = 18$ [bins: experimental data, solid lines: fitting by the log-logistic distribution (17) (p value is 0.38), inset: the distribution tail in the log-log coordinates]. (c) Cumulative distribution of normalized avalanches in log-log coordinates [circles: experimental data, solid lines: the graph of Eq. (18); the squared coefficient of correlation between the data and theoretical line is equal to $R^2 = 0.9549$].

This indicates that observed behavior of crude oil price volatility can be interpreted in terms of scale-free avalanches, which define the intrinsic horizon scale and build up long-range correlations in the price volatility.

FRACTAL DYNAMICS OF SUPPLY CHAIN VOLATILITY IN TELECOM INDUSTRY

Pushed by competitive pressures and market demands for efficient services, providers (carriers) are pursuing to use Telecom Equipment Manufacturers (TEM) after-sales services, to minimize operational and capital expenditures, as well as the impact of an outage in their network. But now, TEM face different challenges to provide the service (Cohen, 2006).

The maintenance service that concernsthis section of this chapter is related to spare parts of repairable circuit packs. Carriers and TEM have established an advance and exchange (AE) spare part service scope through an agreement (Hartley, 2005). Typically, this contract is designed to give support to critical network elements that put the availability of the network at risk. The AE service is the trigger when a critical network element fails.The TEM must send to the Carrier a good circuit pack from their stock under a determined Service Level Agreement.Once received, the Carrier must return the faulty unit back to TEM's warehouse, so that it can be repaired and returned to the pool of good stock (see numbers 1, 2, 3 and 4 in Figure 11). The defective collectprocess plus the repair process is called the recovery process. It wasanalyzed datathat encompasses only the recovery process (seenumbers 2,3 and 4 in Figure 11).

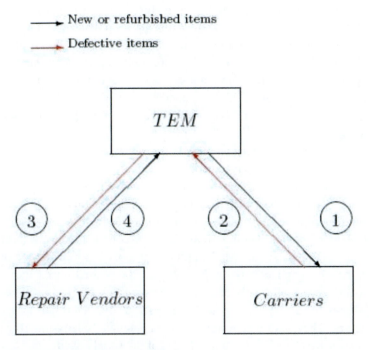

Figure 11. Closed loop supply chain of repairable items.

Most research on supply chain volatility hasfocused on the amplification of upstream ordervariability, namely the "bullwhip effect" (Lee et al., 1997).

The bullwhip effect has been analyzed for some time. This phenomenon suggests that demand variability increases as one moves upstream in a supply chain. Forrester (1961) observed that factory production rate often fluctuates more widely than does the actual consumer purchase rate and stated that this was consequence of industrial dynamics. Sterman

(1989) reported an experiment of a simulated inventory distribution system played by four people, who make independent inventory decisions without consultation with other chain members, just relying on orders from the other players instead. This experiment was called "Beer Distribution Game" and shows that the variance of orders is amplified as one moves up in the supply chain, i.e. bullwhip effect. Sterman attributes this phenomenon to a misperception of feedback from the players.

Lee *et al.* (1997) analyzed the demand information flow in a supply chain and identified four causes of the bullwhip effect: demand signal processing, rationing game, order batching and price variations. By identifying these causes, the authors concluded that the *"combination of sell through data, exchange of inventory status information, order coordination and simplified pricing schemes can help mitigate the bullwhip effect"*. Chen *et al.* (2000) quantified the bullwhip effect in a simple supply chain of two stages. The model includes the demand forecasting and order lead time, which are commonly factors that cause the phenomenon. The work is extended to multiple stage centralized and decentralized supply chains. The study demonstrates that the bullwhip effect can be mitigated but not eliminated.

Daganzo (2003, 2004) studied the bullwhip effect in the frequency domain. He argued that this is trigger with all operational inventory control policies, independent of demand process, but showed that advance demand information in future order commitments can eliminate the bullwhip effect without giving up efficiency under a family of order-up-to policies. Dejonckheere*et al.* (2003, 2004) used control theory to analyze and illustrate the bullwhip effect for a generalized family of order-up-to policies. The work of Kim and Springer (2008) analyzed the volatility of the supply chain with a different focus as previous works: the cyclical oscillation of on-hand and on-order inventories about their target value.

The study of supply chain from the point of view of complex dynamical systems theory has started only recently (Helbing, 2008). Concepts from statistical physics and nonlinear dynamics have recently been used for the investigation of supply networks (Radons and Neugebauer, 2004).

Helbing (2003) generalized concepts from traffic flow to describe instabilities of supply chains. This section of the chapter remarks how small changes in the supply network topology can have enormous impact on the dynamics and stability of supply chains. In order to stabilize the supply chain, some strategies are mention on Radons and Neugebauer (2004).

By simulating a supply chain model, Larsen *et al.* (1999) showed a wide range of non-linear dynamic phenomena that produce an exceedingly complex behavior in the production-distribution chain model. Makui and Madadi (2007) proposed to measure the bullwhip effect by using the Lyanupov exponent. Hwarng and Xie (2008) used chaos theory through the Lyapunov exponent across all levels of a specific supply chain. They showed that chaotic behaviors in supply chain systems can be generated by deterministic exogenous and endogenous factors. They also discovered the phenomenon "chaos-amplification", i.e. the inventory becomes more chaotic at the upper levels of the supply chain.

The fluctuations (volatility) of the queues of inventory in the recovery process into a closed loop supply chain were analyzed. With a different focus as previous works, we characterized the variability by calculating the scaling parameter of the fluctuations and by using a novel algorithm to map the time series into a graph. The latter allows using the tools of complex networks and provides an insight about the emergence of the variability in the queues.

In this section of the chapter there are two different methods used to estimate H : the rescale range (R/S) analysis method (equation 4) and the visibility graph algorithm. These methods allow the calculation of the self-affine parameter H (more details and further estimation methods can be found in Gao*et al.*, 2007; Beran, 1994). Once we determined the H parameter, we map the time series into a graph, in so doing, methods of complex networks analysis are applied to give insight regarding the emergence of volatility in the recovery process queues.

Visibility Graph Analysis

One of the most useful mathematical models for self-affine processes has been the *fractional Brownian motion* (fBm) which is an extension of the central concept of *Brownian motion* (Mandelbrot, 1968; Embrechtc and Maejima, 2002). Self-affine processes such as fBm are currently used to model fractal phenomena of a different nature.

Let (Ω, F, P) be a complete probability space. The fBm $B_H(t)$ is a Gaussian process with mean 0, stationary self-affine increments (fractional Gaussian noise (fGn)), variance $E(B_H(t)^2) = t^{2H}$, that can be characterized by the so-called Hurst exponent, $H \in (0; 1)$. The special value $H = 0.5$ gives the familiar Brownian motion

The visibility graph algorithm is a new method to estimate the H exponent by mapping afBm into a scale-free network according with the following criterion (Lacasa*et al.*, 2008, 2009): two arbitrary data $(t_a; y_a)$ and $(t_b; y_b)$ in the time series have visibility, and consequently become two connected nodes in the associated graph, if any other data $(t_c; y_c)$ such that $t_a < t_c < t_b$ fills

$$y_c < y_a + \frac{(y_b - y_a)(t_c - t_a)}{t_b - t_a} \tag{19}$$

Lacasa*et al.* (2009) showed that the degree distribution of graphs derived from generic fBm follows a power law $P(k) \sim k^{-\gamma}$ where k stands for the degree of a given node. A linear relation between the exponent of the power law degree distribution in the visibility graph and the H of the associated fBm series exist through:

$$\gamma(H) = 3 - 2H \tag{20}$$

And to estimate the exponent γ we plotted the logarithm of the vertex degree k versus the logarithm of the number of vertices of degree k: n_k. The resulting curve should approximate a straight line and the points satisfy the equation

$$\log(n_k) \approx \alpha - \gamma \log(k) \tag{21}$$

Complex Networks

The possibility of mapping a time series into a graph allows us to use the tools from complex networks to understand, with a different perspective, how the volatility in the recovery process queue emerges. Most real networks have the following properties (Chung and Lu, 2006; Chung, 2010):

- Small world phenomenon:it refers to the smallest number of links between the nodes, usually so-called *six degrees of separation* in social networks (Watts, 1998). A network has this property if the average shortage path $\bar{l}$ is comparable with that on a random graph, $\bar{l}/\bar{l}_{RandomGraph} \sim 1$, and the clustering coefficient is much greater than that for a random graph $\bar{C}_{data} \gg \bar{C}_{RandomGraph}$ (Watts, 1998).

- Power law degree distribution:itrefers to networks where the degree distribution of the connected nodes has a power law tail $P(k) \sim k^{-\gamma}$.

- Large: the size of the network typically ranges from hundreds of thousands to billions of vertices.

- Sparse: by defining the density of a graph as a number of existing edges E divided by the maximal possible number N of edges $D = E/\left[\frac{N(N-1)}{2}\right]$, a graph is then sparse if $D \ll 1$ (Barrat*et al.*, 2008).

The exponent γ of graphs that exhibit a power law degree distribution falls into different ranges. Barabási and Albert (1999) found that $\gamma = 3$, and that these networks are the consequence of two generic mechanisms: (i) networks expand continuously by the addition of new vertices, and (ii) new vertices attach preferentially to sites that are already well connected. The so-called *preferential attachment scheme*, that can be described as "the rich get richer", produces power law graphs with aγ exponent from 2 to 1 (Dorogovtsev*et al.*, 2000). The range $1 < \gamma < 2$, is produced by a partial duplication model which is motivated on the duplication of information in the genome in biological networks. The model is described by Chung *et al.* (2003) as:

Let $z_{0\,0}$ be a constant and G_{z_0} be a graph on z_0 vertices. For$z > z_0$, G_z is constructed by partial duplication from G_{z-1} as follows: a Random vertex u of G_{z-1} is selected, then a new vertex v is added to G_{z-1} in such a way that for each neighbor $w$of u, with a probability p, a new edge $v - w$ is added. The exponent γ and the probability p are associated as

$$p(\gamma - 1) = 1 - p^{\gamma - 1} \tag{22}$$

Equation (22) can be linear approximated by

$$\gamma \approx 9.45 - 14.9p \tag{23}$$

Finally, to characterize the network structure obtained by the time series we use the following definitions ((Barrat *et al.*, 2008):

The average degree of an undirected graph is defined as the average value of k over all the vertices in the network,

$$\bar{k} = \frac{1}{N}\sum_i k_i \equiv \frac{2E}{N}$$

(24)

If the degree of a node i is k_i and if these nodes have e_i edges between them, we have the clustering coefficient,

$$C(i) = \frac{e_i}{\dfrac{k_i(k_i - 1)}{2}}$$

(25)

where e_i can be computed in terms of the adjacency matrix X for vertices i, j and l as

$$e_i = \frac{1}{2}\sum_{jl} x_{ij} x_{jl} x_{li}$$

(26)

and the average clustering coefficient is simply,

$$\bar{C} = \frac{1}{N}\sum C(i)$$

(27)

The measure is normalized and bounded to be between 0 and 1. The concept of path lies at the basis of the definition among vertices. The natural distance measure between two vertices i and j is defined as the number of edges traversed by the shortage connecting path is called the shortage path length, denoted by $l_{i,j}$. Then the diameter of a graph is defined as

$$d_g = max \; l_{i,j}$$

(28)

and the average shortage path length is

$$\bar{l} = \frac{1}{N(N-1)}\sum_{ij} l_{ij}$$

(29)

where by definition $\bar{l} \le d_G$. The fact that any pair of nodes is connected by a small shortage path and a combination of high clustering constitutes the so-called small-world phenomenon (Watts, 1998).

Spare Part Service Process

It was analyzed an Advance and Exchange (AE) spare part service process which happens between three stakeholders: the TEM, the Carrier and the Repair Vendor (Figure 11). The AE service is triggered when a critical element of the telecom network from the Carrier failed. At time t_1, a goodunit is delivered (i.e. delivery process (DP(t)))at Carrier site. At time t_2 the Carrier returns thedefective unit to the TEM (i.e. defective collectprocess (DCP(t))), where$t_2 \geq t_1$. At time t_3 the defective unit arrived at the Repair Vendor site (i.e. inbound repair process (IRP(t))), where$t_3 \geq t_2$. Finally at time t_4, once repaired, the unit outbound the repair process (ORP(t)) and returns to the pool of good units, where$t_4 \geq t_3$.

The dynamics of items that flow in the closed loop supply chain described above can be visualized by cumulative plots as shown in Figure 12 (Daganz, 2003). The vertical difference between two curves represents the queue $Q(t)$ of material pending to be processed and the horizontal separation between the curves represents the lead time L each item experiences between consecutive echelons.

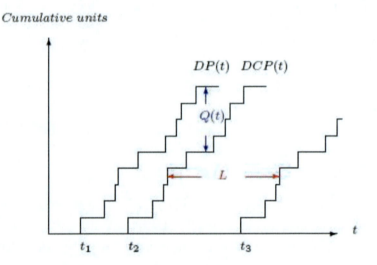

Figure 12. Cumulative data of each process in the closed loop supply chain.

Within this system a conservation principle applies: what comes in must go out. In general the queues can be represented as

$$Q_{kl}(t) = Curve_k(t) \geq 0 \qquad (30)$$

For $k > l$, where $k = 1,2,3,...,n-1$ and $l = 1,2,3,...,n$, and n is the number of echelons in the closed loop supply chain.

Similar to the measure of the bullwhip effect (Cachon*et al.*, 2007), here we calculate the amplification of variability in the queues as

$$\text{Amp. radio} = \frac{V[\text{queue echelon}_{i+1}]}{V[\text{queue echelon}_i]} \qquad (31)$$

for $i = 2, 3, \ldots, n$.

Empirical Findings

The time series encompassed one year of failures (demand of spare parts) of 4,217 units. Unfortunately not all defective units were collected and/or repaired at the moment we began the analysis. Then, only 3,617 units completed the entire process, i.e. since they were demanded until repaired. Figures 13 and 14 show the time series of each process of the supply chain, i.e. (a) delivery process, (b) defective collect process, (c) inbound repair process, and (d) outbound repair process. The demand of the 3,617 units happened during 365 days, the defective collect process and the inbound repair process took 434 days, and finally the outbound repair process took 464 days. Although there is a constant amount of units processed in each echelon, the number of days each process required increased going upstream in the supply chain.

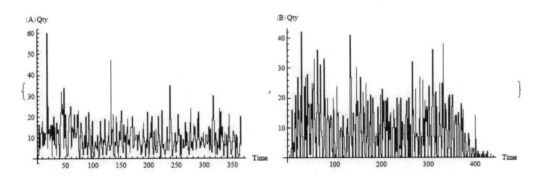

Figure 13. Time series of (a) delivery process and (b) defective collection process.

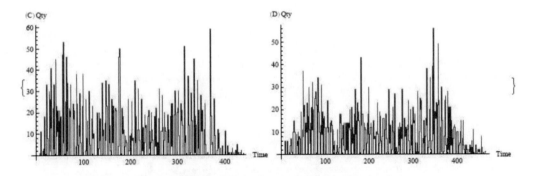

Figure 14. Time series (c) Inbound Repair and (d) Outbound Repair.

Table 1 shows simple statistics of the time series. It is notorious to see in this statistics the increase in variability between *DCP* and *DP*, and between *IRP* and *DCP* which confirm the presence of the bullwhip effect.

By applying equation (32) in last four time series, Figure 15 shows the three queue functions $Q(t)$.

Table 1. Simple statistics of the time series of actual deliveries of the closed loop supply chain

	DP	DCP	IRP	ORP
Average	9.9025	8.3341	8.3341	7.7952
Std. Deviation	7.0877	8.5929	11.468	9.1287
Variance	50.23	73.83	131.51	83.3338

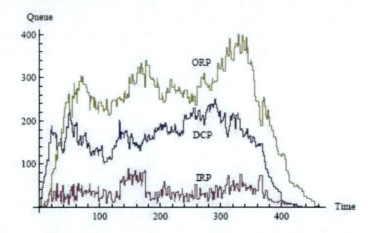

Figure 15. Inventory queues in the recovery process.

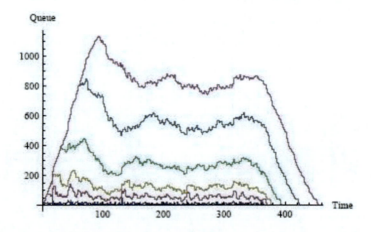

Figure 16. Inventory queues with different constant lead times.

In order to verify the impact of lead time in the volatility of the queues, we also built different queues with lead time equals to: 1, 7, 14, 30, 60 and 90 days (Figure 16).

In Table 2 we can see simple statistics of the queue time series.

Table 2. Simple statistics of queues of the closed loop supply chain

	AVERAGE	STD.DEVIATION	VARIANCE	CV
DCPqueue	147.7	60.91	3710.57	0.41
IRPqueue	32.48	19.49	379.94	0.6
ORPqueue	223.42	105.03	11032.17	0.47
L=1queue	9.9	7.08	50.23	0.71
L=7queue	58.65	20.61	425.14	0.35
L=14queue	124.72	37.12	1378.4	0.29
L=30queue	266.9	81.89	6707.14	0.3
L=60queue	504.49	178.05	31704.03	0.35
L=90queue	710.62	273.52	74817.45	0.38

Fractal Results

The H exponent measures the intensity of long-range dependence in a time series. Table 3 shows the estimated values of the H parameter.

Both methods: R/S and Visibility Graph Algorithm are consistent due they yielded values with the H parameter higher than 0.5. Therefore, the time series show persistence correlation, i.e. long-range dependence.

Table 3. The Hurst exponent calculation

	H R/S	$H_{(\text{Visibility G.A.})}$
DCPqueue	1.00	0.85
IRPqueue	0.89	0.84
ORPqueue	0.95	0.96
L=1queue	0.74	0.78
L=7queue	0.81	0.81
L=14queue	0.89	0.88
L=30queue	0.94	0.90
L=60queue	0.89	0.91
L=90queue	0.90	0.96

Complex Networks Results

In this point we mapped each time series into undirected networks by applying the visibility graph algorithm, and then used some simple formulas from the complex network theory to characterize the system (we used the Network Workbench tool software (NWB Team, 2006)).

The network of each process can be visualized in Figures 17-25. All the networks appear to be sparse due the density value $D \ll 1$ (Table 4), which is one of the characteristics of complex networks. To identify the presence of the small world effect, we ran a random graph per each network constructed on the same vertex set, where the probability that two vertices are connected is $P = \bar{k}/n$ (Newman *et al.*, 2006).

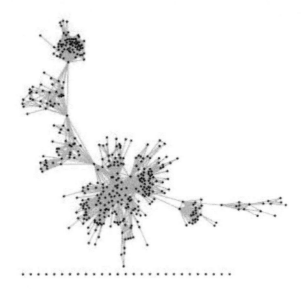

Figure 17. Queue VG Defective Collect.

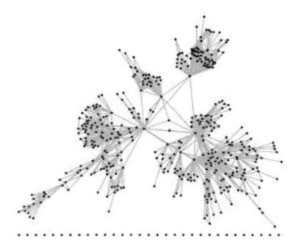

Figure 18. Queue VG Inbound Repair.

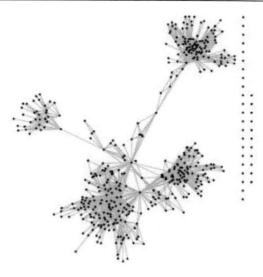

Figure 19. Queue VG Outbound Repair.

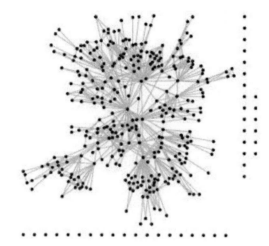

Figure 20. Queue VG L=1.

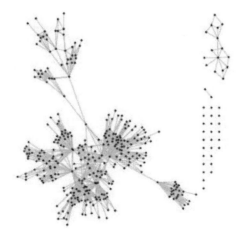

Figure 21. Queue VG L=7.

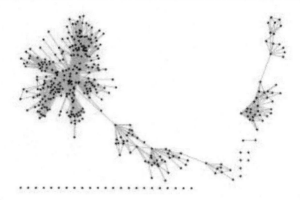

Figure 22. Queue VG L=14.

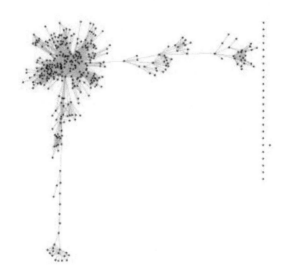

Figure 23. Queue VG L=30.

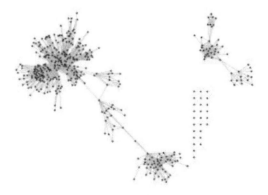

Figure 24. Queue VG L=60.

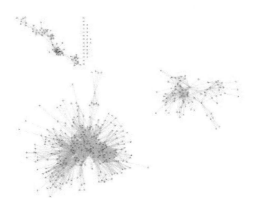

Figure 25. Queue VG L=90.

The random graphs were generated in the NWB tool (NWB Team, 2006). As in all cases the $\bar{C}_{data} \gg \bar{C}_{RandomGraph}$ and the average shortest path of data is comparable to that of the one generated by a random graph, $\bar{l}/\bar{l}_{RandomGraph} \sim 1$ (Tables 4 and 5), then we concluded that the networks are considered small worlds (Watts, 1998).

The power law property of these networks was also confirmed by the value of γ shown in Tables 4 and 5. As the value of γ is always in the range $1 < \gamma < 2$, then the Partial Duplication Model describes how the network emerged.

We can say that the lead time variablerepresents a key factor in mitigating the volatilityin the closed loop supply chain analyzed in this section of the chapter.

Table 4. Networks analysis of recovery process queues

			DC_{queue}	IR_{queue}	OR_{queue}
n			433	431	458
Edges			1795	1183	2220
Averagedegree			8.29	5.48	9.69
Isolatednodes			27	31	33
Density			0.0191	0.0127	0.0212
Average closteringcoef.(from data)			0.4875	0.5484	0.5261
Average closteringcoef.(random graph)			0.0202	0.0138	0.0202
Diameter			10	10	12
Average Shortest path (from data)			4.3325	4.0373	4.9235
Average Shortest path (random graph)			3.1567	3.8783	2.9662
γ			1.2854	1.3069	1.0763
P			0.5479	0.5465	0.5663

Table 5. Networks analysis of lead times

			L=1	L=7	L=14	L=30	L=60	L=90
N			365	370	377	393	423	453
Edges			838	1265	1442	2026	2390	2467
Averagedegree			4.39	6.83	7.64	10.31	11.3	10.89
Isolatednodes			40	26	29	25	26	26
Density			0.0126	0.0185	0.0203	0.0263	0.0267	0.0241
Average closteringcoef.(from data)			0.5995	0.4675	0.4898	0.4742	0.4258	0.4518
Average closteringcoef.(random graph)			0.0177	0.0219	0.0238	0.0266	0.0253	0.0227
Diameter			7	9	10	17	9	9
Average Shortest path (from data)			3.7439	4.0429	3.7606	4.7307	4.0064	2.9688
Average Shortest path (random graph)			4.1832	3.3913	3.2262	2.8181	2.7629	2.83
γ			1.4251	1.3627	1.2309	1.1883	1.0673	1.172
P			0.5385	0.5427	0.5516	0.5544	0.5625	0.5555

CONCLUSION

The fractal dynamics of complex systems has captured the attention of the scientific community to the extent where its proponents consider it a dominant scientific trend because the fractal analysis of time series from complex systems allows us to check for the presence of fractal behavior (or scaling explained by power laws) and in particular long-term correlations.

In the three applications of fractal analysis presented in this chapter we find a transient from anti-persistent to persistent behavior of the volatility in the horizon scale. In the case of fractal analysis of epilepsy, this transient could be interpreted as the beginning and end of seizures, and it could help locate the epileptic focus (the place where the seizures originated).

For the oil crude price market volatility case, this transition isaccompanied by the change in the type of volatility distribution, which is light-tailed for short horizons and is heavy-tailed for long horizons. The findings shown in this chapter have potentially wide-ranging implications in statistical physics of complex systems. Specifically, it is expected that the crossover from anti-persistent to persistent behavior should be observed in a wide variety of systems displaying generalized scaling dynamics.

The existence of a ''universal'' mechanism which gives rise to crossover from anti-persistent to persistent behavior in systems of different natures could provide a new insight into the physics of complex systems governed by avalanche dynamics leading to generalized scaling dynamics with continuously varying exponents.

Fractal Dynamics of Complex Systems

The scaling form (14) is valid for a large variety of systems far from equilibrium, as well as for critical phenomena. Specifically, the Family-Viscek dynamic scaling ansatz is commonly applied to describe the kinetic roughening of growing interfaces.

Examples of systems displaying the general scaling dynamics (14) include certain self-organized critical sand pile models, ion sputtering of surfaces and DLA-related growth processes. The continuously varied scaling exponents were found in some experiments in turbulence, in paper wetting experiments, in numerical analysis of the Kuramoto-Sivashinsky equation, and also were observed in Monte Carlo simulations of ion etch-front roughening.

In the third application of fractal analysis we also decided to model the time series fluctuations as a complex network because we had a few data. We found that all graphs show the presence of thesmall world phenomena and the power law property $P(k) \sim k^{-\gamma}$ with $1 < \gamma < 2$ .According withthe value of γ found in this case, the graphsemerged through the Partial Duplication Model.In conclusion, the fluctuation in the queues ofthe recovery process increases when H is closer to 1 and γ is also closer to 1 as a result of anincrement of the lead time. So, the lead time variablerepresents a key factor in mitigating the volatilityin the closed loop supply chain analyzed in thispaper.

Finally, more work will have to be invested in studying the practicalconsequences of fractal dynamics in time series. Studies should particularlyfocus on predictions of future values and behavior of time series and wholecomplex systems. This is very relevant, not only in hydrology and climateresearch, where a clear distinguishing of trends and natural fluctuations iscrucial, but also for predicting dangerous medical events on-line in patientsbased on the continuous recording of time series.

REFERENCES

Balankin A. 1991. *Synergetics of Deformed Solid*. Ministry of Defense USSR Press, Moscow.

Balankin A, Morales O, Gálvez E, and Pérez A. 2004. Crossover from antipersistent to persistent behavior in time series possessing the generalized dynamic scaling law.*Physical Review*E 69, 036121: 1-7.

Balankin A. 1997. Physics of fracture and mechanics of self-affine cracks, *Engineering Fracture Mechanics*.57 (2/3): 135-203.

Barabási A L. and Albert R. 1999.Emergenceof scaling in random networks.*Science*286: 509-512.

Barrat A, Barthélemy M. and Vespignani A. 2008.*Dynamical processes on complex networks*. Cambridge University Press.

Beran J. 1994. *Statistics for Long-Memory Processes*.Chapman and Hall, US.

Bunge M. 2003. *Emergence and Convergence, Qualitative Novelty and the Unity of Knowledge*.University of Toronto Press.

Cachon G, Randall T. and Schmid, G. 2007.Manufacturing and Service.*Operations Management*9(4): 457-49.

Chave J. and Levin S. 2002.*Scale and scaling in ecological and economic systems*.GuyotHall.Princeton University, Princeton NJ 08544-1003, U.S.

Chen F, Drezner Z, Ryan J. and Simchi-Levi, D. 2000.Quantifying the bullwhip effect in a single supply chain: The impact of forecasting, lead times, and information.*Management Science*46(3):436-443.

Chung F and Lu L. 2006.*Complex Graphs and Networks*.American Mathematical Society.

Chung F, Lu L, Dewey TG. and Galas DJ. 2003. Duplication models for biological networks, *Journal of Computational Biology* 10(5): 677-687.

Chung F. 2010. Graph Theory in the information age. *Notices of the AMS*57(6):726-732.

Cohen M, Agrawal N. and Agrawal V. 2006.Winning in the Aftermarket.*Harvard Business Review*, May: 129-138.

Contreras TI, Morales O. and Tejeida R. 2007. Epilepsy as a Dynamic Complex System, Journals ISSS http:/ /journals.isss.org/ (10 August 2009).

Cuniberti G. 2001. Effects of regulation on a self-organized market. *Quantitative Finance*1: 332-335.

Daganzo C. 2003. *A Theory of Supply Chains*.Springer, New York.

Daganzo C. 2004. On the stability of supply chains.*Operations Research*52(6): 909-921.

Dejonckheere J, Disney S, Lambrecht M, and Towill D. 2003. Measuring and avoiding the bullwhipeffect: A control theoretic approach. *European Journal of Operational Research*147: 567-590.

Dejonckheere J, Disney S, Lambrecht M. and Towill, D. 2004. The impact of informationenrichment on the bullwhip effect in supply chains: A control engineering perspective. *European Journal of Operational Research*153(3): 727-750.

Dorogovtsev S, Mendes J. and Smukhin A. 2000.Structure of growing networks with preferential linking.*Physical Review Letters*85(21): 4632-4633.

Embrechts P. and Maejima M. 2002.*Self-similar processes*.Princeton University Press.

Foote R. 2007. Mathematics and Complex Systems. *Science*318 : 410-412.

Forrester J. 1961. *Industrial Dynamics*.MIT Press and John Wiley and Sons, U.S.

Gao J, Cao Y, Tung W. and Hu J. 2007. *Multiscale analysis of complex time series*, Wiley,UK.

Hartley, KL. 2005. Defining Effective Service Level Agreements for Network Operation and Maintenance. *Bell Labs Technical Journal*9(4): 139-143.

Helbing D. (Ed). 2008. *Managing complexity: insights, concepts, applications*. Springer, New York.

Hwarng H. and Xie N. 2008. Understanding supply chain dynamics: A chaos perspective. *European Journal of Operational Research*184: 1163-1178.

Ivanova VS, Balankin AS, Bunin A, and Oksogoev A. 1994.*Synergetics and Fractals in Material Science*.Nauka, Moscow.

Kim I. and Springer M. 2008. Measuring endogenous supply chain volatility: Beyond the bullwhip effect. *European Journal of Operational Research* 189 (1):172-193.

Klonowsk W. 2000.*Signal and Image Analysis Using Chaos Theory and Fractal Geometry*. Polish Academy of Sciences.

Lacasa L, Luque B, Ballesteros F, Luque J. and Nu~no J. 2008. From time series to complex networks: The visibilitygraph. *Proc. Natl. Acad. Sci. U.S.A.*105 (13):4972-4975.

Lacasa L, Luque B, Luque J. and Nu~no J. 2009. The visibility graph: A new method forestimating the Hurst exponent of fractionalBrownian motion, *Europhysics Letters*86: 30001.

Larsen E, Morecroft J. and Thomsen J. 1999.Complex behavior in a production-distribution model.*European Journal of Operational Research*9: 61-74.

Lee HL, Padmanabhan V. and Whang S. 1997. The Bullwhip Effect in Supply Chains.*Sloan Management Review*38 (3): 93-102.

Makui A. and Madadi A. 2007. The bullwhip effect and Lyapunov exponent. *Applied Mathematics and Computation*189: 35-40.

Mandelbrot B. and Van Ness JW. 1968. Fractional Brownian Motion, Fractional Noises and Applications. *SIAM Review*10(4): 422-437.

Morales O, Gálvez E, Balankin A. and Pérez A. 2003. Dinámica Fractal y predicción de los precios del mercado petrolero. *Científica*3:139-154.

Morales O. 2004. *Modelos mecánicos de la dinámica fractal del mercado petrolero*; PhD thesis. Instituto Politénico Nacional, Mexico.

Newman M, Barabasi A. and Watts D. 2006.The structure and dynamics of networks.Princeton University Press.

NWB Team. 2006. *Network Workbench Tool*. Indiana University, NortheasternUniversity, and University of Michigan, http://nwb.slis.indiana.edu (June 30 2009).

Radons G. and Neugebauer R. (Ed). 2004. *Nonlinear dynamics of production systems*. Wiley-VCH,UK.

Rodriguez J. and Pandolfelli, V.C. 1998.Insights on the Fractal-Fracture Behavior Relationship.*Materials Research,*1(1): 47-52.

Scafetta N, Grigolini P, Hamilton P. and Westet B. 2002.*Non-extensive diffusion entropy analysis: non-stationarity in teen birth phenomena*, arXiv:cond-mat/0205524.

Sterman J. 1989. Modeling managerial behavior: Misperceptions of feedback in a dynamicdecision making experiment, *Management Science*35 (3): 321-339.

Watts D. and Strogatz S. 1998.Collective dynamics of small world networks.*Nature*393:440-442.

Zmeskal O. 2001. *Fractal Analysis of Image Structures*. Institute of Physical and Applied Chemistry, Czech Rep.

In: Classification and Application of Fractals
Editor: William L. Hagen

ISBN 978-1-61209-967-5
© 2012 Nova Science Publishers, Inc.

Chapter 3

FRACTAL ANALYSIS OF ELECTROMAGNETIC EMISSIONS IN POSSIBLE ASSOCIATION WITH EARTHQUAKES

M. Hayakawa[1,2,3], N. Yonaiguchi[2,4], Y. Ida[2,4,5], S. Masuda[6], and Y. Hobara[2,4]

[1]The University of Electro-Communications (UEC),
Advanced Wireless Communications research Center, Chofugaoka, Chofu Tokyo, Japan
[2]UEC, Research Station on Seismo Electromagnetics, Chofu Tokyo, Japan
[3]Information Systems Inc., Earthquake Analysis Laboratory,
Minami-Aoyama, Minato-ku Tokyo, Japan
[4]UEC, Graduate School of Informatics and Engineering, Chofu Tokyo, Japan
[5]Japan Radio Co., Ltd, Shimorenjaku, Mitaka Tokyo, Japan
[6]Tohoku Intelligent Communications Co., Ltd., Sendai, Japan

ABSTRACT

An earthquake (EQ) is known to be a large-scale fracture phenomenon in the Earth's crust and a vital problem in the short-term EQ prediction is the identification of precursors of EQs. When a heterogeneous crust is strained, its nonlinear evolution toward the final rupture is characterized by self-organization toward the critical point including the local nucleation and coalescence of microcracks (i.e., self-organized criticality (SOC)). Both acoustic as well as electromagnetic emissions in a wide frequency range from DC, ULF (Ultra-low-frequency) up to VHF are produced by those microcracks during the preparatory phase of EQs. This nonlinear dynamics in the lithosphere can be extensively investigated with the use of fractal analysis. This paper deals with the reviews on those fractal analyses especially on the two types of seismogenic emissions; one is ULF electromagnetic electromagnetic emissions and the second is VHF electromagnetic noises. Significant changes in the fractal properties of those electromagnetic emissions are found mainly prior to an EQ, which provides a rather promising candidate for predicting EQs.

Keywords: VHF electromagnetic emissions, ULF emissions, self-organized criticality, fractal analysis, earthquakes

1. INTRODUCTION

An earthquake (EQ) is a large-scale fracture phenomenon in the Earth's crust and a vital problem in the short-term EQ prediction is the identification of precursors of EQs (e.g., Hayakawa and Hobara, 2010). When a heterogeneous material like Earth's crust is strained, its nonlinear evolution toward the final rupture is characterized by self-organization toward the critical point including the local nucleation and coalescence of microcracks (i.e., self-organized criticality (SOC) (Bak, 1997; Turcott, 1997; Kapiris et al., 2004b)). Both acoustic as well as electromagnetic emissions in a wide frequency range from DC, ULF (Ultra low frequency, frequency less than ~ 1 Hz) up to VHF (30-300 MHz), are produced by microcracks, which can be considered as the so-called precursors of the general rupture of EQs (see the monographs by Hayakawa and Fujinawa (Eds.) (1994), Hayakawa (Ed.) (1999), Hayakawa (2001), Hayakawa and Molchanov (Ed.) (2002) and Molchanov and Hayakawa (2008)).

Such a nonlinear dynamics taking place in the Earth's crust toward the rupture (EQ) can be satisfactorily investigated with the use of fractal analysis. Hayakawa et al. (1999) performed the first attempt of fractal analysis to ULF emissions based on their spectral slope, and found that a significant change was observed just before the EQ. Later, different kinds of analysis methods have been developed by Smirmova et al. (2001, 2004), Gotoh et al. (2003, 2004), and Ida and Hayakawa (2006). Recently, Ida et al. (2005, 2006) have extended the previous mono-fractal analysis to multifractal one, in order to know the fine structure changes taking place just before the EQ. Then, this kind of fractal analysis has been rather popular in other frequency ranges as well, including DC emissions (SES (seismic electric signals)) (Varotsos, 2005), and VHF/VLF emissions (e. g. Eftaxias et al., 2002; Kapiris et al., 2002, 2004). The first part of this paper deals with the review on the previous fractal analysis on seismogenic ULF emissions and a comparison of this fractal analysis result to the corresponding result by means of another independent method of flicker noise spectroscopy, in order to find the consistency of the results between the two methods. In the second part, we will pay attention to VHF electromagnetic emissions recorded at multiple stations for a particular EQ. Our main observational tool is to monitor the microfractures, which occur in the prefocal area before the final breakup, by recording VHF emissions. In our previous work on ULF, we have found that the preseismic signature occurs from a few months before the EQ, with significant changes just before the EQ. Unlike the case of ULF emissions, higher frequency (VHF) emissions have exhibited the precursors only a few days to a few hours before the EQ (Yonaiguchi et al., 2007a). So that, Yonaiguchi et al. (2007a) have performed a much more extensive study on the VHF electromagnetic emissions by means of multi-stationed network than the Greek work (e. g., Kapiris et al., 2002), the essential results of which will be reviewed.

2. FRACTAL ANALYSIS OF SEISMOGENIC ULF ELECTROMAGNETIC EMISSIONS

2.1. General Introduction

The presence of precursory signature of EQs has been clearly identified in the ULF range for a few large (magnitude greater than 7) EQs such as Spitak, Loma Prieta, Guam, Biak etc. (Fraser-Smith et al., 1990; Molchanov et al., 1992; Kopytenko et al., 1993; Hayakawa et al., 1996, 2000). See recent reviews on seismogenic ULF emissions by Hayakawa and Hattori (2004), Hayakawa et al. (2007), Molchanov and Hayakawa (2008), Fraser-Smith (2009) and Kopytenko et al. (2009). ULF emissions are generally considered to be produced by current systems of different kinds including microfracturing (Molchanov and Hayakawa (1995, 1998), Vallianatos and Tzanis (1999)), which can be considered as the so-called precursors of general rapture of an EQ.

2.2. Fractal Analysis on Seismogenic ULF Emissions

Since the pioneering paper by Hayakawa et al. (1999) on the basis of mono-fractal analysis for ULF electromagnetic emissions using the spectrum slope, there have been published a lot of papers on the use of fractal analysis not only in the ULF range (Smirnova et al., 2001, 2004; Gotoh et al., 2003, 2004; Ida et al., 2005; Ida and Hayakawa, 2006), but also in other frequency ranges (Kapiris et al., 2002, 2004; Varotsos et al., 2005). Also, besides mono-fractal analysis, the analysis is extended to multifractal analysis (such as Ida et al. (2005)) in order to investigate the detailed fractal features of SOC in the lithosphere.

As taking an example of a huge EQ, the 1993 Guam EQ with magnitude of 8.2, Hayakawa and Ida (2008) have published a review by summarizing the previous fractal analysis for ULF emissions.

2.3. A Comparison of Fractal Analysis Result to that Analyzed by Flicker Noise Spectroscopy

In order to have some confidence on the fractal results, we have compared those with the corresponding results by any other method based on the different principle (here we use the flicker noise spectroscopy) (Hayakawa and Timashev, 2006). Again we use the same 1993 Guam EQ in the following.

(a) Experimental ULF Data and Guam EQ

The details of the ULF data for the Guam EQ have already been given in Hayakawa et al. (1999), but we have to repeat only the important points as follows. The period of data analysis is from January 1992 to the end of 1994 (total three years). The Guam EQ with magnitude Ms = 8.2, occurred on 8 August 1993 at 08:34UT suddenly and without any foreshocks. Its epicenter was located in the sea near the Guam island (geographic coordinates: 12.89°N, 1444.80 °E) as shown in Figure 1, and its depth was 60km. The Guam observatory where the

ULF data were recorded, is located at ~65km from the epicenter in Figure 1. We here comment on the sensitivity distance of seismogenic ULF emissions. Hayakawa and Hattori (2004), Hayakawa et al. (2004), Hayakawa et al. (2007) and Molchanov and Hayakawa (2008) have summarized all of the previous ULF emissions, who have concluded that the distance of sensitivity of ULF emissions is approximately 100 km for an EQ with magnitude 7.0. Figure 1 illustrates the relative location of our ULF observatory with respect to the EQ epicenter. A regular magnetic observation is maintained there using a three-axis ring-core-type fluxgate magnetometer (Hayakawa et al., 1996). Three components of magnetic variations are usually recorded on a digital cassette tape with a sampling rate of 1 s.

We analyze the data during the whole period, and we analyze the data during daytime (LT=14:00-15:00), because Gotoh et al. (2004) have found that the most significant change in the mono-fractal dimension was observed for the Guam EQ during daytime. One hour data are treated, so that the number of data is 3600 point per day.

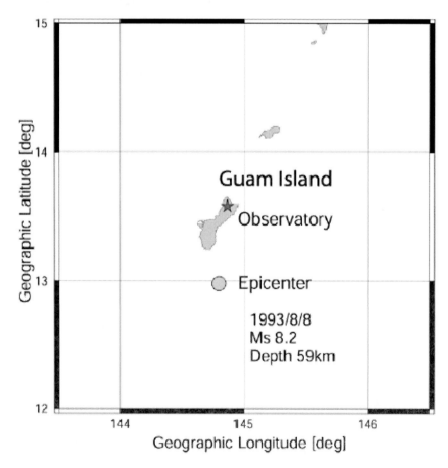

Figure 1. Relative location of our ULF observing station and the epicenter of the Guam EQ, together with the EQ characteristics.

(b) Fractal (Mono) Analysis

The method of analysis has already been described in details in Ida and Hayakawa (2006). Different kinds of methods of fractal behavior of time series data have been proposed, but we have used the Higuchi (1988) method because Gotoh et al. (2003, 2004) have compared different methods (Burlaga and Klein, 1986; Higuchi, 1988) extensively and have come to the conclusion that the Higuchi method is superior to others. Here we explain briefly the Higuchi's method. The estimation of fractal dimension (D) of the time series data (in our case ULF data with sampling of 1 sec) is based on the estimation of the length of the curve X(t). The fractal time series X(t) of our concern is used to define the following new time series.

$$\tilde{X}_m(k); X(m), X(m+k), X(m+2k), \cdots\cdots, X\left(m + \left[\frac{N-m}{k}\right]\cdot k\right)(m = 1,2,\cdots,k)$$

We estimate the curve length for each $\tilde{X}_m(k)$, and we call it $L_m(k)$ $(m = 1,2,\cdots,k)$.

$$L_m(k) = \frac{\left\{\sum_{i=1}^{\left[\frac{N-m}{k}\right]}|X(m+ik) - X(m+(i-1)\cdot k)|\frac{N-1}{\left[\frac{N-m}{k}\right]\cdot k}\right\}}{k}$$

where the term, $(N-1)\Big/\left[\dfrac{N-m}{k}\right]\cdot k$ is the normalizing factor. Finally, the curve length is defined as the arithmetic average as follows.

$$<L(k)>= \frac{\sum_{m=1}^{k} L_m(k)}{k}$$

Then, we plot $< L(k) >$ versus k $(k = 1,2,\cdots,10)$ and we estimate the slope by fitting, leading to the estimation of D.

Figure 2 is the summary of the temporal evolution of mono-fractal dimension of the result for H component in the bottom panel. One result is obtained for one day, so that the thin line is the connection of one-day results. While, the full line is the running average over ± 5 days. In order to indicate the statistical significance of the peaks in the fractal dimension, D, we have plotted the average value for the whole period (as a horizontal line) and the $\pm\sigma$ (σ: standard deviation) lines. For the sake of comparison, the top panel illustrates the temporal evolution of geomagnetic activity expressed by Ap index.

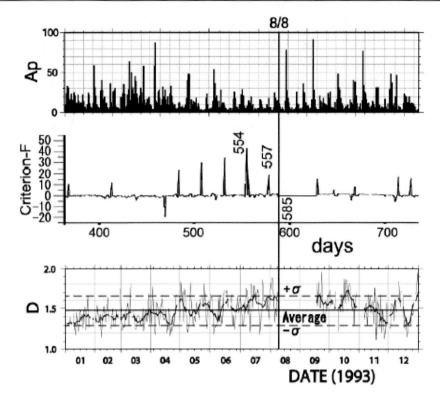

Figure 2. Comparison of the results by the two analysis methods; Top panel refers to the geomagnetic activity (Ap), 2nd panel is the result by the flicker noise spectroscopy and the bottom panel, that by the mono-fractal analysis (fractal dimension D).

(c) Flicker Noise Spectroscopy Method

This flicker noise spectroscopy is a new phenomenological method for the retrieval of information contained in chaotic time signals, the details of which have already been described in Timashev (2001) and Hayakawa and Timashev (2006). According to this phenomenological approach, the main information hidden in a chaotic signal at an interval T is provided by sequences of distinguishing types of irregularities-spikes, jumps, and discontinuities of derivatives of different orders at all space-time hierarchical levels of systems. It is possible to introduce different types of information. The ability to distinguish the irregularities means that the parameters or patterns characterizing the totality of properties of the irregularity sequences are extracted from the following power spectra $S(f)$ (f, frequency).

$$S(f) = \left| \int_{-T/2}^{T/2} \langle X(t) X(t + t_1) \cdot exp(2\pi i f t_1) dt_1 \rangle \right|,$$

$$\langle (\cdots) \rangle = \frac{1}{T} \int_{-T/2}^{T/2} (\cdots) dt,$$

and the difference moments $\Phi^{(2)}(\tau)$ of the 2nd order,

$$\Phi^{(2)}(\tau) = \left\langle X(t) - X(t + \tau)\right\rangle^2 = \left\langle \left[\int\limits_{t}^{t+\tau} \frac{dX(x)}{dx}\right]^2 \right\rangle,$$

where τ is time delay.

In this case, $\Phi^{(2)}(\tau)$ is formed exclusively by jumps of the dynamic variable different space-time hierarchical levels of the system under consideration, and $S(f)$ is formed by spikes and jumps.

In other words, the power spectra and difference moments of the 2nd order carry different information, which complement each other. The "passport parameters" are characteristic quantities, which are the correlation times, parameters characterizing the loss of "memory" for these correlation times, and characterizing the sequences of "spikes", "jumps" and discontinuities of derivatives of different orders.

In the most real cases, $S(f)$ and $\Phi^{(2)}(\tau)$ dependences manifest their complexity and non-stationary behavior. The behavior of the real $S(f)$ and $\Phi^{(2)}(\tau)$ is very specific and individual to each study case. These dependences have a definite physical sense and characterize the sequences of spikes, jumps and discontinuities of derivatives of different orders. That is why these dependences can be considered as "characteristic passport patterns" of the evolution under study.

While studying non-stationary processes, dynamics of the $S(f)$ and $\Phi^{(2)}(\tau)$ variations is being analyzed at sequential shift of the averaging interval $[k\Delta T, t_k]$ with the extension T, where $k = 0, 1, 2, 3 \cdots$ and $t_k = T + k\Delta T$, for the value ΔT along the whole time interval T_{tot} ($T + \Delta T < T_{tot}$) of the available experimental data. The time intervals T and ΔT should be selected as based on the physical sense of the problem considered - revealing the typical time of a process which determines the most important internal structural reconstruction of the evolution studied.

It is obviously natural to associate a phenomenon of "precursor" occurrence with the sharper variations of the relations $S(f)$ and $\Phi^{(2)}(\tau)$ at the approach of the upper boundary of the time interval of averaging t_k to a moment t_c of a catastrophic event when reconstruction takes place at all the possible spatial scales in the system. It is also expected (in this case we may speak about a "precursor") that the time of the "precursor's" manifestation t_k should stand from the moment t_c not less than at an interval ΔT, i.e. $\Delta T_{cn} = t_c - t_k \geq \Delta T$, at realization of the inequality $\Delta T_{cn} \ll T_{tot}$. When revealing a "precursor", it is important to distinguish cases when sharp variations in $S(f)$ and $\Phi^{(2)}(\tau)$ at averaging interval T shift are caused by significant signal variations on the "front" or "back" boundary of the interval T by approaching the "front" boundary t_k to a moment t_c of the expected event. A given problem is being solved by the analysis of the time behavior of the corresponding criteria at the T variations: it is obvious that when T increases in a value ΔT_1 the non-stationary effects associated with the signal behavior at the "back" boundary should be displayed with the same time delay ΔT_1, when the factor display caused by sharp signal variations in the area of the front boundary does not depend so strongly on the average interval value. Next we consider the "precursors" that are defined by the difference moments $\Phi^{(2)}(\tau)$. These functions can be reliably calculated only for a delay τ in the range $[0, \alpha T]$ with $\alpha \le 0.5$. Let us introduce the dimensionless quantities:

$$C(t_{k+1}) = 2 \cdot \frac{Q_{k+1} - Q_k}{Q_{k+1} + Q_k} \left/ \frac{\Delta T}{T} \right. ;$$

$$Q_k = \int_0^{\alpha T} \left[\Phi^{(2)}(\tau) \right]_k d\tau .$$

Here $t_{k+1} = k\Delta T \, (k = 0, 1, 2, \cdots)$ and subscripts of square brackets show that $\Phi^{(2)}(\tau)$ dependence was calculated for time interval $[k\Delta T, k\Delta T + T]$. The introduced quantities characterize a measure or a factor of non-stationarity of the signals, as the averaging interval T moves along time axis by a step ΔT, in particular, when the "forward" boundary of the averaging interval t_k approaches the catastrophic event at time t_c. Evidently $C(t_{k+1}) = 0$ for the stationary processes at $T \rightarrow \infty$.

We use the "high-frequency" components of the signals (C_F), because the criterion factors $C_F(t_k)$ demonstrated clearer results as compared to the low-frequency $C_G(t_k)$. The problem is to reveal the non-stationary factors and to understand whether these factors could be considered as precursors of the EQ.

The initial second data were used to get the initial minute data; every 60th second data were taken. Then we formed the hourly as well as the daily time series (formed by every 24th reading of the hour time series). At first, we began to analyze the daily data. It is well known that large EQs are prepared during several years. That is why we began our analysis by considering large T-intervals for finding a precursor for our case. We processed the daily time series and calculated the $C_F(t_k)$ factors by choosing $T = 550, 500, 400, 300, 200$ and 100 days as well as $\Delta T = 1$ day for all cases. The results are calculated for $T = 550$ (500, 400, 300, 200, and 100) days, and $\Delta = 1$ day, but we present only one example among them in Figure 2 with $T = 300$ days and $\Delta T = 1$ day. Many large peaks are found to be present in the calculated dependences in Figure 2. The appearance of every peak at different meanings of the current time means that the state of geophysical medium is changed at these times. If anyone considers an ordinary time series the appearance of any peak means that the measured value drops after rising, and that is all. But in the case of the noise spectroscopy non-stationary criteria the appearance of every peak means that the state of medium is different before and after the peak. It means that the seismo-active medium could be reconstructed (changes its state) several times before the EQ. It is interesting to study the dynamics of the realized peaks during the whole time (3 years in our case). We can notice five significant peaks in C_F in Figure 2 before the large EQ located on the day of 585 (Guam EQ), which will be our greatest concern. It is possible to think that the 5 peaks many reflect the 5 stages of the complex processes of the medium rearrangement before the coming catastrophic EQ. These precursors appeared at 101, 78, 54, 31 and 8 days before the EQ.

(d) Comparison of Results by Fractal Analysis and by Flicker Noise Spectroscopy Method

As is already shown by Smirnova and Hayakawa (2007), the main part of the ULF signals is of magnetospheric origin, so that we have to pay attention to the temporal evolution of Ap index in the top panel of Figure 2. Smirnova and Hayakawa (2007) have shown that the SOC process is also taking place before a major geomagnetic storm, just in the case of EQs. The geomagnetic activity during the period when we notice several peaks in C_F and D plots (during 4 months before the EQ), is found to be relatively quiet, so that we do not expect any SOC process in the Earth's magnetosphere.

The result by the flicker noise spectroscopy is summarized in the second panel of Figure 2 in the form of temporal evolution of C_F (criterion-F) factor just around the Guam EQ (August 8, 1993; Day = 585d). This analysis is done with $T = 300$ days and $\Delta T = 1$ day. It is possible for us to identify significant peaks in the C_F factor evolution; that is, these peaks appeared 101, 78, 54, 31 and 8 days before the EQ. This is indicative of 5 stages of the complex process of the medium rearrangement before the coming catastrophic EQ. The result by the flicker noise spectroscopy provides us with spiky results, while the mono-fractal analysis (in the bottom of Figure 2) gives us the continuous variation. However, as you can see from the bottom panel of Figure 2, we note that the fractal dimension (D) exhibits a significant change from the end of April, 1993 in such a way that the fractal dimension is notably increased from the end of April. This transition time is coincident with the time of the first spike in the C_F factor (101 days before the EQ), and the period for which the mono-fractal dimension D is enhanced, is found to be overlapping with the period of 5 peaks identified by the flicker noise spectroscopy. The next important point from the comparison by two methods is that the running curve of the mono-fractal dimension exhibits several maxima, whose positions are generally coincident with the peaks found by the flicker noise spectroscopy. Another point we have to pay attention is the period between the 4th peak (31 days before the EQ) and the EQ. While, the fractal dimension is seen to remain at a rather high value with large fluctuations. After the EQ, the fractal dimension is seen to show a tendency of decrease toward the pre-EQ level in December, 1993 or so. The general conclusion as based on the extensive comparison of results by the two methods, can be drawn as follows;

1) Only one year data (January to December, 1993) are presented in Figure 2, though we have analysed ±1.5 years around the EQ. It is clear from the results by both methods that some significant effects have definitely taken place just around the EQ.

2) Some precursory effects seem to start about 3 months before the EQ, and re-arrangement of the lithospheric medium seems to be taking place 3 months before the EQ and a few months even after the EQ.

3) Both analysis methods have yielded very consistent results on the precursory behavior taking place in the lithosphere in the period from 3 months before the EQ up to the EQ. The flicker noise spectroscopy indicates that there are 5 peaks before the EQ (101, 78, 54, 31 and 8 days before the EQ), and during this period of those peaks the mono-fractal dimension, D is found to be significantly enhanced.

4) Several peaks in the flicker noise spectroscopy result might indicate the step-like changes in the lithosphere just before the EQ due to the self-organization criticality process.

3. FRACTAL ANALYSIS OF VHF ELECTROMAGNETIC EMISSIONS

3.1. EQ Treated

The EQ for our analysis took place at 11h 46m 25s (L.T.) on 16 August, 2005, with the name of Miyagi-ken oki EQ (Off-sea of Miyagi prefecture). Its magnitude was 7.2 (a rather large EQ in this area) with the depth of 42 km. Its epicenter is located at the geographic coordinates (38.1 °N, 142.4 °E) as indicated by a cross in Figure 3.

Seven observing stations are set up for VHF natural electromagnetic emissions in the Tohoku area around Sendai, but only three representative VHF stations (Dairokuten, Kurihara and Kunimi) are illustrated in Figure 3. The distance between any pair of two stations is about 50 km, and the distance between Dairokuten and the EQ epicenter is approximately 90 km. We present the time series data of VHF electromagnetic emissions during a sufficiently long period (about one month before and one month after the EQ). The VHF observing system is briefly described here. The observing antenna is a discon-type (wideband characteristics from 25 to 1300 MHz) installed on the top of a tower at each station, but the observing frequency is tuned to 49.5MHz in the radio receiver (Yonaiguchi et al., 2007a). The intensity recorded at each station, is transmitted to the master station in Sendai.

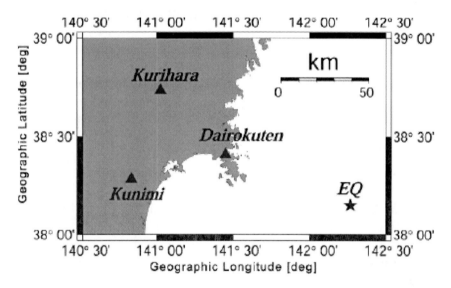

Figure 3. Relative location of the EQ epicenter (as indicated by a cross in the sea) and three representative VHF observing stations (Dairokuten, Kurihara and Kunimi). The distance between each two of 3 stations is approximately 50 km, and the epicentral distance from Dairokuten is about 90 km.

3.2. Multifractal Analysis for VHF Emissions

The VHF emission data as a time series are sampled every 100 sec (before the EQ) or 72 sec (after the EQ), and we apply the multifractal analysis for those VHF data. Multifractal analysis is known to provide us with much more information than the simple mono-fractal analysis (Ida et al., 2005, 2006).

The most important point of this VHF observation is (1) multi-stationed network and (2) longer period of analysis than previous works. First of all, we have to point out one important aspect of the data comparing the VHF data at seven stations. It is found that the VHF data at the three stations of Dairokuten, Kurihara and Kunimi are completely different from each other, which is importantly indicative of high locality nature of VHF emissions. Also, there is no diurnal variation in the VHF intensity variation, being very different from lower frequency emissions (such as ULF, ELF and VLF). There have already been proposed several methods for the multifractal analysis; extended MF-DFA (Multifractal detrended fluctuation analysis) (Kantelhardt et al., 2002), WTMM (Wavelet transform modulus maxima) (Muzy et al., 1993; Mallat and Hwang, 1992) etc. In order to confirm the reliability of the analysis results, we have utilized the above-mentioned two methods and have confirmed that those two methods have yielded consistent results (Yonaiguchi et al., 2007b). So, in this section we present only the result by the MF-DFA (not standard, but extended). The way of this analysis can be seen in Kantelhardt et al. (2002) and Ida et al. (2006).

We show one example of very conventional multifractal $f(\alpha)$ curve for a particular time interval from 15:54 on 28 July to 20:19 on 29 July before the EQ and Figure 4 is the result. The sampling before the EQ is 100 s (as one point), so that we took 1024 points (this corresponds to the above time interval of about 28 hours). By using these data, the full line in Figure 4 is obtained by the extended MF-DFA method.

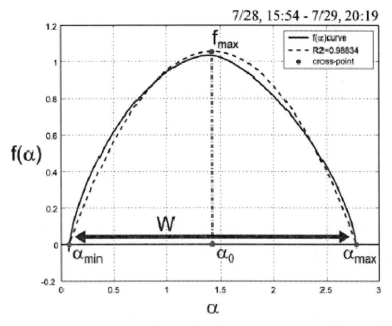

Figure 4. An example of obtained $f(\alpha)$ curve of the VHF emissions during a period of 15:54 L.T. on 28 July to 20:19 L.T. on 29 July at the station of Kunimi. The full line is observed $f(\alpha)$ and the broken line is its corresponding approximated $f(\alpha)$. The definitions of α_{max}, α_{min}, w ($\equiv \alpha_{max} - \alpha_{min}$), α_0 and f_{max} are given.

The general form of this $f(\alpha)$ is very regular, so that we can estimate very easily the multifractal parameters. However, when we deduce the $f(\alpha)$ curve, we have sometimes some irregularity in the $f(\alpha)$ curve and we wonder whether we have to accept such a $f(\alpha)$ curve or not. So we have made the following criterion: When the obtained $f(\alpha)$ curve as in a full line in Figure 4 is approximated by a quadratic equation of α with correlation function exceeding 0.95, we accept the $f(\alpha)$ curve and we use the approximated $f(\alpha)$ curve to deduce the multifractal parameters. First, α_{max} and α_{min} are obtained from the relation of $f(\alpha) = 0$, and the width w is defined as w ($\equiv \alpha_{max} - \alpha_{min}$). $f(\alpha)$ takes a maximum value f_{max} at a specific α_0. In Figure 4, the correlation is found to be 0.988, well above 0.95.

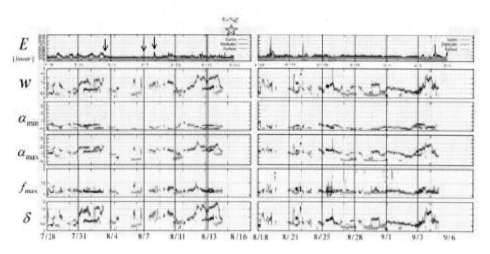

Figure 5. Temporal evolutions of VHF radio emissions and their corresponding multifractal quantities. EQ means the time of the EQ or 16 August, 2005. The left part refers to the period before the EQ, while the right, after the EQ. The top panel indicates the real VHF emissions intensity at three stations (Blue Dairokuten, green Kurihara and red Kunimi). Below the parameters of multifractality are given from the top, w ($\equiv \alpha_{max} - \alpha_{min}$), α_{max}, α_{min}, α_0 and δ ($\equiv w / f_{max}$) (non-uniformity factor). The same color is valid for each station.

Figure 5 is the overall summary of the multifractal analysis of VHF emissions at the three stations. The occurrence time of the EQ is indicated by a EQ star; the left part of Figure 5 refers to the period before the EQ, while the right part, the period after the EQ. The color corresponds to a station: blue line refers to Dairokuten, green, Kurihara and red, Kunimi as shown in Figure 3. The top panel in Figure 5 illustrates the electric field intensity of observed VHF emissions. Below we illustrate the parameters of multifractality; from the top, w ($\equiv \alpha_{max} - \alpha_{min}$), α_{max}, α_{min}, α_0 and finally the non-uniformity factor δ ($= w / f_{max}$) (Goltz, 1997). The color is the same as in the top panel. First of all we look at the temporal evolutions of VHF emission intensity. The VHF activity at all of the three stations are seen to be very stable after the EQ. While, we observe relatively enhanced VHF activity before the EQ. But, 3 stations exhibited considerably different features. The VHF emissions at Kurihara (in green) is still at a low level, without any significant enhancement. On the contrary, the VHF intensity at Dairokuten (in blue) is considerably enhanced with some modulations during the period of 28 July to 3 August. And we have a quiet period from 4 August to 8 August, Then, again we have a prolonged period of high activity from 8 August to just before the EQ. The intensity at Kunimi in red, showed the similar general tendency. When we look at the parameters of multifractality in the lower panels in Figure 5, we notice that the points are not continuous with time. We have to explain why we have such discontinuities in the multifractal parameters. When we present one example in Figure 4, we reject the $f(\alpha)$ curve obtained when the $f(\alpha)$ curve is not good enough to have the correlation coefficient less than 0.95. The computation of multifractal parameters is performed for one data with 1024 points, with shifting one point (100s before the EQ and 72 s after the EQ). Where we have no point in the multifractal analysis in Figure 5, we have not good $f(\alpha)$ curves.

We examine, in Figure 5, the temporal behavior of the multifractal parameters at each station, one by one. The first three parameters (w ($\equiv \alpha_{max}$ - α_{min}), α_{max} and α_{min}) are inter-related, so that we deal with these together. The most important and seemingly significant point at Dairokuten (in blue) is that even though the VHF emission is considerably enhanced for some periods, both of α_{min} and α_{max} (correspondingly w) are seen to be stable during the whole period before and after the EQ. Also, the value of α_o with f_{max} (5th panel) remains just around $\alpha_o \approx 1.25$ during the whole period. The most significant parameter (δ (= w / f_{max})); non-uniformity factor remains stable in a range of δ ⁿ⁽ʳ⁾ 0.5 – 1.0 . Then, we consider the data at Kurihara (in green). The VHF emission there is extremely low, without any significant enhancement (though the close inspection indicates the enhancement before the EQ as compared with that after the EQ). The multifractal parameters (w, α_{max}, α_{min} and α_o) are seen to exhibit no significant changes, together with the δ (at the bottom). If we suppose that the VHF emissions at Dairokuten (in blue) were due to the EQ, we would expect significant precursory effects before the EQ, which is in complete discrepancy with the results at Dairokuten. This suggests that the enhanced VHF noises observed at Dairokuten during the period of 28 July to 11 August, are not seismogenic effects, but they may be local noises having nothing to do with the EQ. Finally, we go to the VHF data observed at Kunimi (in red). As compared with the situation after the EQ, the VHF emission intensity is considerably more intense. The temporal evolutions of multifractality parameters indicate that there are a few periods with significantly enhanced value of w only before the EQ; that is, 31 July to 3 August, 8-10 August and 12 to 15 August just before the EQ. During these periods, α_{max} is considerably increased α_{max} ⁿ⁽ʳ⁾ 2.5 – 3.0, suggesting that the Earth's crust appears more complex, with a wider range of possible fractal exponents. This effect is most clearly recognized in the plot of the most sensitive parameter, δ (non-uniformity factor). The VHF emissions during these periods with enhancement are likely to reflect the SOC process in the lithosphere near Kunimi. Furthermore we note that this δ value has relaxed to the normal (or background) value after the EQ. This result appears to be consistent with the well-known result on the transition from mono- to multifractal behavior in the percolation model of breakdown at the last stage of the fracture process. This is strongly indicating that the VHF emissions observed at Kunimi (in red) may be electromagnetic precursor of the EQ, which is reflecting the nonlinear SOC behavior taking place only just around the station of Kunimi.

We here comment on some other factors, including (1) lightning and (2) magnetospheric (solar-terrestrial) effect. In the top panel in Figure 5 the occurrence or detection of lightning discharges (detected by lightning detector) is indicated by downward arrows, so that the effect of lightning is not found to be so influential. As for the second effect, (ii) magnetic activity, we looked at the temporal variation of Dst and ΣKp, and have found that the magnetic activity was not so active during the whole period, except a large geomagnetic storm on 24 August. Even this storm has not made any significant effect on VHF noise.

The presence of precursory VHF emissions only around Kunimi and the absence of seismic effects at Dairokuten and Kurihara, can be closely related with the geological structures (including faults information) around those stations. The seismological observation after the EQ has indicated in Figure 6 that the strongest EQ intensity was observed at the EQ shock around the station of Kunimi, and Dairokuten had the smallest seismic intensity because of the strong rock structure below the station. We have also indicated the fault region in this area in Figure 6, which indicates that there are fault regions named Nagamachi-Rifu fault regions, being coincident with the distribution of higher EQ intensity. Because of this geological structure beneath the station of Kumini, VHF electromagnetic emissions tend to easily appear in the atmosphere, to be observed only at the station.

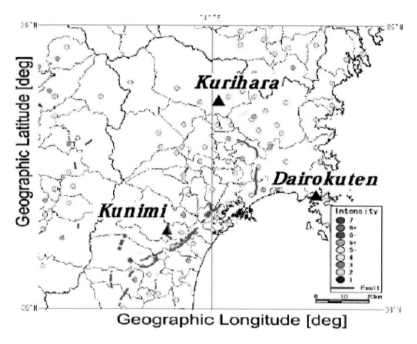

Figure 6. Distribution of intensity of the EQ, with the strongest seismic intensity around the station of Kunimi. Also, there is illustrated the spatial distribution of faults around this station.

3.3. Summary on Fractal Properties of VHF Noise

The nonlinear process taking place in the Earth's heterogenous crust is investigated by monitoring the VHF electromagnetic noise, and we identify the VHF precursory signature of an EQ. Our method is based on monitoring the microfractures, which occur in the prefocal area before the final breakup of EQ, by recording their VHF emissions. We have chosen a large EQ in the off-sea of Sendai, which took place on 16 August, 2005 and had a magnitude of 7.2 (depth 40 km).

The VHF data are obtained by means of a network in the Tohoku area, consisting of seven observing stations and also we have used a much longer period of observations than any previous VHF studies. A comparison of raw data at different stations and multifractal analysis performed to the VHF time series data observed at three stations, have yielded the following findings.

1) The VHF time series data are very different at the three stations, which is indicative of a high locality of the VHF emissions.

2) When we look only at the VHF intensity , the VHF intensity at Dairokuten is very variable and considerably stronger than at two other stations (Kurihara and Kunimi). However, the multifractal parameters at Dairokuten did not show any significant change (especially before the EQ), indicating the absence of self-organization in the VHF noise at Dairokuten. The Kurihara station is characterized by the smallest and stable VHF intensity before and after the EQ. No changes in multifractal parameters are seen there either.

3) While, the multifractality parameters (w and δ) at Kunimi where the VHF intensity is relatively enhanced before the EQ, indicated the presence of a few periods with significant changes (increase in w and δ) 2-3 weeks before and a few days before the EQ. But, these parameters are found to be completely relaxed after the EQ. This sharp contrast before and after the EQ, may be indicative of the SOC behavior toward the final rupture only at this station of Kunimi.

4) The significant effect of nonlinear process only around Kunimi, is highly likely to be related with the geological structure there (fault region).

Unlike the seismogenic ULF emissions which are believed to be generated near the focal region of an EQ, VHF emissions are expected to be heavily damped during the propagation in the lithosphere, and so they are considered to be generated near the Earth's surface or in the atmosphere as a secondary effect of precursory lithospheric signatures. As summarized above, we have found significant changes only at a station of Kunimi. A sharp contrast of multifractal parameters before and after the EQ (Summary (3) and (2)), is strongly suggesting that some SOC process is going on only at Kunimi, but not at the other two stations. A few days before the EQ we have observed seemingly significant changes in multifractal parameters (especially δ), which would be a precursor to the EQ. The change in multifractality before the EQ is characterized by (1) a decrease in the value of α_{min}, and (2) a drastic increase in the value of α_{max} (correspondingly a significant increase in w). These result in a drastic enhancement in the value of δ (non-uniformity factor). Although δ can, of course, not completely describe the $f(\alpha)$ spectra, it is very convenient for the study. This value of δ is found to exhibit a significantly high value approaching 2.0 before the EQ (i.e., a support that the system showed a transition from homogenous to heterogeneous (from monofractal to multifractal)). This may be an important precursor to this EQ.

CONCLUSION

A typical rapture phenomenon of EQs can be investigated by means of fractal analysis of electromagnetic emissions in a wide frequency range (ULF and VHF) because it is recently confirmed that electromagnetic emissions are useful for short-term EQ prediction. The fractal analysis provides us with the information on the nonlinear evolution toward the final rupture characterized by self-organization toward the critical point including the local nucleation and coalescence of microcracls closely related with the generation of electromagnetic emissions. As the result of fractal analysis for ULF electromagnetic emissions, some significant changes in mono-fractal dimension are found to exist prior to an EQ, which is supported by a comparison with the corresponding result by another independent method by the flicker noise spectroscopy. Then, the multifractal analysis is applied to a VHF network in the Tohoku area with a target EQ (Miyagi-ken oki EQ). A sharp contrast of multifractal parameters of VHF electromagnetic noise before and after the EQ is strongly suggesting that some SOC process is taking place only at a particular station.

ACKNOWLEDGMENTS

This work is, in part, supported by NICT (National Institute of Information and Communications Technology), Japan, to which we are grateful. The original ULF data were made available to us by Prof. K. Yumoto of Kyushu University, whom we would like to thank.

REFERENCES

Bak, P., How Nature works (The Science of Self-organized Criticality), Oxford University Press, 201 pp, 1997.

Burlaga, L. F., and L. W. Klein, Fractal structure of the interplanetary magnetic field, *J. Geophys. Res.*, 91, 347-350, 1986.

Eftaxias, K., P. Kapiris, E. Dologlou, J. Kapanas, N. Bogris, G. Antonopoulos, A. Peratzakis, and V. Hadjicontis, EM anomaly before the Kozani earthquake: A study of their behavior through laboratory experiments, *Geophys. Res. Lett.*, 29, 1228, doi:10.1029/2001GL013786, 2002.

Fraser-Smith, A. C., The ultralow-frequency magnetic fields associated with and preceding earthquakes, in "Electromagnetic Phenomena Associated with Earthquakes", Ed. by M. Hayakawa, Transworld research Network, Trivandrum (India), 1-20, 2009.

Fraser-Smith, A. C., A. Bernardi, P. R. McGill, M. E. Ladd, R. A. Helliwell, and O. G. Jr. Villard, Low-frequency magnetic field measurements near the epicenter of the Ms 7.1 Loma Prieta earthquake, *Geophys. Res. Lett.*, 17,1465-1468, 1990.

Goltz, C., Fractal and Chaotic Properties of Earthquakes, Springer Verlag, Berlin, 1997.

Gotoh, K., M. Hayakawa, and N. Smirnova, Fractal analysis of the geomagnetic data obtained at Izu peninsula, Japan in relation to the nearby earthquake swarm of June-August 2000, *Natural Hazards Earth System Sci.*, 3, 229-236, 2003.

Gotoh, K., M. Hayakawa, N. A. Smirnova, and K. Hattori, Fractal analysis of seismogenic ULF emissions, *Phys. Chem. Earth,* 29, 419-424, 2004

Hayakawa, M. (Editor), Atmospheric and Ionospheric Electromagnetic Phenomena Associated with Earthquakes, *Terra Sci. Pub. Comp.*, Tokyo, 996p, 1999.

Hayakawa, M., NASDA's Earthquake Remote Sensing Frontier Research, Seismo-electromagnetic Phenomena in the Lithosphere, Atmosphere and Ionosphere, Final Report 228p, *Univ. of Electro-Communications*, March, 2001.

Hayakawa, M. and Y. Fujinawa (Editors), Electromagnetic Phenomena Related to Earthquake Prediction, *Terra Sci. Pub*. Comp., Tokyo, 677p, 1994.

Hayakawa, M. and K. Hattori, Ultra-low-frequency electromagnetic emissions associated with earthquakes, Inst. Electr. Engrs. Japan, Trans. Fundamentals and Materials, 124, No. 12, 1101-1108, 2004.

Hayakawa, M., and Y. Hobara., Current status of seismo-electromagnetics for short-term earthquake prediction, Geomatics, *Natural Hazards and Risk*, 1, No. 2, 115-155, 2010.

Hayakawa, M., and Y. Ida, Fractal (mono- and multi-) analysis for the ULF data during the 1993 Guam earthquake for the study of prefracture criticality, *Current Development in Theory and Applications of Wavelets*, 2(2), 159-174, 2008.

Hayakawa, M., and O. A. Molchanov (Editors), Seismo Electromagnetics : Lithosphere – Atmosphere – Ionosphere Coupling, *TERRAPUB*, Tokyo, 477p, 2002.

Hayakawa, M. and S. F. Timashev, An attempt to find precursors in the ULF geomagnetic data by means of flicker noise spectroscopy, *Nonlinear Processes Geophys.*, 13, 255-263, 2006.

Hayakawa, M., K. Hattori, and K. Ohta, Monitoring of ULF (ultra-low-frequency) geomagnetic variations associated with earthquakes, *Sensors,* 7, 1108-1122, 2007.

Hayakawa, K., R. Kawate, O. A. Molchanov and K. Yumoto, Results of ultra-low-frequency magnetic field measurements during the Guam earthquake of 8 August 1993, *Geophys. Res. Lett.*, 23, 241-244, 1996.

Hayakawa, M., T. Itoh, and N. Smirnova, Fractal analysis of ULF geomagnetic data associated with the Guam earthquake on August 8, 1993, *Geophys. Res. Lett.,* 26, No. 18, 2797-2800, 1999.

Hayakawa, M., O. A. Molchanov and NASDA/UEC Team, Summary report of NASDA's earthquake remote sensing frontier project, in Special Issue on Seismo Electromagnetic and Related Phenomena, Ed. by M. Hayakawa, O. A. Molchanov, P. Biagi and F. Vallianatos, *Phys. Chem. Earth*, 29, Issues 4-9, 617-626, 2004.

Hayakawa, M., T. Itoh, K. Hattori and K. Yumoto, ULF electromagnetic precursors for an earthquake at Biak, Indonesia on February 17, 1996, *Geophys. Res. Lett.*, 27, 1531-1534, 2000.

Higuchi, T., Approach to an irregular time on the basis of fractal theory, *Physica D*, 31, 277-283, 1988.

Ida, Y., and M. Hayakawa, Fractal analysis for the ULF data during the 1993 Guam earthquake to study prefracture criticality, *Nonlinear Processes Geophys.*, 13, 409-412, 2006.

Ida, Y., M. Hayakawa, and K. Gotoh, Multifractal analysis for the ULF geomagnetic data during the Guam earthquake, *IEEJ Trans. Fundamentals and Materials*, 126, No.4, 215-219, 2006.

Ida, Y., M. Hayakawa, A. Adalev, and K. Gotoh, Multifractal analysis for the ULF geomagnetic data during the 1993 Guam earthquake, *Nonlinear Processes Geophys.*, 12, 157-162, 2005.

Kantelhardt, J. W., S. A. Zschiegner, E. K. Bunde, A. Bunde, S. Havlin, and H. E. Stanley, Multifractal detrended fluctuation analysis of nonstationary time series, *Physica A*, 316, 87-114, 2002.

Kapiris, P. G., K. A. Eftaxias, and T. L. Chelidze, Electromagnetic signature prefracture criticality in heterogeneous media, *Phys. Rev. Lett.*, 92, No. 6, 065702, 2004a.

Kapiris, P., J. Polygiannakis, A. Peratzakis, K. Nomicos, and K. Eftaxias, VHF-electromagnetic evidence of underlying pre-seismic critical stage, *Earth Planets Space*, 54, 1237-1246, 2002.

Kapiris, P. G., G. T. Balasis, J. A. Kopanas, G. N. Antonopoulos, A. S. Pertezakis and K. A. Eftaxias, Scaling similarities of multiple fracturing of solid materials, *Nonlinear Process Geophys.*, 11, 137-151, 2004b.

Kopytenko, Yu. A., V. S. Ismaguilov and L. V. Nikitina, Study of local anomalies of ULF magnetic disturbances before strong earthquakes and magnetic fields induced by tsunami, in "Electromagnetic Phenomena Associated with Earthquakes", Ed. by M. Hayakawa, Transworld research Network, Trivandrum (India), 21-40, 2009.

Kopytenko, Y. A., T. G.. Matishvili, P. M. Voronov, E. A. Kopytenko and O. A. Molchanov, Detection of ultra-low-frequency emissions connected with the Spitak earthquake and its aftershock activity, based on geomagnetic pulsations data at Dusheti and Vatdzia observations., *Phys. Earth Planet. Inter.*, 77, 85-95, 1993.

Mallat, S., and W. L. Hwang, Singularity detection and processing with wavelets, *IEEE Trans. Information Theory*, 38, No. 2, 617-643, 1992.

Molchanov, O.A., and M. Hayakawa, Generation of ULF electromagnetic emissions by microfracturing, *Geophys. Res. Lett.* 22, 3091-3094, 1995.

Molchanov, O. A., and M. Hayakawa, Seismo Electromagnetics and Related Phenomena: History and latest results, *TERRAPUB,* Tokyo, 189 p., 2008.

Molchanov, O. A., Y. A. Kopytenko, P. M. Voronov, E. A. Kopytenko, T. G. Matiashvill, A. C. Fraser-Smith, A. Bernardi, Results of ULF magnetic field measurements near the epicenters of the Spitak (ms=6.9) and Loma Prieta (Ms=7.1) earthquakes: Comparative analysis, *Geophys. Res. Lett.*, 19, 1495-1498, 1992.

Muzy, J. F., E. Bacry, and A. Arneodo, Multifractal formalism for fractal signals: The structure-function approach versus the wavelet-transform modulus-maxima method, *Phys. Rev.*, E 47, No.2, 875-884, 1993.

Smirnova, N. A., M. Hayakawa, Fractal characteristics of the ground-observed ULF emissions in relation to geomagnetic and seismic activities, *J. Atmos. Solar-terr. Phys.,* 69, 1833–1841, 2007.

Smirnova, N., M. Hayakawa, and K. Gotoh, Precursory behavior of fractal characteristics of the ULF electromagnetic fields in seismic active zones before strong earthquakes, *Phys. Chem. Earth*, 29, 445-451, 2004.

Smirnova, N., M. Hayakawa, K. Gotoh, and D. Volobuev, Scaling characteristics of ULF geomagnetic field at the Guam seismoactive area and their dynamics in relation to the earthquake, *Natural Hazards Earth System Sci.*, 1, 119-126, 2001.

Timashev, S.F., Flicker-noise spectroscopy as a tool for analysis of fluctuations in physical system, Noise in Physical Systems and 1/f Fluctuations, World Scientific, Ed. by G. Bosman, *World Scientific*, 775-778, 2001.

Turcott, D.L., Fractal and Chaos in Geology and Geophysics, 2nd Edition, Cambrige Univ. Press, (ambridge) 1997.

Vallianatos, F. and A. Tzanis, A model for the generation of precursory electric and magnetic fields associated with the deformation rate of the earthquake focus, in "Atmospheric and Ionospheric Electromagnetic Phenomena Associated with Earthquakes", Ed. by M. Hayakawa, *TERRAPUB*, Tokyo, 287-306, 1999.

Varotsos, P., The Physics of Seismic Electric Signals, *TERRAPUB*, Tokyo, 338p, 2005.

Yonaiguchi, N., Y. Ida, M. Hayakawa, and S. Masuda, Fractal analysis for VHF electromagnetic noises and the identification of preseismic signature of an earthquake, *J. Atmos. Solar-terr. Phys.*, 69, 1825–1832, 2007a.

Yonaiguchi, N., Y. Ida, M. Hayakawa, and S. Masuda, A comparison of different fractal analyses for VHF electromagnetic emissions and their self-organization for the off-sea Miyagi-prefecture earthquake, *Natural Hazards Earth System Sci.*, 7, 485–493, 2007b.

In: Classification and Application of Fractals
Editor: William L. Hagen

ISBN 978-1-61209-967-5
© 2012 Nova Science Publishers, Inc.

Chapter 4

DESIGN OF FREQUENCY SELECTIVE SURFACE USING FRACTALS GEOMETRIES

Antonio Luiz Pereira de Siqueira Campos[1] *and Paulo Henrique da Fonseca Silva*[2]

[1]Federal University of Rio Grande do Norte (UFRN)
Comunications Engineering Department (DCO)
Caixa Postal, CEP: Natal, RN, Brazil
[2]Group of Telecommunications and Applied Electromagnetism
IFPB - Federal Institute of Education, Science and Technology of Paraiba
Av. Primeiro de Maio, Jaguaribe, CEP, João Pessoa, PB, Brazil

ABSTRACT

This chapter describes applications of fractal geometries to design Frequency Selective Surfaces (FSSs), where we emphasize the use of fractal geometries to improve their frequency responses. Initially, we do a review about fractals and fractal geometries. After this, we describe the L-system and the Iterated Function Systems – IFS to generate fractal geometries. Thereafter, we show a review of literature about the use of fractal geometries in FSS design. Then, we present some numerical results of FSS designing using fractal geometries. The obtained numerical results are compared with the experimental ones. Finally, conclusions are listed.

INTRODUCTION

Typical frequency selective surface (FSS) comprise a bi-dimensional periodic array of elements (patches or apertures) in a conducting screen, which must be either freestanding or etched on supporting dielectric substrates. These structures are designed to reflect or transmit electromagnetic waves with frequency discrimination and have contributed significantly toward advancing of the modern telecommunication systems. Early researches have been carried through the application of FSS as subreflectors of parabolic dish Cassegrain antennas.

Several authors proposed the design of frequency selective surfaces using fractals. The use of space-filling properties of the Minkowski loop and the Hilbert curve was proposed in order to reduce the overall size of the FSS elements. In particular, the attractive features of certain self-similar fractals have received attention of microwave engineers to design multiband FSS. A dual-band FSS was designed using Sierpinski gasket dipole, with attenuation in excess of 30 dB. Fractal tree elements were considered to design tri-band FSS. Many others self-similar geometries have been explored in the design of dual-band and dual polarized FSS. The self-similarity property of these fractals enables the design of multiband fractal elements or fractal screens. Gosper fractal elements allow the design of FSS, which have dual-band rejection in a short band of frequencies. Furthermore, as the number of fractal iterations increases, the resonant frequencies of these periodic structures decrease, allowing the construction of FSS compact.

The use of Minkowski fractal may allow the design of frequency selective surfaces, with band-stop characteristics, enabling reductions in unit cell dimensions and metal areas of the FSS. The FSS with Minkowski elements have bandwidth less than 1 GHz, for some applications it may be desirable.

Fractal elements of the type of Dürer's pentagon can be used to design FSS with dual-band response type. The proposed structures act as spatial band-stop filters and have interesting properties for the design of FSS compact broadband or dual-band responses.

The aim of this chapter is to demonstrate how the use of fractal geometry can allow the enhancement of FSS response, allowing FSS with dual-band response, increasing the bandwidth of the designed structures, enabling the miniaturization of the dimensions of FSS, etc. In this chapter, we will consider different fractal geometries, among which we can cite: Minkowski, Dürer's pentagon, Koch, etc.

REVIEW OF FRACTALS AND FRACTAL GEOMETRIES

The idea of fractals is relatively new, but some of the fractals and their descriptions date back to classical mathematicians of 19th century. The mathematics behind fractals can be traced back to 1872, when Karl Weierstrass studied functions with the non-intuitive property of being everywhere continuous but nowhere differentiable. In 1883, Georg Cantor gave examples of subsets of the real line with unusual properties. Subsequently, Giuseppe Peano (1890) and David Hilbert (1891) constructed space-filling curves whose range contains the entire 2-dimensional unit square. Helge von Koch (1904) gave a more geometric definition of a function everywhere continuous but nowhere differentiable. Wacław Sierpiński constructed his triangle in 1915. Gaston Julia investigated iterated functions in the complex plane in 1918. In 1938, Paul Pierre Lévy developed the idea of self-similar curves.

In the 1960s, Benoît Mandelbrot started investigating self-similarity. Finally, in 1975 Mandelbrot coined the word "fractal" to denote an object whose Hausdorff–Besicovitch dimension is greater than its topological dimension. In 1977, Mandelbrot inspired by the Gaston Julia work published in 1918 used computers to explore it and discovered (quite by accident) the most famous fractal of all, which now bears his name: the Mandelbrot set [1]. In 1982, Mandelbrot demonstrated that these early mathematical fractals in fact have many features in common with shapes found in nature [2]. The creations of these mathematicians

Design of Frequency Selective Surface Using Fractals Geometries   105

played a key role in Mandelbrot's concept of a new geometry, but Mandelbrot is often recognized as the father of fractal geometry [3].

The word 'fractal' (from the Latin 'fractus' meaning broken) introduced by Benoit Mandelbrot in his foundational essay in 1975 was given to "a rough or fragmented geometric shape that can be split into parts, each of which is (at least approximately) a reduced-size copy of the whole". Before Mandelbrot coined this term, the common name for such structures was monster curve [4]. Fractal objects, as defined by Mandelbrot have certain special properties such as self-similarity (containing replicas/copies of themselves), underivability at every point, and a Hausdorff-Besicovitch (HB) dimension strictly greater than its topological dimension. [1].

A fractal is a mathematically generated pattern that is reproducible at any magnification or reduction and the reproduction looks just like the original, or at least has a similar structure. The fractals contain their own scale down, rotate and skew replicas embedded in them. Aside the fact that the object looks fractioned, the term fractal was given due to the strange fact that these objects have a dimension, which is not a whole number but a fraction. A mathematical fractal is based on an equation that undergoes iteration, a form of feedback based on recursion [5].

There are many definitions of a fractal. Possibly the simplest way to define a fractal is as an object which appears self-similar under different scales. In effect, a fractal possesses symmetry across scale, with each small part of the object replicating the structure of the whole. This is perhaps the loosest of definitions, however, it captures the essential, defining characteristic, that of self-similarity [4].

These definitions proved to be unsatisfactory in that it excluded a number of sets that clearly ought to be regarded as fractals [6]. Various attempts have been made to give a mathematical definition of a fractal, but such definitions have not proved satisfactory in a general context [7]. Falconer proposed that there is no hard and precise definition, but just a list of properties characteristic for fractals and preferring to consider a set E in Euclidean space to be a fractal if it has all or most of the following features [6, 7]:

1) E has a fine structure, that is irregular detail at arbitrarily small scales,
2) E is too irregular to be described by calculus or traditional geometrical language, either locally or globally,
3) Often E has some sort of self-similarity or self-affinity, perhaps in a statistical or approximate sense,
4) Usually the 'fractal dimension' of E (defined in some way) is strictly greater than its topological dimension,
5) In many cases of interest E has a very simple, perhaps recursive, definition,
6) Often E has a 'natural' appearance.

Fractal geometry is the study of sets with properties such as (i)-(vi). Fractal geometry is concerned with the properties of fractals. Fractal geometry has become a common tool to describe objects or phenomena in which a scale invariance of some sort exists [1]. Fractals may be found in nature or generated using a mathematical recipe. Mandelbrot proposed that fractals and fractal geometry could be used to describe real objects, such as trees, lightning, river meanders and coastlines, to name but a few [4]. Many natural objects and man-made processes exhibit intricate detail and scale invariance. Given some underlying geometry, one can study these objects and processes under a branch of mathematics known as fractal geometry. Benoit Mandelbrot christened this field in 1970s. Mandelbrot's "The fractal Geometry of Nature", was the first "text book" in the field [1].

There are an infinite variety of fractals and types. It is convenient to resort to conventional classification for representation of all the variety of fractals [1]. Examples of fractals abound, but certain classes have attracted particular attention, such as: geometrical fractals, algebraic fractals and stochastic fractals [1]. Fractals that are invariant under simple families of transformations include self-similar, self-affine, approximately self-similar and statistically self-similar fractals. Figure 1 shows some examples of classical fractals. Certain self-similar fractals are especially well known: the middle-third Cantor set, the von Koch curve, the Sierpinski triangle (or gasket) and the Sierpinski carpet. Fractals that occur as attractors or repellers of dynamical systems, for example the Julia sets resulting from iteration of complex functions, have also received wide coverage [7].

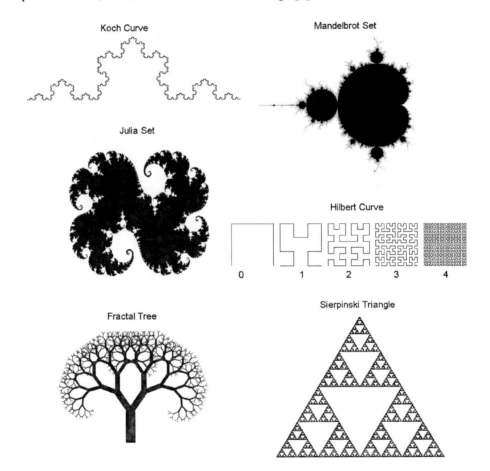

Figure 1. Examples of fractals.

Fractals have been generated or represented by different means, like recursive mathematical families of equations, recursive transformations (generators) applied to an initial shape (the initiator), Brownian movements, etc. [1]. Methods used for fractal generation include: Iterated Function Systems (IFS), Lindenmayer Systems, dynamic systems, Newton-Raphson iteration, etc.

Fractal geometry has attracted widespread and sometimes controversial attention. The subject has grown on two fronts: on the one hand many "real fractals" of science and nature have been identified. On the other hand, the mathematics that is available for studying fractal sets, much of which has its roots in geometric measure theory, has developed enormously with new tools emerging for fractal analysis [7].

In recent years, the science of fractal geometry has grown into a vast area of knowledge, with almost all branches of science and engineering gaining from the new insights it has provided [4]. The variety of scientific applications is enormous; as such structures exist from physics to astrophysics, from biology to chemistry, and even in market fluctuations analysis. However, till recently, fractal concepts were used more to understand than to build. The period is now appropriate to stress the importance of the practical applications of fractal objects and fractal concepts in many fields of direct importance to industry, communications, environment, and physiology [1].

L-System

Lindenmayer systems (or L-systems) are a mathematical formalism for parallel grammars conceived by the biologist Aristid Lindenmayer in 1968 as a theoretical framework for studying the biological structure and development of simple multicellular organisms, which is well adapted to the modeling of growth phenomena [8]. Initially, L-systems development was motivated to model and study biological forms and growth processes and geometric aspects were beyond the scope of the theory. This new class of grammars is similar to Chomsky grammars [9]. Both kinds of grammars handle an initial string of symbols (the axiom) and include a set of production rules that may be applied to the symbols to generate new strings, but they differ in the way in which production rules are applied. Chomsky grammars change a symbol at a time sequentially, while Lindenmayer grammars apply many rules at the same time in parallel.

In order to describe growth of living organisms, Lindenmayer introduced the notion of a parallel rewriting system. The basic idea is to define a complex organism by successively replacing parts of a simple organism using a set of rewriting rules or productions carried out recursively. Productions are applied in parallel, which is intended to capture the simultaneous progress of time in all parts of the growing organism. The whole organism is treated as an assembly of discrete units, called modules. Each module is represented by a symbol (a letter of the L-system alphabet), which specifies the module's type. L-system models are inherently dynamic, which means that the form of an organism is viewed as a result of development: an event in space-time, and not merely a configuration in space [10].

This formalism was closely related to abstract automata and formal languages, and attracted the immediate interest of theoretical computer scientists. In fact, the applications for L-systems increased after 1984 with the use of L-Systems for computer graphics developed by Smith [11]. More recently, L-systems have found several applications in computer graphics to synthesize realistic images. In particular, geometric interpretations of L-systems were proposed in order to turn them into a versatile tool for fractal and plant modeling.

Prusinkiewicz et al [12] describe a graphic interpretation of strings based on turtle geometry. This interpretation may be used to produce fractal images through of deterministic L-systems. Other extensions were introduced for L-systems in various ways: bracketed L-systems, context sensitive L-systems, parametric L-systems, stochastic L-systems, etc. [13 – 16].

Prusinkiewicz and Hammel have presented equivalence between L-systems and IFS [17] through of presentation of L-system models, its equivalent IFS and the resulting geometric forms. The two formalisms describe structures with self-similarity between the partial and the global forms. Early studies led to the concise L-system description of linear fractals, including classic space-filling curves. The geometric interpretations of L-systems used in these studies were based on turtle geometry or chain coding. More recently, L-systems with an interpretation based on affine geometry have been demonstrated to provide a concise description of the subdivision curves used in geometric modeling [18]. A comparison between L-systems and the traditional specification of subdivision algorithms using indexed points and/or matrices reveals the advantages of L-systems.

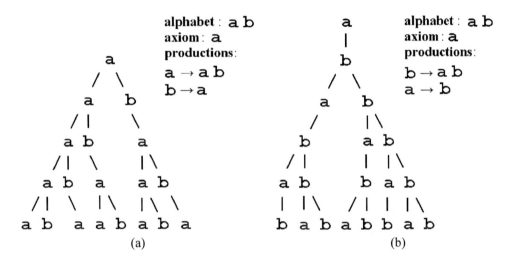

Figure 2. Examples of derivations in a D0L-system: (a) growth of algae, (b) Fibonacci numbers.

The recursive nature of the L-system rules leads to self-similarity and thereby fractal-like forms which are easy to describe with an L-system. The simplest class of L-systems is termed D0L-systems (D0L stands for deterministic and context-free) [1, 16]. D0L-system is defined as a triple G = (V, ω, P), where V is the alphabet of the system, ω is a nonempty word called the axiom and P is a finite set of productions. A production (a, χ) $\in$ P is written as $a \to \chi$. The letter a and the word χ are called the predecessor and the successor of this production, respectively. It is assumed that for any letter $a \in V$, there is at least one word $\chi \in V^*$ such that $a \to \chi$. If no production is explicitly specified for a given predecessor a, the identity production $a \to a$ is assumed to belong to the set of productions P. An 0L-system is deterministic (noted D0L-system) if and only if for each $a \in V$ there is exactly one $\chi \in V^*$ such that $a \to \chi$. Formal definitions of D0L-systems and their operation can be found in [1, 16]. The classical examples presented in Figure 2 provide an intuitive understanding of the main idea behind D0L-systems.

To produce fractals, strings generated by L-systems must contain the necessary information about figure geometry. A graphic interpretation of strings, based on turtle geometry (similar to those in the Logo programming language), is described in [1, 16]. This interpretation may be used to produce fractal images.

A state of the turtle is defined as a triplet *(x, y, a)*, where the Cartesian coordinates *(x, y)* represent the turtle's position, and the angle *a*, called the heading, is interpreted as the direction in which the turtle is facing. Given the initial state of turtle *(x_0, y_0, a_0)*, the step size *d* and the angle increment *b*, the turtle can respond to the commands in response to the string *v* represented by the following symbols [1, 16]:

F → Move forward a step of length d. The state of the turtle changes to (x', y', a), where $x' = x + d \cos(a)$ and $y' = y + d \sin(a)$. A line segment between points (x, y) and (x', y') is drawn.

f → Move forward a step of length d without drawing a line. The state of the turtle changes as above.

+ → Turn left by angle b. The next state of the turtle is $(x, y, a+b)$.

− → Turn left by angle b. The next state of the turtle is $(x, y, a-b)$.

All other symbols are ignored by the turtle (the turtle preserves its current state). The above description gives us a rigorous method for mapping strings to pictures, which may be applied to interpret strings generated by L-systems. Figure 3 presents four approximations ($k = 0,1,2,3$) of the Minkowski fractals. The Matlab code used for DOL-system generation of fractals in Figure 3 is shown below.

```
level=2; % FRACTAL LEVEL
axiom='k'; % FRACTAL CURVE
axiom='k+k+k+k'; % FRACTAL ISLAND
rule='k-k+k+kk-k-k+k';
scale_factor=4;
for N=0:level,
reply='';
if N==0,
font=axiom; M=length(font);
x=0; y=1; %theta=0;
end
if axiom=='k', x=0; y=0; end
% REWRITING
if N>0
for m=1:M
character=font(m);
if character=='k'
reply=[reply rule];
else
reply=[reply character];
end
end
font=reply;
M=length(font);
```

```
reply='';
end
% TURTLE GEOMETRY
theta=0;
dtheta=pi/2;
d=(scale_factor^-N);
for f=1:M
if font(f)=='k'
nx=x+d*cos(theta);
ny=y+d*sin(theta);
line(0.5+[x nx],0.5+[y ny]);
x=nx; y=ny;
end
if font(f)=='+', theta=theta-dtheta; end
if font(f)=='-', theta=theta+dtheta; end
end
end
```

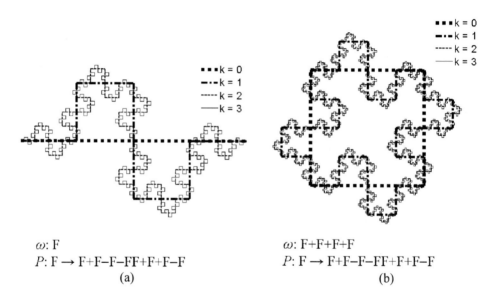

ω: F $\qquad\qquad\qquad\qquad$ ω: F+F+F+F
P: F $\rightarrow$ F+F−F−FF+F+F−F $\qquad$ P: F $\rightarrow$ F+F−F−FF+F+F−F
$\qquad\qquad$ (a) $\qquad\qquad\qquad\qquad\qquad\qquad$ (b)

Figure 3. (a) Minkowski sausage and (b) Minkowski sland.

ITERATED FUNCTION SYSTEMS – IFS

One of the most common ways of generating fractals is as the fixed attractor set of an iterated function system. Iterated Function Systems were conceived in their present form by John E. Hutchinson in 1981 [19] and popularized by Michael Barnsley's book *Fractals Everywhere* [20]. In mathematics, iterated function systems or IFSs are a method of constructing self-similar fractals. IFS fractals, as they are normally called, are commonly computed and drawn in 2D. The fractals are constructed by creating copies of a geometrical object that are transformed by transformation functions (thus, "function systems"). The results of the transformation are combined with the original image and then transformed again

Design of Frequency Selective Surface Using Fractals Geometries   111

and again (thus, "iterated"). The fractal is the limit object arising as the result of the IFS. Interestingly, the result depends only on the IFS itself and not on the initial object. Thus, when rendering IFS fractals, a point is taken as the initial object for simplicity. The functions are normally contractive which means they bring points closer together and make shapes smaller. Hence the shape of an IFS fractal is made up of several possibly-overlapping smaller copies of itself, each of which is also made up of copies of itself, ad infinitum. This is the source of its self-similar fractal nature.

Formally, an iterated function system is a finite set of contraction mappings on a complete metric space [20]. Symbolically, $\{f_i : X \to X \mid i = 1, 2, \ldots, N\}$ is an iterated function system if each f_i is a contraction on the complete metric space X. Although different kinds of transformations can be used in IFS, *affine transformations* are used most commonly. In this case, IFS consists of affine transformations involving rotations, scalings by a constant ratio, and translations. For the two dimensional case, an affine transformation of a point (x_n, y_n) to the point (x_{n+1}, y_{n+1}) can be described as

$$\begin{cases} x_{n+1} = ax_n + by_n + e \\ y_{n+1} = cx_n + dy_n + f \end{cases} \tag{1}$$

Thus, a transformation can be described through of coefficients *(a, b, c, d, e, f) that represent* the IFS code. A given image will normally require multiple transformations, each with their own set of coefficients. In the random iteration algorithm each transformation is assigned a probability P. With each round of iteration one of the transformations is chosen randomly, using the probability as factor in the choice, and the transformed point is plotted on the graphic plane. As the points are plotted the image emerges. An IFS is then a collection of m transformations ($m \geq 2$):

$$\begin{cases} T_1 : (a_1, b_1, c_1, d_1, e_1, f_1, P_1) \\ T_2 : (a_2, b_2, c_2, d_2, e_2, f_2, P_2) \\ \ldots \\ T_m : (a_m, b_m, c_m, d_m, e_m, f_m, P_m) \end{cases} \tag{2}$$

The additional parameter P_m for each transformation is the transformation probability. Normally, the IFS is initialized with a random point $p_0 = (x_0, y_0)$ in the plain. A transformation T_m is then randomly chosen with the probability (P_m). The resulting point $p_1 = T_m(p_0)$ is calculated and rendered. Then the next transformation is chosen randomly with a probability specified by the transformation parameters. The next point is obtained using this transformation and rendered. This process is then continued *ad infinitum* (for an ideal fractal) or until the required saturation is reached (for a rendering approximation).

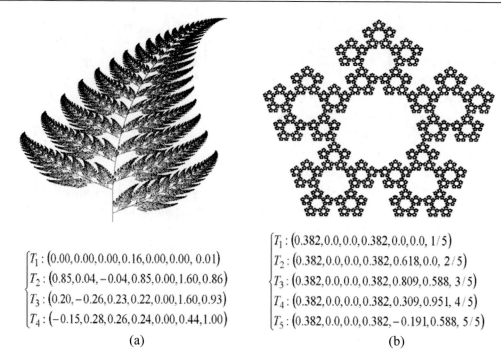

Figure 4. IFS fractals and transformations employed: (a) Barnsley fern and (b) Siepinski pentagon.

The IF method has the ability to create realistic images with very small sets of numbers. It can encode a scene of almost any level of complexity and detail as a small group of numbers, thereby achieving amazing compression ratios of images. The most common algorithm to compute IFS fractals is called the chaos game. It consists of picking a random point in the plane, then iteratively applying one of the functions chosen at random from the function system and drawing the point. The IFS algorithm for generating the image is simply this:

1) Start with an arbitrary point in the plane (x_0, y_0).
2) Pick a random transformation, according to the probabilities (P_m).
3) Transform the point and plot it.
4) Go to step 2.

Figure 4 presents two examples of IFS fractals: fern and Sierpinski pentagon. The Matlab code used for implementation of IFS method used in generation of the Sierpinski pentagon is presented below.

```
N=500000;
X=zeros(N,2);
X(1,:)=[0.5,0.5];
U=sqrt(3)/3.5;
for k=1:N-1
r=rand;
if r<1/5
```

```
X(k+1,1)=0.382*(X(k,1)); X(k+1,2)=0.382*(X(k,2));
elseif r<2/5
X(k+1,1)=0.382*(X(k,1))+0.618; X(k+1,2)=0.382*(X(k,2));
elseif r<3/5
X(k+1,1)=0.382*(X(k,1))+0.809; X(k+1,2)=0.382*(X(k,2))+0.588;
elseif r<4/5
X(k+1,1)=0.382*(X(k,1))+0.309; X(k+1,2)=0.382*(X(k,2))+0.951;
else
X(k+1,1)=0.382*(X(k,1))-0.191; X(k+1,2)=0.382*(X(k,2))+0.588;
end
end
plot(X(:,1),X(:,2),'K.','markersize',1)
```

FSS DESIGN USING FRACTAL GEOMETRIES

Frequency selective surfaces (FSSs) are used in many commercial and military applications. Usually, conducting patches and isotropic dielectric layers are used to build these FSS structures. Single and multiple dielectric layer structures can be used in FSS design. In several applications, the FSS design specifications are very stringent. These structures are composed of arrays of conducting patches or aperture elements on a dielectric substrate and have been the subject of many studies during the last two decades. A typical frequency selective surface is constructed from a 2D-periodic planar structure consisting of one or more metallic patterns, each backed by a dielectric substrate with a frequency response that is entirely determined by the geometry of the structure in one period called a unit cell. Figure 5 illustrates the considered array in this chapter. The structure has periodicities T_x and T_y, in x and y directions, respectively.

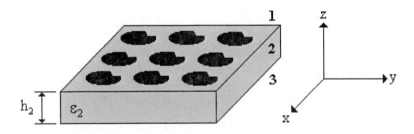

Figure 5. FSS with one dielectric isotropic layer.

The first fractal geometry chosen in this chapter was the self-similar geometry shown in Figure 6 [21]. To obtain this element, the original geometry is divided into ones of smaller scales, but they are identical copies of the original element. If there are n copies of the original geometry, reduced by a factor of scale, s, the size D is defined as [22]:

$$D = \frac{log(n)}{log(1/s)} \quad (3)$$

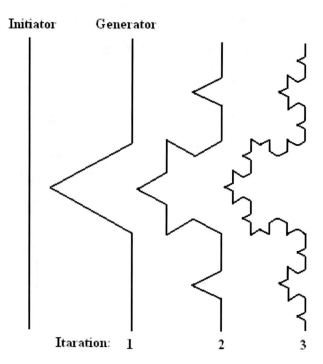

Figure 6. Koch curves.

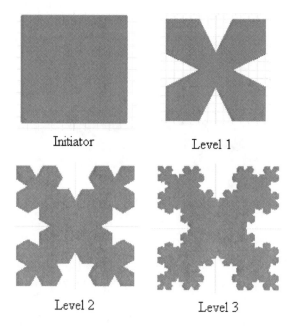

Figure 7. Koch fractal elements.

In generation of fractal Koch proposed in this paper, we beginning with a rectangular patch with dimensions W and L, in x and y directions, respectively. Then we applied the scale factor of 1/3 getting a fractal Koch level 1 element. Applying again the scale factor in the fractal Koch level 1 element, we obtain the fractal Koch level 2 and applying again the scale factor in the fractal Koch level 2 elements, we obtain the fractal Koch level 3. The considered element geometries are shown in Figure 7.

The second considered fractal geometry is the Minkowski Island [23]. The generation of this fractal geometry is carried out using a recursive procedure, which is characterized by two fractal parameters: iteration number (or level) and iteration factor. From the initiator square at each recursion, an 8-side generator is applied to the pre-fractal at the previous iteration. For the Minkowski fractal, the iteration factor ($1/r$) is four and the number of copies (N) is eight. Thus, at a given iteration, the Minkowski island perimeter is duplicated, but its area remains the same. The fractal dimension, $D = \log(N)/\log(1/r)$, is 1.5.

The Lindenmayer system (or L-system) and turtle algebra are used to generate and visualize the Minkowski fractals shown in Figure 1. For this particular case we used an iteration factor equal to 1/4 of the segment.

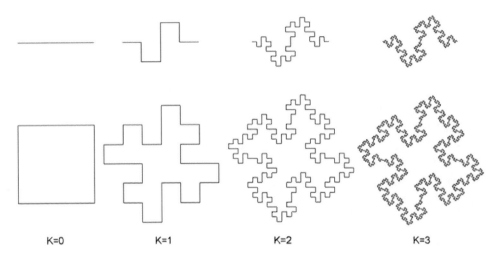

Figure 8. Minkowski sausage and Minkowski island fractal iterations for levels K=0, 1, 2, 3.

The third fractal geometry considered in this chapter, is the spiral fractal. The FSSs were designed to present multiband resonances. The generation of the fractal-like spiral element (FSE) is carried out through a recursive procedure, which is characterized by two fractal parameters: iteration number or level (N_S) and scale factor (K_f). Starting from a conventional thin dipole of the height L generic segments with L_n length are calculated by [24]

$$L_n = K_f^n L \qquad (4)$$

where $n=1, 2, 3, ..., N_S$, L is the initial length, and K_f is the scale factor ($0 < K_f < 1$).

Orthogonal segments are added successively to the conventional thin dipole of length L. Figure 1 shows the process of successive iterations for this fractal element. An example of a FSE is given in Figure 9.

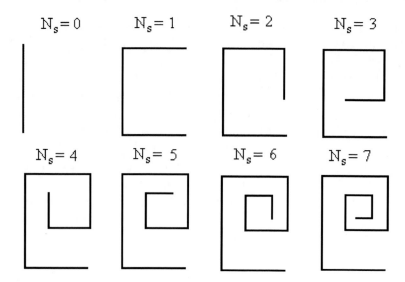

Figure 9. Examples of successive iterations of the FSEs.

Finally, the last proposed fractal geometry, to design FSS, consists of patch elements based on Dürer's pentagon fractal geometry generated with the application of iterated function system [25]. From a regular pentagon of side L, that corresponds to the generator element (or level 0), we use a scale factor of 0.382 for the generation of Dürer's pentagon at levels $k = 1, 2$, and 3 with $L_k = L \cdot 0.382^k$. At each iteration fractal, five copies are generated in small-scale, it generating the fractal element (see Figure 10). In 1525, the artist Albrecht Dürer illustrated various ways for drawing geometric figures. One section is on "Tile Patterns Formed by Pentagons". Dürer's description forms the basis for the attractor of an iterated function system [26].

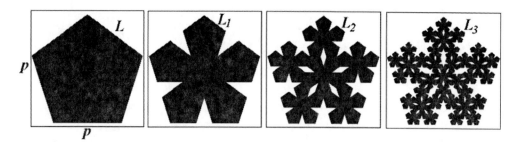

Figure 10. Proposed Dürer's pentagon pre-fractal patch elements for FSS design.

RESULTS

In this section, the influences of the different elements on FSS performance are characterized. To perform this study, the full wave analysis tool Ansoft DesignerTM was used. The four different fractal geometries presented in the last section were considered.

For the first fractal geometry, we design a periodic array of rectangular conducting patch on a dielectric substrate. The dimensions of the structure are $W = L = 9$ mm (area equal to

81mm^2); and $T_x = T_y = 10$ mm. Applying the scale factor of 1/3 we obtain the fractal Koch level 1 and 2. We simulated the FSSs in Ansoft Designer^{TM} software and the transmissions in dB for both structures are shown in Figure 11.

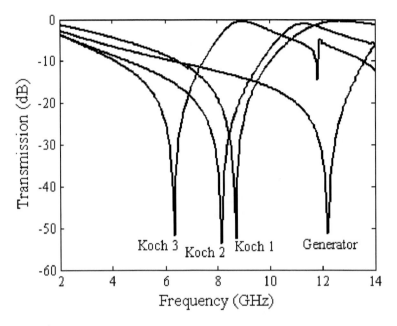

Figure 11. Simulated results for the four different elements.

The FSS initiator obtained a resonance frequency of 12.191 GHz with a bandwidth of – 20 dB equals to 1.83 GHz. The Koch fractal level 1 FSS obtained a resonance frequency of 9.115 GHz with a bandwidth of – 20 dB equals to 1.951 GHz. So, we can see a frequency reduction of 25.22 per cent proximally. We obtained with the Koch fractal level 2 FSS a resonance frequency of 8.333 GHz with a bandwidth of – 20 dB equals to 1.618 GHz. So, we can see a resonance frequency reduction of 33.14 per cent proximally. For the Koch fractal level 3 FSS we obtained a resonance frequency of 6.794 GHz with a bandwidth of – 20 dB equals to 1.047 GHz. So, we can see a resonance frequency reduction of 44.27 per cent proximally. So, we can observe a decrease of resonant frequency when the fractal level increases. Thus, we can reduce the size of fractals to increase the resonant frequency.

After the simulations, the prototypes of FSS Koch fractal level 1 and 2 were built to validation propose of the simulated results. The equipment used in measurements was the Hewlett Packard vector network analyzer model N5230A from Agilent. In figures 12 and 13 we can see the comparison between the simulated and measured results, for FSS built prototypes. A good agreement is observed. For the Koch fractal level 1 FSS the simulated results point to a resonance frequency of 9.115 GHz with a bandwidth of – 20 dB equals to 1.913 GHz. The measured results point to a resonance frequency of 9.032 GHz with a bandwidth of – 20 dB equals to 1.500 GHz. For the FSS Koch fractal level 2 the simulated results point to a resonance frequency of 8.333 GHz with a bandwidth of – 20 dB equals to 1.688 GHz. The measured results point to a resonance frequency of 8.332 GHz with a bandwidth of – 20 dB equals to 1.350 GHz.

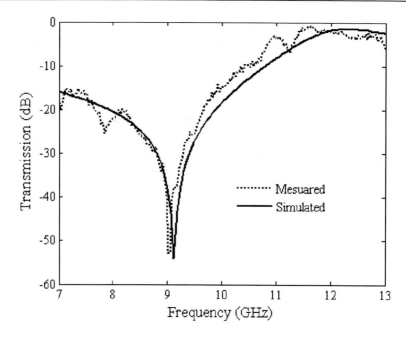

Figure 12. Comparison between measured and simulated results for the Koch fractal level 1.

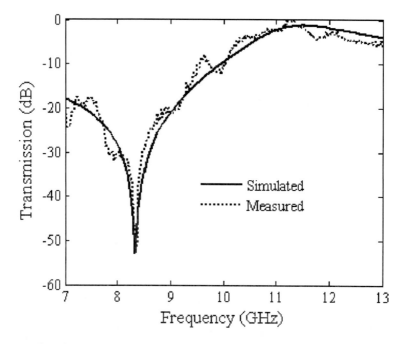

Figure 13. Comparison between measured and simulated results for the Koch fractal level 2.

For the Minkowski geometry FSSs we used a dielectric isotropic layer. The substrate used was the RT-Duroid 3010, with a height of 1.27 mm and relative permittivity of 10.2. The distance between the centers of any two adjacent patches is 15.2 mm.

We simulated the FSSs in Ansoft Designer™ software. The proposed structures were arranged in a periodic array, in which a normal incident plane wave was considered, preserving the far-field condition for analysis of the structure investigated. A vertical polarization was considered for the incident wave. Minkowski fractal level 1 FSS obtained a resonance frequency of 9.84 GHz with a − 10 dB bandwidth of 797 MHz. With Minkowski fractal level 2 FSS, we obtained a resonance frequency of 9.00 GHz with a − 10dB bandwidth of at 670 MHz. We can observe a decrease in resonant frequency, whereas the bandwidth showed little change when the fractal level increased. At resonant frequency the rejection was close to 35 dB for the two structures. Figures 14 and 15 show simulated and measured results, respectively, for the fabricated FSSs. Good agreement can be observed. Resonance frequency and bandwidth errors were less than 1% and 6%, respectively.

For the fractal spiral geometry, periodic arrays of were mounted on a dielectric isotropic layer. The substrate used was the FR-4. It has 1.57mm of height and a relative permittivity of 4.4. The initial length (L) was 13.5 mm and (w) is 1.5 mm. We use a reduction factor (K_f) equals to 0.8 and an iteration level (N_S) equals to 3, 4, 5, and 6. The distance between the centers of any two adjacent patches was 20 mm. The fractal spiral element levels considered were 3, 4, 5, and 6.

In figures 16 and 17 we can see the comparison between the simulated and measured results, for the FSS with level 4 and 6 elements. A good agreement is observed. Two resonant bands are obtained for two structures.

Finally, the Dürer's pentagon geometry was used to design a FSS consisting of a periodic array of patch elements mounted on a fiberglass substrate (FR-4), with thickness and dielectric constant equal to 1.57 mm and 4.4, respectively. The patch elements were designed based on Dürer's pentagon fractal geometry generated with the application of iterated function system.

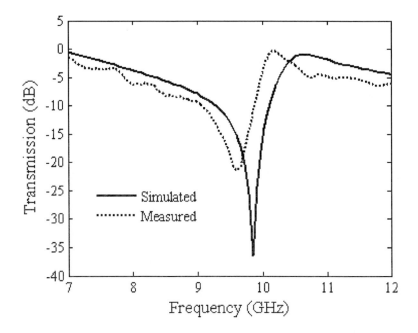

Figure 14. Comparison between measured and simulated results for Minkowski fractal level 1.

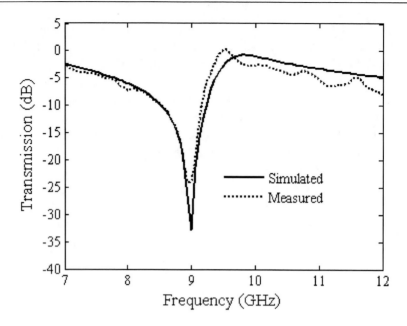

Figure 15. Comparison between measured and simulated results for Minkowski fractal level 2.

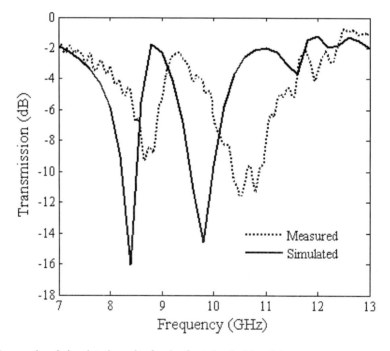

Figure 16. Measured and simulated results for the fractal spiral level 4.

From a regular pentagon of side L, that corresponds to the generator element (or level 0), we use a scale factor of 0.382 for the generation of Dürer's pentagon at levels $k = 1$, 2, and 3 with $L_k = L \cdot 0.382^k$. At each iteration fractal, five copies are generated in small-scale, it generating the fractal element.

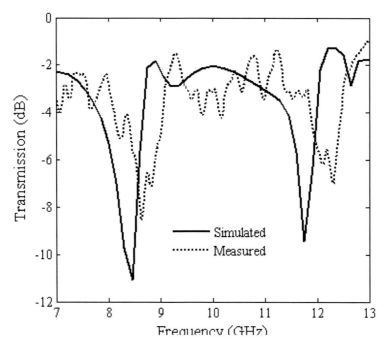

Figure 17. Measured and simulated results for the fractal spiral level 6.

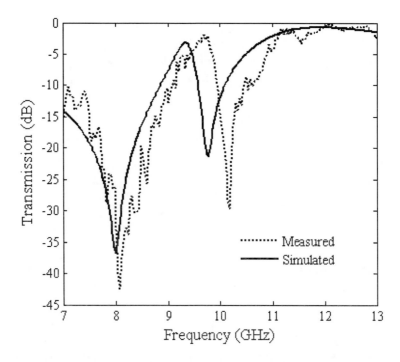

Figure 18. Comparison between the measured and simulated results for the FSS of level 1 with p = 16.5 mm and L = 10 mm.

Simulated and measured results for the FSS with Durer's pentagon fractal level 1 are shown in Figure 18. As we can see, two resonances occur. The first resonance occurs at 8.00 GHz with a bandwidth equals to 0.79 GHz and the second resonance occurs at 9.77 GHz with a bandwidth equals to 0.1 GHz, for simulated results. The measured results point to a first resonance at 8.07 GHz with a bandwidth equals to 0.76 GHz and it points to a second at 10.19 GHz with a bandwidth equals to 0.15 GHz. Again, a good agreement between the results is observed. We can verify a reduction of resonance frequency and bandwidth to the first rejection band.

For the FSS with Durer's pentagon fractal of level 2, we also observe two resonances as we can see in Figure 19. The first resonance occurs at 7.02 GHz with a bandwidth equals to 0.66 GHz and the second resonance occurs at 8.53 GHz and it did not reach – 20 dB, for simulated results. The measured results point to a first resonance at 7.29 GHz with a bandwidth equals to 0.76 GHz and it points to a second at 9.24 GHz with a bandwidth equals to 0.085 GHz.

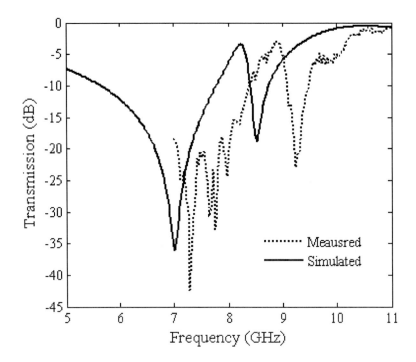

Figure 18. Comparison between the measured and simulated results for the FSS of level 2 with p = 16.5 mm and L = 10 mm.

CONCLUSION

This chapter described the use of fractal geometries to design Frequency Selective Surfaces (FSS) on isotropic dielectric substrates, emphasizing the use of fractal geometries to miniaturize FSS dimensions and to obtain FSS with dual band response. In designs were considered four different fractal geometries: Koch, Minkowski, Spiral and Dürer's Pentagon.

Due to the bigger electrical length, the Koch fractal geometry enables a considerable reduction in the unit cell and in the metallic area of a FSS compared with a rectangular element. As the Koch fractal geometry, we should note that the use of the Minkowski island fractal in FSS also enables a reduction in the unit cell and in the metallic area. The Minkowski fractal element exhibited a bandwidth at – 10 dB below 1 GHz, and for some applications this narrow bandwidth may be ideal Compared with a square patch FSS, the bandwidth obtained with the Minkowski fractal elements were larger, with a 36% reduction in unit cell area for fractal level 1 and 52.3 per cent for fractal level 2. The fractal spiral element enable us t obtain fractal geometries with high levels. This is very difficult with other geometries, like Koch or Minkowski curves. Also, we should note that the use of fractal spiral element enable a multiband response. The Durer's pentagon fractal geometries were used to design FSS stop-band filters with dual-band responses. A single substrate has been used to generate stop-band characteristics. It has been shown that the use of Durer's pentagon fractal elements allows the design of FSSs with dual-band responses. The structures can be redesigned to obtain a four band response, rejecting two bands and transmitting other two bands.

REFERENCES

[1] Mishra, J. and Mishra, S. L-Systems Fractals. Amsterdam, The Netherlands: Elsevier, 2007. 274p.

[2] Mandelbrot, B. The Fractal Geometry of Nature. New York: W.H. Freeman and Company. 1983. 420p.

[3] Peitgen, H. O., Jurgens, H. and Saupe, D. Chaos and fractals: new frontiers of science. New York: Springer-Verlag, 1992. 864p.

[4] Addison, P. S. Fractals and Chaos: An Illustrated Course. CRC Press, 1997. 256p.

[5] Briggs, J. Fractals: The Patterns of Chaos. London : Thames and Hudson, 1992. 148p.

[6] Falconer, K. Fractal Geometry: Mathematical Foundations and Applications. John Wiley and Sons. 2003. 310p.

[7] Falconer, K. Techniques in fractal geometry, New York: John Wiley and Sons. 1997. 256p.

[8] Lindenmayer, A. Mathematical models for cellular interaction in development, Parts I and II. *Journal of Theoretical Biology* 18, 1968, p. 280–315.

[9] Chomsky, N. Three models for the description of language. *IRE Transactions on Information Theory*, 2(3), 1956, p. 113–124.

[10] Prusinkiewicz, P. Introduction to modeling with L-systems. In: L-systems and beyond SIGGRAPH'2003 Course Notes, 2003, p. 9–25, ACM, ACM Press, 2003.

[11] Smith, A. R. Plants, Fractals and Formal Languages. *Computer Graphics*, 18(3), 1984, p. 1–10.

[12] Prusinkiewicz, P., Hammel, M., Mech R., and Hanan, J. The artificial life of plants. In: Artificial Life for Graphics. Animation, and Virtual Reality, SIGGRAPH'95 Course Notes. p. l-38. ACM, ACM Press, 1995.

[13] P. Prusinkiewicz. Graphical applications of L-systems. In Proceedings of Graphics Interface – Vision Interface '86, p. 247–253. CIPS, 1986.

[14] Prusinkiewicz, P. and Hanan, J. Visualization of botanical structures and processes using parametric L-systems. In D. Thalmann, editor, Scientific Visualization and Graphics Simulation, pages 183–201. J. Wiley and Sons, 1990.

[15] Eichhorst, P. and Savitch, W. J. Growth functions of stochastic Lindenmayer systems. Information and Control, 45:217–228, 1980.

[16] Prusinkiewicz, P. and Lindenmayer, A. The Algorithmic Beauty of Plants, Springer Verlag. 2004, 228 p.

[17] Prusinkiewicz, P. and Hammel, M., Automata, languages and iterated function systems, Lecture Notes for the SIGGRAPH '91 course: Fractal Modeling in 3D Computer Graphics and Imagery, 1991.

[18] Prusinkiewicz, P., Samavati, F., Smith, C. and Karwowski, R. L-system description of subdivision curves, *International Journal of Shape Modeling* 9 (1), pages 41-59, 2003.

[19] Hutchinson, J. E. Fractals and self similarity. *Indiana Univ. Math.* 1981, J. 30: 713–747.

[20] Barnsley, M. Fractals Everywhere. San Diego: Academic Press, Inc., 1988. 394 p.

[21] Campos, A. L. P. S., Oliveira, E. E. C., and Silva, P. H. F., Miniaturization of frequency selective surfaces using fractal Koch curves. *Microwave and Optical Technology Letters*, v. 51, p. 1983-1986, 2009.

[22] K. J. Vinoy, J. K. Abraham, and V. K. Varadan, On the Relationship Between Fractal Dimension and the Performance of Multi-Resonant Dipole Antennas Using Koch Curves, *IEEE Transactions on Antennas and Propagation*, Vol. 51, No. 9 (2003), 2296 – 2303.

[23] Campos, A. L. P. S., Oliveira, E. E. C., and Silva, P. H. F., Design of Miniaturized Frequency Selective Surfaces Using Minkowski Island Fractal. *Journal of Microwaves and Optoelectronics*, v. 9, p. 43-49, 2010.

[24] Campos, A. L. P. S., Manicoba, Robson H. C., Cavalcante, G. A., d'Assunção, A. G., Fractal-Like Spiral Element Used to Design Multiband Frequency Selective Surfaces. In: MOMAG 2010 - 14° SBMO - Simpósio Brasileiro de Microondas e Optoeletrônica e 9° CBMAG - Congresso Brasileiro de Eletromagnetismo, 2010, Vila Velha. *Annals of MOMAG* 2010, 2010. p. 147-150.

[25] Trindade, J. I., Fonseca, P. H. S., Campos, A. L. P. S., d'assunção, A. G., Analysis of Stop-Band Frequency Selective Surfaces With Dürer s Pentagon Pre-Fractals Patch Elements. In: 14th Biennial IEEE Conference on Electromagnetic Field Computation, 2010, Chicago. Annals of 14th Biennial IEEE Conference on Electromagnetic Field Computation, 2010.

[26] R. Lück, Dürer–Kepler–Penrose, "The Development of Pentagon Tilings," *Materials Science and Engineering*, Vol. 294-296, pp. 263-267, December 2000.

In: Classification and Application of Fractals
Editor: William L. Hagen

ISBN 978-1-61209-967-5
© 2012 Nova Science Publishers, Inc.

Chapter 5

FRACTAL-BASED MATHEMATICAL MODELS OF CANCER IN THE ERA OF SYSTEMS BIOLOGY

Fabio Grizzi[*1] *and Irene Guaraldo*[2]

[1]IRCCS Istituto Clinico Humanitas, Rozzano, Milan, Italy
[2]Dipartimento di Metodi e Modelli Matematici per le Scienze Applicate,
University of Rome "La Sapienza", Rome, Italy

ABSTRACT

Cancer research has undergone radical changes in the past few years. Amount of information both at the basic and clinical levels is no longer the issue. Rather, how to handle this information has become the major obstacle to progress. System biology is the latest fashion in cancer biology, driven by advances in technology that have provided us with a suite of "omics" techniques. It can be seen as a conceptual approach to biological research that combines "reductionist" (parts) and "integrationist" (interactions) research, to understand the nature and maintenance of entities. In geometrical terms, cancerous lesions can be depicted as fractal entities mainly characterized by their irregular shape, self-similar structure, scaling relationship and non-integer or fractal dimension. It is indubitable that The Fractal Geometry of Nature has provided an innovative paradigm, a novel epistemological approach for interpreting the anatomical world. It is also known that mathematical methods and their derivatives have proved to be possible and practical in oncology. Viewing cancer as a system that is dynamically complex in time and space will probably reveal more about its underlying behavioral characteristics. It is encouraging that mathematicians, biologists and clinicians contribute together towards a common quantitative understanding of cancer complexity.

INTRODUCTION

Despite progresses in our biological and clinical knowledge, human cancer remains one of the major public health problems throughout the world. Cancer is today recognized as a

---

[*]E-mail: fabio.grizzi@humanitasresearch.it.

highly heterogeneous disease: more than 100 distinct types of human cancer have been described, and various tumor subtypes can be found within specific organs. It encompasses various pathological entities and a wide range of clinical behaviors, and is underpinned by a complex array of gene alterations that affect supra-molecular processes. This genetic and phenotypical *variability* is what primarily determines the *self-progression* of neoplastic disease and its response to therapy [1,2]. Additionally, the *asynchrony* and self-progression of a cancer cell population suggests that the extent to which each neoplastic cell shares the properties of a natural cell may differ in *time* and in *space*. Individual cells from a clonal cell population respond differently to the same stimulus, some not responding at all.

The complexity of alterations in cancer presents a daunting problem with respect to treatment: how can we effectively treat cancers arising from such varied perturbations? A tumor consists of genetically distinct subpopulations of cancer cells, each with its own characteristic sensitivity profile to a given therapeutic agent. Each cancer therapy can be viewed as a filter that remove a subpopulation of cancer cells that are sensitive to this treatment while allowing other insensitive subpopulations to escape. It is also indubitable that the conception of anatomical entities as a hierarchy of graduated forms and the increase in the number of observed anatomical sub-entities and structural variables has generated a growing complexity [3], thus highlighting new properties of normal cells and their tumoral counterpart. The need to tackle system complexity has become even more evident since completion of the various genome projects [4-7]. The still unsolved central question is how to transform molecular knowledge into an understanding of complex phenomena in *cells*, *tissues*, *organs* and *organisms*. Almost all the anatomical entities display hierarchical forms: their component sub-entities at different spatial scales or their process at different time scales are related to each other [3]. One of the pre-eminent characteristics of the entire living world is its tendency to form multi-level structures of "systems within systems", each of which forms a Whole in relation to its parts and is simultaneously part of a larger Whole. Anatomical entities, when viewed at microscopic as well as macroscopic level of observation, show a different degree of complexity [3]. In addition, complexity can reside in the *structural organization* of the anatomical system or in its *behavior*, and often complexity in structure and behavior go together.

The need to find a new way of classifying natural as well tumoral anatomical entities, and objectively quantifying their different structural changes, prompted us to investigate the Fractal geometry and the theories of Complex Systems, and to apply their concepts to modeling human cancer.

SYSTEM BIOLOGY: FROM THE PAST TO THE PRESENT

It is indubitable that recent advances in molecular biology, genetics and imaging have demonstrated important insights about the complexity of human diseases. However, a comprehensive understanding of their pathogenesis remains still incomplete. Although *Reductionism* (*i.e.* the theory that every complex phenomenon can be explained by analyzing the simplest, most basic physical mechanisms that are in operation during the phenomenon) has been successful in enumerating and characterizing the component parts of most living organisms, it has failed to generate knowledge on how these parts interact in complex arrangements to allow and sustain two of the most fundamental properties of the organism as a Whole: its *fitness*, also termed its *robustness* (*i.e.* the ability of a system to adapt to changes in its environment), and its capacity to evolve. Although a formal definition of the term "System Biology" has not yet been widely accepted, most researchers agree in that it represents an *integrative approach* that attempts to understand higher-level operating principles of living organisms, including humans [7-11]. Under the reductionist paradigm, a positive correlation between a single biological parameter and the occurrence of a disease is often considered a major success, even though the complete pathogenic mechanisms may remain largely unknown. Notably, this approach has been followed by the pharmaceutical industry, despite a remarkably low rate of success (only 11% of new therapeutic targets reach the market as new drugs) [10].

System biology focuses on understanding not only the component parts of a given system, thus complementing the reductionist approach, but also the effect of *interactions* among them and the interaction of the system with its *environment* (Figure 1). Like physiology, system biology is deeply rooted in the principle that the Whole is more than the sum of its component parts. Systemist biologists first indicated the existence of different organisational levels governed by different laws. They stressed that a fundamental characteristic of the structural organisation of living organisms is their *hierarchical nature*. Systemism was born in the first half of the twentieth century as a reaction to the previous mechanistic movement (*i.e.* Reductionism). It was based on the awareness that classical causal/deterministic schemata are not sufficient to explain the variety of interactions characterising living systems. Advances in the fields of *cybernetics* [defined by Norbert Wiener (1894-1964) as the interdisciplinary study of the structure of regulatory systems) and *biology* [from *bios* (Greek for "life") and *logos* (Greek for "reasoned account")] led to the proposition of new interpretative models that were better suited to identifying and describing the complexity of phenomena that could no longer be seen as abstractly isolated entities divisible into parts or explicable in terms of temporal causality, but needed to be studied in terms of the dynamic interactions of their parts. The word *system* means "putting together". A system has been defined as "an entity that maintains its existence through the mutual interaction of its parts" [11,12]. In keeping with this concept, research into systems therefore must combine: *a)* the identification and *b)* detailed characterisation of the parts, with the *c)* investigation of their interaction with each other and *d)* with their wider environment, to *e)* elucidate the maintenance of the entity [11]. Systemic understanding literally means putting things in a context and establishing the nature of their interactions, and implies that the phenomena observed at each level of organisation have properties that do not apply lower or higher levels. According to systemic thinking, the essential properties of a living being belong to the Whole and not to its component parts. This led to the fundamental discovery that, contrary to the belief of René Descartes (1596-1650), biological systems cannot be understood by means of reduction [3-6]. The properties of the individual component parts can only be understood in the context of the wider Whole. The biologist and epistemologist Ludwig von Bertalanffy (1901-1972) provided the first theoretical construction of the complex organisation of living systems [12]. Like other organic biologists, he firmly believed that to understand biological phenomena, new modes of thought that went beyond the traditional methods of the physical sciences were required [3, 13,14]. According to von Bertalanffy, living beings should be considered as complex systems with specific activities to which the principles of the thermodynamics of "closed" systems studied by physicists do not apply. Unlike *closed systems* (in which a *state of equilibrium* is established), *open systems* remain in a stationary state far from equilibrium and are characterised by the *input* and *output* of *matter*, *energy* and *information* [15]. James Grier Miller (1916-2002) first introduced the *Living System Theory* (LST) about how living systems "work", how they maintain themselves and how they develop and change [16].

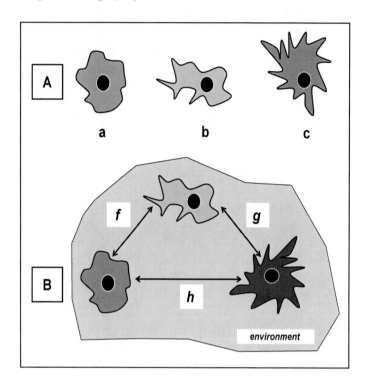

Figure 1. Cancer environment consists of many interacting components characterized by their *number*, *species* and *interactions*. A. Each cell *(a, b, c)* can be considered as an isolated anatomical entity with a proper structural and functional organization. B. Three different cell types interconnected each other making-up a complex system, *i.e.* cancer. System biology focuses on understanding not only the component parts of a given system, but also the effect of *interactions (f, g, h)* among them and the interaction of the system with its environment.

By definition, living systems are *open, self-organizing* systems that have the peculiar characteristics of life and interact with their environment. This takes place by means of information, matter and energy exchanges. The term self-organization defines an evolutionary process where the effect of the environment is minimal, *i.e.* where the generation of new, complex structures takes place fundamentally in and through the system itself [17-19]. In open systems, it is the continuous flow of matter, energy and information that allows the system to self-organize and to exchange entropy with the environment. Supported by a plethora of scientific data, LST asserts that all the great variety of living systems that evolution has generated are complexly structured open systems [20]. They maintain thermodynamically improbable energy states within their boundaries by continuous interactions with their environments [19-21]. LST indicates that living systems exist at eight levels of increasing complexity: *cells, organs, organisms, groups, organizations, communities, societies,* and *supranational systems* [16, 19-21]. All living systems are organized into critical *subsystems*, each of which is a structure that performs an essential life process, *i.e.* a subsystem is identified by the process it carries out. LST is resulted an integrated approach to studying biological and social systems, the technology associated with them, and the ecological systems of which they are all parts [22,23]. It can be said that Systems biology forms a logical juxtaposition to the recently prevailing "reductionist" drive, serving as the "post-genomic" manifestation of the need to balance *dissection* and *synthesis*. The concept of complexity in biology and how to assess the links between information at the molecular level to that at the living organism (*i.e.* genomics, proteomics, metabolomics) is the actual foundation of systems biology (Table 1). As suggested by Butcher *et al.* [9] the "-omics" *(bottom-up)* approach focuses on the identification and global measurement of molecular components. Modelling (the *top-down approach*) attempts to form integrative (across *scales*) models of human physiology and disease, although with current technologies, such modelling focuses on relatively specific questions at particular scales (*i.e.* at the pathway or organ levels). An intermediate approach, with the potential to bridge the two, is to generate profiling data from high-throughput experiments designed to incorporate biological complexity at multiple levels: *multiple interacting active pathways*, *multiple intercommunicating cell types*, and *multiple environments*.

Table 1. Omics approach focuses on the identification and global measurement of biological systems

BIOLOGICAL DATA SUBSET	OMIC DESIGNATION	APPLICABLE TECHNOLOGY
DNA	Genomics	Nucleotide sequencing
RNA	Functional genomics	Gene Expression Profiling
Protein	Proteomics	2D Gels; Mass spectrometry
Metabolites	Metabolomics	Mass spectrometry; NMR
Cells	Cellomics	Fluorescence Probe Digital Imaging; Cytology; Immunocytochemistry
Tissues	Pathology	Microscope; Immunohistochemistry
Organism	Clinical informatics	Relational Databasing; Support technology

THE COMPLEX ORGANIZATION OF HUMAN SYSTEMS

Human beings are complex hierarchical systems consisting of a number of hierarchical levels of anatomical organization that interrelate differently with each other to form networks of growing complexity [3]. It is accepted that in the experimental sciences, observed patterns can often be conceptualized as *macro-scale* manifestations of *micro-scale* processes. However, in many cases, a more typical situation involves observed patterns or system states that are created or influenced by *multiple processes* and *controls* [24]. This *multiple scale causality* not only recognizes multiple processes and controls acting at multiple scales but, in contrast to a strictly reductionist approach, may also recognize that relevant "first principles" may reside at scales other than the smallest micro-scales. In other words, the observed phenomenon at each scale has structural and behavioural properties that do not exist at lower or higher organizational levels. Unlike an anatomical entity, and despite the fact that it has a unique shape, a crystal has no unequivocally defined size that can be used for classification; a small crystal of a given substance will always have the same general structure as a large crystal of the same type [3]. Any fragment of a crystal has the same physical and chemical characteristics as the Whole crystal, but this is not true of any fragment of a living organism because the chemical compositions and physical properties of the individual parts do not correspond with the composition of the Whole. Furthermore, the various components of a living system are characterised by the integration of precise functional criteria that form a Whole. Returning once again to crystals, their macroscopic structures can easily be predicted on the basis of their microscopic structures [3]; they lack what are called *emergent properties*: *i.e.* those that strictly depend on the level of organisation of the material being observed. *Emergence* is a seminal concept in system theory, where it denotes the principle that the "emergent" global properties defining higher order systems or "Wholes" cannot generally be reduced to the properties of the lower order subsystems or "parts". We shall here use the word emergence to mean the appearance of unexpected structures and/or the occurrence of surprising behaviours in large systems consisting of microscopic (physical or biological) non-identical parts: *i.e.* structures and behaviours that are not intuitive or simply predictable [25,26].

THE FRACTAL DESIGN OF HUMAN SYSTEMS

It is widely accepted that anatomical systems can be geometrically depicted as fractal objects [27,28]. In the human body, many apparatus, organs, histological and cytological sub-entities (Figure 2) have been defined as fractal objects [27-41] Fractals are objects mainly characterized by its *self-similar structure* (*i.e.* its parts resemble the Whole), its *scaling* (which means that measured properties depends on the scale at which they are measured), and its *non-integer or fractal dimension* [42,43]. The most important hallmark of the fractal objects is that the schemes that define them are found again continually in orders of greatness decreasing, so that their component parts have a form similar to the Whole system. This peculiar characteristic is called *self-similarity* and can be *geometrical* or *statistical*. A geometrical object can be defined as self-similar, when every smaller piece of the object is an exact duplicate of the Whole object [*i.e.* the "curve" and the "snowflake", from Niels Fabian Helge von Koch (1870-1924) or the "Sierpinski's triangle", described in 1915 by the mathematician Waclaw Sierpinski (1882-1969)]. The *statistical self-similarity* concerns all the natural systems, including anatomical entities. In fact, small pieces making up the objects are rarely identical copies of the Whole object. Usually, the pieces are "kind of like" the Whole. Thus, the statistical properties of the pieces are proportional to the statistical properties of the Whole.

Dimension is a numerical attribute that has been defined in two ways: the topological or Euclidean dimension *(Dγ)*, which assigns *0* to a *point* (defined as that which has no part), *1* to a *straight line* (defined as a length without thickness), *2* to a *plane surface* (defined as having length and thickness, but no depth), and *3* to a *three-dimensional figure* (a volume defined by length, thickness and depth); and the dimension *(D)* introduced by the mathematicians Felix Hausdorff (1868-1942) and Abram S. Besicovitch (1891-1970), who attributed a real number to every natural object between the topological dimensions of *0* and *3*. *Dγ* and *D* coincide *(Dγ = D)* in the case of all Euclidean figures, but remain unequal in the case of all anatomical fractal objects *(D > Dγ)* because no anatomical object corresponds to a regular Euclidean figure [42,43].

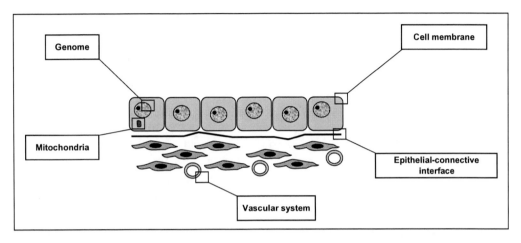

Figure 2. A number of anatomical systems have been depicted as fractals. Among them the cell membrane [36], DNA sequence and chromatin pattern distribution [37,38], mitochondria [39], vascular system [29,30,33,40] and the epithelial-connective interface [41].

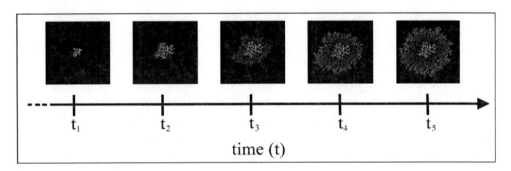

Figure 3. Computer-aided fractal model simulating the growth and invasive behavior of cancer.

Scaling relationships and *power-law distributions* are commonly reported in ecological systems [40]. Living entities are embedded in and constituted by, *networks* at any level of organization, from cells to ecosystems. Biological networks typically consist of a large number of non-identical component parts whose interaction are usually localized, although their effects are not and whose *emergence, maintenance* and *dynamics* represent a challenge to understanding let alone prediction. Biological networks represent the most complex physical system in the universe and yet, as most complex systems they can be described by simple relationships [44]. These relationships are of the form P_{in} where Y is the dependent variable, x represents an independent or explanatory variable, $\square$ is a normalization constant and α is the scaling exponent. Depending on the value of the exponent these relationships are called *allometric ($\alpha \neq 1$)* or *isometric ($\alpha = 1$)* [44].

MATHEMATICAL MODELING OF CANCER

Mathematics is a powerful instrument to study a lot of physical phenomena. During the last years, we have seen a development of a type of mathematical research in which scientists study physical and biological phenomena using some types of partial or ordinary differential equations, through which they can find models which describes these phenomena. If we have a mathematical model which describes a physical phenomenon, we can understand it in a complete way because we can know how and in which way this phenomenon will evolve itself, using the *initial data* and the *boundary conditions* given by the phenomenon and put in the mathematical model. We have just said that also the study of a biological phenomenon can be done using a mathematical model. To reach this goal, there are a lot of difficulties: in fact biological phenomena are complex phenomena because they involve different mechanisms which take place at different scales, from the *microscopic* scale, to the *mesoscopic*, and the *macroscopic* scale. In particular, when we look at the tumors, usually we can't find a specific behavior but often a chaotic behavior, very difficult to understand and to model. To find a mathematical model to describe a biological phenomenon, at first we have to understand the interactions of mechanisms in the different scales, and then find models that describe each of these mechanism and each of this interaction between two (or more) different scales. In other words, we can say that the phenomenon must be divided in some "sub-phenomena", each of these describes the mechanisms in a specific scale (*i.e.* microscopic or macroscopic scale); for each of these "sub-phenomenon" we have to find a mathematical model which describe it. So, at the end, we have not only one model but also more models, and all of these describe the biological phenomenon in its totality. One of the most important problems in this way of work is that the models in each of scale involve different quantities and parameters. For example, if we want to describe the microscopic sub-phenomenon, we have to consider and to work with cells and their interactions, while if we want to describe the macroscopic phenomenon we have to look at the entire macroscopic tissue and at its changes caused by biological phenomenon we are observing, and so we have to work with different variables and parameters (for example the density of the tissue, or its composition). We have also to observe that often each "sub-phenomenon" should be described by different *differential equations* (each scale has a precise type of equation, different from the others). Although these obstacles, during the last years a lot of mathematical models have been founded to describe biological phenomena and some kind of tumors. But we have to say that sometimes these models are simpler than the original biological phenomenon, especially if they had to reproduce the development of a tumor. In fact, the principal difficult to modeling a tumor invasion is that generally tumor has not a precise way of growth; it's difficult to understand its behavior from both biological and (in consequence) mathematical point of view. So sometimes models are not very precise but they can predict and solve some biological problems and for this reason they are used to do simulations in which scientists can see the development of the biological phenomenon, or the tumor, and its changes if they change the values of parameters in the mathematical model. To be more precise, in the model there are parameters and variables that, for example, represent the density of cells, blood vessels and the drug provided from scientists. If they change the values of some of these parameters in the model, they can see through the simulation the kind of development of the tumor. It is indubitable that the future goal of the research, which combines medical and mathematical studies, is that it will be possible to have more precise mathematical models.

CONCLUSION

It is indubitable that human cancer emerges from multiple alterations that induce changes in expression patterns of genes and proteins that function in networks controlling critical cellular events. A primary task of the tumor research is the translation of molecular *biomarkers* candidates into clinical practice. However, there is still not agreement with regard to the sequence and nature of steps that need to be taken to warrant efficient translation of prognostic and/or predictive biomarkers into clinical use and to the introduction of novel, effective and less toxic therapeutic strategies to diminish human suffering and cure life-threatening diseases. Individual cells from a clonal cell population respond differently to the same stimulus, some not responding at all. Variability in cell response can have important implications. It is now known that in a heterogeneous population, patients may display a multiplicity of genetic variations that respond differently to a given medical intervention. The same treatment could be of benefit to some patients yet harmful to others. Each cancer therapy can be viewed as a filter that remove a subpopulation of cancer cells that are sensitive to this treatment while allowing other insensitive subpopulations to escape. The above considerations, in conjunction with the complexity of tumor-host interactions within the tumor microenvironment caused by temporal changes in tumor phenotypes and an array of immune mediators expressed in the tumor microenvironment might partially explain the limited reliability and applicability of current immunotherapeutic approaches. Human carcinogenesis is a dynamical process that depends on a large number of variables and is regulated at multiple spatial and temporal scales and whose behavior does not follow clearly predictable and repeatable pathways. This multiple scale causality not only recognizes multiple processes and controls acting at multiple scales but, unlike a strict reductionist approach, may also recognize the fact that relevant "first principles" may reside at scales other than the smallest micro-scales. In other words, the observed phenomenon at each scale has structural and behavioral properties that do not exist at lower or higher organizational levels.

In mathematical terms, carcinogenesis is a non-linear process. In linear systems, the relationship between an environmental factor increases, the system behavior changes linearly in response to it. In contrast, non-linear systems are mainly characterized by three basic properties: *(a)* they do not react proportionally to the magnitude of their inputs; *(b)* they depend on their initial conditions. Small changes in the initial conditions may generate very different end points. *(c)* Their behavior is not deterministic, *i.e.* periods of inactivity may be punctuated by sudden change, apparent patterns of behavior may disappear and new patterns surprisingly emerge. Such behavior emerges in complex systems, and is permanently sensitive to small perturbations. Over the last decade, many mathematical models of tumor growth, both temporal and spatio-temporal, have appeared in the research literature (Figure 3). Much of the experimental data that exist on the growth kinetics of avascular tumors have been integrated into mathematical models using various growth laws such as *Gompertzian growth*, *logistic growth* and *exponential growth*. Deterministic *reaction-diffusion equations* have been used to model the spatial spread of tumors both at an early stage in its growth and at the later invasive stage. Typical solutions observed in all these models appear as invading traveling waves of cancer cells. While these models are able to capture the tumor structure at the tissue level, they fail to describe the tumor at the cellular level and, subsequently, the sub-cellular level. This failure is critical, because it hampers connections with experimental biologists who perform the studies on cells. Therefore, although these models may be mathematically valid, they are difficult to test experimentally and, consequently, of limited impact. Additionally, mathematical models of tumor growth range from simple fitting of experimental data on the growth kinetics of tumor spheroids using various growth laws to more complex simulations of tumor-induced angiogenesis and capillary network formation, and tumor spreading at early and later invasive stages. In order to understand cancer as a complex system that involves so many interacting components, we also need to determine the type of data that needs to be collected at each level of organization, the boundary conditions to use when describing the disease (*i.e.* a perturbed system), and the technologies and approaches best suited to reveal its underlying biological behavior. Critical analysis of traditional clinical concepts is needed, as is reinterpretation of the clinical significance of failed therapies from the perspective of complexity. Two main concepts, *multi-scale causality* and *heterogeneity* need to be considered when generating new medical interventions. It is known that mathematical methods have proved to be possible and practical in oncology, but the current models are often simplifications that ignore vast amounts of knowledge; furthermore, most models struggle to resolve the *10-12* order-of-magnitude span of the timescales of systemic events, be they molecular (*ion channel gating: 10^{-6} sec*), cellular (*mitosis: 10^2-10^3 sec*) or physiological (*cancer progression: 10^8 sec*). Viewing cancer as a system that is dynamically complex in time and space will however probably reveal more about its underlying behavioral characteristics. It is encouraging that mathematicians, biologists and clinicians contribute together towards a common quantitative understanding of *cancer complexity*. This multi-disciplinary approach may help to clarify old concepts, categorize the actual knowledge, and suggest an alternative approach to discover biomarkers with potential clinical value.

REFERENCES

[1]   Grizzi, F; Chiriva-Internati, M. Cancer: looking for simplicity and finding complexity. *Cancer Cell Int*, 2006, 6, 4.

[2]   Grizzi, F; Di Ieva, A; Russo, C; Frezza, EE; Cobos, E; Muzzio, PC; Chiriva-Internati, M. Cancer initiation and progression: an unsimplifiable complexity. *Theor. Biol. Med. Model*, 2006, 3, 37.

[3]   Grizzi, F; Chiriva-Internati, M. The complexity of anatomical systems. *Theor. Biol. Med. Model*, 2005, 2, 26.

Fractal-Based Mathematical Models of Cancer in the Era of Systems Biology   135

[4]   Noble, D. The music of life. Biology beyond genes. Oxford, Oxford University Press; 2006.

[5]   Nurse, P. Reductionism. The ends of understanding. *Nature*, 1997, 387, 657.

[6]   Rose, S. What is wrong with reductionist explanations of behaviour? *Novartis Found Symp.*, 1998, 213, 176–86.

[7]   Westerhoff, HV; Palsson, BO. The evolution of molecular biology into systems biology. *Nat Biotechnol*, 2004, 22, 1249-52.

[8]   Villoslada, P; Steinman, L; Baranzini, SE. Systems biology and its application to the understanding of neurological diseases. *Ann. Neurol.*, 2009, 65, 124-39.

[9]   Butcher, EC; Berg, EL; Kunkel, EJ. Systems biology in drug discovery. *Nat. Biotechnol.*, 2004, 22, 1253-9.

[10] Wehling, M. Assessing the translatability of drug projects: what needs to be scored to predict success? *Nat. Rev. Drug Discov.*, 2009, 8, 541-6.

[11] Kohl, P; Noble, D. Systems biology and the virtual physiological human. *Mol. Syst. Biol.* 2009, 5, 292.

[12] Bertalanffy, LV. General system theory. New York, Braziler; 1968.

[13] Chang, L; Ray, LB. Whole-istic biology. *Science*, 2002, 295, 1661.

[14] Kitano, H. Computational systems biology. *Science*, 2002, 295, 1662-64.

[15] Bertalanffy, LV. The theory of open systems in physics and biology. *Science*, 1950, 111, 23-29.

[16] Miller, GJ. Living systems. New York, McGraw-Hill; 1978.

[17] Kauffman, AS. The origins of order: Self-Organization and Selection in Evolution. New York, Oxford University Press; 1993.

[18] Haken, H. Information and Self-Organization: A Macroscopic Approach to Complex Systems. Berlin, Springer; 2000.

[19] Miller, GJ; Miller, LJ. Introduction: the nature of living systems. *Behav. Sci.*, 1990, 35, 157-63.

[20] Miller, GJ. Living systems: basic concepts. *Behav. Sci.*, 1965, 10, 193-237.

[21] Miller, GJ. Living systems: structures and processes. *Behav. Sci.*, 1965, 10, 337-79.

[22] Braham, MA. A general theory of organization. *Gen. Systems*, 1973, 18, 13-24.

[23] Laszlo, E. The Systems View of the World: A Holistic Vision for Our Time. Cresskill, NJ, Hampton Press; 1996.

[24] Phillips, JD. Entropy analysis of multiple scale causality and qualitative causal shifts in spatial systems. *The Professional Geographer*, 2005, 57, 83-93.

[25] Lumsden, CY; Brandts, WA; Trainor, LEH. Physical theory in biology. Foundations and explorations. London: World Scientific; 1997.

[26] Casti, JL. Complexification: explaining a paradoxical world through the science of surprise. New York: HarperCollins; 1994.

[27] Losa, GA. The fractal geometry of life. *Riv. Biol.* 2009, 102, 29-59.

[28] Losa, GA. Fractal morphometry of cell complexity. *Riv. Biol.*, 2002, 95, 239-58.

[29] Taverna, G; Colombo, P; Grizzi, F, Franceschini, B; Ceva-Grimaldi, G; Seveso, M; Giusti, G; Piccinelli, A; Graziotti, P. Fractal analysis of two-dimensional vascularity in primary prostate cancer and surrounding non-tumoral parenchyma. *Pathol. Res. Pract.* 2009, 205, 438-44.

[30] Di Ieva, A; Grizzi, F; Ceva-Grimaldi, G; Russo, C; Gaetani, P; Aimar, E; Levi, D; Pisano, P; Tancioni, F; Nicola, G; Tschabitscher, M; Dioguardi, N; Baena, RR. Fractal

dimension as a quantitator of the microvasculature of normal and adenomatous pituitary tissue. *J. Anat.*, 2007, 211, 673-80.

[31] Grizzi, F; Russo, C; Franceschini, B; Di Rocco, M; Torri, V; Morenghi, E; Fassati, LR; Dioguardi, N. Sampling variability of computer-aided fractal-corrected measures of liver fibrosis in needle biopsy specimens. *World J. Gastroenterol.*, 2006, 12, 7660-5.

[32] Grizzi, F; Muzzio, PC. The morphologic complexity of the peripheral solitary pulmonary nodule. *Radiology*, 2005, 237, 373-4.

[33] Grizzi, F; Russo, C; Colombo, P; Franceschini, B; Frezza, EE; Cobos, E; Chiriva-Internati, M. Quantitative evaluation and modeling of two-dimensional neovascular network complexity: the surface fractal dimension. *BMC Cancer*, 2005, 5, 14.

[34] Torres Munoz, I; Grizzi, F; Russo, C; Camesasca, FI; Dioguardi, N; Vinciguerra, P. The role of amino acids in corneal stromal healing: a method for evaluating cellular density and extracellular matrix distribution. *J. Refract. Surg.*, 2003, 19, S227-30.

[35] Grizzi, F; Muzzio, PC; Di Maggio, A; Dioguardi, N. Geometrical analysis of benign and malignant breast lesions. *Radiol. Med.*, 2001, 101, 432-5.

[36] Nonnenmacher, TF; Baumann, G; Barth, A; Losa, GA. Digital image analysis of self-similar cell profiles. *Int. J. Biomed. Comput.*, 1994, 37, 131-8.

[37] Spinelli, G. Heterochromatin and complexity: a theoretical approach. *Nonlinear Dynamics Psychol. Life Sci.*, 2003, 7, 329-61.

[38] Losa, GA; Castelli, C. Nuclear patterns of human breast cancer cells during apoptosis: characterisation by fractal dimension and co-occurrence matrix statistics. *Cell Tissue Res.*, 2005, 322, 257-67.

[39] Paumgartner, D; Losa, G; Weibel, ER. Resolution effect on the stereological estimation of surface and volume and its interpretation in terms of fractal dimensions. *J. Microsc.*, 1981, 121, 51-63.

[40] Cross, SS; Start, RD; Silcocks, PB; Bull, AD; Cotton, DW; Underwood, JC. Quantitation of the renal arterial tree by fractal analysis. *J. Pathol.*, 1993, 170, 479-84.

[41] Abu Eid, R; Landini, G. Quantification of the global and local complexity of the epithelial-connective tissue interface of normal, dysplastic, and neoplastic oral mucosae using digital imaging. *Pathol. Res. Pract.*, 2003, 199, 475-82.

[42] Mandelbrot, BB. Les objects fractales: forme, hazard et dimension. Paris, Flammarion; 1975.

[43] Mandelbrot, BB. The Fractal Geometry of Nature. San Francisco, Freeman; 1983.

[44] Marquet, PA; Quiñones, RA; Abades, S; Labra, F; Tognelli, M; Arim, M; Rivadeneira, M. Scaling and power-laws in ecological systems. *J. Exp. Biol.*, 2005, 208, 1749-69.

In: Classification and Application of Fractals
Editor: William L. Hagen

ISBN 978-1-61209-967-5
© 2012 Nova Science Publishers, Inc.

Chapter 6

FRACTAL ANALYSIS OF SOIL STRUCTURE

S. H. Anderson

Department of Soil, Environmental and Atmospheric Sciences,
University of Missouri, Columbia, MO, US

ABSTRACT

Fractal analysis has been a very useful characterization tool to assist in understanding physical properties and processes of earth systems. This analysis has been applied to soil systems to help in assessing management and landscape effects on soil properties. Applications include fractal analysis of soil pores, hydraulic properties, solute transport properties, and cracking processes. Fractal dimensions of X-ray computed tomography (CT)–measured porosity vary depending upon land management and landscape position. Investigations have found that fractal dimension of soil macropores increases 19% under vegetative buffers as compared to row crop management. Similar increases (26%, 9% and 18%) in fractal dimension were found for agroforestry (tree/grass) buffers compared to row crop management, agroforestry buffers compared to pasture management, and native prairie compared to row crop management, respectively. Fractal dimension of macropores was found to be highly correlated with saturated hydraulic conductivity (r = 0.87). CT-measured solute pore-water velocity and dispersivity were found to be fractal, and fractal dimension increased with average grain size in soil cores. Soil cracking patterns have also been characterized by both mass fractal dimension and crack edge fractal dimension; mass fractal dimension was found to be a function of soil landscape position, while crack edge fractal dimension values did not vary with landscape. Fractal analysis has been shown to be a useful characterization tool to differentiate management influences on critical soil physical properties and processes.

INTRODUCTION

Soils are a valuable resource in the natural landscape and are essential for life on earth. Soils are mapped throughout the landscape to understand how properties change and influence plant growth (Brady and Weil, 2008). Soils store essential nutrients and water to sustain plant and animal life. Soils are also vital for filtering nutrients and other chemicals from water which enhances the quality of both surface and ground water resources (Brady and Weil, 2008).

The soil fabric which is described by soil texture and structure is a vital part of the system for supporting roots and allowing fluid transport in the system. Coarse-textured soils which include large fractions of sand-sized particles allow rapid transport of water and air. However, these soils do not retain sufficient water and nutrients for sustained plant growth. Fine textured soils with larger fractions of clay-sized particles do retain water and nutrients but restrict water and air movement in soils. Moderate textured soils with higher fractions of silt-size particles provide sufficient water and nutrient retention for good plant growth while also allowing for moderate transport of water and air.

In addition to texture, soil structure plays a significant role in water and air transport in soils. Evaluation of transport of chemicals dissolved in water (solutes) through soils is of great importance in evaluating and preventing degradation of water quality from possible contamination. Chemicals, which include fertilizers, pesticides, antibiotics, heavy metals, and wastes, often move from the soil surface through the vadose zone of the soil towards groundwater and water bodies, thus resulting in the deterioration of soil and water quality. Such deterioration challenges scientists to better understand transport processes of chemicals through the soil. Transport processes are complex, and groundwater contamination through the vadose zone needs to be investigated (Onsoy et al., 2005). Soil structure plays a significant role in these complex transport processes. In fact, soils are seldom homogeneous; their properties usually vary spatially on both a small and large scale due to textural and structural changes (Kazemi et al., 2008).

FRACTAL ANALYSIS OF SOIL PROPERTIES

Understanding contaminant transport in porous media is a key area of study to aide in the protection of the quality of water resources. Fluid transport properties and the characterization of their associated spatial variability are essential for accurate prediction of water and chemical solute transport behavior. A useful method for quantitatively describing spatial changes in these transport properties employs fractal analysis (Baveye et al., 1998b). The fractal dimension can be used in explaining the variation of structural properties of a porous media (Dathe and Thullner, 2005). This fractal dimension can be used to assess macropore structure measured using CT methods (Rachman et al., 2005; Udawatta and Anderson, 2008).

The inner quality of the soil structure and distributions of soil pore sizes can be assessed using the fractal dimension. Fractal theory can be used to compare structural complexities and values can be used as an index for macroporosity and water retention in porous media (Tyler and Wheatcraft, 1989; Rasiah, 1995; Perret et al., 2003). Several investigators have measured the fractal dimension of pores in different textured soils from sand to silty clay loam and found values varying between 1.011 and 1.485 (Tyler and Wheatcraft, 1989). Riparian buffer management effects on the fractal dimension of soil macroporosity under grass and tree buffers showed values of 1.21 to 1.34 (Udawatta and Anderson, 2008). These values are in a similar range to those found by Tyler and Wheatcraft (1989). The fractal dimension of CT-measured macroporosity has been found to be highly correlated with saturated hydraulic conductivity (r = 0.87; Udawatta and Anderson, 2008).

Computed tomography using X-rays has been used as a diagnostic tool in soil science (Anderson et al., 2003). This method has significant potential for advancing dynamic solute transport research. Only a few studies have quantified the macropore-scale spatial structure of solute transport parameters. These parameters include the solute pore-water velocity and dispersivity which can be measured using these CT techniques (Anderson et al., 2003). Developing an understanding of the macropore-scale variability of solute transport parameters will play an important role in predicting the fate and transport of chemicals through heterogeneous porous media systems. Since CT methods utilize imaging techniques to determine fractal properties such as fractal dimension and lacunarity (Zeng et al., 1996), the contrast, accuracy, and precision of these estimates depend on the resolution of the images, threshold values used to transform the image, and techniques for estimating the fractal dimension (Baveye et al., 1998a; Udawatta et al., 2008b).

Soil physical properties such as particle size distribution, soil pore size distribution, and hydraulic properties can be evaluated in terms of the fractal dimension. Perrier and Bird (2003) reviewed the work on fractal analysis in soils and proposed the pore solid fractal (PSF) model in which the soil matrix can be divided into pores, solids and other areas. This method requires higher resolution in order to determine whether an area is pore or solid (thus fractal). This system can be useful in utilizing soil particle size data, soil pore size data, soil aggregate size data, and solid-pore interface area scaling. Thus, the PSF model can allow a powerlaw soil pore size distribution and a powerlaw soil particle size distribution. The exponents of these distributions involve the same fractal D parameter. This model approach can also be useful for evaluating links between soil structural and soil hydraulic properties.

Evaluation of images of resin-impregnated soil was reviewed by Tarquis et al. (2003). These images provide an opportunity to evaluate pore and solid spaces and their interface. Useful fractal and multifractal methods for evaluating these images include the box-counting method, dilation method, random walk method, and the singularity spectrum method. When an object exhibits multifractal scaling, the mass of the object is distributed in a hierarchical pattern as in the fractal case (Tarquis et al., 2003). For example, the whole object can be formed by a union of similar subsets; however, the subsets are related to the whole by different scaling factors (Tarquis et al., 2003).

MANAGEMENT EFFECTS ON SOIL MACROPORE FRACTAL DIMENSION

Assessments have shown that row crop and pasture management systems can have an effect on soil structure. Fractal analysis comparisons among native prairie, restored prairie and row crop management (Udawatta et al., 2008a) have been found to influence the fractal dimension of macropores (pores > 1000 μm in effective diameter). Pasture management systems compared with agroforestry and grass buffer management have also been found to influence the fractal dimension of macropores (Kumar et al., 2010).

Native prairie, restored prairie, conservation reserve program (CRP), and row crop management treatments have been compared for claypan soils (Mexico silt loam; fine, smectitic, mesic Aeric Vertic Epiaqualf) relative to soil macropore characteristics (>1000 μm diameter; Udawatta et al., 2008a). Claypan soils have a hydraulically restrictive subsoil horizon with a predominance of smectitic clay (Jamison et al., 1968; Blanco-Canqui et al.,

2002 and 2004; Anderson, 2011). Pore properties were shown to be significantly affected by row crop soil management for over 100 years compared to undisturbed native prairies (Udawatta et al., 2008a). Soils under a corn (*Zea mays* L.) –soybean (*Glycine max* (L.) Merr.) rotation showed macropores with a fractal dimension of 1.29 and restored prairie and CRP areas had fractal dimensions of 1.31, where the undisturbed prairie had a fractal dimension of 1.52. These significant differences among treatments suggest soil structure is highly influenced by management. Undisturbed prairies had significantly higher numbers of macropores. In addition, the pores under native prairie were more space filling compared to pores under row crop areas. Fractal dimension decreased with increasing soil depth; however, the decrease was smaller for the native prairie compared to the other treatments. This indicates better soil structure was preserved even for deeper soil depths under the native prairie soils.

Udawatta et al. (2006) and Udawatta and Anderson (2008) evaluated macropore parameters for a Putnam silt loam (fine, smectitic, mesic Vertic Albaqualf) to compare row crop, agroforesty buffer and grass buffer areas. Fractal dimension was significantly different among the three treatments and five depth zones evaluated. Fractal dimension decreased with soil depth with a correlation coefficient of -0.83. The agroforestry buffer treatment had the highest fractal dimension of 1.34 averaged across all depths with the grass buffer treatment having a value of 1.21 and the row crop area having a value of 1.06. Fractal dimension values were highly correlated with measured macroporosity and coarse mesoporosity (200 to 1000 μm diameter). Comparing fractal dimension of macropores with saturated hydraulic conductivity, Udawatta and Anderson (2008) found a very strong correlation (r = 0.87). Thus, 76% of the variation in this important fluid transport property was explained by the fractal dimension of macropores. These grass and agroforestry buffer systems significantly affect macroporosity which strongly influences soil hydraulic properties (Seobi et al., 2005).

Cheng et al. (2001) and Anderson et al. (2007) used multiple resolution blankets to evaluate the fractal nature of X-ray CT-measured macroporosity. These blankets are theoretical and evaluate the change in the fractal properties of images as a function of resolution. Cheng et al. (2001) were able to predict soil saturated hydraulic conductivity with these properties using fuzzy logic.

Pasture management affects soil structure. Kumar et al. (2010) evaluated pores using X-ray CT for rotational and continuous pasture systems compared to agroforestry and grass buffers on a well drained Menfro silt loam soil (fine-silty, mixed, superactive Typic Hapludalf). Continuously grazed pasture management systems do not restrict cattle grazing; cattle have access to all areas of the pasture. Rotationally grazed pasture management systems restrict cattle grazing to specific areas with cattle being moved every few days. Rotationally grazed pasture systems have been found to maximize cattle production and to be better for conserving soil resources. They found that CT-measured soil macroporosity was 13 times higher (0.053 m^3 m^{-3}) for the buffer treatments compared to the pasture treatments (0.004 m^3 m^{-3}) within the upper 10 cm. Kumar et al. (2010) also found a fractal dimension of macropores was 1.08 for the continuously grazed pastures compared to 1.14 for the rotationally grazed pastures, 1.21 for agroforestry buffers and 1.41 for grass buffers. Fractal dimension values decreased with soil depth for this study. These fractal dimension parameters were highly correlated with measured macroporosity.

Tillage of soils has been found to significantly affect soil properties such as soil structure. Gantzer and Anderson (2002) evaluated a chisel-disk tillage treatment compared to a no-till treatment on macropore properties for a Mexico silt loam claypan soil. They found that the tillage treatment had a significantly higher macropore area (11%) compared to the no-till treatment (5%). The tillage treatment also had a higher fractal dimension (1.44) compared to the no-till treatment (1.26). This reflects the greater space-filling nature of the chisel-disk tillage treatment for this soil.

Like tillage of soils, compaction also significantly affects soil structure and thus, fractal properties of soils. Kim et al. (2010) evaluated compacted and non-compacted treatment effects on macropore properties for a Mexico silt loam soil. They found fractal dimension of soil macropores decreased with a compaction treatment (fractal dimension of 1.11 from 1.17) within the 0 to 30 cm soil depth. As with prior studies, fractal dimension decreased with increasing soil depth.

MACROPORES UNDER VEGETATIVE BUFFERS

Macropores have been evaluated using X-ray CT under different permanent vegetative grass hedges compared to traditional row crop areas (Rachman et al., 2005). Estimates of the box-counting fractal dimension, D, for CT-measured macroporosity in three landscape positions (row crop area or erosional area, sediment deposition area, and grass hedge area) and two depths were determined (Rachman et al., 2005). Fractal dimension was found to be related to the number of CT-measured macropores and their size distribution; hence, it measures the space-filling nature of the macropores. Similar to macroporosity data, the fractal D in the grass hedge position was significantly higher compared to values for the other positions. The macropores under the permanent grass hedges were more space filling as indicated by the higher fractal dimension. Fractal dimensions were 1.70, 1.49 and 1.16 for the grass hedge, row crop and depositional areas, respectively. This increase in macropores has a significant beneficial influence on soil hydraulic properties (Rachman et al., 2004) which can reduce runoff and sediment transport from watersheds with these management systems (Rachman et al., 2008).

Rachman et al. (2005) also found that the fractal dimension decreased with increasing soil depth. This is due in part to the decrease in macroporosity with increasing soil depth. Correlation coefficients between fractal dimension and macroporosity were 0.74, 0.90, and 0.97 for the deposition zone, row crop area, and grass hedge area, respectively.

SOLUTE TRANSPORT PROPERTIES

Soil systems have a complex heterogeneous structure which presents serious challenges in estimating parameters needed for solute transport models. Past research has shown the variability of solute transport parameters using both laboratory (Lennartz, 1999; Strock et al., 2001) and field approaches (Kazemi et al., 2008). Pore-water velocity and dispersivity are significant parameters used in models which are influenced by measurement scale. Recently, the importance of macropore-scale heterogeneities in influencing solute transport through porous media has been recognized (Strock et al., 2001). These macropores (greater than 1000 μm in diameter) are defined as pores or structural features in porous media.

Methods for characterization of solute transport parameters in porous media in time and space on a macropore-scale are laborious and time consuming. In addition, obtaining simple, efficient parameters using non-invasive techniques to obtain data on a macropore-scale is difficult with traditional laboratory procedures (Liu et al., 2008).

In the past few decades, X-ray CT systems have been developed as diagnostic tools in medicine for rapid and non-destructive assessment of density inside opaque objects in three-dimensions. These instruments have been utilized to measure density and water content in soils and to characterize soil macropores in terms of size and spatial distribution (Rachman et al., 2005). CT techniques have also been used to measure chemical breakthrough curves in undisturbed soil cores and to characterize solute transport parameters such as solute dispersivity (Peyton et al., 1994; Clausnitzer and Hopmans, 2000). Other researchers, such as Perret et al. (2000), have employed single photon emission CT to visualize preferential flow in soils.

Liu et al. (2008 and 2010) hypothesized that fractal properties such as fractal dimension and lacunarity of solute transport properties on a macropore-scale would provide better indicators to differentiate solute movement through porous media. Assessment of transport properties in cores with selected sizes of glass beads was conducted by Liu et al. (2008). Liu et al. (2010) evaluated whether CT-measured pore-water velocity and dispersivity in intact soil core samples were fractal. They found these properties to be fractal and estimated the fractal dimension and lacunarity of these solute transport properties.

Porosity, pore-water velocity, and solute dispersivity were evaluated in cores with selected sizes of glass beads (Liu et al., 2008). CT-measured breakthrough curve experiments were conducted in columns of glass beads (1.4 to 8.0 mm diam.). CT-measured porosity, pore-water velocity, and dispersivity were found to be fractal. Fractal dimensions of these parameters decreased with the logarithm of glass bead diameter. Results of the study indicated that both fractal dimension and lacunarity are required to discriminate spatial distributions of the solute transport parameters among different porous media. If fractal dimensions are the same for different fractal sets, lacunarity analysis may reveal different spatial patterns or fractal structures for such fractal sets.

Undisturbed soil cores were taken from the surface horizon of a Sarpy loamy sand (Typic Udipsamment), an alluvial soil (Liu et al., 2010). CT-measured breakthrough curve experiments were conducted through eight intact soil cores. Breakthrough experiments with a dilute solution of KI (potassium iodide) were conducted using a medical X-ray CT scanner. Based on the breakthrough curve for each pixel, solute pore-water velocity and dispersivity distributions were determined. CT-measured pore-water velocity and dispersivity were found to be fractal. Fractal dimensions of pore-water velocity ranged from 2.16 to 2.43 and for dispersivity from 2.38 to 2.66. Results indicated that both fractal dimension and lacunarity are required to discriminate spatial distributions of the solute transport parameters among samples. Again, when fractal dimensions are the same for different data sets, lacunarity analysis may be useful in revealing different spatial patterns or fractal structures for these data sets.

FRACTAL DIMENSION TO QUANTIFY SOIL CRACKING

Fractal dimension has been utilitized for quantifying soil desiccation cracks (Baer et al. 2009). Rapid transport of water can occur through dessication cracks which may lead to ground water contamination. Baer et al. (2009) evaluated fractal dimension of crack edges as well as the mass fractal dimension of soil cracks for a claypan soil prone to dessication cracking during dry periods of the summer. Crack edge fractal dimensions were approximately 1.08 while mass fractal dimensions of desiccation cracks ranged from 1.44 to 1.64. Similar crack edge fractal dimensions indicated similar crack edge features among locations with significantly different levels of cracking. Higher mass fractal dimension occurred when the claypan argillic soil horizon was closer to the soil surface and thus produced more surface cracks.

Locations in the landscape with a high degree of cracking may provide more rapid transport of water and solutes into subsurface soil horizons and the groundwater (Baer and Anderson, 1997). With appropriate soil properties and crack descriptions, crack networks can be generated to more successfully model water and chemical transport in cracking soils (Baer et al., 2009).

LACUNARITY AS A FRACTAL DIAGNOSTIC TOOL

Fractal lacunarity, $C(L)$, comes from the word *lacuna* which is Latin for gap. The parameter, L, is the length of the box size used to estimate the fractal lacunarity. Hence, $C(L)$ describes the uniformity of gaps or voids in the fractal object or in this study, the highs and lows of the CT-measured solute parameters. The estimate of lacunarity described in this article is similar to a coefficient of variation.

Work discussed in this article used a point-distribution method (PDM) which counts the number of data points after discretizing the data set in a box. The PDM method facilitates calculation of the fractal lacunarity, $C(L)$. The trace is computed using the PDM as follows:

$$N(L) = H \sum_{m=1}^{K} \frac{1}{m} P(m, L)$$

[1]

where H is the total number of points in the trace, K is the maximum number of points that can be contained in a box of side length L, and $P(m,L)$ is the probability that m points fall within a box of side length L centered at an arbitrary point along the trace. For each value of L, the relationship can be described as follows: $\sum_{m=1}^{K} P(m,L) = 1$. The fractal dimension, D, can be estimated with a least-squares fit of $\log[N(L)]$ vs. $\log(L)$. The $P(m,L)$ is calculated by sequentially centering an imaginary box of side length L at each point along the trace and counting the number of points in the box at each position. The points represent discrete solute transport property values in the image. The frequency of occurrence for m points is recorded. The $P(m,L)$ is the frequency of occurrence divided by H. The $N(L)$ is then calculated using Equation [1]. This is repeated for a range of box side lengths.

The lacunarity was estimated as follows:

$$C(L) = \frac{M_2(L) - [M_1(L)]^2}{[M_1(L)]^2} \qquad [2]$$

where $C(L)$ is lacunarity as a function of box side length L, $M_1(L)$ is the first moment of $P(m,L)$, and $M_2(L)$ is the second moment of $P(m,L)$. These moments can be defined as follows:

$$M_1(L) = \sum_{m=1}^{K} mP(m, L) \qquad [3]$$

and

$$M_2(L) = \sum_{m=1}^{K} m^2 P(m, L) \qquad [4]$$

The computer code discussed by Liu et al. (2008 and 2010) was used in this analysis to calculate D and $C(L)$ for the parameters.

Liu et al. (2008) conducted solute breakthrough curve experiments using a potassium iodide tracer for cores with selected glass bead sizes. These were conducted in columns of glass beads ranging from 1.4 to 8.0 mm in diameter. The CT-measured macropore properties were found to be fractal. Lacunarity at both 1 mm and 20 mm box sizes were highly correlated with glass bead size ($r^2 > 0.90$). Lacunarity at 1 mm box size was negatively correlated with glass bead diameter for all three parameters while lacunarity at 20 mm box size was positively correlated with glass bead diameter. Results of the study indicated that both fractal dimension and lacunarity are required to discriminate spatial distributions of the solute transport parameters among different porous media.

Liu et al. (2010) conducted experiments with undisturbed soil cores with a predominant texture of loamy sand. Results indicated that both fractal dimension and lacunarity are required to discriminate spatial distributions of the solute transport parameters among the undisturbed soil samples (Liu et al., 2010). If fractal dimensions are the same for different fractal sets, lacunarity analysis may reveal different spatial patterns or fractal structures for such fractal sets.

Lacunarity of soil cracks for a Mexico silt loam claypan soil was determined by Baer et al. (2009). They found that fractal lacunarity of soil cracking measured by imaging increased over time and was affected by landscape position. Fractal lacunarity is an assessment of the space-filling pattern of the crack area. High lacunarity indicates more space between crack areas and low lacunarity indicates a more dense pattern of crack areas (Baer et al., 2009).

Yakimenko et al. (2009) evaluated lacunarity and fractal dimension for Russian soils under different management: forest, meadow, and a forest-meadow transition. They found that lacunarity measured at 5 mm box size was significantly different among treatments for moist soils (-10 kPa soil water pressure). Forest soils were significantly higher compared to soils under meadow. Yakimenko et al. (2009) also found that the slope of the lacunarity curve as a function of box size was negatively correlated with saturated hydraulic conductivity. This implies that a stronger decrease in lacunarity was associated with higher saturated soil hydraulic conductivity.

Lacunarity of CT-measured soil macropores was evaluated for a Mexico silt loam soil by Zeng et al. (2009). They found that fractal lacunarity was more significantly related to soil depth than fractal dimension. Lacunarity was significantly correlated with saturated hydraulic conductivity (r = 0.66) and CT-measured macroporosity (r = 0.74). Thus, fractal lacunarity is useful for discriminating relative to soil property changes.

CONCLUSION

Evaluation and comparison of fractal dimension of soil properties is helpful in discriminating the effects of management on soil structure. Fractal analysis of X-ray CT-measured porosity obtained fractal dimensions which vary depending upon land management and landscape position. Studies have found that the fractal dimension of soil macropores increases under vegetative buffers compared to row crop management. Increases in fractal dimension of macropores were found for agroforestry (tree/grass) buffers (26%) compared to row crop management, agroforestry buffers (9%) compared to pasture management, and native prairie (18%) compared to row crop management. Significantly higher fractal dimension for macropores in soils under permanent management indicates the higher space-filling nature of macropores. This fractal dimension of macropores has been useful in predicting saturated hydraulic conductivity (r^2 = 0.76). Solute pore-water velocity and dispersivity measured using X-ray CT were found to be fractal, and the fractal dimension increased with average grain size of the porous media. The crack edge fractal dimension of soil desiccation cracks was similar among landscape positions, while the mass fractal dimension was dependent upon landscape position primarily due to differential depths of the claypan relative to the soil surface. Fractal analysis of soil structural features has been shown to be a useful characterization tool to differentiate management influences on critical soil physical properties and processes.

REFERENCES

Anderson, S.H. 2011. Claypan and its environmental effects. In Jan Glinski, Jozef Horabik, Jerzy Lipiec (Eds.), *Encyclopedia of Agrophysics*, Springer. (in press).

Anderson, S.H., Cheng, Z., and Gantzer, C.J. 2007. MRB signatures of X-ray CT images for estimating soil hydraulic conductivity. *Intelligent Engineering Systems Through Artificial Neural Networks* 17:131-136.

Anderson, S.H., Wang, H., Peyton, R.L., and Gantzer, C.J. 2003. Estimation of porosity and hydraulic conductivity from X-ray CT-measured solute breakthrough. *In* F. Mees, R. Swennen, M. Van Geet, and P. Jacobs (eds.) *Applications of X-ray Computed Tomography in the Geosciences.* Geological Society of London. Special Publication 215:135-149.

Baer, J.U. and Anderson, S.H. 1997. Landscape effects on desiccation cracking in an Aqualf. *Soil Science Society of America Journal* 61:1497-1502.

Baer, J.U., Kent, T.F., and Anderson, S.H. 2009. Image analysis and fractal geometry to characterize soil desiccation cracks. *Geoderma* 154:153-163.

Baveye, P., Boast, C.W., Ogawa, S., Parlange, J., and Steenhuis, T. 1998a. Influence of image resolution and thresholding on the apparent mass fractal characteristics of preferential flow patterns in field soils. *Water Resources Research* 34:2783-2796.

Baveye, P., Parlange, J.Y., and Stewart, B.A. (eds.) 1998b. Fractals in soil science. *Advances in Soil Science*, CRC Press, Boca Raton, FL.

Blanco-Canqui, H., Gantzer, C.J., Anderson, S.H., Alberts, E.E., Ghidey, F. 2002. Saturated hydraulic conductivity and its impact on simulated runoff for claypan soils. *Soil Science Society of America Journal* 66:1596-1602.

Blanco-Canqui, H., Gantzer, C.J., Anderson, S.H., Alberts, E.E., and Thompson, A.L. 2004. Grass barrier and vegetative filter strip effectiveness in reducing runoff, sediment, nitrogen, and phosphorous loss. *Soil Science Society of America Journal* 68:1670-1678.

Brady, N.C. and Weil, R.R. 2008. *The Nature and Properties of Soils*. 14th ed. Prentice Hall, Upper Saddle River, New Jersey.

Cheng, Z., Anderson S.H., Gantzer, C.J., and Chu, Y. 2001. Fuzzy logic for predicting soil hydraulic conductivity using CT images. *Intelligent Engineering Systems Through Artificial Neural Networks* 11:307-312.

Clausnitzer, V. and Hopmans, J.W. 2000. Pore-scale measurements of solute breakthrough using microfocus X-ray computed tomography. *Water Resources Research* 36:2067–2079.

Dathe, A. and Thullner, M. 2005. The relationship between fractal properties of solid matrix and pore space in porous media. *Geoderma* 129:279-290.

Gantzer, C.J. and Anderson, S.H. 2002. Computed tomographic measurement of macroporosity in chisel-disk and no-tillage seedbeds. *Soil and Tillage Research* 64:101-111.

Jamison, V.C., Smith, D.D., and Thornton, J.F. 1968. Soil and water research on a claypan soil. *USDA Technical Bulletin* 1379. U.S. Gov. Print. Office, Washington, DC.

Kazemi, H.V., Anderson, S.H., Goyne, K.W., and Gantzer, C.J. 2008. Spatial variability of bromide and atrazine transport parameters for a Udipsamment. *Geoderma* 144:545-556.

Kim, H.M., Anderson, S.H., Motavalli, P.P., and Gantzer, C.J. 2010. Compaction effects on soil macropore geometry and related parameters for an arable field. *Geoderma* 160:244-251.

Kumar, S., Anderson, S.H., and Udawatta, R.P. 2010. Agroforestry and grass buffer influences on macropores measured by computed tomography under grazed pasture systems. *Soil Science Society of America Journal* 74:203-212.

Lennartz, B. 1999. Variation of herbicide transport parameters within a single field and its relation to water flux and soil properties. *Geoderma* 91:327-345.

Liu, X., Anderson, S.H., and Udawatta, R.P. 2010. Fractal dimension and lacunarity of CT-measured solute pore-water velocity and dispersivity. *Intelligent Engineering Systems Through Artificial Neural Networks* 20:421-428.

Liu, X., Anderson, S.H., and Udawatta, R.P. 2008. Fractal analysis of CT-measured solute transport parameters. *Intelligent Engineering Systems Through Artificial Neural Networks* 18:171-178.

Onsoy, Y.S., Harter, T., Ginn, T.R., and Horwath, W.R. 2005. Spatial variability and transport of nitrate in a deep alluvial vadose zone. *Vadose Zone Journal* 4:41-55.

Perrier, E.M.A. and Bird, N.R.A. 2003. The PSF model of soil structure: A multiscale approach. pp. 1-18. *In* Y. Pachepsky, D.E. Radcliffe, and H.M. Selim (eds.) *Scaling Methods in Soil Physics*, CRC Press, Boca Raton, FL.

Perret, J.S., Prasher, S.O., and Kacimov, A.R. 2003. Mass fractal dimension of soil macropores using computed tomography: from the box-counting to the cube-counting algorithm. *European Journal of Soil Science* 54:569-579.

Perret, J., Prasher, S.O., Kantzas, A., Hamilton, K., and Langford, C. 2000. Preferential solute flow in intact soil columns measured by SPECT scanning. *Soil Science Society of America Journal* 64:469-477.

Peyton, R.L., Anderson, S.H., Gantzer, C.J., Wigger, J.W., Heinze, D.J., and Wang, H. 1994. Soil-core breakthrough measured by X-ray computed tomography. pp. 59-71. *In* S.H. Anderson and J.W. Hopmans (ed.) *Tomography of Soil-Water-Root Processes.* Soil Science Society of America Special Publication No. 36, Madison, WI.

Rachman, A., Anderson, S.H., Alberts, E.E., Thompson, A.L., and Gantzer, C.J. 2008. Predicting runoff and sediment yield from a stiff-stemmed grass hedge system for a small watershed. *Transactions of the American Society of Agricultural and Biological Engineers* 51:425-432.

Rachman, A., Anderson, S.H., and Gantzer, C.J. 2005. Computed-tomographic measurement of soil macroporosity parameters as affected by stiff-stemmed grass hedges. *Soil Science Society of America Journal* 69:1609-1616.

Rachman, A., Anderson, S.H., Gantzer, C.J., and Thompson, A.L. 2004. Influence of stiff-stemmed grass hedge systems on infiltration. *Soil Science Society of America Journal* 68:2000-2006.

Rasiah, V. 1995. Fractal dimension of surface connected macropore count-size distribution. *Soil Science* 159:105-108.

Seobi, T., Anderson, S.H., Udawatta, R.P., and Gantzer, C.J. 2005. Influence of grass and agroforestry buffer strips on soil hydraulic properties for an Albaqualf. *Soil Science Society of America Journal* 69:893-901.

Strock, J.S., Cassel, D.K., and Gumpertz, M.L. 2001. Spatial variability of water and bromide transport through variably saturated soil blocks. *Soil Science Society of America Journal* 65:1607-1617.

Tarquis, A.M., Gimenez, D., Saa, A., Diaz, M.C., and Gasco, J.M. 2003. Scaling and multi-scaling of soil pore systems determined by image analysis. pp. 19-34. *In* Y. Pachepsky, D.E. Radcliffe, and H.M. Selim (eds.) *Scaling Methods in Soil Physics*, CRC Press, Boca Raton, FL.

Tyler, S.W. and Wheatcraft, S.W., 1989. Application of fractal mathematics to soil water retention estimation. *Soil Science Society of America Journal* 53:987-996.

Udawatta, R.P., Anderson, S.H., Gantzer, C.J., Garrett, H.E. 2006. Agroforestry and grass buffer influence on macropore characteristics: A computed tomography analysis. *Soil Science Society of America Journal* 70:1763-1773.

Udawatta, R.P. and Anderson, S.H. 2008. CT-measured pore characteristics of surface and subsurface soils influenced by agroforestry and grass buffers. *Geoderma* 145:381-389.

Udawatta, R.P., Anderson, S.H., Gantzer, C.J., and Garrett, H.E. 2008a. Influence of prairie restoration on CT-measured soil pore characteristics. *Journal of Environmental Quality* 37:219-228.

Udawatta, R.P., Gantzer, C.J., Anderson, S.H., and Garrett, H.E. 2008b. Agroforestry and grass buffer effects on high resolution X-ray CT-measured pore characteristics. *Soil Science Society of America Journal* 72:295-304.

Yakimenko, E.Y., Anderson, S.H., and Udawatta, R.P. 2009. Assessment of CT-measured porosity in Russian soils using fractal dimension and lacunarity. *Intelligent Engineering Systems Through Artificial Neural Networks* 19:147-154.

Zeng, Y., Gantzer, C.J., Anderson, S.H., and Udawatta, R.P. 2009. Fractal analysis of CT-measured porosity for claypan soils. *Intelligent Engineering Systems Through Artificial Neural Networks* 19:115-122.

Zeng, Y., Gantzer, C.J., Peyton, R.L., and Anderson, S.H. 1996. Fractal dimension and lacunarity of bulk density determined with X-ray computed tomography. *Soil Science Society of America Journal* 60:1718-1724.

In: Classification and Application of Fractals
Editor: William L. Hagen

ISBN 978-1-61209-967-5
© 2012 Nova Science Publishers, Inc.

Chapter 7

DYNAMICS OF MISCELLANEOUS FRACTAL STRUCTURES IN HIGHER-DIMENSIONAL EVOLUTION MODEL SYSTEMS

Victor K. Kuetche, Thomas B. Bouetou and Timoleon C. Kofane

[1]National Advanced School of Engineering, University of Yaounde I, Cameroon
[2]Department of Physics, Faculty of Science, University of Yaounde I, Cameroon

ABSTRACT

Throughout the present chapter, based upon the viewpoint of Weiss–Tabor–Carnevale formalism [J. Weiss, M. Tabor, and G. Carnevale, J. Math. Phys. **24**, 522 (1983); **25**, 13 (1984)], we study the integrability properties of a set of higher-dimensional evolution model systems, namely, the coupled nonlinear extension of the reaction-diffusion equation modeling the development of highly complex organisms based upon nonlinear interactions between common genes, the two-coupled nonlinear Schrödinger equation arising in the description of dynamics of miscellaneous Bose–Einstein condensate mixtures confined within a time-independent anisotropic parabolic trap potential mapped onto the higher-dimensional time-gated Manakov system up to a first-order of accuracy, the dynamics of bulk polaritons in ferromagnetic slab through the single-oscillation two-dimensional soliton system, and the three-coupled Gross–Pitaevskii type nonlinear equations arising in the context of spinor Bose–Einstein condensates of atomic hyperfine spin $f = 1$ species. As a result, due to the arbitrariness of some functions stemming from the Laurent expansion up to a suitable truncation, we unearth an interesting family of fractal structures of miscellaneous patterns and dynamics worthy to the understanding of many physical phenomena occurring in the nature.

PACS 02.30.Ik, 05.45.Yv, 11.10.Lm, 42.81.Dp, 03.75.Lm, 42.50.Md, 42.65.Tg.

Keywords: Weiss–Tabor–Carnevale formalism, nonlinear extension of the reaction-diffusion equation, nonlinear Schrödinger equation, bulk polaritons, spinor Bose–Einstein condensates, fractal structures.

---

*E-mail address: vkuetche@yahoo.fr
†E-mail address: tbouetou@yahoo.fr
‡E-mail address: tckofane@yahoo.com

1. Introduction

There are many ordered/desordered complex features in nature, with similar/dissimilar patterns observed in various fields. In such patterns, the formation of branching structures in open systems is often observed [1–7]. For example, branching patterns can be observed in neurons, protoplasmic streaming tubes of slime molds, trees, bacterial colonies, among others [1–7]. In these branching structures, propagation of signals or transportation of substances is observed, particularly in the structures of living organisms. Such branching structures are interesting from the viewpoint of a transportation system. Several models have been proposed to describe complex branching pattern formations [1–7]. Alongside the above real natural intricate phenomena, there are folded structures such as the folded protein, folded brain and skin surfaces, and many other kinds of folded biologic systems, just to name a few [8–12]. Getting a rough satisfactory analytic survey of such complicated folded phenomena is actually nontrivial. Nonetheless, many attempts in this line are currently in the route of a performing task.

Nowadays, research in physics devotes much attention to nonlinear phenomena. The reason for this is undoubtedly that most physical phenomena such as those mentioned above are intrinsically nonlinear. As further illustrations, there are Newton's gravitational attraction law involving a spatial dependency $1/r$ (r being the position magnitude), the transcendent pendulum equation, among others. Advances in nonlinear science have been plentiful in recent years. In particular, interests in nonlinear wave propagation continue to grow, stimulated by new applications such as fiber-optics communication systems, as well as the many classical unresolved issues of fluid dynamics. Owing to the existence of many complicated physical phenomena in real natural world, one important tool in characterizing such media is the nonlinear partial differential evolution model (NLPDEM) equations. Studying the solutions to the associated model equations should help one better understand the physical mechanism of these problems. Among such evolution equations, there are many coupled systems which have been proposed in the wake of the great discovery of the soliton theory and its remarkable related features. In fact, the coupled systems arguably arise in the context of mixtures, but also as an indicator of anisotropy in a given medium. Because of the rich structures of most coupled equations, both mathematicians and physicists have paid attention to the above structures. For example, an infinite number of conservation laws, multiHamiltonian structures, symmetry reductions, just to name a few, have been found in such systems [13–18].

It has been shown that some simplest NLPDEM equations with soliton solutions possess an infinite set of conservation laws [19–21]. However, the full details showing the relationship of infinitely conserved quantities with soliton solutions have been shown to be unclear [22–24] until Wahlquist and Estabrook [25] developed a technique known as "prolongation structure" which has been generalized to nonlinear evolution equations. As applied to the Korteweg- de Vries equation, this powerful method which is a set of interrelated potentials and pseudopotentials for NLPDEM equations in two independent variables, generates infinite conservation laws leading directly to the soliton solutions, Bäcklund transformation (BT) between solutions, and inverse scattering transform (IST) to the initial value problem [24]. Following such results, Zhai et al. [26] recently investigated the integrable (2+1)-dimensional (modified) Heisenberg ferromagnet model [27] using the prolon-

Dynamics of Miscellaneous Fractal Structures ... 151

gation structure theory. The corresponding geometrical equivalent counterparts, such as the $(2+1)$-dimensional nonlinear Schrödinger equation and the coupled $(2+1)$-dimensional integrable equations, presented through the motion of Minkowski space curves endowed with an additional spatial variable, have been constructed. These last coupled $(2+1)$-dimensional integrable equations are given by [26, 28]

$$\psi_t + \psi_{xy} + \gamma\psi = 0, \quad \phi_t - \phi_{xy} - \gamma\phi = 0, \quad \gamma_x + (\phi\psi)_y = 0, \tag{1}$$

where ψ, ϕ and γ are physical observables and subscripts denote partial differentiation. Owing to the miscellaneous geometrical and physical applications of Eq. (1), it is worth investigating such a system both from the viewpoint of its integrability properties and from the viewpoint of the existence of stable fractal excitations. In fact, as the aforementioned system appears in combined reaction and diffusion processes such as the development of highly complex organisms based upon nonlinear interactions between common genes, it is arguably dubbed throughout this chapter as the coupled nonlinear extension of the reaction-diffusion (CNLERD) system.

The slowing of atoms by use of cooling apparatus produces a singular quantum state known as Bose–Einstein condensate (BEC) [29, 30]. This phenomenon was predicted in 1925 by generalizing Bose's work [29] on the statistical mechanics of (massless) photons to (massive) atoms. The result of the efforts of Bose and Einstein [29, 30] is the concept of a Bose gas, governed by Bose–Einstein statistics, which describes the statistical distribution of identical particles with integer spin, now known as bosons. Since then, Bose–Einstein condensation has long been a key element of macroscopic quantum phenomena such as superconductivity and superfluidity. Investigations with two-species Bose–Einstein condensates [31] are actually useful in performing experiments with two-component matter-wave solitons [32–34]. As a matter of fact, second-harmonic generation with two-component BEC has been subject to much interests [35, 36]. In this configuration, the interaction between the two species plays a significant role in determining the dynamics of the clouds [37, 38]. Such experimental realization of a two-component BEC [32, 33] has also stimulated considerable attention in a quasi one-dimensional regime [39, 40] when the Gross–Pitaevskii (GP) equations for two interacting BEC reduce to coupled nonlinear Schrödinger (CNLS) equations in an extended form of the Manakov system [41]. In the wake of such interests, Schumayer and Apagyi [42] have discussed the possible forms of the external potentials of a binary BEC while investigating the complete integrability of the one-dimensional coupled GP system. In our investigation, we extend the concept of Bose–Einstein solitons [43–45] to the case of multi-species condensates within a higher-dimensional manifold. We show the possibility of producing higher-dimensional Manakov fractal patterns [41, 46–48] with matter waves, and of using them in the design of wave-switching devices. As an illustration, the dynamics of binary mixture species are modeled by the following normalized coupled system [49]

$$\imath\bar{\psi}_t + \Delta\bar{\psi} + \bar{\varrho}\left(|\bar{\psi}|^2 + \bar{\alpha}|\bar{\phi}|^2\right)\bar{\psi} = 0, \quad \imath\bar{\phi}_t + \Delta\bar{\phi} + \bar{\varrho}\left(\bar{\alpha}|\bar{\psi}|^2 + |\bar{\phi}|^2\right)\bar{\phi} = 0, \tag{2}$$

where $\bar{\psi}$ and $\bar{\phi}$ are complex-valued physical observables, functions of the independent real-valued variables t, x and y. The operator Δ stands for the Laplacian in two-dimensional space expressed as $\Delta = \partial_x^2 + \partial_y^2$. The quantity $\bar{\varrho} = \pm 1$ is the parameter which in nonlinear optics distinguishes between self-focusing or self-defocusing Kerr nonlinearity whereas in BEC, it represents the attractive or repulsive interactions between the species. The value of the constant $\bar{\alpha}$ varies over a wide range [50, 51]. In BEC, the variable t is the time, while in optics it is the propagation distance. For BEC or spatial optical solitons, the independent variables x and y are transverse coordinates; for spatiotemporal optical solitons in a two-dimensional waveguide with anomalous chromatic dispersion, the variable y is the "local time". In the context of BEC, Eq. (2) is usually called the (2+1)-dimensional coupled GP equation [52–54].

In the wake of theoretical [55–57] and experimental [58–60] investigations of soliton propagation of magnetic polaritons in ferromagnetic media, many interests are being paid to the further survey of the dynamics of such a highly nonlinear system (see refs. [61–66] and references therein). The theoretical investigations are made around two main considerations: the longwave and the modulational asymptotic models [61–66]. The latter are wave envelope soliton equations which represent nonlinear modulation of a wave train under the main assumption that the wave number of the wave envelope is much smaller than that of the carrier wave, meaning a slowly varying envelope approximation. This approximation in a lower-dimensional system leads to the (1+1)-dimensional NLS equation [67, 68]. Extending such a study to higher-dimensional systems, it has been shown that the short-wave approximation accounts in particular for the propagation of line solitons which stability has been studied for certain values of the soliton parameter [69]. Taking into account of both damping and demagnetizing fields in real ferromagnets, and regarding the instability of the constant background of the line solitons, the previous model [69] is rather improved to the following (2+1)-dimensional most accurate system given by

$$\varphi_{xt} - \varphi\chi_x - \varphi_{yy} + \chi_y + s\varphi_x = 0, \quad \chi_{xt} + \varphi\varphi_x - \chi_{yy} - \varphi_y = 0, \tag{3}$$

where quantities φ and χ are physical observables, parameter s is a damping constant. Besides, system (3) actually paves a route for a survey of restabilizing the polariton short solitary waves against transverse damping perturbations. Following the above interests, we look forward to studying the integrability properties of Eq. (3) and discuss the possible unwrapping of magnetic polaritons with fractal structure.

Moreover, waves propagation along the matter, that is, matter waves, can also be found in some kinds of ferromagnetic state matter such as spinor BEC within an optical trap potential. The dynamics of BEC is actually a subject of great interests to experimentalists and theorists alike. The mean-field theoretical description of BEC is based on GP equations which correctly predict matter-wave solitons in various configurations of condensates with attractive and repulsive interactions [70]. In the presence of an optical trap, one-dimensional spinor condensates exhibit ferromagnetic and polar stationary states, whose stability depends on the scattering lengths of atomic collisions in different angular-momentum channels [71–73]. Matter-wave solitons in a new field of atom optics [74] are expected to be useful for applications in atom laser, atom interferometry, and coherent atom transport. One interesting feature of atomic BEC is the many internal degrees of freedom of atoms liberated under an optical trap [75–77] which enrich the multiplicity of signals. Such properties have been investigated recently [71–73, 75–80]. Combining the above fascinating properties of atomic BEC to hyperfine spin state $f = 1$, an integrable model of a multidimensional spinor BEC in one dimension allowing an exact description of the dynamics of bright solitons has been proposed by Ieda, Miyakawa and Wadati [81] and given by

$$
\begin{aligned}
\imath\hbar\Phi_{\pm1,t} &= -\hbar^2\Phi_{\pm1,xx}/2m + (c_0 + c_2)(|\Phi_{\pm1}|^2 + |\Phi_0|^2)\Phi_{\pm1} \\
&\quad +(c_0 - c_2)|\Phi_{\pm1}|^2\Phi_{\pm1} + c_2\Phi_{\pm1}^\star\Phi_0^2, \tag{4a} \\
\imath\hbar\Phi_{0,t} &= -\hbar^2\Phi_{0,xx}/2m + c_2|\Phi_0|^2\Phi_0 + (c_0 + c_2)(|\Phi_{+1}|^2 + |\Phi_{-1}|^2)\Phi_0 \\
&\quad +2c_2\Phi_0^\star\Phi_{+1}\Phi_{-1}, \tag{4b}
\end{aligned}
$$

where $\Phi_{\pm1}$ and Φ_0 represent the three hyperfine states corresponding to $f = 1$. Parameters c_0 and c_2 are the coupling constants. When the mean-field interaction is attractive ($c_0 < 0$) and the spin-exchange interaction is ferromagnetic ($c_2 < 0$), the system (4) possesses a completely integrable point. By considering a reduction of results from the IST, the exact multibright soliton solutions to the system with the spin-exchange interaction is unearthed [81]. In fact, the spin mixing within condensates [78–80] happening during the head-on collision is generated from the spin-exchange interaction between these components, which is absent in the system owing to the frozen spin degrees of freedom under additional magnetic fields [44, 82]. Analyzing the collision law for two-soliton solutions, it is found that the soliton dynamics can be explained in terms of the spin precession [81]. Generalizing the above equations to a higher-dimensional system, we aim at investigating some particular solutions with fractal support satisfying to these generalized coupled equations.

Despite the difficulty caused by the lack of superposition principles, the last few decades have seen revolutionary progress in solving nonlinear systems, guided by advances in experiments, phenomenal success in the computer simulation of nonlinear systems, and new mathematical analytical tools. There is a rich theory of evolution equations which is mostly devoted to the problem of integration and to the study of the underlying algebraic and analytic structures. Whether a given equation is integrable or not, whether the powerful machinery developed for integrable equations can be applied to an equation of our particular interest in a concrete problem, constitute a very challenging problem to establish. The investigation of the exact solutions to nonlinear evolution equations plays an important role in the study of nonlinear physical phenomena. For example, the wave phenomena observed in fluid dynamics, plasma and elastic media are modeled by the bell-shaped sech-solutions and the kink-shaped tanh-solutions. The exact solutions, if available, to those nonlinear equations facilitate the verification of numerical solvers and aid in the stability analysis of solutions. In the past few decades, there has been significant progress in the development of various methods. Among them are IST, Darboux transformation, BT, Hirota's bilinearization (HB) method, mapping and deformation approaches, sine-cosine method, standard and extended truncated Painlevé (P)-analysis, just to name a few [83–88]. This last method, particularly useful in unearthing integrability properties of nonlinear evolution systems, has been performed to NLPDEM equations by Weiss, Tabor and Carnevale (WTC) [89, 90] through a formalism stating that a NLPDEM equation possesses the P-property if its solutions are single-valued about a movable singularity manifold. The remarkable feature of the P-analysis, particularly for soliton solutions, is that a natural connection exists between Lax pairs, BT, HB and Miura transformation, which can be constructed through the expansion of the solutions about the singularity manifold.

Following the development of the WTC-formalisms [89, 90] to a set of evolution systems (see refs. [91, 92] and references therein), many interests have been paid to the construction of corollaries for solving such equations, such as the multilinear variable separation approach (MLVSA) which has this peculiarity that it deals with arbitrary functions endowing integrable MLVSA solvable models quite rich localized excitations such as the solitoffs, dromions, lumps, breathers, instantons, ring solitons, peakons, compactons, localized chaotic and fractal patterns [93]. According to the MLVSA, the solutions to solvable models are expressed in terms of a field called potential U which expression is given by

$$U = \frac{-2\Delta q_y p_x}{(a_0 + a_1 p + a_2 q + a_3 pq)^2}, \quad \Delta = a_0 a_3 - a_1 a_2, \tag{5}$$

where the quantities $p \equiv p(x, t)$ and $q \equiv q(y, t)$ are arbitrary functions with a_k ($k = 0, \ldots, 3$) being arbitrary constants. In the "universal" formula (5), the appearance of the arbitrary functions p and q for some MLVSA solvable models, is closely related to the arbitrary boundary conditions of some types of the quantities for the related models. The effects of arbitrary boundary conditions have been considered for some (2+1)-dimensional integrable systems. As illustration, the exact solutions to the Davey–Stewartson system with arbitrary boundary data have been studied by Fokas and Santini [94–96]. Such investigation has also been carried out through the (2+1)-dimensional sine-Gordon system by Konopelchenko and Dubrovsky [97, 98].

In another hand, scrutinizing the geometrical structure of the aforementioned pattern formations, it sometimes appears that these systems split into parts, each of which is (at least approximately) a reduced-size copy of the whole [99]. Such a property is called "self-similarity" and is fundamental in characterizing a kind of matter with rough or fragmented geometric shape, known as "fractal". In fact, this term was coined by Mandelbrot in 1975 [99] as derived from the Latin *fractus* meaning "broken" or fractured. A mathematical fractal is based on an equation that undergoes iteration, a form of feedback based on recursion [100]. By fractal, we mean the following features [100–102]

- Fine structures at arbitrary small scales.

- Too irregular structure to be easily described in traditional Euclidean geometry language.

- Self-similar patterns (at least approximately or stochastically).

- Hausdorff dimension greater than the topological dimension of the system (although this requirement is not met by space-filling curves such as the Hilbert curve).

- Simple and recursive definition.

Because they appear similar at all levels of magnification, fractals are often considered to be infinitely complex. Since the inception of the concept of "fractal" proposed by Mandelbrot [99], it has seen the most remarkable developments in mathematical science and has had important influences on many fields such as biology, physics, social science, just to name a few [103]. For instance, in physics, this concept has been widely used through fractal growth phenomena including diffusion-limited aggregation, viscous fingering, cracks, and the physical properties such as diffusion, flow, vibration magnetism of fractal structures with percolation clusters, polymers, porous media, among others [103]. Alongside the above acquaintance of fractals to the realms of mathematics and computer graphics, they also exist nearly everywhere in nature, such as in tree branching, cloud structures, galaxy clustering, fern shapes, human veins, leaves, music, coastlines, fluid turbulence, crystal growth patterns [104–106]. Although fractals are found in many natural structures through various features, one important query rely on how to provide a rough analytical characterization of such patterns. In the wake of such query, many attempts on the subject can be performed by selecting different types of lower-dimensional fractal models while constructing beautiful higher-dimensional fractal patterns which can be useful in many designs such as costume design, architecture, system modeling, and others. In addition, music can be composed by choosing appropriate forms of the arbitrary functions such as p and q. In the future, perhaps the most famous artists will also be the most famous physicists and mathematicians. Usually, solitons, chaos and fractals are the most important three parts of nonlinear science [107–113]. Though solitons are the representatives of integrable systems while the two others are on the behalf of nonintegrable systems, Tang, Chen and Lou [114] have found chaotic and fractal structures for the lump and dromion solutions to the (2+1)-dimensional dispersive long wave system which is integrable by IST. Thus, the question of what on earth the integrability definition is, arguably casts on one's mind.

In general, there are four main techniques for generating fractals as follows.

- *Escape-time fractals.* Sometimes known as "orbits" fractals, these structures are defined by a formula or recurrence relation at each point in a space such as the complex plane. As examples, we have the Mandelbrot set, the Julia set, the Burning Ship fractal, the Nova fractal and the Lyapunov fractal, just to name a few. The two-dimensional vector fields that are generated by one or two iterations of escape-time formulae also give rise to a fractal form when points are passed through this field repeatedly.

- *Iterated function systems.* They are characterized by a fixed geometric replacement rule. As illustration, we have the Cantor set, Sierpinski carpet, Sierpinski gasket, Peano curve, Koch snowflake, Harter-Highway dragon curve, T-Square, Menger sponge, among others.

- *Random fractals.* Such systems are generated by stochastic rather than deterministic processes. For example, we have trajectories of the Brownian motion, Lévy flight, fractal landscapes and the Brownian tree yielding mass- or dendritic fractals such as diffusion-limited aggregation or reaction-limited aggregation clusters.

- *Strange attractors.* These systems are generated by iteration of a map or the solution to a system of initial-value differential equations that exhibit chaos.

Taking advantage of the previous generating techniques, we aim at unearthing quite number of typical fractal structures to the set of physical systems described above, while performing the MLVSA to these solvable models. As mentioned above, following the recent investigation of many types of solitons and other coherent structures with chaotic (sensitive dependence on the initial conditions) and fractal (self-similar structures) behaviors in integrable or nonintegrable (2+1)-dimensional systems [115–126], additionally, as lower-dimensional arbitrary functions are present in the exact or approximate solutions to these model systems, we can use lower-dimensional fractal solutions in view of obtaining higher-dimensional fractal pattern formations to such models. Throughout this chapter, we classify our fractal patterns through three considerations based upon the expressions of a lower-dimensional arbitrary function Θ as follows [120–124].

(a) *Nonlocal fractal excitations.* In this case, the lower-dimensional arbitrary function Θ reads

$$\Theta(\xi, t) = \prod_{j=1}^{2} \lambda_j \theta_j |\theta_j| \left\{ \sin[\ln(\theta_j^2)] - \cos[\ln(\theta_j^2)] \right\}, \tag{6}$$

with constant λ_j being an arbitrary parameter and $\theta_j \equiv k_j \xi - v_j t + \theta_{0j}$. Quantities ξ, k_j, and v_j refer to generalized space-like variable, wave number, and velocity of the j-wave component, respectively. Constant θ_{0j} is an arbitrary parameter.

(b) *Fractal dromion and lump excitations.* In such a configuration, the dromion-like (lump-like) structure is exponentially (algebraically) localized in large scale and possesses self-similar structure near the center of the pattern. The function Θ can be expressed as

$$\Theta(\xi, t) = \exp \left\{ -c\theta \left\{ \theta + \sin[\ln(\theta^2)] - \cos[\ln(\theta^2)] \right\} \right\}, \tag{7}$$

with $\theta \equiv k\xi - vt + \theta_0$, θ_0 being an arbitrary parameter, and constant c is a real negative-definite parameter, while describing fractal dromion. But also, we can find

$$\Theta(\xi, t) = |\theta| \left\{ \sin[\ln(\theta^2)] - \cos[\ln(\theta^2)] \right\}^2 / (1 + \theta^4), \tag{8}$$

for fractal lump solution.

(c) *Stochastic fractal excitations.* Well-known is the stochastic fractal property of the continuous but nowhere differentiable Weierstrass function $\wp$ defined as

$$\wp(\xi; t) = \sum_{j=0}^{N} \alpha^{-j/2} \sin(\beta^j \theta), \quad N \to \infty, \tag{9}$$

with constants α and β being arbitrary parameters. A stochastic fractal excitation can be expressed as

$$\Theta(\xi, t) = \sum_{i,j} R_i(\theta_i) R_j(\theta_j), \tag{10}$$

where $R_i = \wp(\theta_i) + \theta_i^2 + \mu_i$, $\mu_i \in \mathbb{R}$.

- *Stochastic fractal dromion/solitoff excitations.* Such structures are obtained by including the Weierstrass function into the dromion solution as follows

$$\Theta(\xi, t) = \kappa + \sum_{j=0}^{M} \eta_j \wp(\theta_j) \tanh^{\nu_j}(\theta_j), \tag{11}$$

with θ_j defined above and, κ, η_j and $\nu_j > 0$ being arbitrary parameters.

- *Stochastic fractal lump excitations.* Such kinds of patterns are obtained by including the Weierstrass function into the lump solution. Alternatively, Eq. (10) is reduced as

$$\Theta(\xi, t) = \sum_{j=0}^{\bar{N}} \varrho_j R_j(\theta_j), \tag{12}$$

with R_j defined above, and ϱ_j being an arbitrary parameter.

In the wake of the above considerations, the structure of the present chapter which sections are treated independently, is organized as follows. In Sec. II, we present the integrability properties of the CNLERD equation, followed by the construction of its fractal composite solutions. In Sec. III, we discuss the different fractal structures, solutions to the $(2 + 1)$-dimensional GP coupled system, while carrying out its singularity structure analysis from the viewpoint of WTC-formalism. Based upon this formulation, in Sec. IV, we also construct solutions with fractal supports to the $(2 + 1)$-dimensional polariton equation. In Sec. V, extending this concept to the spinor condensates, we investigate the dynamics of a $(2+1)$-dimensional spinor model system of $f = 1$ hyperfine spin state while unwrapping its integrability properties by means of the P-analysis. Finally, we end this work with a brief summary through paving the ways to further and underlying surveys.

2. The Coupled Nonlinear Extension of the Reaction-Diffusion System and Its Fractal Patterns

First of all, it is worth noting that with the transformation $\partial_x = \partial_y$, Eq. (1) straightforwardly reduces to a $(1+1)$-dimensional coupled NLPDEM equation of diffusion type investigated by Nakayama [127] while surveying from the viewpoint of a geometrical approach the motion of curves in hyperboloid in the Minkowski space [127].

The development of highly complex organisms is an intriguing and fascinating problem. The genetic material is the same in each cell of an organism. One question that is worth investigating, therefore, is formulated as follows: How do cells produce spatial patterns under the influence of their common genes? To provide an answer based upon nonlinear interactions of at least two chemicals and on their diffusion, regarding autocatalysis and long-range inhibition as fundamental phenomena, Koch and Meinhardt [128] investigated some simple models of isotropic systems that describe the generation of patterns out of an initially nearly homogeneously state. Owing to the complexity of such biological systems, we aim at shedding light on the genuine anisotropic nature of these systems which would be very helpful in the understanding of the dynamical behavior of these patterns.

Following the pioneer work of Turing [129] on the interactions of two substances with different rates, Gierer and Meinhardt [130] and independently Seger and Jackson [131] depicted two important features playing a central role in pattern formation: the local self-enhancement and long-range inhibition. Such features give rise to some important typical classification of biological apparatus, namely; the activator-inhibitor systems describing at least stripelike patterns in monkey and zebra, faceted eye of drosophila flies, just to name a few [132]; the activator-substrate systems describing at least reticulated dragonflies, animal coat patterns and also Brusselator systems [133–135]; and finally the biochemical switches [136–139]. Alongside such classifications, other kinds of interactions are possible, mediated for instance by mechanical forces [140, 141], by electric potentials [142, 143] or surface contact between cell membrane [144]. Cellular automata are also often used to explain the emergence of inhomogeneous patterns [134]. Nevertheless, chemical interactions coupled with the exchange of molecules, are the main motor of primary pattern genesis in biological systems.

Following the seminal work of Koch and Meinhardt [128], taking into account the anisotropic properties of such systems, the models for complex biological pattern formation are described by the following reaction-diffusion equation

$$U_{kt} = \nabla \cdot \mathbf{J}_k + R_k(U),\tag{13}$$

where the flow $\mathbf{J}_k$ is given by

$$\mathbf{J}_k = \bar{\bar{\mathbf{D}}}_k \nabla U_k.\tag{14}$$

The quantity $\bar{\bar{\mathbf{D}}}_k$ $(k \in \mathbb{N})$ stands for the diffusion-stress which components represent the diffusion cœfficients with respect to a specified direction. The quantities R_k $(k \in \mathbb{N})$ are functions of the dynamical fields $U = \{U_k\}_{k\in\mathbb{N}}$ characterizing the nonlinear reactions among them. The gradient operator ∇ is expressed as $\nabla = (\partial_1, \partial_2, \partial_3, \cdots)$ with $\partial_i = \partial/\partial x_i$, x_i $(i \in \mathbb{N})$ being the spacelike coordinates. It is noted that in some cases where all the diffusion cœfficients are different, the system can straightforwardly evolve towards an instability known as Turing instability [129] and complicated patterns related to morphogenesis, reaction front dynamics, self-organization henceforward emerge [145–147]. It is also worth noting that negative diffusion cœfficients deserve to be studied both mathematically [148, 149] and physically [150]. Indeed, by considering a simple case consisting of two (1+1)-dimensional interacting chemicals with the diffusion-stresses given by $\bar{\bar{\mathbf{D}}}_1 = \begin{pmatrix} 1 & 0 \\ 0 & 0 \end{pmatrix}$ and $\bar{\bar{\mathbf{D}}}_2 = \begin{pmatrix} -1 & 0 \\ 0 & 0 \end{pmatrix}$, and the nonlinear functions $R_1 = 2aU_1 - 2U_1^2 U_2$ and $R_2 = 2U_1 U_2^2 - 2aU_2$, the constant a being a fixed parameter, this system is shown to be gauge equivalent to a (1+1)-dimensional gravity [149], but also emerges as a good model for complex biological organisms such as zebra stripes or butterfly stripes where the nonlinear interacting cells are behaving as damped-oscillators from the viewpoint of a thermo-field approach [150]. Extending such a study to a (2+1)-dimensional system where the diffusion-stresses $\bar{\bar{\mathbf{D}}}_1$ and $\bar{\bar{\mathbf{D}}}_2$ are given by

$$\bar{\bar{\mathbf{D}}}_1 = \begin{pmatrix} 1 & -1 \\ 1 & -1 \end{pmatrix}, \quad \bar{\bar{\mathbf{D}}}_2 = \begin{pmatrix} -1 & -1 \\ 1 & 1 \end{pmatrix},\tag{15}$$

and considering the nonlinear functions R_1 and R_2 as

$$R_1 = [\partial_2\chi - \partial_1\chi - a(t)]U_1, \quad R_2 = [a(t) + \partial_1\chi - \partial_2\chi]U_2, \tag{16}$$

where the spaceless quantity $a \equiv a(t)$ stands for a time-dependent function while the pseudopotential quantity χ is introduced in order to characterize a typical nonlinear interaction between the two observables U_1 and U_2 as $U_1U_2 = \partial_1\chi + \partial_2\chi - b(t)$ (the spaceless quantity $b \equiv b(t)$ depending on time), we derive a model valuable for a better understanding of the properties of some natural biological organisms such as zebra stripes, reticulated dragonflies, butterfly stripes, faceted eye of drosophila flies, and constituting typical nonlinear systems of different kinds such as activator-substrate, activator-inhibitor, just to name a few. Following a variable transformation of the form $x = (x_1 + x_2)/2$ and $y = (x_2 - x_1)/2$, the system consisting of Eqs. (13)–(16) reduces to Eq. (1), provided $U_1 = \psi$, $U_2 = \phi$ and $\gamma = a(t) + \partial_1\chi - \partial_2\chi$. Furthermore, the extension towards a (2+1)-dimensional manifold is interesting both from the viewpoint of the investigation of its integrability properties and from the viewpoint of the existence of stable pattern formations.

2.1. Singularity structure analysis of the (2+1)-dimensional coupled nonlinear extension of the reaction-diffusion system

According to the Painlevé-analysis generalized to NLPDEM equations by Weiss et al. [89, 90], a NLPDEM equation is said to possess the Painlevé-property if the only singularities of the general integral which can be found on arbitrary non-characteristic ('movable') hypersurfaces are poles. The Painlevé-test has proved to be a useful criterion for the identification of completely integrable NLPDEM system.

In fact, the Painlevé-analysis essentially consists of a local analysis of the structure of the solution to a NLPDEM equation in the neighborhood of a movable singularity in the complex space and identifies conditions under which it is free from movable critical singularities especially of the logarithmic branch-point type or certain complex branchpoint and irrational types (so called dense branching). This investigation follows three fundamental steps:

- Determination of the leading-order behavior of the Laurent expansion of the solution to the NLPDEM equation under interests. It is required that the leading orders being negative-valued integers for the system to pass the Painlevé-test ("resonance" values).

- Determination of the "resonances" expressing the powers at which arbitrary functions can enter into the Laurent expansion while satisfying to some compatibility relations ("resonance" conditions).

- Verification that a "sufficient" number of arbitrary functions exist in the Laurent series solution without the introduction of movable critical points. The word "sufficient" meaning in this context that the number of arbitrary functions is at least equal to the number of "resonances" and eventually their multiplicities ("sufficiency" condition).

In other words, if the singular manifold is determined by

$$g(x, y, t) = 0, \quad g_{xy}(x, y, t) \neq 0, \tag{17}$$

and $u(x, y, t)$ is a solution to a NLPDEM equation, it is required that

$$u = \sum_{k=0}^{\infty} u_k g^{k+s},\tag{18}$$

where $u_0 \neq 0$, $g = g(x, y, t)$ and $u_k = u_k(x, y, t)$, (k being a nonzero integer) are functions of (x, y, t) in a neighborhood of the manifold given by equation (17), and s is a negative integer.

Concretely, if Eq. (1) is P-integrable, then all the available solutions to the model is written in the full Laurent series as follows

$$\psi = \sum_{k=0}^{\infty} \psi_k g^{k+\alpha}, \quad \phi = \sum_{k=0}^{\infty} \phi_k g^{k+\beta}, \quad \gamma = \sum_{k=0}^{\infty} \gamma_k g^{k+\varsigma},\tag{19}$$

with sufficient arbitrary functions among ψ_k, ϕ_k, γ_k and g. The constants α, β and ς should be negative integers. This means that the previous solutions are written as single-valued expressions among the arbitrary singularity manifold.

The formal way to find the constant α, β and ς is known as the standard leading order analysis. Thus, truncating the previous series given by Eq. (19) to the zeroth order, and then replacing them into Eq. (1) such as to compare the leading order terms for $g \sim 0$, we find only one possible branch

$$\alpha = \beta = -1, \quad \varsigma = -2,\tag{20}$$

and

$$\psi_0 \phi_0 = 2g_x^2, \quad \gamma_0 = -2g_x g_y.\tag{21}$$

This implies that one of the three functions ψ_0, ϕ_0 and γ_0 is arbitrary. In general, there is no restriction on the valuedness of these functions. In fact, they can be real-or complex-valued expressions. Nonetheless, throughout the section, the arbitrary functions shall be regarded as real-valued expressions.

In order to obtain the recursion relations to determine the functions ψ_k, ϕ_k and γ_k, we substitute Eqs. (19), (20) and (21) into (1). This leads us to the following algebraic system

$$\mathcal{M}_k \mathcal{V}_k = \mathcal{T}_k,\tag{22}$$

where $\mathcal{M}_k$ is a square matrix, $\mathcal{V}_k = (\psi_k, \phi_k, \gamma_k)^T$ and $\mathcal{T}_k = (P_k, Q_k, U_k)^T$ with,

$$
\begin{aligned}
P_k = {}& -\sum_{j=1}^{k-1} \gamma_{k-j} \psi_j - \psi_{k-2,xy} - \psi_{k-2,t} \\
& -(k-2)\left(\psi_{k-1}g_t + \psi_{k-1,x}g_y + \psi_{k-1,y}g_x + \psi_{k-1}g_{xy}\right),
\end{aligned}\tag{23}
$$

$$
\begin{aligned}
Q_k = {}& -\sum_{j=1}^{k-1} \gamma_{k-j} \phi_j - \phi_{k-2,xy} + \phi_{k-2,t} \\
& -(k-2)\left(-\phi_{k-1}g_t + \phi_{k-1,x}g_y + \phi_{k-1,y}g_x + \phi_{k-1}g_{xy}\right),
\end{aligned}\tag{24}
$$

and

$$U_k = -\left[\sum_{j=0}^{k-1}(\psi_{k-j-1}\phi_j)_y + (k-2)g_y\sum_{j=1}^{k-1}\psi_j\phi_{k-j} + \gamma_{k-1,x}\right].\tag{25}$$

The matrix $\mathcal{M}_k$ is given by

$$\mathcal{M}_k = \left[\begin{array}{ccc} A_{1k} & A_{2k} & A_{3k} \\ B_{1k} & B_{2k} & B_{3k} \\ C_{1k} & C_{2k} & C_{3k} \end{array}\right],\tag{26}$$

with

$$A_{1k} = k(k-3)g_xg_y, \quad A_{2k} = 0, \quad A_{3k} = \psi_0,\tag{27}$$

$$B_{1k} = 0, \quad B_{2k} = A_{1k}, \quad B_{3k} = \phi_0,\tag{28}$$

$$C_{1k} = (k-2)\phi_0g_y, \quad C_{2k} = (k-2)\psi_0g_y, \quad C_{3k} = (k-2)g_x.\tag{29}$$

Thus, the determinant Δ_k of the matrix $\mathcal{M}_k$ is given by

$$\Delta_k = k(k-2)(k-3)(k-4)(k+1)g_y^2g_x^3.\tag{30}$$

The resonances are found at

$$k = -1, 0, 2, 3, 4.\tag{31}$$

The resonance at $k = -1$ corresponds to that of the singularity manifold g being arbitrary. If the model is P-integrable, we require four resonance conditions at $k = 0, 2, 3, 4$, which are satisfied identically such that the other four arbitrary functions among ψ_k, ϕ_k and γ_k can be introduced into the general series expansion given by Eq. (19). From the leading order analysis, we know that the resonance at $k = 0$ is satisfied identically and one of ψ_0, ϕ_0 and γ_0, is arbitrary.

For $k = 1$, ψ_1, ϕ_1 and γ_1 are explicitly found as follows

$$\psi_1 = -\frac{2g_xg_t\psi_0 + 2g_xg_y\psi_{0,x} - \psi_0^2\phi_{0,y}/2 + \psi_{0,y}g_x^2}{4g_yg_x^2},\tag{32a}$$

$$\phi_1 = -\frac{2g_xg_y\phi_{0,x} - \phi_0^2\psi_{0,y}/2 + \phi_{0,y}g_x^2 - 2g_xg_t\phi_0}{4g_yg_x^2},\tag{32b}$$

$$\gamma_1 = 2g_{xy}.\tag{32c}$$

For $k = 2$, one of ψ_2, ϕ_2 and γ_2 is arbitrary.

For $k = 3$, we also find that one of ψ_3, ϕ_3 and γ_3 is arbitrary.

Finally, for the resonance at $k = 4$, one of ψ_4, ϕ_4 and γ_4 is arbitrary. Consequently, the (2+1)-dimensional coupled NLERD Eq. (1) possesses a sufficient number of arbitrary functions. We conclude that this system is P-integrable. Its complete integrability will be established if some essential properties such as BT, Lax-Pairs, Hirota's bilinearization [85, 151], just to name a few, are derived.

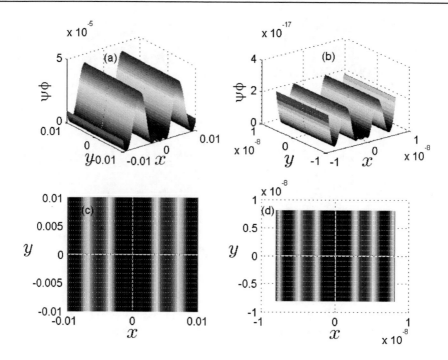

Figure 1. Nonlocal fractal excitations depicted at $t = 0$ by the observable $\psi\phi$ which expression is given by Eqs. (42) and (43). In this case, the parameters are selected as $a_0 = 1$, $a_1 = 1$, $a_2 = 1$, $a_3 = 1/2$, and $\gamma_2 = 0$ such that: -For $p(x,t) = \Theta(x,t)$, $\lambda_1 = 1/4$, $\theta_{01} = 0$ $\lambda_2 = 0$, $k_1 = 1$, and $v_1 = 1$. -For $q(y,t) = \Theta(y,t)$, $\lambda_1 = 1/4$, $\lambda_2 = 0$, $\theta_{02} = 0$, $k_2 = 1$, and $v_2 = 0$. Panels (a) and (b) represent the pattern formations depicted in 3D-perspective, and the the two others (c) and (d) are their corresponding densities represented within the square regions $[-1 \cdot 10^{-2}, 1 \cdot 10^{-2}] \times [-1 \cdot 10^{-2}, 1 \cdot 10^{-2}]$ and $[-8 \cdot 10^{-9}, 8 \cdot 10^{-9}] \times [-8 \cdot 10^{-9}, 8 \cdot 10^{-9}]$, respectively.

2.2. Bäcklund transformation and Hirota's bilinearization of the (2+1)-dimensional coupled nonlinear extension of the reaction-diffusion system

It is well known that the P-analysis is also useful in searching for other interesting properties [89, 90] of a given system. In this section, we use the truncated P-expansion to derive the BT and Hirota's bilinearization [85, 151] of the (2+1)-dimensional coupled NLERD Eq. (1). Thus, setting

$$\psi_k = \phi_k = \gamma_{k+1} = 0, \quad k \geq 2, \tag{33}$$

Eq. (19) is transformed as follows

$$\psi = \psi_0/g + \psi_1, \quad \phi = \phi_0/g + \phi_1, \quad \gamma = \gamma_0/g^2 + \gamma_1/g + \gamma_2. \tag{34}$$

Substituting Eq. (34) into (1) yields,

$$\psi_{0,xy} + \psi_{0,t} + \psi_1\gamma_1 = -\psi_0\gamma_2, \quad \phi_{0,xy} - \phi_{0,t} + \phi_1\gamma_1 = -\phi_0\gamma_2, \tag{35}$$

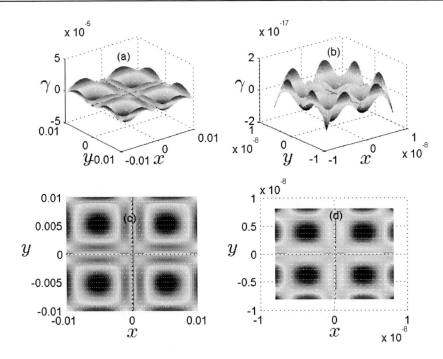

Figure 2. Nonlocal fractal excitations depicted at $t = 0$ by the observable γ which expression is given by Eqs. (42) and (43). In this case, the parameters are selected as $a_0 = 1$, $a_1 = 1$, $a_2 = 1$, $a_3 = 1/2$, and $\gamma_2 = 0$ such that: -For $p(x,t) = \Theta(x,t)$, $\lambda_1 = 1/4$, $\lambda_2 = 0$, $\theta_{01} = 0$, $k_1 = 1$, and $v_1 = 1$. -For $q(y,t) = \Theta(y,t)$, $\lambda_1 = 1/4$, $\lambda_2 = 0$, $\theta_{02} = 0$, $k_2 = 1$, and $v_2 = 0$. Panels (a) and (b) represent the pattern formations depicted in $3D$-perspective, and the the two others (c) and (d) are their corresponding densities represented within the square regions $[-1 \cdot 10^{-2}, 1 \cdot 10^{-2}] \times [-1 \cdot 10^{-2}, 1 \cdot 10^{-2}]$ and $[-8 \cdot 10^{-9}, 8 \cdot 10^{-9}] \times [-8 \cdot 10^{-9}, 8 \cdot 10^{-9}]$, respectively.

and

$$\psi_{1,t} + \psi_{1,xy} + \gamma_2 \phi_1 = 0, \quad \phi_{1,t} - \phi_{1,xy} - \gamma_2 \phi_1 = 0, \quad \gamma_{2,x} + (\phi_1 \psi_1)_y = 0. \quad (36)$$

From Eq. (36), it follows that $\{\psi_1, \phi_1, \gamma_2\}$ is a solution of the (2+1)-dimensional coupled NLERD Eq. (1). In order words, the truncated expansion (34) actually stands for a BT. A seed solution is written as follows

$$\psi_1 = \phi_1 = 0, \quad \gamma_2 \equiv \gamma_2(y,t). \quad (37)$$

This seed solution is a simple one and is actually useful for constructing many other solutions. For other existing seed solutions, many other classes of solutions are derived. It is that property of the P-method for constructing various kinds of solutions by means of arbitrary functions that makes it potentially and powerfully underlying. The solutions are given by the Eq. (34) expressed in a truncated form. Due to the arbitrariness of these functions, many solutions are constructed in a straightforward way, provided to solve analytically or

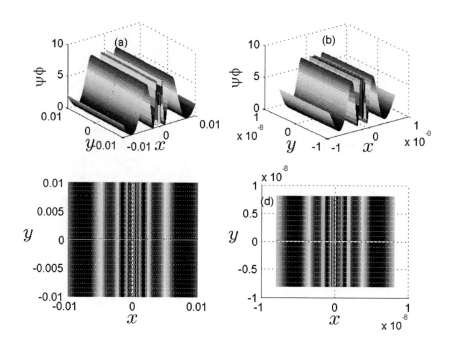

Figure 3. Typical fractal excitations depicted at $t = 0$ by the observable $\psi\phi$ which expression is given by Eqs. (42) and (43). In this case, the parameters are selected as $a_0 = 1$, $a_1 = 1$, $a_2 = 1$, $a_3 = 2$, $\theta_0 = 0$, $k = 1$, and $\gamma_2 = 0$, and lower-dimensional fractal dromion functions are chosen as follows: -For $p(x,t) = \Theta(x,t)$, $c = 1$, and, $v = 1$. -For $q(y,t) = \Theta(y,t)$, $c = 1$, and $v = 0$. Panels (a) and (b) represent the pattern formations depicted in $3D$-perspective, and the the two others (c) and (d) are their corresponding densities represented within the square regions $[-1 \cdot 10^{-2}, 1 \cdot 10^{-2}] \times [-1 \cdot 10^{-2}, 1 \cdot 10^{-2}]$ and $[-8 \cdot 10^{-9}, 8 \cdot 10^{-9}] \times [-8 \cdot 10^{-9}, 8 \cdot 10^{-9}]$, respectively.

numerically some NLPD constraint equations. Many examples will be given in the next subsection while studying the interactions between such structures. Using the seed solution given by Eq. (37), Eqs. (32a) and (32b) are written in the following compact form

$$\mathcal{A}\mathcal{V}_{0,x} + \mathcal{B}\mathcal{V}_{0,y} + \mathcal{C}\mathcal{V}_0 = 0, \tag{38}$$

where $\mathcal{V}_0 = (\psi_0, \phi_0)^T$ and

$$\mathcal{A} = \begin{bmatrix} 2g_x g_y & 0 \\ 0 & 2g_x g_y \end{bmatrix}, \quad \mathcal{B} = \begin{bmatrix} g_x^2 & -\dfrac{\psi_0^2}{2} \\ -\dfrac{\phi_0^2}{2} & g_x^2 \end{bmatrix}, \quad \mathcal{C} = \begin{bmatrix} 2g_x g_t & 0 \\ 0 & -2g_x g_t \end{bmatrix}. \tag{39}$$

Thus, solving Eq. (38) by means of the characteristics method, it yields

$$\mathcal{V}_0 = \mathcal{G}_0\left(x - \int \frac{dy}{\mathcal{A}^{-1}\mathcal{B}}\right)\exp\left(\int \frac{dx}{\mathcal{C}^{-1}\mathcal{A}}\right), \tag{40}$$

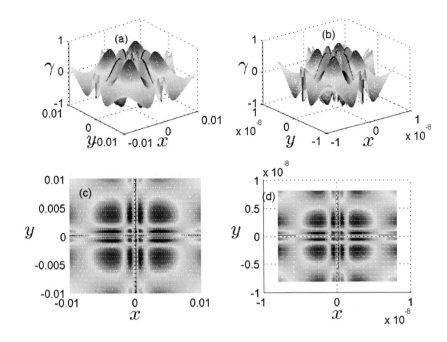

Figure 4. Fractal dromion excitations depicted at $t = 0$ by the observable γ which expression is given by Eqs. (42) and (43). In this case, the parameters are selected as $a_0 = 1$, $a_1 = 1$, $a_2 = 1$, $a_3 = 2$, $\theta_0 = 0$, $k = 1$, and $\gamma_2 = 0$ such that: -For $p(x,t) = \Theta(x,t)$, $c = 1$, and, $v = 1$. -For $q(y,t) = \Theta(y,t)$, $c = 1$, and $v = 0$. Panels (a) and (b) represent the pattern formations depicted in $3D$-perspective, and the the two others (c) and (d) are their corresponding densities represented within the square regions $[-1 \cdot 10^{-2}, 1 \cdot 10^{-2}] \times [-1 \cdot 10^{-2}, 1 \cdot 10^{-2}]$ and $[-8 \cdot 10^{-9}, 8 \cdot 10^{-9}] \times [-8 \cdot 10^{-9}, 8 \cdot 10^{-9}]$, respectively.

where $\mathcal{G}_0$ stands for an arbitrary array-function of $\left(x - \int \frac{dy}{A^{-1}B}\right)$ to be determined.

Now, substituting Eqs. (34) and (37) into (1), the following bilinear system is derived as follows

$$(D_t + D_x D_y - \nu)\psi_0 \cdot g + \psi_0 \gamma_1 = 0, \quad (41a)$$

$$(-D_t + D_x D_y - \nu)\phi_0 \cdot g + \phi_0 \gamma_1 = 0, \quad (41b)$$

$$(D_x D_y - \mu) g \cdot g - \gamma_0 = 0, \quad (41c)$$

$$D_x \gamma_0 \cdot g + D_y (\psi_0 \phi_0) \cdot g + g D_x \gamma_1 \cdot g = \gamma_0 g_x + \psi_0 \phi_0 g_y, \quad (41d)$$

where $\nu - \gamma_2 - \mu = 0$ and μ, ν, δ and ϱ stand for arbitrary quantities to be determined. We note that the symbols D_x, D_y and D_t represent the Hirota's operators [85,151] with respect to the variables x, y and t. By expanding the functions g, ψ_0, ϕ_0, γ_0, γ_1 and γ_2 as formal power series, and using them in the system (41), the one-, two- and N-soliton solutions (N being an integer) to the system (1) can straightforwardly be constructed. However, such solutions will be studied in detail in a separate paper. Knowing that the BT of the system

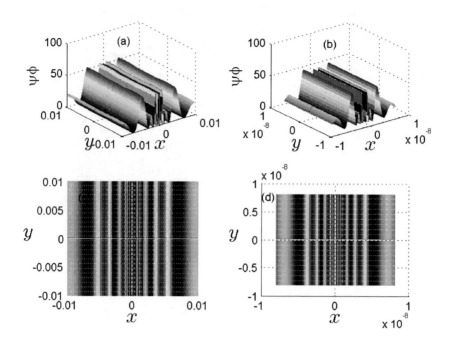

Figure 5. Fractal Lump excitations depicted at $t = 0$ by the observable $\psi\phi$ which expression is given by Eqs. (42) and (43). In this case, the parameters are selected as $a_0 = 1$, $a_1 = 1$, $a_2 = 1$, $a_3 = 2$, $\theta_0 = 0$, $k = 1$, and $\gamma_2 = 0$ such that: -For $p(x,t) = \Theta(x,t)$, $v = 1$. -For $q(y,t) = \Theta(y,t)$, $v = 0$. Panels (a) and (b) represent the pattern formations depicted in 3D-perspective, and the the two others (c) and (d) are their corresponding densities represented within the square regions $[-1 \cdot 10^{-2}, 1 \cdot 10^{-2}] \times [-1 \cdot 10^{-2}, 1 \cdot 10^{-2}]$ and $[-8 \cdot 10^{-9}, 8 \cdot 10^{-9}] \times [-8 \cdot 10^{-9}, 8 \cdot 10^{-9}]$, respectively.

(1) has been found and its related Hirota's bilinearization derived, we conclude that the (2+1)-dimensional coupled NLERD Eq. (1) is completely integrable. This confirms the power of the 'prolongation structure' coined by Wahlquist and Estabrook [25] establishing the integrability properties of a given NLPDEM equation. Nonetheless, the usefulness of the WTC-formalism also stems from its allowance to construct interesting solutions based upon the previous results of the P-analysis.

Now, owing to the arbitrariness of some functions derived from the P-analysis, there are many interesting solutions to investigate. Thus, we convey our attention to solutions for which the quantities $\psi\phi$ and γ are expressed as follows

$$\psi\phi = 2\left(\partial_x \ln|g|\right)^2, \quad \gamma = \gamma_2 + \frac{D_x D_y g \cdot g}{g^2}, \qquad (42)$$

and which stem from Eqs. (21) and (34). From a seed solution to system (36), as given by Eq. (37), it is seen that this solution does not depend on whether γ_1 is different from zero or not. Thus, by setting $\gamma_1 = 0$ and using Eq. (32c), one gets $g_{xy} = 0$ which shows that g is the sum of two arbitrary functions $g_1 \equiv g_1(x,t)$ and $g_2 \equiv g_2(y,t)$. Now, considering

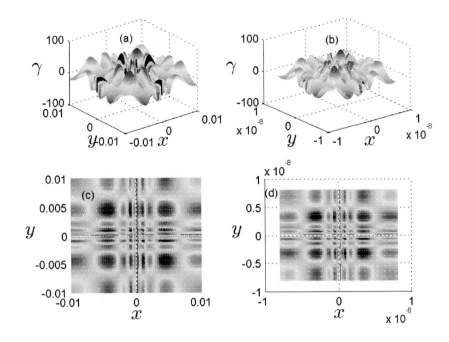

Figure 6. Fractal lump excitations depicted at $t = 0$ by the observable γ which expression is given by Eqs. (42) and (43). In this case, the parameters are selected as $a_0 = 1$, $a_1 = 1$, $a_2 = 1$, $a_3 = 2$, $\theta_0 = 0$, $k = 1$, and $\gamma_2 = 0$ such that: -For $p(x,t) = \Theta(x,t)$, $v = 1$. -For $q(y,t) = \Theta(y,t)$, $v = 0$. Panels (a) and (b) represent the pattern formations depicted in $3D$-perspective, and the the two others (c) and (d) are their corresponding densities represented within the square regions $[-1 \cdot 10^{-2}, 1 \cdot 10^{-2}] \times [-1 \cdot 10^{-2}, 1 \cdot 10^{-2}]$ and $[-8 \cdot 10^{-9}, 8 \cdot 10^{-9}] \times [-8 \cdot 10^{-9}, 8 \cdot 10^{-9}]$, respectively.

the case where $\gamma_1 \neq 0$, and searching for a class of solutions generalizing the previous ones such that $\gamma_1 = f_1(x,t)f_2(y,t)$ with $f_1(x,t)$ and $f_2(y,t)$ being arbitrary functions, from Eq. (32b), the function g is expressed as follows

$$g = a_0 + a_1 p + a_2 q + a_3 pq, \qquad (43)$$

where $p \equiv p(x,t)$ and $q \equiv q(y,t)$ stand for arbitrary functions and the parameters a_i ($i = 0, 1, 2, 3$) are arbitrary constants. With the above form of g given by Eq. (43), we combine the two Eqs. (35) and (40) in order to get a nonlinear system expressed in terms of $\mathcal{G}_0$ and γ_2 which can be solved analytically or numerically. For simplicity, it is interesting to take $\gamma_2 = 0$ as considered in the next section while investigating the scattering behavior of some localized excitations.

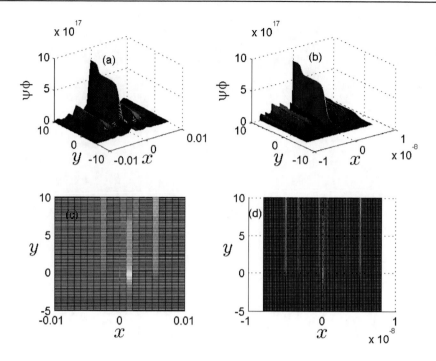

Figure 7. Typical Stochastic fractal solitoff excitations depicted at $t=0$ by the observable $\psi\phi$ which expression is given by Eqs. (42) and (43). In this case, the parameters are selected as $a_0 = 1$, $a_1 = 1$, $a_2 = 1$, $a_3 = 2$, $\alpha = 3/2$, $\beta = 3/2$, and $\gamma_2 = 0$ such that $p(x,t) = \Theta(x,t)$ and $q(y,t) = \tanh(y)/5 + \tanh(2y-15)/4$, with $\kappa = 2$, $M = 1$, $\eta_0 = 0$, $\eta_1 = 1/2$, $\eta_m = 0$ $(m \geq 2)$, $\nu_1 = 1$, $k_1 = 4$, $v_1 = -1$, $\theta_{01} = -20$. Panels (a) and (b) represent the pattern formations depicted in $3D$-perspective, and the the two others (c) and (d) are their corresponding densities represented within the square regions $[-1 \cdot 10^{-2}, 1 \cdot 10^{-2}] \times [-5, 10]$ and $[-8 \cdot 10^{-9}, 8 \cdot 10^{-9}] \times [-5, 10]$, respectively.

2.3. Discussion of the localized and stochastic excitations with fractal structures

From the quantum mechanics's viewpoint, when a body moves from one state level to another one, it can gain or loose energy in terms of quantum quantities $nh\nu$, n being a nonzero integer, h the Planck's constant and ν the frequency of the radiation. During the process of absorbing or emitting radiation, it can generate miscellaneous pattern formations such as the local deformation of the manifold in which it is embedded, its shape-changing during the interaction with its neighbors, just to name a few. Classically, the word "excitation" is related to the different features of a system which is submitted to perturbations from its "stationary" state. Owing to the different properties of the manifolds in which these features are embedded, there mainly exist two kinds of "excitations", i.e., the lower-dimensional "excitations" and the higher-dimensional "excitations". More significantly, by lower-dimensional and higher-dimensional excitations, we mean excitations derived from evolution systems in (1+1)-dimensional and $(N+1)$-dimensional $(N \geq 2)$ manifolds, respectively. It is important to point out the genuine difference between the "excitations" of a system and the solutions to a model equation governing the spatio-temporal dynamical behavior of the system. In fact, "excitations" use to appear in the perturbed system as induced pattern formations which can be expressed in terms of the solutions to the model equations governing the dynamics of the system. The powerfulness of the Painlevé-analysis is twofold specifically for coupled systems. In fact, although it stands for a method particularly useful in assessing the integrability properties of a given system, the Painlevé-analysis also paves the way for construction of miscellaneous "excitations" amongst nontravelling or travelling "localized" and periodic waves in spite of their physical meaning to be clarified for applications.

When we talk of "Localized excitations", we refer to physical structures which either have vanishing tails or possess compact supports. The physical interests of such structures stem from the "localized" features of their energy density which ascribes to the system its particle nature giving rise to interesting applications. Recently, great interests have been paid to many (2+1)-dimensional equations ([152–154] and references therein) and a rich diversity of coherent structures such as the folded solitary waves (FSW) have been found. These structures are more abundant than those of (1+1)-dimensional cases because some types of arbitrary functions have been included in the solution expressions [155–157].

Basically, the general procedure of searching for miscellaneous solutions to an evolution systems passing the Painlevé-analysis relies on the arbitrariness of the singularity manifold and the functions entering the Laurent expansion (see [91, 92] and references therein). Following the pioneering work of Tang and Lou (see [152] and references therein) on the subject, one fundamental step is to express the above functions in terms of arbitrary quantities by means of a "variable separation approach" (see [152] and references therein). The next interesting step is to make advantage of these arbitrariness while constructing miscellaneous excitations from the known solutions to some lower-dimensional integrable models, naturally, provided to solve analytically or numerically some related conditions expressed in the form of NLPDEM equations.

By selecting appropriate parameters and arbitrary functions entering into the expression of the manifold g above, we can construct an interesting set of pattern formations. As illustration, in Fig. 1, the $\psi\phi$-observable at $t = 0$ varies self-similarly. Indeed, in a $3D$-representation, the features presented in panel $1(a)$ within the space region $[-1 \cdot 10^{-2}, 1 \cdot 10^{-2}]^2 \times \psi\phi$ and those depicted in $1(b)$ within region $[-8 \cdot 10^{-9}, 8 \cdot 10^{-9}]^2 \times \psi\phi$ are close to each other. Such a similarity in the profile is clearly shown in panels $1(c)$ and $1(d)$ representing their density plots, respectively. By pursuing further with different scales of x and y to the reducing orders of 10^{-14} and lesser, we obtain similar pictures. The same observations are made when representing the features of the γ-observable both in planar and $3D$-perspectives. The nonlocal structures depicted in panels $2(a)$ and $2(b)$ both showing four holes are similarly represented in blue colors in their density plots, respectively. The dynamics of the aforementioned pattern formations can be investigated merely by depicting their profiles at different evolving times. In fact, through such dynamical behaviors, the structures describe traveling waves with unit x-velocity. Interestingly, one can survey the interactions among these waves and see how elastically such traveling structures behave during the scattering process. Such study will constitute a worth investigating issue to carry out within a separate work.

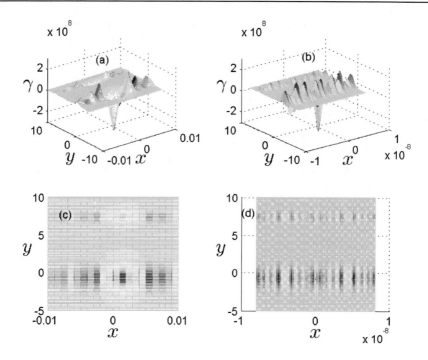

Figure 8. Stochastic fractal dromion excitations depicted at $t = 0$ by the observable γ which expression is given by Eqs. (42) and (43). In this case, the parameters are selected as $a_0 = 1$, $a_1 = 1$, $a_2 = 1$, $a_3 = 2$, $\alpha = 3/2$, $\beta = 3/2$, and $\gamma_2 = 0$ such that $p(x,t) = \Theta(x,t)$ and $q(y,t) = \tanh(y)/5 + \tanh(2y - 15)/4$, with $\kappa = 2$, $M = 1$, $\eta_0 = 0$, $\eta_1 = 1/2$, $\eta_m = 0$ ($m \geq 2$), $\nu_1 = 1$, $k_1 = 4$, $v_1 = -1$, $\theta_{01} = -20$. Panels (a) and (b) represent the pattern formations depicted in $3D$-perspective, and the the two others (c) and (d) are their corresponding densities represented within the square regions $[-1 \cdot 10^{-2}, 1 \cdot 10^{-2}] \times [-5, 10]$ and $[-8 \cdot 10^{-9}, 8 \cdot 10^{-9}] \times [-5, 10]$, respectively.

Alongside the previous structures, by choosing appropriately lower-dimensional arbitrary dromion functions and suitable parameters as presented in the caption of Fig. 3, we obtain a typical pattern formation which is exponentially localized in a large scale of x and y. Nonetheless, in order to better present the self-similarity in structure of such fractal patterns, we enlarge a small region near the centre of the previous figure. For instance, by reducing the region of panel 3(a) $(x, y) \in [-1 \cdot 10^{-2}, 1 \cdot 10^{-2}]^2$ to $[-8 \cdot 10^{-9}, 8 \cdot 10^{-9}]^2$ of panel 3(b), we obtain a totally similar structure with density plots represented in panels 3(c) and 3(d), respectively. The process can continue further with lesser reducing orders of x and y independent space variables. Such fractal dromion excitation is more expressive in Fig. 4 where features of γ-observables are depicted in both planar and $3D$-perspectives. From these pictures, one can see how the four smooth peaks distributed around the center in panels 4(a) and 4(b) are projected similarly in the xy-plane such as to obtain a density plot as shown in panels 4(c) and 4(d). Globally, it appears that the fractal dromions have relatively higher amplitudes than those of the nonlocal fractal patterns.

The lump excitation (algebraically localized in all directions) is another type of signifi-

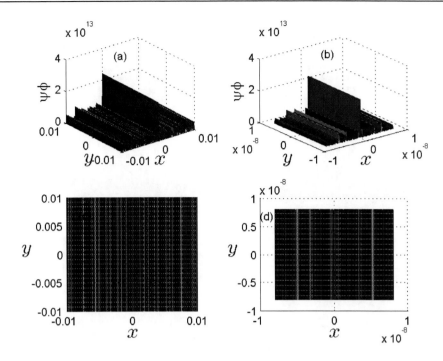

Figure 9. Typical Stochastic fractal excitations depicted at $t=0$ by the observable $\psi\phi$ which expression is given by Eqs. (42) and (43). In this case, the parameters are selected as $a_0 = 1$, $a_1 = 1$, $a_2 = 1$, $a_3 = 2$, $\alpha = 3/2$, $\beta = 3/2$, $\bar{N} = 2$, and $\gamma_2 = 0$, and lower-dimensional stochastic fractal lump functions are chosen as follows: -For $p(x,t) = \Theta(x,t)$, $k_1 = 1$, $v_1 = 1$, $\theta_{01} = 0$, $\varrho_0 = 0$, $\varrho_1 = 1$, $\varrho_2 = 0$, and $\mu_1 = 1000$. -For $q(y,t) = \Theta(y,t)$, $k_2 = 1$, $v_2 = 0$, $\theta_{02} = 0$, $\varrho_0 = 0$, $\varrho_1 = 0$, $\varrho_2 = 1$, and $\mu_2 = 1000$. Panels (a) and (b) represent the pattern formations depicted in $3D$-perspective, and the the two others (c) and (d) are their corresponding densities represented within the square regions $[-1 \cdot 10^{-2}, 1 \cdot 10^{-2}] \times [-1 \cdot 10^{-2}, 1 \cdot 10^{-2}]$ and $[-8 \cdot 10^{-9}, 8 \cdot 10^{-9}] \times [-8 \cdot 10^{-9}, 8 \cdot 10^{-9}]$, respectively.

cant localized solution in high dimensions. With the details presented in the caption of Fig. 5, we obtain a typical pattern formation which is algebraically localized in a large scale of x and y. In fact, near the centre, there are infinitely many peaks which are distributed in a fractal manner. In order to see the fractal structure of the previous pattern, we look at the structure more carefully. Panel $5(a)$ presents the $3D$-representation of the fractal features of $\psi\phi$-observable and its corresponding density plot at region $(x,y) \in [-1 \cdot 10^{-2}, 1 \cdot 10^{-2}]^2$ is depicted in panel $5(c)$. Reducing such a region to $[-8 \cdot 10^{-9}, 8 \cdot 10^{-9}]^2$ shows self-similar structures in panels $5(b)$ and $5(d)$. More detailed studies show the self-similar structure of the previous fractal pattern. For instance, if we reduce the region of $5(b)$ to $[-8 \cdot 10^{-15}, 8 \cdot 10^{-15}]^2$, $[-8 \cdot 10^{-22}, 8 \cdot 10^{-22}]^2$, ..., we retrieve a totally similar structure to that plotted in $5(a)$. Such a fractal lump excitation where there are an infinite distribution of many peaks near the centre is better expressed in Fig. 6. These structures which look like fractal dromion patterns are actually localized in a large scale although for some

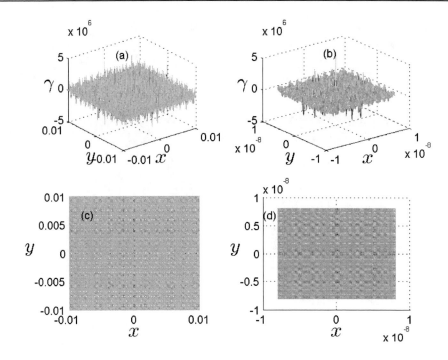

Figure 10. Stochastic fractal lump excitations depicted at $t = 0$ by the observable γ which expression is given by Eqs. (42) and (43). In this case, the parameters are selected as $a_0 = 1$, $a_1 = 1$, $a_2 = 1$, $a_3 = 2$, $\alpha = 3/2$, $\beta = 3/2$, $\bar{N} = 2$, and $\gamma_2 = 0$ such that: -For $p(x,t) = \Theta(x,t)$, $k_1 = 1$, $v_1 = 1$, $\theta_{01} = 0$, $\varrho_0 = 0$, $\varrho_1 = 1$, $\varrho_2 = 0$, and $\mu_1 = 1000$. -For $q(y,t) = \Theta(y,t)$, $k_2 = 1$, $v_2 = 0$, $\theta_{02} = 0$, $\varrho_0 = 0$, $\varrho_1 = 0$, $\varrho_2 = 1$, and $\mu_2 = 1000$. Panels (a) and (b) represent the pattern formations depicted in $3D$-perspective, and the the two others (c) and (d) are their corresponding densities represented within the square regions $[-1 \cdot 10^{-2}, 1 \cdot 10^{-2}] \times [-1 \cdot 10^{-2}, 1 \cdot 10^{-2}]$ and $[-8 \cdot 10^{-9}, 8 \cdot 10^{-9}] \times [-8 \cdot 10^{-9}, 8 \cdot 10^{-9}]$, respectively.

convenience we have only considered some selected regions. As it can be observed, the amplitudes of the aforementioned fractal lump pattern formations are greater than those of the previous fractal depictions.

In addition to the self-similar regular fractal dromion and lump excitations, the lower-dimensional stochastic fractal functions can also be used to construct high-dimensional stochastic fractal dromions and lump solutions. The most well known stochastic function is the so-called Weierstrass function which, when included in the dromion or lump solutions leads to stochastic fractal dromions and lumps. Figure 7 presents a typical stochastic fractal solitoff pattern derived from a lower-dimensional stochastic arbitrary fractal dromion. In fact, by expanding the planar region of Fig. 7 to larger scale shows that the previous structure is actually localized. Investigating the self-similarity in detail, we reduce the region $(x, y) \in [-1 \cdot 10^{-2}, 1 \cdot 10^{-2}] \times [-5, 10]$ of panel 7(a) to $(x, y) \in [-8 \cdot 10^{-22}, 8 \cdot 10^{-22}] \times [-5, 10]$ as shown in panel 7(b). Their densities presented at the bottom in panels 7(c) and 7(d), respectively, clearly present such a similarity in structure. In the wake of such analysis, the depiction of the features of the γ-observables at time $t = 0$ is also presented in Fig. 8, and as it can be witnessed, the stochastic fractal nature of this structure is conserved even in the very small scale of representation. In such a configuration, panels $8(a)$ and $8(b)$ actually describe fractal dromion excitations which amplitudes vary following a stochastic manner. These features would also be more expressive if the Weierstrass function was included in the expression of q. In Figs. 9 and 10, with the selecting parameters and suitable choices of lower-dimensional arbitrary stochastic fractal lump functions as presented in the captions of these figures, we obtain higher-dimensional stochastic lump excitations. Through the panels $9(a)$ and $9(b)$ depicting the variations of the $\psi\phi$-observable at $t = 0$, and panels $10(a)$ and $10(b)$ related to the γ-observable at the same initial time, the self-similarity in structure of these observables shows how the peaks are distributed stochastically within the regions $(x, y) \in [-1 \cdot 10^{-2}, 1 \cdot 10^{-2}]^2$ and $(x, y) \in [-8 \cdot 10^{-22}, 8 \cdot 10^{-22}]^2$. Globally, we can see that the aforementioned stochastic fractal structures have amplitudes greater than the previous ones.

In the wake of the previous surveys, it is important to see how fractal can the structure of some biological organisms be modeled. In fact, the observables depicted previously merely provide more information on that the nonlinear interactions between common genes of an organism satisfying to a reaction-diffusion equation can produce intriguing and fascinating pattern formations, particularly underlying in the understanding of the dynamical properties of some highly natural complex organisms such as zebra stripes, reticulated dragonflies, and butterfly stripes, just to name a few. Nonetheless, further attention should be payed to the scattering behavior of such structures for a better understanding of the appearance of some other kinds of patterns resulting probably from a scattering process between the aforementioned structures.

3. Theoretical Foundation for Multi-Species Condensates and Basic Equations

At very low temperatures, when the self-consistent field theory approximation is applicable [52], the evolution of interacting multi-species BEC can be described by a higher-dimensional coupled GP system [33, 39, 47, 48] from the viewpoint of two-body interactions between the condensates. Considering a (N+1)-dimensional manifold spanned by the N-spatial coordinates σ_k $(k = 1, \ldots, N)$ and time coordinate τ, the previous coupled GP system [33, 39, 47, 48] is given by

$$
\imath\hbar\Psi_{j\tau} = \left[-\hbar^2 \sum_{k=1}^{N} \partial_{\sigma_k}\partial_{\sigma_k}/2m_j + V_j(\sigma) + 2\pi\hbar^2 \sum_{l=1}^{M} a_{jl}|\Psi_l|^2/\mu_{jl} \right] \Psi_j, \tag{44}
$$

where the quantities m_j, V_j and a_{jl} $(j = 1, \ldots, M)$ are the mass of species, the external trap potential as a function of the vector-position σ of the j condensates and the scattering lengths of the respective atomic interactions. The function Ψ_j stands for an order-parameter representing the macroscopic wave function of the states of j component. The integer M represents the number of species. The constant μ_{jl} represents the reduced mass of a two-species system given by $\mu_{jl} = m_j m_l/(m_j + m_l)$. Considering the case where the potential consists of a superposition of an external magnetic trap providing a cigar-shaped condensate, the trapping potential V_j can be expressed in an asymmetric form as

$$V_j = m_j \Omega_j^2 \sum_{k=1}^{N} \lambda_k^2 \sigma_k^2 / 2, \qquad (45)$$

where the cœfficients λ_k are constants describing the anisotropy of the parabolic trap [158]. The parameter Ω_j is the linear oscillator frequency of the j component of the condensates. Indeed, different oscillator frequencies infer different locations of the corresponding condensates in the parabolic potential and thus, the effective densities of these components are different even in the case of an equal number of atoms. Consequently, even at approximately equal s-wave scattering length, the multicomponents will experience different nonlinearities proportional to the atomic densities.

What is also worth mentioning is the mode structure of the cigar-shaped BEC viewed as a waveguide for matter-waves. Compared with the nonlinear optical waveguides [159], the intrinsic nonlinearity of BEC results in energy distribution among modes. In the case of a weak nonlinearity, the main state of the condensate can be considered as a weakly modulated ground state, as it is clear that for a two-component BEC, the corresponding small parameter ϵ is the ratio of the energy of two-body interactions to the kinetic energy. Such parameter also expresses the ratio of the square linear oscillator length a_j of each j component to their corresponding square healing length ξ_j $(a_j \ll \xi_j)$. Assuming that the linear oscillator length a_j and the s-wave scattering lengths a_{jl} between atoms are all of the same order, a self-consistent transformation of the original higher-dimensional GP system to an effective coupled equations can be proceeded by means of the multiscale technique [160, 161].

Although nowadays different atomic species are used to form a two-component BEC, it is always of special interest both mathematically and physically [32–34, 39] to investigate the dynamical behavior of mixtures BEC trapped within a given potential in which these condensates have very close masses identical in the first-order of approximation. Physically, such situation can be found in atomic system of isotopes of the same element, but also it can correspond to a system of different spin states with the same mass m without any violation of the extended concept of species characterized fundamentally only by their order parameter Ψ_j $(j = 1, \dots, M)$. In the wake of such consideration, we introduce the following dimensionless variables

$$\sigma_k' = \sigma_k / a \quad (k = 1, \dots, N), \quad \tau' = \omega \tau, \quad \Phi_j = \sqrt{\frac{M a^N}{\mathcal{N}}} \Psi_j \quad (j = 1, \dots, M), \quad (46)$$

where constant $\mathcal{N}$ refers to the total number of atoms in the trap defined by $\mathcal{N} = \sum_{j=1}^{M} \int |\Psi_j|^2 d\sigma$, quantities a and ω are arbitrary constants. While avoiding any confusion with indices, it is important to note that throughout the paper, the indices k $(k = 1, \dots, N)$ refer to spatial coordinates of the $(N + 1)$-dimensional manifold whereas integers i $(i = 1, \dots, M)$, j $(j = 1, \dots, M)$ and l $(l = 1, \dots, M)$ represent the species of ith,

Dynamics of Miscellaneous Fractal Structures ... 175

j^{th} and l^{th} condensates, respectively. Thus, with Eq. (46), the system (44) is written as

$$\imath\Phi_{j\tau'} = \left[-\Delta' + \omega_j^2 \sum_{k=1}^{N} \lambda_k^2 \sigma_k'^2 + \sum_{l=1}^{M} g_{jl}|\Phi_l|^2 \right] \Phi_j, \quad (j = 1, \ldots, M) \tag{47}$$

where $\omega_j = \Omega_j/2\omega$, $g_{jl} = 4\pi\hbar\mathcal{N}a_{jl}/Mmwa^N \equiv 8\pi\mathcal{N}a_{jl}/Mwa^{N-2}$ and $\Delta' = \sum_{k=1}^{N} \partial_{\sigma_k'}\partial_{\sigma_k'}$, provided

$$a = \sqrt{\hbar/2m\omega}, \tag{48}$$

with the constant m representing the mass of the condensates. Taking into account the fact that ϵ is a small parameter and that the nonlinearity disappears in the limit $\epsilon \to 0$, we construct a solution to Eq. (44) perturbatively as a modulation of the solution to the linear problem associated to Eq. (47):

$$\mathcal{L}_{jn}\psi_{jn} = \mathcal{E}_{jn}\psi_{jn}, \tag{49}$$

where

$$\mathcal{L}_{jn} = -\Delta_0' + \omega_j^2 \sum_{k=1}^{N} \lambda_k^2 \sigma_{kn}'^2. \tag{50}$$

Equation (49) is the equation for a linear quantum oscillator whose solutions ψ_{jn} express the wave function of the oscillator associated to the energy $\mathcal{E}_{jn}$ in the state characterized by the three quantum numbers referred by the constant n [162].

On the particular statement of the problem, we concentrate on the most interesting case from practical point of view where ψ_{jn} is the ground state of the higher-dimensional linear oscillator such that [162]

$$\psi_{jn} \equiv \psi_{j0}(\sigma) = \left(\frac{\omega_j^N \prod_{k=1}^{N} \lambda_k}{\pi^N} \right)^{1/4} \exp\left(-\omega_j \sum_{k=1}^{N} \lambda_k \sigma_{k0}'^2/2 \right), \tag{51a}$$

$$\mathcal{E}_{jn} \equiv \mathcal{E}_{j0} = \omega_j \sum_{k=1}^{N} \lambda_k. \tag{51b}$$

We note that we have chosen the basis ψ_{jn} to be orthonormal:

$$\langle \psi_{jn}, \psi_{jn'} \rangle = \int_{-\infty}^{+\infty} \psi_{jn}^{\star}(\sigma_0)\psi_{jn'}(\sigma_0)d\sigma_0 = \delta_{nn'}, \tag{52}$$

where the star $(\star)$ refers to complex conjugation and the symbol δ is the Kronecker delta. The next steps are conventional for the multiple scale expansion [160, 161] where we introduce scaled independent variables $x_{kn} = \epsilon^n \sigma_k'$, $t_n = \epsilon^n \tau'$ (n integer). Looking for solutions in the form

$$\Phi_j = \alpha_j \sum_{k=1}^{N} \epsilon^n \Phi_j^{(n)}, \tag{53}$$

where the constants $\alpha_j (j = 1, \ldots, M)$ are arbitrary parameters, we collect all terms of the same order in ϵ after substituting Eq. (53) into (47) as follows:

- First order of ϵ:

$$\imath \Phi_{jt_0}^{(1)} - \mathcal{L}_{j0} \Phi_j^{(1)} = 0. \tag{54}$$

- Second order of ϵ:

$$\imath \Phi_{jt_0}^{(2)} - \mathcal{L}_{j0} \Phi_j^{(2)} = -\imath \Phi_{jt_1}^{(1)} - 2 \sum_{k=1}^{N} \Phi_{jx_{k0}x_{k1}}^{(1)}. \tag{55}$$

- Third order of ϵ:

$$\imath \Phi_{jt_0}^{(3)} - \mathcal{L}_{j0} \Phi_j^{(3)} = -\imath \Phi_{jt_2}^{(1)} - \sum_{k=1}^{N} \Phi_{jx_{k1}x_{k1}}^{(1)} - 2 \sum_{k=1}^{N} \Phi_{jx_{k0}x_{k2}}^{(1)}$$

$$+ \Phi_j^{(1)} \sum_{l=1}^{M} \tilde{g}_{jl} |\Phi_l^{(1)}|^2 - \imath \Phi_{jt_1}^{(2)} - 2 \sum_{k=1}^{N} \Phi_{jx_{k0}x_{k1}}^{(2)}, \tag{56}$$

where $\tilde{g}_{jl} = g_{jl}\alpha_l$ and $\mathcal{L}_{j0} = -\Delta_0' + \omega_j^2 \sum_{k=1}^{N} \lambda_k^2 x_{k0}^2$.

Since the stationary ground state ψ_{j0} function satisfies to the equation $\mathcal{L}_{j0} \psi_{j0} = \mathcal{E}_{j0} \psi_{j0}$, it is quite natural and straightforward while solving the system (54) to search for a weakly modulated linear ground state wave function $\Phi_j^{(1)}$ in the form

$$\Phi_j^{(1)} = Q_j(x_{k1}, t_1) \psi_{j0}(x_{k0}) \exp\left(-\imath \mathcal{E}_{j0} t_0\right) \quad (k = 1, \ldots, N), \tag{57}$$

where $Q_j(x_{k1}, t_1)$ is the modulating amplitude of the background state due to nonlinearity. In fact, the argument (x_{k1}, t_1) tells that Q_j is a function of a set of spatial $(x_{11}, x_{21}, x_{31}, \ldots)$ and temporal $(t_1, t_2, t_3, \ldots)$ variables in such a way that in the arguments of the modulating amplitude, only the most "rapid" variables are explicitly shown.

Besides, while looking forward solving the system (55), its solutions can be sought in the form of superposition of weakly modulating linear ground states ψ_{jn} $(n \neq 0)$ such as $\Phi_j^{(2)} = \sum_{n \neq 0} B_n(x_{k1}, t_1) \psi_{jn}(x_{k0}) \exp\left(-\imath \mathcal{E}_{j0} t_0\right) (k = 1, \ldots, N)$ which can be rewritten as

$$\Phi_j^{(2)} = \sum_n (1 - \delta_{n0}) B_n(x_{k1}, t_k) \psi_{jn}(x_{k0}) \exp\left(-\imath \mathcal{E}_{jn} t_0\right) \quad (k = 1, \ldots, N), \tag{58}$$

with the initial condition $\Phi_j^{(2)}(x_{k0}, 0)$ at $t_n = 0$ (n integer≥ 0). It should be noted that the correction $\Phi_j^{(2)}$ is orthogonal to ψ_{j0} according to Eq. (52) (since this is a generic property

of a perturbative expansion [162]) and consequently, $\Phi_j^{(1)}$ and $\Phi_j^{(2)}$ are orthogonal, i.e.
$\left\langle \Phi_j^{(1)}, \Phi_j^{(2)} \right\rangle = 0$. Applying the operator $\int\limits_{-\infty}^{+\infty} dx_{k0}\psi_{jn}(x_{k0})$ to Eq. (55), it comes

$$Q_{jt_1} = 0, \quad \Phi_j^{(2)} \equiv 0. \tag{59}$$

Finally, requiring orthogonality between the right-hand-side of Eq. (56) and the kernel of operator $i\partial_{t_0} - \mathcal{L}_{j0}$, and taking into account of the expression of $\Phi_j^{(1)}$ and $\Phi_j^{(2)}$ given above, the following coupled equation is derived:

$$iQ_{jt_2} + \sum_{k=1}^{N} Q_{jx_{k1}x_{k1}} + Q_j \left(\sum_{l=1}^{M} \beta_{jl}|Q_l|^2 \right) = 0, \tag{60}$$

where

$$\beta_{jl} = - \left(\frac{\prod\limits_{k=1}^{N} \lambda_k \frac{\omega_j\omega_l}{\omega_j+\omega_l}}{\pi^N} \right)^{1/2} \tilde{g}_{jl}. \tag{61}$$

Setting $t_2 = t$, $x_{11} = x$, $x_{21} = y$, $x_{31} = z,\ldots$, it comes

$$iQ_{jt} + \Delta Q_j + \left(\sum_{l=1}^{M} \beta_{jl}|Q_l|^2 \right) Q_j = 0, \tag{62}$$

where $\Delta = \sum\limits_{k=1}^{N} \partial_{x_{k1}}\partial_{x_{k1}} \equiv \partial_x^2 + \partial_y^2 + \partial_z^2 + \ldots$. Eq. (62) is regarded as the higher-dimensional CNLS system appearing as the parameter-dependent higher-dimensional time-gated Manakov system. Recently, Desyatnikov et al. [163] have investigated the multi-component vortex solutions to the system (62) in the context of physical experiments in photorefractive crystals and BEC.

If one considers condensates of, for instance, sodium 23 and rubidium 87, the non-linear interactions are due to elastic s-wave scattering amongst the atoms, and are effectively repulsive for both systems in which multicomponent condensates have been realized [164]. Very recently, there has been a tremendous interest in studying the dynamics of two-component BEC coupled to the environment using both theoretical and experimental means [165]. In the wake of such interests, for a such binary mixture of two-species condensates from the standpoint of a cigar-shaped condensate elongated in the planar space by the coordinates x and y, Eq. (62) reduces to the system (2) provided the following settings $Q_1 = \bar{\psi}$, $Q_2 = \bar{\phi}$ and $\bar{\varrho} = \beta_{11} = \beta_{22}$ hold. In this situation, we assume that $a_{12} = a_{21}$ which leads to $\beta_{12} = \beta_{21}$. We also note that if $\bar{\alpha} = 1$, then Eq. (2) is mapped onto the two-dimensional time-gated Manakov system [41]. Due to the arbitrary values of the parameters a, ω, ω_j and α_j, the quantities $\bar{\varrho}$ and $\bar{\alpha}$ can be properly controlled corresponding to a physical purpose of the problem statement. Since quantities $|\bar{\phi}|^2$ and $|\bar{\psi}|^2$ describe the local density of atoms, condensate excitations governed by Eq. (2) can be called matter wave excitations.

For further convenience, throughout the analytical survey of this section, we remove the bars over the dependent variables and parameters of Eq. (2).

3.1. Weiss–Tabor–Carnevale formalism to the (2+1)-dimensional CNLS equations

In order to investigate the integrability properties of Eq. (2), we rewrite this system in terms of four real-valued functions p, q, u and v defined by $\psi = p + \imath q$ and $\phi = u + \imath v$. Consequently, we derive the following equations

$$p_t + \Delta q + \varrho \left[p^2 + q^2 + \alpha \left(u^2 + v^2\right)\right] q = 0, \tag{63a}$$

$$-q_t + \Delta p + \varrho \left[p^2 + q^2 + \alpha \left(u^2 + v^2\right)\right] p = 0, \tag{63b}$$

$$u_t + \Delta v + \varrho \left[\alpha \left(p^2 + q^2\right) + u^2 + v^2\right] v = 0, \tag{63c}$$

$$-v_t + \Delta u + \varrho \left[\alpha \left(p^2 + q^2\right) + u^2 + v^2\right] u = 0. \tag{63d}$$

Following thee performed P-formalism established for NLPDEM equations by Weiss et al. [89, 90], in view of finding the powers at which some arbitrary functions could enter into the series, we consider the full Laurent series as follows

$$p = \sum_{k=0}^{\infty} p_k f^{k+\beta}, \quad q = \sum_{k=0}^{\infty} q_k f^{k+\gamma}, \quad u = \sum_{k=0}^{\infty} u_k f^{k+\kappa}, \quad v = \sum_{k=0}^{\infty} v_k f^{k+\lambda}. \tag{64}$$

Let us assume that the leading order behaviors are of the forms

$$p \sim p_0 f^{\beta}, \quad q \sim q_0 f^{\gamma}, \quad u \sim u_0 f^{\kappa} \quad v \sim v_0 f^{\lambda}, \tag{65}$$

where β, γ, κ and λ are constants to be determined, and $f(x, y, t)$ is the singularity manifold. Substituting Eq. (65) into (63) and equating the most dominant terms, we obtain the unique choice

$$\beta = \gamma = \kappa = \lambda = -1, \tag{66}$$

and

$$\varrho \left[p_0^2 + q_0^2 + \alpha \left(u_0^2 + v_0^2\right)\right] + 2 \left(f_x^2 + f_y^2\right) = 0, \tag{67a}$$

$$\varrho \left[\alpha \left(p_0^2 + q_0^2\right) + u_0^2 + v_0^2\right] + 2 \left(f_x^2 + f_y^2\right) = 0. \tag{67b}$$

This implies that two of the four functions p_0, q_0, u_0 and v_0 are arbitrary without any parametric constraints. Indeed, the two equations of the system (67) are equivalent if $\alpha = 1$ or $p_0^2 + q_0^2 = u_0^2 + v_0^2$. The other parameter ϱ takes arbitrary values. In a previous study, Zhang et al. [49] have shown that in the case of $\varrho = 1$ referring to the self-focusing Kerr nonlinearity, there are solutions with "bright" background having elastic and nonelastic properties. In the next, we shall consider the following values $\varrho = -1$ and $\alpha = 1$, referring to a physical system with self-defocusing Kerr nonlinearity. We aim at showing that alongside the class of 'dark' solitons satisfying to the (1+1)-dimensional NLS equations [166] with self-defocusing Kerr nonlinearity, the (2+1)-dimensional counterparts enrich the above class of solutions with miscellaneous patterns.

Substituting Eq. (64) into (63) and vanishing all the cœfficients of the powers f^k, we can obtain the recursion relations to determine the functions p_k, q_k, u_k and v_k as follows

$$\mathcal{M}_k \mathcal{V}_k = \mathcal{T}_k, \tag{68}$$

where $\mathcal{M}_k$ is a square matrix, $\mathcal{V}_k = (p_k, q_k, u_k, v_k)^T$ and $\mathcal{T}_k = (P_k, Q_k, U_k, V_k)^T$ with

$$
\begin{aligned}
P_k = {} & \Delta p_{k-2} - q_{k-2,t} - (k-2)q_{k-1}f_t + (k-2)[2p_{k-1,x}f_x + 2p_{k-1,y}f_y + p_{k-1}\Delta f] \\
& - \sum_{j=1}^{k-1} p_j \left(p_0 p_{k-j} + q_0 q_{k-j} + u_0 u_{k-j} + v_0 v_{k-j} \right) \\
& - \sum_{j=1}^{k-1} \sum_{i=0}^{j} p_i \left(p_{k-j}p_{j-i} + q_{k-j}q_{j-i} + u_{k-j}u_{j-i} + v_{k-j}v_{j-i} \right), \quad (69)
\end{aligned}
$$

$$
\begin{aligned}
Q_k = {} & \Delta q_{k-2} + p_{k-2,t} + (k-2)p_{k-1}f_t + (k-2)[2q_{k-1,x}f_x + 2q_{k-1,y}f_y + q_{k-1}\Delta f] \\
& - \sum_{j=1}^{k-1} q_j \left(p_0 p_{k-j} + q_0 q_{k-j} + u_0 u_{k-j} + v_0 v_{k-j} \right) \\
& - \sum_{j=1}^{k-1} \sum_{i=0}^{j} q_i \left(p_{k-j}p_{j-i} + q_{k-j}q_{j-i} + u_{k-j}u_{j-i} + v_{k-j}v_{j-i} \right), \quad (70)
\end{aligned}
$$

$$
U_k = P_k \bigg|_{\substack{p_k \leftrightarrow u_k \\ q_k \leftrightarrow v_k}} , \quad V_k = Q_k \bigg|_{\substack{p_k \leftrightarrow u_k \\ q_k \leftrightarrow v_k}} . \quad (71)
$$

Eq. (71) merely tells that the couples (U_k, V_k) and (P_k, Q_k) are equivalent through the substitution of p_k by u_k, and q_k by v_k, and reciprocally.

The matrix $\mathcal{M}_k$ is given by

$$
\mathcal{M}_k = \begin{bmatrix} A_{1k} & A_{2k} & A_{3k} & A_{4k} \\ A_{2k} & B_{1k} & B_{2k} & B_{3k} \\ A_{3k} & B_{2k} & C_{1k} & C_{2k} \\ A_{4k} & B_{3k} & C_{2k} & D_k \end{bmatrix}, \quad (72)
$$

with

$$
\begin{aligned}
A_{1k} &= 2p_0^2 - k(k-3)\left(f_x^2 + f_y^2\right), & A_{2k} &= 2p_0 q_0, & A_{3k} &= 2u_0 p_0, & (73a) \\
B_{1k} &= 2q_0^2 - k(k-3)\left(f_x^2 + f_y^2\right), & B_{2k} &= 2q_0 u_0, & A_{4k} &= 2v_0 p_0, & (73b) \\
C_{1k} &= 2u_0^2 - k(k-3)\left(f_x^2 + f_y^2\right), & C_{2k} &= 2u_0 v_0, & & & (73c) \\
B_{3k} &= 2q_0 v_0, & D_k &= 2v_0^2 - k(k-3)\left(f_x^2 + f_y^2\right). & & & (73d)
\end{aligned}
$$

Thus, the determinant Δ_k of the matrix $\mathcal{M}_k$ is given by

$$
\Delta_k = -(k-4)(k-3)^3(k+1)k^3 \left(f_x^2 + f_y^2\right)^4. \quad (74)
$$

The "resonances" are then found at

$$
k = -1, 0, 0, 0, 3, 3, 3, 4. \quad (75)
$$

The "resonance" at $k = -1$ corresponds to that of the singularity manifold f being arbitrary. If the model is P-integrable, it is required three "resonance" conditions at $k = 0, 3, 4$, which are satisfied identically such that the other seven arbitrary functions among p_k, q_k, u_k and v_k can be introduced into the general series expansion given by Eq. (64).

From the leading order analysis, we know that the "resonance" at $k = 0$ is satisfied identically and three of p_0, q_0, u_0 and v_0 are arbitrary.

For $k = 1$, we find that

$$p_1 = q_1 = u_1 = v_1 = 0, \tag{76}$$

provided that the system

$$p_0 f_t + q_0 \Delta f + 2 \left(q_{0x} f_x + q_{0y} f_y \right) = 0, \quad q_0 f_t - p_0 \Delta f - 2 \left(p_{0x} f_x + p_{0y} f_y \right) = 0, \tag{77a}$$
$$u_0 f_t + v_0 \Delta f + 2 \left(v_{0x} f_x + v_{0y} f_y \right) = 0, \quad v_0 f_t - u_0 \Delta f - 2 \left(u_{0x} f_x + u_{0y} f_y \right) = 0, \tag{77b}$$

holds.

We also see that the functions p_2, q_2, u_2 and v_2 vanish identically while p_0, q_0, u_0 and v_0 satisfy

$$p_{0t} + \Delta q_0 = 0, \quad q_{0t} - \Delta p_0 = 0, \quad u_{0t} + \Delta v_0 = 0, \quad v_{0t} - \Delta u_0 = 0. \tag{78}$$

On the other hand, solving the case $k = 3$, the following "resonance" condition is derived

$$p_0 p_3 + q_0 q_3 + u_0 u_3 + v_0 v_3 = 0, \tag{79}$$

so that three of the four functions p_3, q_3, u_3 and v_3 are arbitrary.

In the similar way, we easily check that any one of the functions p_4, q_4, u_4 and v_4 is arbitrary provided that the system

$$p_3 f_t + q_3 \Delta f + 2 \left(q_{3x} f_x + q_{3y} f_y \right) = 0, \quad q_3 f_t - p_3 \Delta f - 2 \left(p_{3x} f_x + p_{3y} f_y \right) = 0, \tag{80a}$$
$$u_3 f_t + v_3 \Delta f + 2 \left(v_{3x} f_x + v_{3y} f_y \right) = 0, \quad v_3 f_t - u_3 \Delta f - 2 \left(u_{3x} f_x + u_{3y} f_y \right) = 0, \tag{80b}$$

holds.

Let us briefly recap the above results. For $k = -1$ with multiplicity one, there is one arbitrary function. For $k = 0$ with multiplicity three, there are three arbitrary functions, similar to $k = 3$. And for $k = 4$ with multiplicity one, there is only one arbitrary function. The straightforward and easily way to check the existence of these arbitrary functions is to consider the full Laurent expansion as given by Eq. (64) while solving the system (68) under the conditions derived from each of the above "resonances". Since the number of arbitrary functions (eight) entering into the Laurent expansion is equal to the number of "resonances" with "degeneracy" (eight) as given by Eq. (75), according to the Painlevé-formalism, we can undoubtedly say that Eq. (63) admits "sufficient" number of arbitrary functions. Thus, this equation is Painlevé-integrable for the parametric restriction $(\varrho = -1, \alpha = 1)$ provided that Eqs. (77), (78), and (80) hold. It is well known that the Painlevé-analysis can also be used to obtain other interesting properties [89, 90]. In this paper, we use the truncated Painlevé-expansion to construct an interesting set of excitations.

Setting

$$p_k = q_k = u_k = v_k = 0, \quad k \geq 2, \tag{81}$$

Equation (64) becomes

$$p = p_0/f + p_1, \quad q = q_0/f + q_1, \quad u = u_0/f + u_1, \quad v = v_0/f + v_1, \tag{82}$$

where $\{p_1, q_1, u_1, v_1\}$ is also a solution of the (2+1)-dimensional CNLS equations. Thus, the truncated expansion given above is also a BT. Generally, in order to construct a typical family of solutions to Eq. (63) in a simple manner, it is useful to consider very simple expressions of p_1, q_1, u_1 and v_1. Thus, for convenience later, we select the seed solution

$$p_1 = q_1 = u_1 = v_1 = 0. \tag{83}$$

All the remainder equations to solve the functions p_0, q_0, u_0, v_0 and f are simplified to Eqs. (77) and (78) provided

$$p_0^2 + q_0^2 + u_0^2 + v_0^2 = 2\left(f_x^2 + f_y^2\right), \tag{84}$$

holds.

Using Eqs. (82) and (83) into (63), we derive

$$\left[\left(D_x^2 + D_y^2\right) f \cdot f + \mathcal{V}_0^T \mathcal{V}_0\right] \mathcal{V}_0 - f\left(\mathcal{A}D_t + D_x^2 + D_y^2\right) \mathcal{V}_0 \cdot f = 0, \tag{85}$$

where the operators D_x and D_t are defined by [85, 151, 167, 168]

$$D_t^m D_x^n(G \cdot F) = (\partial_t - \partial_{t'})^m (\partial_x - \partial_{x'})^n G(x, t) F(x', t')|_{x=x';t=t'}. \tag{86}$$

The constant matrix $\mathcal{A}$ is given by

$$\mathcal{A} = \begin{bmatrix} 0 & -1 & 0 & 0 \\ 1 & 0 & 0 & 0 \\ 0 & 0 & 0 & -1 \\ 0 & 0 & 1 & 0 \end{bmatrix}, \tag{87}$$

such that $\mathcal{A}\mathcal{A} = -\mathcal{I}$, $\mathcal{I}$ being the identity matrix.

Eq. (85) can be decoupled to the following bilinear equations

$$\left(D_x^2 + D_y^2 - \mu f D_x D_y\right) f \cdot f + \mathcal{V}_0^T \mathcal{V}_0 = 0, \tag{88a}$$
$$\left(\mathcal{A}D_t + D_x^2 + D_y^2\right) \mathcal{V}_0 \cdot f - \mu \mathcal{V}_0 D_x D_y f \cdot f = 0, \tag{88b}$$

where μ stands for an arbitrary nonzero parameter.

In particular, from Eq. (88), we have

$$p^2 + q^2 + u^2 + v^2 = 2\left(\mu f \partial_{xy}^2 - \Delta\right) \ln f. \tag{89}$$

Now, expanding the functions f and $\mathcal{V}$ as power series of an arbitrary perturbative parameter ε, and using these expressions in Eq. (88), we can construct the N-soliton solution according to the standard perturbative procedure due to Hirota's formalism [85, 151, 167, 168].

Nonetheless, the investigation of these N-soliton solution does not constitute the purpose of this paper. We shall briefly present the general scheme to derive such structures and we shall illustrate the simple case of one-soliton solution.

The P-test is actually useful in finding the BT and hence the Lax pairs such that the spectral problem and also the Hirota's bilinearization of the system can straightforwardly be written down. As illustration, even though the (1+1)-dimensional NLS equation has been solved by means of the IST method, its Painlevé-integrability has been established by Gibbon et al. [169] and its connection with the Hirota's bilinearization has been shown [169]. In the same way, in a previous study, Sahadevan et al. [170] have investigated the (1+1)-dimensional CNLS equations. Avoiding to consider the problem in full generality, they have found conditions at which these equations are integrable, and the BT and Hirota's bilinearization have been derived. Thus, whether a system does pass the Painlevé-test or partially passes this test under some conditions, the result from the Painlevé-test can be used to find out other integrability properties such as BT, Lax pairs, just to name a few, so that the question of integrability should be clearly investigated.

Now, before exploiting the bilinear form of the (2+1)-dimensional CNLS equations to construct the one-soliton solution, it is useful to make an emphasis: in the wake of the Painlevé-analysis of this coupled system, without any truncation procedure, the general solutions $\mathcal{S} \equiv (p, q, u, v)^T$ are written as follows

$$\mathcal{S} = \left(\mathcal{V}_0 + \sum_{k=1}^{\infty} \mathcal{V}_k f^k \right) / f. \tag{90}$$

In the Hirota's bilinearization, we use this result to write the solution $\mathcal{S} = \mathcal{Y}/f$ where $\mathcal{Y}$ is an arbitrary matrix expressed as

$$\mathcal{Y} = \mathcal{V}_0 + \sum_{k=1}^{\infty} \mathcal{V}_k f^k, \tag{91}$$

satisfying the bilinear Eq. (88), provided to replace $\mathcal{V}_0$ by $\mathcal{Y}$. The form of the general solution $\mathcal{S} = \mathcal{Y}/f$ given by Eqs. (90) and (91) looks like we have truncated the Painlevé-expansion to $\mathcal{V}_k = 0$, $k \geq 1$. Looking forward avoiding any confusion, we shall consider the bilinear Eq. (88) to be expressed in terms of $\mathcal{Y}$ instead of $\mathcal{V}_0$. Thus, in order to find the soliton solution to this bilinear system by means of the Hirota's method [85, 151, 167, 168], we consider the following expressions

$$\mathcal{Y} = \varepsilon \mathcal{Y}_{(1)} + \varepsilon^3 \mathcal{Y}_{(3)} + \varepsilon^5 \mathcal{Y}_{(5)} + \cdots, \quad f = 1 + \varepsilon^2 f_{(2)} + \varepsilon^4 f_{(4)} + \varepsilon^6 f_{(6)} + \cdots, \tag{92}$$

where $\mathcal{Y}_{(j)}$ and f_{j+1} $(j = 1, 3, 5)$ satisfy the following equations

1. order of ε,

$$(\Delta + \mathcal{A}\partial_t) \, \mathcal{Y}_{(1)} = 0, \tag{93}$$

2. order of ε^2,

$$(\mu \partial_{xy}^2 - \Delta) \, f_{(2)} - \frac{1}{2} \mathcal{Y}_{(1)}^T \mathcal{Y}_{(1)} = 0, \tag{94}$$

Dynamics of Miscellaneous Fractal Structures ...

3. order of ε^3,

$$(\Delta + \mathcal{A}\partial_t)\,\mathcal{Y}_{(3)} + \left(\mathcal{A}D_t + D_x^2 + D_y^2\right)\mathcal{Y}_{(1)} \cdot f_{(2)} - 2\mu\partial_{xy}^2 f_{(2)} = 0, \tag{95}$$

4. order of ε^4,

$$\left(\mu\partial_{xy}^2 - \Delta\right) f_{(4)} + \frac{1}{2}\left(\mu D_x D_y - D_x^2 - D_y^2\right) f_{(2)} \cdot f_{(2)}$$
$$+\mu f_{(2)}\partial_{xy}^2 f_{(2)} - \mathcal{Y}_{(1)}^T \mathcal{Y}_{(3)} = 0, \tag{96}$$

5. order of ε^5,

$$(\Delta + \mathcal{A}\partial_t)\,\mathcal{Y}_{(5)} + \left(\mathcal{A}D_t + D_x^2 + D_y^2\right)\left(\mathcal{Y}_{(1)} \cdot f_{(4)} + \mathcal{Y}_{(3)} \cdot f_{(2)}\right)$$
$$-\mu\mathcal{Y}_{(1)}\left(D_x D_y f_{(2)} \cdot f_{(2)} + 2\partial_{xy}^2 f_{(4)}\right)$$
$$-2\mu\mathcal{Y}_{(3)}\partial_{xy}^2 f_{(2)} = 0, \tag{97}$$

6. order of ε^6,

$$\left(\mu\partial_{xy}^2 - \Delta\right) f_{(6)} + \left(\mu D_x D_y - D_x^2 - D_y^2\right) f_{(2)} \cdot f_{(4)}$$
$$+\mu\left(f_{(2)}\partial_{xy}^2 f_{(4)} + f_{(4)}\partial_{xy}^2 f_{(2)} + \frac{1}{2}D_x D_y f_{(2)} \cdot f_{(2)}\right)$$
$$-\frac{1}{2}\left(\mathcal{Y}_{(3)}^T \mathcal{Y}_{(3)} + 2\mathcal{Y}_{(1)}^T \mathcal{Y}_{(5)}\right) = 0, \tag{98}$$

Now, in order to derive the one-soliton solution, it is useful to write $\mathcal{Y}_{(1)}$ as follows

$$\mathcal{Y}_{(1)} = \theta \exp(\eta) + \theta^\star \exp(\eta^\star), \tag{99}$$

where $\eta = k_1 x + k_2 y - \omega t + \eta_0$ and $\theta = (a, b, c, d)^T$, $k_1, k_2, \omega, \eta_0, a, b, c, d \in \mathbb{C}$ with the star symbol referring to complex conjugation. Thus, solving the Eq. (93) leads to the following system

$$\begin{bmatrix} k_1^2 + k_2^2 & \omega \\ -\omega & k_1^2 + k_2^2 \end{bmatrix}\begin{bmatrix} a \\ b \end{bmatrix} = \begin{bmatrix} 0 \\ 0 \end{bmatrix}, \quad \begin{bmatrix} k_1^2 + k_2^2 & \omega \\ -\omega & k_1^2 + k_2^2 \end{bmatrix}\begin{bmatrix} c \\ d \end{bmatrix} = \begin{bmatrix} 0 \\ 0 \end{bmatrix}. \tag{100}$$

Looking for nontrivial solutions to Eq. (100), the following dispersion equation is satisfied,

$$\left(k_1^2 + k_2^2\right)^2 + \omega^2 = 0. \tag{101}$$

The phase η now becomes

$$\eta = k_1 x + k_2 y - \bar{\epsilon}\imath\left(k_1^2 + k_2^2\right)t + \eta_0, \tag{102}$$

with $\bar{\epsilon} = \pm 1$. Besides, θ takes the following form

$$\theta = (a, -\bar{\epsilon}\imath a, c, -\bar{\epsilon}\imath c)^T. \tag{103}$$

Solving Eq. (94), it can easily be found that $f_{(2)}$ takes the following form,

$$f_{(2)} = \delta \exp\left(\eta + \eta^{\star}\right), \tag{104}$$

where

$$\delta = \frac{1}{2} \frac{|a|^2 + |c|^2}{\mu k_{1R} k_{2R} - k_{1R}^2 - k_{2R}^2}, \tag{105}$$

provided

$$\mu \neq \frac{k_{1R}^2 + k_{2R}^2}{k_{1R} k_{2R}}. \tag{106}$$

We note that k_{1R} and k_{2R} stand for real parts of the complex-valued wave numbers k_1 and k_2, respectively. We pay interests to the quantity $\mathcal{W} = \sqrt{p^2 + q^2 + u^2 + v^2}$. Two conditions are regarded as follows

1.

$$\mu < \frac{k_{1R}^2 + k_{2R}^2}{k_{1R} k_{2R}}, \tag{107}$$

$$W = \frac{\sqrt{2\left(k_{1R}^2 + k_{2R}^2 - \mu k_{1R} k_{2R}\right)}}{|\sinh\left(\eta_R + \varrho\right)|}, \tag{108}$$

where

$$\varrho = \frac{1}{2} \ln\left[\frac{|a|^2 + |c|^2}{2\left(k_{1R}^2 + k_{2R}^2 - \mu k_{1R} k_{2R}\right)}\right]. \tag{109}$$

From Eq. (108), it is seen that the solution is undefined at some particular points meaning that it takes infinite values at these singular points.

2.

$$\mu > \frac{k_{1R}^2 + k_{2R}^2}{k_{1R} k_{2R}}, \tag{110}$$

$$W = \sqrt{2\left(\mu k_{1R} k_{2R} - k_{1R}^2 - k_{2R}^2\right)} \, sech\left(\eta_R + \varpi\right), \tag{111}$$

with

$$\varpi = \frac{1}{2} \ln\left[\frac{|a|^2 + |c|^2}{2\left(\mu k_{1R} k_{2R} - k_{1R}^2 - k_{2R}^2\right)}\right]. \tag{112}$$

From Eq. (111), since the argument of the sech-function is real-valued, the solution has no singular points and takes the form of a hump. The amplitude of this soliton solution is given by $\sqrt{2\left(\mu k_{1R}k_{2R} - k_{1R}^2 - k_{2R}^2\right)}$. In order to get symmetry in the (x,y,t)-space, it is useful to take

$$\eta_{0R} = -\varpi, \tag{113}$$

such that η now reads

$$\eta_R = k_{1R}x + k_{2R}y - \omega_R t - \frac{1}{2}\ln\left[\frac{|a|^2 + |c|^2}{2\left(\mu k_{1R}k_{2R} - k_{1R}^2 - k_{2R}^2\right)}\right]. \tag{114}$$

Looking forward getting more information about the interactions between such structures, the further subsequent equations shown above should be solved. Even N-soliton solutions ($N \geq 2$) can be investigated if full series instead of truncated ones are considered. Even though the localized soliton-like solutions displayed in Eqs. (108) and (111) can be obtained by symmetry reduction of the (2+1)-dimensional CNLS equations, it actually seems important to show that the Hirota's bilinearization of this system, which is found after a preliminary P-analysis of the system, holds very well such that the one-, two-, or N-soliton ($N \geq 3$) can be constructed. This constitutes another interesting investigation. However, in this chapter, we would like to focus our interests to some types of excitations that exist owing to the arbitrariness of some functions derived from the Painlevé-analysis. But, in the next, we first discuss the complete integrability of the previous system while providing its general Lax-representation.

3.2. General Lax-representation of the (2+1)-dimensional CNLS equations

In order to check the complete integrability of the above coupled system (2), we now follow the Zakharov–Shabat (ZS)-scheme ([171–173]) which generalizes the Lax-method for higher-dimensional systems. Making use of the Matrix operators, we define two integral operators J_F and $J_\pm$ as follows

$$J_F = \int_{-\infty}^{\infty} dz F(x, z), \quad J_\pm = \int_{-\infty}^{\infty} dz K_\pm(x, z), \tag{115}$$

where the functions F and $K_\pm$ are $N \times N$ (N integer) matrices such that

$$K_+(x, z) = 0 \quad if \quad z < x, \tag{116a}$$
$$K_-(x, z) = 0 \quad if \quad z > x. \tag{116b}$$

The operators J_F and $J_\pm$ are related through the following identity

$$(I + J_+)(I + J_F) = (I + J_-), \tag{117}$$

where the quantity I stands for the unit matrix. Following the operator identity given above, the matrix Marchenko equations ([171–173]) for $K_\pm$ can be expressed in terms of the function F leading to the constructions of the solutions to the investigated system. However, reaching such a goal requires the introduction of two pairs of operators Δ_0 and Δ referred to as "undressed" and "dressed" operators, respectively. A typical choice is given by

$$\Delta_0^{(1)} = I\lambda\partial_t - M_0, \quad \Delta_0^{(2)} = I\mu\partial_y + L_0, \tag{118a}$$

$$\Delta^{(1)} = I\lambda\partial_t - M, \quad \Delta^{(2)} = I\mu\partial_y + L, \tag{118b}$$

where the quantities λ and μ are arbitrary constants, the matrices M_0, L_0, M and L are differential operators in x only. We note that $\partial_t \equiv \partial/\partial_t$ and $\partial_y \equiv \partial/\partial_y$. The undressed and dressed operators satisfy the following relations

$$[\Delta_0^{(i)}, J_F] = 0, \quad \Delta^{(i)}(I + J_+) = (I + J_+)\Delta_0^{(i)}, \tag{119}$$

in such a way that it easily comes

$$\lambda L_t + \mu M_y + [L, M] = 0. \tag{120}$$

This equation stands for the generalization of the Lax-equation, representing the system of nonlinear evolution equations. The procedure for solving the above system is now described. The variable cœfficients which arise in the dressed operators L and M constitute some functions satisfying the system of evolution equation. These functions are related to the kernel K_+, solution to the linear matrix Marchenko equations ([171–173]) where the function F satisfies to a pair of equations. In the wake of the above development, let us consider the following operators

$$\Delta_0^{(1)} = I(\imath 3\partial_t - \partial_{xx}^2), \quad \Delta_0^{(2)} = I\partial_y + \begin{pmatrix} 2 & 0 \\ 0 & 1 \end{pmatrix}\partial_x/3, \tag{121}$$

from which the operators M_0 and L_0 defined in Eq. (118) can easily be explicited. Thus, following the commutation relations (see Eq. (119)) given above, the following equations are obtained

$$\imath 3F_t + F_{zz} - F_{xx} = 0, \quad 3F_y + \begin{pmatrix} 2 & 0 \\ 0 & 1 \end{pmatrix}F_x + F_z\begin{pmatrix} 2 & 0 \\ 0 & 1 \end{pmatrix} = 0. \tag{122}$$

Following the second equation of the system (119), it comes

$$\Delta^{(1)} = I\left[\imath 3\partial_t - \partial_{xx}^2 - W(x, y, t)\right] - 2d\hat{K}_+/dx, \tag{123a}$$

$$\Delta^{(2)} = I\partial_y + \begin{pmatrix} 2 & 0 \\ 0 & 1 \end{pmatrix}\partial_x/3 + V(x, t), \tag{123b}$$

from which the operators M and L defined in Eq. (118) can obviously be found. The matrices V and K_+ are expressed as

$$3V = \begin{pmatrix} 2 & 0 \\ 0 & 1 \end{pmatrix}\hat{K}_+ - \hat{K}_+\begin{pmatrix} 2 & 0 \\ 0 & 1 \end{pmatrix}, \tag{124a}$$

$$\begin{pmatrix} 2 & 0 \\ 0 & 1 \end{pmatrix}K_{+x} + K_{+z}\begin{pmatrix} 2 & 0 \\ 0 & 1 \end{pmatrix} + 3VK_+ = 0, \tag{124b}$$

with $\hat{K}_+ = K(x, x; t)$. It is worth noting that the matrix $W(x, y, t)$ stands for an arbitrary differential operator which in general satisfies to some conditions in the derivation of an evolution equation solvable by the ZS-scheme. Let us express the matrix $\hat{K}_+$ as follows

$$\hat{K}_+ = \begin{pmatrix} A & B \\ C & D \end{pmatrix}. \tag{125}$$

Eq. (124) gives

$$3V = \begin{pmatrix} 0 & B \\ -C & 0 \end{pmatrix}, \quad 2A_x + BC = 0, \quad D_x - BC = 0. \tag{126}$$

By setting $B = \Phi$ and $C = \pm\Phi^*$ (star referring to complex conjugation) where Φ stands for a complex-valued observable which components ϕ and ψ span a two-dimensional manifold generated by the basis unit vectors $\mathbf{e}_\phi$ and $\mathbf{e}_\psi$ such that

$$\Phi = \phi\mathbf{e}_\phi + \psi\mathbf{e}_\psi. \tag{127}$$

The dressed operators are then expressed as follows

$$\Delta^{(1)} = I\left[\imath3\partial_t - \partial_{xx}^2 - W(x, y, t)\right] - \begin{pmatrix} \mp|\Phi|^2 & 2\Phi_x \\ \pm2\Phi_x^* & 2|\Phi|^2 \end{pmatrix}, \tag{128a}$$

$$\Delta^{(2)} = I\partial_y + \begin{pmatrix} 2 & 0 \\ 0 & 1 \end{pmatrix}\partial_x/3 + \begin{pmatrix} 0 & \Phi/3 \\ \mp\Phi^*/3 & 0 \end{pmatrix}. \tag{128b}$$

By setting

$$W = \begin{pmatrix} p & q \\ r & s \end{pmatrix}, \tag{129}$$

and using the generalized Lax system (120), we straightforwardly derive

$$\imath\Phi_t + \Delta\Phi \pm |\Phi|^2\Phi = 0, \tag{130}$$

where the differential operators p, q, r and s satisfy to the following relations

$$q_y + 2q_x/3 + q\partial_x/3 + \Phi s/3 - p\Phi/3 + 2\Phi_{xy} - \Phi_{yy} = 0, \tag{131a}$$

$$r_y + r_x/3 - r\partial_x/3 \pm s\Phi^*/3 \mp \Phi^*p/3 \pm 2\Phi_{xy}^* - \Phi_{yy}^* = 0, \tag{131b}$$

$$p_y + 2p_x/3 + \Phi r/3 \mp (|\Phi|^2)_y = 0, \tag{131c}$$

$$s_y + s_x/3 \mp \Phi^*q \pm 2(|\Phi|^2)_y = 0. \tag{131d}$$

The previous development of the ZS-scheme leading to the derivation of the CNLS system (130) shows that the system (2) under our study actually possesses a generalized Lax-representation in the particular case consisting of $\varrho = -1$ and $\alpha = 1$. Consequently, the system can be solved by means of the matrix Marchenko equations ([171–173]). But, it is not the main purpose of the present work. It is however useful to mention that following the present results, there actually exist some (2+1)-dimensional CNLS equations which

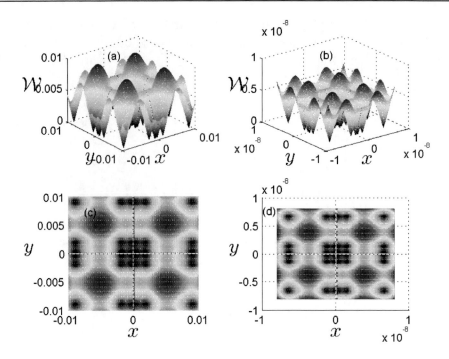

Figure 11. Nonlocal fractal excitations depicted at $t = 0$ by the observable $\mathcal{W}$ which expression is given by Eq. (132). In this case where $a_0 = 0$, the other parameters are selected as follows: -For $f_1(x,t) = \Theta(x,t)$, $\lambda_1 = 1/4$, $\lambda_2 = 0$, $\theta_{01} = 0$, $k_1 = 1$, and $v_1 = 1$. -For $f_2(y,t) = \Theta(y,t)$, $\lambda_1 = 1/4$, $\lambda_2 = 0$, $\theta_{02} = 0$, $k_2 = 1$, and $v_2 = 0$. Panels (a) and (b) represent the pattern formations depicted in $3D$-perspective, and the the two others (c) and (d) are their corresponding densities represented within the square regions $[-1 \cdot 10^{-2}, 1 \cdot 10^{-2}] \times [-1 \cdot 10^{-2}, 1 \cdot 10^{-2}]$ and $[-8 \cdot 10^{-9}, 8 \cdot 10^{-9}] \times [-8 \cdot 10^{-9}, 8 \cdot 10^{-9}]$, respectively.

are solvable, at least by the ZS-scheme and possess miscellaneous soliton solutions which stabilities against various types of perturbations can properly be discussed.

From the previous sections to the present one, we have shown that the system (2) is P-integrable and possesses both bilinear forms and generalized Lax-representation under the arbitrary choice $\varrho = -1$ and $\alpha = 1$. We can now conclude that the system is completely integrable. In the next, we now discuss physically the different pattern formations that such equations in (2+1)-dimensional space can support.

3.3. Localized and stochastic fractal excitations to the (2+1)-dimensional CNLS equations

Within present work, we focus our interests either on the intensity of the waves which is proportional to the square of their amplitude as it is done in nonlinear optics propagation or on the density of atoms also proportional to the square of the amplitude of the wave as investigated in matter-wave interactions in systems such as BEC. However, in the different plots

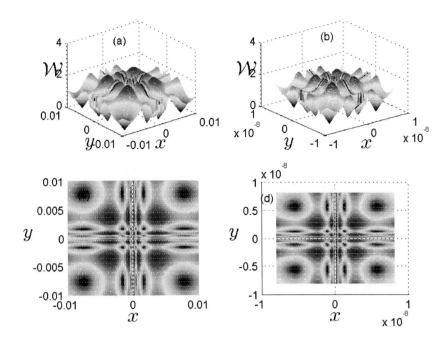

Figure 12. Fractal dromion excitations depicted at $t = 0$ by the observable $\mathcal{W}$ which expression is given by Eq. (132). In this case where $a_0 = 1$, $\theta_0 = 0$, and $k = 1$, the other parameters are selected as follows: -For $f_1(x,t) = \Theta(x,t)$, $c = 1$, and $v = 1$. -For $f_2(y,t) = \Theta(y,t)$, $c = 1$, and $v = 0$. Panels (a) and (b) represent the pattern formations depicted in $3D$-perspective, and the the two others (c) and (d) are their corresponding densities represented within the square regions $[-1 \cdot 10^{-2}, 1 \cdot 10^{-2}] \times [-1 \cdot 10^{-2}, 1 \cdot 10^{-2}]$ and $[-8 \cdot 10^{-9}, 8 \cdot 10^{-9}] \times [-8 \cdot 10^{-9}, 8 \cdot 10^{-9}]$, respectively.

presented in the present section, we consider the square root of the previous observables. Thus, actually, there is no need to search for the explicit individual solutions necessarily owing to the fact that the sum of the square of their amplitudes are explicitly found. However, recently, through a controllable scattering of vector Bose–Einstein solitons, Babarro et al. [174] have shown the possibility of producing matter-wave switching devices by using Manakov interactions between vector matter-wave solitons of two-species BEC. For such an attempt, the initial Gaussian trial functions have been used. Performing such a study to the higher-dimensional time-gated Manakov system, one can explicitly found the expressions of the observables ϕ and ψ while deeply surveying the soliton structure of the system. This is currently a matter of active investigations.

We define a quantity $\mathcal{W} = \sqrt{|\phi|^2 + |\psi|^2}$ expressed as follows

$$\mathcal{W}^2(x,y,t) = \frac{\mathcal{V}_0^T \mathcal{V}_0}{f^2} = 2\frac{(f_x^2 + f_y^2)}{f^2}, \quad (132)$$

which actually constitutes the nonlinear excitations of the system under investigation, and which physical meaning will be examined according to their different features. We note

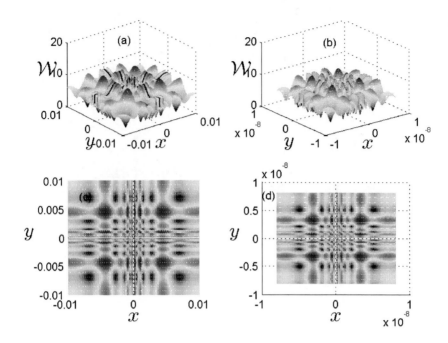

Figure 13. Fractal lump excitations depicted at $t = 0$ by the observable $\mathcal{W}$ which expression is given by Eq. (132). In this case where $a_0 = 0$, $\theta_0 = 0$, and $k = 1$, the parameters are selected as follows: -For $f_1(x,t) = \Theta(x,t)$, $v = 1$. -For $f_2(y,t) = \Theta(y,t)$, $v = 0$. Panels (a) and (b) represent the pattern formations depicted in $3D$-perspective, and the the two others (c) and (d) are their corresponding densities represented within the square regions $[-1 \cdot 10^{-2}, 1 \cdot 10^{-2}] \times [-1 \cdot 10^{-2}, 1 \cdot 10^{-2}]$ and $[-8 \cdot 10^{-9}, 8 \cdot 10^{-9}] \times [-8 \cdot 10^{-9}, 8 \cdot 10^{-9}]$, respectively.

that Eq. (77) and (78) are rewritten in a compact form as follows

$$2\left(f_x \mathcal{V}_{0,x} + f_y \mathcal{V}_{0,y}\right) + \mathcal{B}\mathcal{V}_0 = 0, \quad \mathcal{V}_{0,t} - \mathcal{A}\Delta\mathcal{V}_0 = 0, \tag{133}$$

where

$$\mathcal{B} = (\Delta f)\mathcal{I} + f_t \mathcal{A}. \tag{134}$$

Solving the first equation of the system (133) by using the characteristic method, it leads to the following set

$$\mathcal{V}_0 = \mathcal{G}_0\left(y - \int \frac{f_y dx}{f_x}\right) \exp\left(\int \frac{dx}{2f_x \mathcal{B}^{-1}}\right), \tag{135}$$

where $\mathcal{G}_0$ is an arbitrary array-function of $\left(y - \int \frac{f_y}{f_x}dx\right)$ to be determined.

From equation (133), it can be shown that

$$f_t + \mathcal{E}(x,t)f_x + \mathcal{F}(y,t)f_y = 0, \quad f_t + \mathcal{H}(x,t)f_x + \mathcal{K}(y,t)f_y = 0, \tag{136}$$

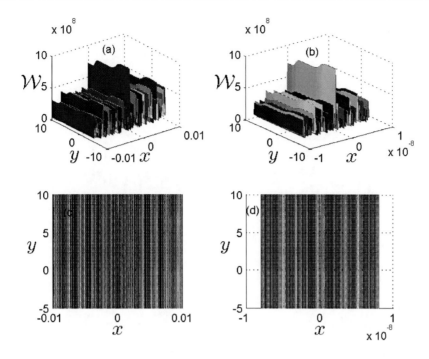

Figure 14. Stochastic fractal solitoff excitations depicted at $t = 0$ by the observable $\mathcal{W}$ which expression is given by Eq. (132). In this case where $a_0 = 0$, $\alpha = 3/2$, and $\beta = 3/2$, the parameters are selected as follows: $f_1(x,t) = \Theta(x,t)$ and $f_2(y,t) = \tanh(y)/5 + \tanh(2y-15)/4$, with $\kappa = 2$, $M = 1$, $\eta_0 = 0$, $\eta_1 = 1/2$, $\eta_m = 0$ $(m \geq 2)$, $\nu_1 = 1$, $k_1 = 4$, $v_1 = -1$, $\theta_{01} = -20$. Panels (a) and (b) represent the pattern formations depicted in 3D-perspective, and the the two others (c) and (d) are their corresponding densities represented within the square regions $[-1 \cdot 10^{-2}, 1 \cdot 10^{-2}] \times [-5, 10]$ and $[-8 \cdot 10^{-9}, 8 \cdot 10^{-9}] \times [-5, 10]$, respectively.

where $\mathcal{E}$ and $\mathcal{H}$ are arbitrary functions of x and t, $\mathcal{F}$ and $\mathcal{K}$ being arbitrary functions of y and t, defined by

$$2\left(\frac{q_0}{p_0}\right)_x = \mathcal{E}(x,t)\left[1 + \left(\frac{q_0}{p_0}\right)^2\right], \tag{137a}$$

$$2\left(\frac{q_0}{p_0}\right)_y = \mathcal{F}(y,t)\left[1 + \left(\frac{q_0}{p_0}\right)^2\right], \tag{137b}$$

$$2\left(\frac{v_0}{u_0}\right)_x = \mathcal{H}(x,t)\left[1 + \left(\frac{v_0}{u_0}\right)^2\right], \tag{137c}$$

$$2\left(\frac{v_0}{u_0}\right)_y = \mathcal{K}(y,t)\left[1 + \left(\frac{v_0}{u_0}\right)^2\right], \tag{137d}$$

such that the compatibility equations $\partial^2_{xy}\left(\frac{q_0}{p_0}\right) = \partial^2_{yx}\left(\frac{q_0}{p_0}\right)$ and $\partial^2_{xy}\left(\frac{v_0}{u_0}\right) = \partial^2_{yx}\left(\frac{v_0}{u_0}\right)$ are satisfied.

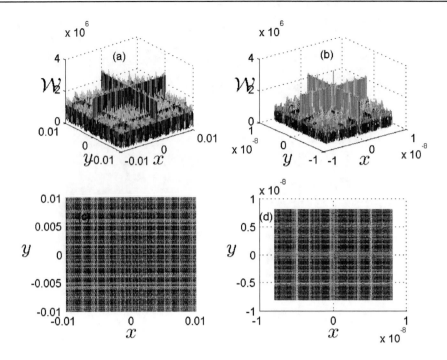

Figure 15. Stochastic fractal lump excitations depicted at $t = 0$ by the observable $\mathcal{W}$ which expression is given by Eq. (132). In this case where $a_0 = 0$, $\alpha = 3/2$, $\beta = 3/2$, and $\bar{N} = 2$, the parameters are selected as follows: -For $f_1(x,t) = \Theta(x,t)$, $k_1 = 1$, $v_1 = 1$, $\theta_{01} = 0$, $\varrho_0 = 0$, $\varrho_1 = 1$, $\varrho_2 = 0$, and $\mu_1 = 1000$. -For $f_2(y,t) = \Theta(y,t)$, $k_2 = 1$, $v_2 = 0$, $\theta_{02} = 0$, $\varrho_0 = 0$, $\varrho_1 = 0$, $\varrho_2 = 1$, and $\mu_2 = 1000$. Panels (a) and (b) represent the pattern formations depicted in $3D$-perspective, and the the two others (c) and (d) are their corresponding densities represented within the square regions $[-1 \cdot 10^{-2}, 1 \cdot 10^{-2}] \times [-1 \cdot 10^{-2}, 1 \cdot 10^{-2}]$ and $[-8 \cdot 10^{-9}, 8 \cdot 10^{-9}] \times [-8 \cdot 10^{-9}, 8 \cdot 10^{-9}]$, respectively.

Expressing $f(x, y, t)$ as follows,

$$f(x, y, t) = f_1(x, t) + f_2(y, t) + a_0, \qquad (138)$$

where a_0 is an arbitrary constant, and solving Eq. (137), it easily comes that

$$\begin{aligned} f_1(x,t) &= \mathcal{L}_1\left(x - \int \mathcal{E}(x,t)dt\right) - \int \varsigma(t)dt \\ &= \mathcal{M}_1\left(x - \int \mathcal{H}(x,t)dt\right) - \int \nu(t)dt, \end{aligned} \qquad (139a)$$

$$\begin{aligned} f_2(y,t) &= \mathcal{L}_2\left(y - \int \mathcal{F}(y,t)dt\right) + \int \varsigma(t)dt \\ &= \mathcal{M}_2\left(y - \int \mathcal{K}(y,t)dt\right) + \int \nu(t)dt, \end{aligned} \qquad (139b)$$

where ς, ν, $\mathcal{L}_j$ and $\mathcal{M}_j$ ($j = 1, 2$) stand for arbitrary functions. Thus, the functions f_1 and f_2 are arbitrary. For this reason, there exists a rich variety of structures due to the different

Dynamics of Miscellaneous Fractal Structures ...   193

expressions that these functions can take. It seems worth noting that the functions $\mathcal{E}$, $\mathcal{H}$, $\mathcal{F}$ and $\mathcal{K}$ are derived from Eq. (137) after solving the second equation of the system (133). Besides, solving Eq. (137), the following system is found

$$\frac{q_0}{p_0} = \tan\left(\frac{1}{2}\int \mathcal{E}(x,t)dx + \frac{1}{2}\int \mathcal{F}(y,t)dy + d_1(t)\right), \tag{140a}$$

$$\frac{v_0}{u_0} = \tan\left(\frac{1}{2}\int \mathcal{H}(x,t)dx + \frac{1}{2}\int \mathcal{K}(y,t)dy + d_2(t)\right), \tag{140b}$$

where d_1 and d_2 stand for arbitrary functions depending on time t. Equation (140) shows that for some arbitrary expressions of $\mathcal{E}$, $\mathcal{H}$, $\mathcal{F}$ and $\mathcal{K}$, $\frac{q_0}{p_0}$ and $\frac{v_0}{u_0}$ are found, provided the second equation of the system (133) be satisfied.

Choosing appropriate parameters and arbitrary functions which enter into the expression of the manifold f above arguably yields an interesting set of pattern formations. For instance, in Fig. 11, the $\mathcal{W}$-observable at $t = 0$ varies self-similarly. Indeed, in a $3D$-representation, the features presented in panel $11(a)$ within the space region $[-1 \cdot 10^{-2}, 1 \cdot 10^{-2}]^2 \times \mathcal{W}$ and those depicted in $11(b)$ within region $[-8 \cdot 10^{-9}, 8 \cdot 10^{-9}]^2 \times \mathcal{W}$ are identical in structure. Such a similarity in the profile is clearly shown in panels $11(c)$ and $11(d)$ representing their density plots, respectively. By pursuing further with different scales of x and y to the reducing orders, we obtain similar pictures. The dynamics of the aforementioned pattern formations can be studied while depicting their profiles at different evolving times.

Besides, selecting appropriately lower-dimensional arbitrary dromion functions and suitable parameters as presented in the caption of Fig. 12, we obtain a typical pattern formation which is exponentially localized in a large scale of x and y. Nonetheless, in order to better present the self-similarity in structure of such fractal patterns, we reduce the region of panel $12(a)$ $(x, y) \in [-1 \cdot 10^{-2}, 1 \cdot 10^{-2}]^2$ to $[-8 \cdot 10^{-9}, 8 \cdot 10^{-9}]^2$ of panel $12(b)$, and we obtain a totally similar structure with density plots represented in panels $12(c)$ and $12(d)$, respectively. The process can continue further with lesser reducing orders of x and y.

With the details presented in the caption of Fig. 13, we obtain a typical pattern formation known as fractal lump excitation, which is algebraically localized in a large scale of x and y. In fact, near the centre, there are infinitely many peaks which are distributed in a fractal manner. In order to see the fractal structure of the previous pattern, we look at the structure more carefully. Panel $13(a)$ presents the $3D$-representation of the fractal features of $\mathcal{W}$-observable and its corresponding density plot at region $(x, y) \in [-1 \cdot 10^{-2}, 1 \cdot 10^{-2}]^2$ is depicted in panel $13(c)$. Reducing such a region to $[-8 \cdot 10^{-9}, 8 \cdot 10^{-9}]^2$ shows self-similar structures in panels $13(b)$ and $13(d)$.

In addition, Fig. 14 presents a typical stochastic fractal solitoff pattern derived from a lower-dimensional stochastic arbitrary fractal dromion. In fact, in large scale, the previous structure is actually localized. Investigating the self-similarity in detail, we reduce the region $(x, y) \in [-1 \cdot 10^{-2}, 1 \cdot 10^{-2}] \times [-5, 10]$ of panel $14(a)$ to $(x, y) \in [-8 \cdot 10^{-22}, 8 \cdot 10^{-22}] \times [-5, 10]$ as shown in panel $14(b)$. Their densities presented at the bottom in panels $14(c)$ and $14(d)$, respectively, clearly show such a similarity in structure. Also, in Fig. 15, with the selecting parameters and suitable choices of lower-dimensional arbitrary stochastic fractal lump functions as presented in the caption of this figure, we obtain higher-dimensional stochastic lump excitations. Through the panels $15(a)$ and $15(b)$

depicting the variations of the $\mathcal{W}$-observable at $t = 0$, the self-similarity in structure of this observable shows how the peaks are distributed stochastically. As mentioned above, globally, it appears that the aforementioned stochastic fractal structures have amplitudes greater than the previous ones.

In the wake of the aforementioned results, we point out that the composite matter-wave pattern $\mathcal{W}$ in which square form is regarded as the total energy density of the two-species condensates, in a fractal manner describes the scattering among two-dimensional dark-bright matter-waves, but also some induced instabilities in the BEC, such as the thermal instability combined to the diffusivity of the atoms. The typical fractal patterns derived from the above observable are also applicable to spatial time-gated optical devices, especially the single-valued excitations referring to the total energy of the interacting defocusing light beam. During the matter-wave propagation, some kinds of defects can occur within the system such as singularities in the propagation distance related to the fractal nature of the matter. In such situations, the fractal patterns can be representative of such a state of the matter. Investigating the dynamical behavior of the previous patterns in the context of propagation of successive perturbations in the BEC system, the nonlinear interaction between some kinds of excitations are likely to describe the presence of some induced undesirable effects in the medium such as wave distortion, transmission characteristic deterioration, and transmission rate distortion. Such issue is left for further interests.

4. Fractal Structure of the Higher-Dimensional Ferromagnetic Material

4.1. Basic equations

Let us consider a saturated nonconducting ferromagnetic slab in which an electromagnetic wave propagates. The evolution of the magnetic field $\mathbf{H}$ is governed by Maxwell equations which reduce to

$$\Delta \mathbf{H} - \nabla(\nabla \cdot \mathbf{H}) = (\mathbf{H} + \mathbf{M})_t/c^2, \tag{141}$$

where $c^2 = 1/\mu_0 \varepsilon_0$ is the speed of light in the vacuum with ε_0 and μ_0 being the permittivity and permeability scalars of the vacuum, respectively. The magnetization density $\mathbf{M}$ obeys the Landau-Lifschitz equation [175] which reads

$$\mathbf{M}_t = -\gamma \mu_0 \mathbf{M} \wedge \mathbf{H}_{eff} + \sigma \mathbf{M} \wedge (\mathbf{M} \wedge \mathbf{H}_{eff})/M_s, \tag{142}$$

where constants γ and σ stand for gyromagnetic ratio and damping parameter, respectively. The quantity M_s represents the saturation magnetization. The effective magnetic field is $\mathbf{H}_{eff} = \mathbf{H} - N \cdot \mathbf{M}$, with N being the demagnetizing factor tensor. In the present study, we consider a ferromagnetic film lying in the xy-plane in such a way that N is diagonal.

By considering bulk polaritons, that is, wavelengths ($\sim 10 - 100 \mu m$) larger with regard to the exchange length, inhomogeneous interaction/exchange can be neglected and the pinning boundary conditions not considered. Also, the slab thickness is assumed to be larger with respect to the wavelength, says, typically $\sim 0.5 mm$. This assumption justifies that the exact boundary conditions are replaced by a mere demagnetizing tensor. Although such an assumption is particularly useful in the study of surface or volume modes in thin films, throughout this section, we won't go further through this consideration. Additionally, we neglect the crystalline and surface anisotropy of the sample. The quantities $\mathbf{M}$, $\mathbf{H}$ and t are rescaled into $c\mathbf{M}/\mu_0\gamma$, $c\mathbf{H}/\mu_0\gamma$ and t/c, respectively, such that constants $\mu_0\gamma/c$ and c are replaced by 1 into Eqs. (141) and (142), while M_s is replaced by m, which is the normalized saturation magnetization.

Let us firstly study the linear regime of perturbation where the sample is supposed to be magnetized to saturation by means of an external uniform field, according to

$$\mathbf{M}_0 = (\text{m}\cos\theta, \text{m}\sin\theta, 0), \quad \mathbf{H}_0 = \alpha\mathbf{M}_0, \tag{143}$$

which reads the static field $\mathbf{H}_0$ lying in the plane of the film and collinear to the magnetization $\mathbf{M}_0$. The angle θ represents the deflection between the dominant propagation direction and the internal magnetic field. The parameter α stands for the strength of this internal field. It is important to consider the following settings

$$\mathbf{M} = \mathbf{M}_0 + \mathbf{M}_1\exp[\imath(\mathbf{k}\cdot\mathbf{r} - \omega t)], \quad \mathbf{H} = \mathbf{H}_0 + \mathbf{H}_1\exp[\imath(\mathbf{k}\cdot\mathbf{r} - \omega t)], \tag{144}$$

where $\mathbf{M}_1 = (M_1^x, M_1^y, M_1^z)$ and $\mathbf{H}_1 = (H_1^x, H_1^y, H_1^z)$. Quantities $\mathbf{k}$, $\mathbf{r}$ and ω stand for wave vector, position vector and frequency of the perturbation, respectively. Substituting Eq. (144) into (141) and (142) leads to

$$\omega^2(\omega^2 - \mathbf{k}^2) + \text{m}^2[(2 + \alpha)\omega^2 - (1 + \alpha)\mathbf{k}^2] \times$$
$$[(k^y\cos\theta - k^x\sin\theta)^2 - (1 + \alpha)\omega^2 + \alpha\mathbf{k}^2] = 0. \tag{145}$$

In the short-wave approximation, the dispersion relation admits the following expansion [176, 177]

$$\omega = \tilde{a}/\epsilon + \tilde{b}\epsilon + \tilde{c}\epsilon + \tilde{d}\epsilon + \cdots, \tag{146}$$

where the small parameter ϵ is related to the magnitude of the wavelength through $k^x = k_0/\epsilon$. It corresponds to short waves with k_0 being an arbitrary constant standing for some reference value of the wave number, that is, $k_0 = \omega_r/c$, where ω_r is the ferromagnetic resonance frequency. Computations of the cœfficients $\tilde{a}$, $\tilde{b}$, $\tilde{c}$, and $\tilde{d}$ show that the expansion holds provided $\theta = \pi/2$.

Since we assume that the wave propagation dynamics are predominant in the x-axis in such a way that the y-variable gives only account of a slow transverse deviation, the component k^y of the wavelength is very small compared to k^x, and we can set $k^y = l_0$ of zeroth-order to ϵ, accordingly. The phase η up to order ϵ is thus given by

$$\eta = (k_0 x - \tilde{a}t)/\epsilon + l_0 y - \tilde{\epsilon b}t, \tag{147}$$

which motivates the introduction of new variables ζ and τ as follows

$$\zeta = (x - vt)/\epsilon, \quad y = y, \quad \tau = \epsilon t, \tag{148}$$

where constant v represents the velocity of the short wave. It appears that the slow time variable τ accounts for the propagation at very long time, on distances very large with regard to the wavelength. Concretely, we take the value of the perturbative parameter as $\epsilon = 1/100$ which yields a length of the solitary wave in the range of $50\mu m$ and propagation distances limited to a few centimeter. It is then useful to consider a slab width of about $0.5cm$ as reference length, and slab thickness should have an intermediary scale between the wavelength and the width, that is, about $0.5mm$.

Eq. (148) allows us to define rescaled space and time operators as follows

$$\partial_x = \partial_\zeta/\epsilon, \quad \partial_y = \partial_y, \quad \partial_t = -v\partial_\zeta/\epsilon + \epsilon\partial_\tau. \tag{149}$$

We then expand the fields $\mathbf{M}$ and $\mathbf{H}$ into power series of ϵ as given below

$$\mathbf{M} = \sum_{j=0} \mathbf{M}_j \epsilon^j, \quad \mathbf{H} = \sum_{j=0} \mathbf{H}_j \epsilon^j, \tag{150}$$

where $\mathbf{M}_j$ and $\mathbf{H}_j$ $(j \in \mathbb{N})$ are functions of ζ, y and τ. These functions satisfy to the following boundary conditions

$$\lim_{\zeta \to -\infty} \mathbf{H}_j = \lim_{\zeta \to -\infty} \mathbf{M}_j = 0, (j \neq 0), \quad \lim_{\zeta \to -\infty} \mathbf{H}_0 = \alpha \lim_{\zeta \to -\infty} \mathbf{M}_0 = \alpha(0, \mathrm{m}, 0). \tag{151}$$

Let us assume that the damping is weak. In fact, in yttrium-iron-garnet (YIG) films, the observed envelope solitons [58, 178] show that the NLS-type model including the damping terms should be accounted. This system can be derived from the Landau–Lifschitz and Maxwell equations above, while assuming that the dimensionless damping constant σ is of second-order to ϵ measuring the amplitude of the magnetic wave pulse [179]. In such YIG films, the parameter σ can take small values as 10^{-4} which actually correspond to perturbative parameter of about 10^{-2} [180]. In order to derive a model system which takes a stronger damping effect into account, it is suitable to transform σ to $\epsilon^p\sigma$, $(p \in \mathbb{N})$. The value $p = 2$ holds for a YIG film with low losses, and $p = 1$ in other cases.

Indeed, taking $p = 1$ and substituting Eqs. (149) and (150) into (141) shows that

- At leading order ϵ^{-2}, the magnetization density $\mathbf{M}_0$ is uniform, the x-component of the magnetic field $\mathbf{H}_0$ vanishes and the velocity v is unity.

- At the order ϵ^{-1}, it comes

$$\mathbf{M}_1 = \mathrm{m} \int_{-\infty}^{\zeta} H_0^z d\zeta' \mathbf{e}_x, \quad H_1^x = - \int_{-\infty}^{\zeta} (H_{0,y}^y + \mathrm{m}H_0^z)d\zeta', \tag{152}$$

where $\mathbf{e}_x$ denotes the unitary vector in the x-direction.

- At order ϵ^0, the damping appears in Eq. (142) such as

$$\mathbf{M}_{2,\zeta} = (\mathrm{m}H_1^z, -M_1^x H_0^z, -\mathrm{m}H_1^x + M_1^x H_0^y + \sigma \mathrm{m}H_0^z). \tag{153}$$

Replacing Eq. (153) into (141) leads to

$$-H_{1,y}^x + 2H_{0,\tau}^y + M_1^x H_0^z = 0, \tag{154}$$

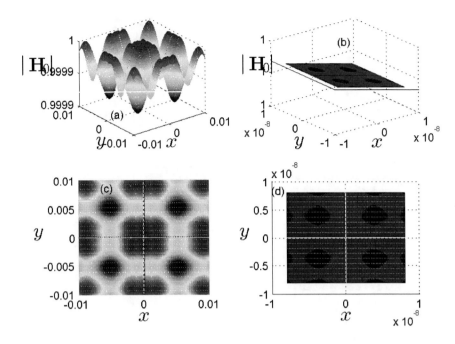

Figure 16. Nonlocal fractal excitations depicted at $t = 0$ by the observable $|\mathbf{H}_0|$ which expression is given by Eq. (190). In this case, the parameters are selected as $a_0 = 1$, $a_1 = 1$, $a_2 = 1$, and $a_3 = 1/2$ such that: -For $p(x,t) = \Theta(x,t)$, $\lambda_1 = 1/4$, $\lambda_2 = 0$, $\theta_{01} = 0$, $k_1 = 1$, and $v_1 = 1$. -For $q(y,t) = \Theta(y,t)$, $\lambda_1 = 1/4$, $\lambda_2 = 0$, $\theta_{02} = 0$, $k_2 = 1$, and $v_2 = 0$. Panels (a) and (b) represent the pattern formations depicted in $3D$-perspective, and the the two others (c) and (d) are their corresponding densities represented within the square regions $[-1 \cdot 10^{-2}, 1 \cdot 10^{-2}] \times [-1 \cdot 10^{-2}, 1 \cdot 10^{-2}]$ and $[-8 \cdot 10^{-9}, 8 \cdot 10^{-9}] \times [-8 \cdot 10^{-9}, 8 \cdot 10^{-9}]$, respectively.

and

$$H^z_{0,yy} + 2H^z_{0,\tau\zeta} + \mathrm{m}H^x_{1,\zeta} - (M^x_1 H^y_0)_\zeta - \sigma \mathrm{m} H^z_{0,\zeta} = 0. \tag{155}$$

Consequently, Eq. (154) and (155) reduce to

$$\int_{-\infty}^{\zeta} H^y_{0,yy} d\zeta' + M^x_{1,y} + 2H^y_{0,\tau} + M^x_1 M^x_{1,\zeta}/\mathrm{m} = 0, \tag{156a}$$

$$-M^x_{1,\zeta yy}/\mathrm{m} - 2M^x_{1,\tau\zeta\zeta}/\mathrm{m} + \mathrm{m}H^y_{0,y} + \mathrm{m}M^x_{1,\zeta} + (M^x_1 H^y_0)_\zeta + \sigma M^x_{1,\zeta\zeta} = 0, \tag{156b}$$

yielding the sought asymptotic model.

On the other hand, as is reasonable in a YIG film, for $p = 2$, it can be shown that from the previous calculation, the first correction due to the damping will appear at the next order in the perturbative scheme only, and hence the asymptotic model obtained above won't contain any damping term at all. It can be recovered by setting $\sigma = 0$ into Eq. (156). In fact,

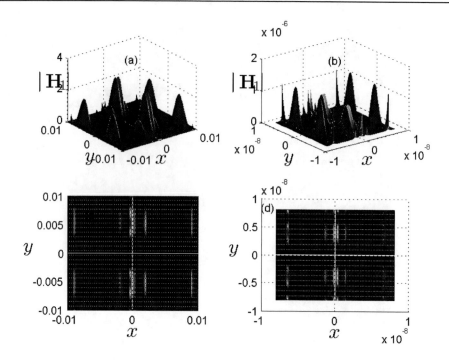

Figure 17. Nonlocal fractal excitations depicted at $t = 0$ by the observable $|\mathbf{H}_1|$ which expression is given by Eq. (190). In this case, the parameters are selected as $a_0 = 1$, $a_1 = 1$, $a_2 = 1$, and $a_3 = 1/2$ such that: -For $p(x,t) = \Theta(x,t)$, $\lambda_1 = 1/4$, $\lambda_2 = 0$, $\theta_{01} = 0$, $k_1 = 1$, and $v_1 = 1$. -For $q(y,t) = \Theta(y,t)$, $\lambda_1 = 1/4$, $\lambda_2 = 0$, $\theta_{02} = 0$, $k_2 = 1$, and $v_2 = 0$. Panels (a) and (b) represent the pattern formations depicted in $3D$-perspective, and the the two others (c) and (d) are their corresponding densities represented within the square regions $[-1 \cdot 10^{-2}, 1 \cdot 10^{-2}] \times [-1 \cdot 10^{-2}, 1 \cdot 10^{-2}]$ and $[-8 \cdot 10^{-9}, 8 \cdot 10^{-9}] \times [-8 \cdot 10^{-9}, 8 \cdot 10^{-9}]$, respectively.

if the damping effect is neglected, the system (156) leads obviously to coupled equations obtained in previous studies [69] where demagnetizing factor was omitted. This rather remarkable result reveals that within the considered approximation, and for the geometry of the system presented above, the demagnetizing field arguably has no effect on wave propagation.

Setting

$$X = -\mathrm{m}\zeta/2, \quad Y = \mathrm{m}y, \quad T = \mathrm{m}\tau, \quad (157a)$$
$$A = -(1 + H_0^y/\mathrm{m}), \quad B = M_1^x/2\mathrm{m}, \quad s = -\sigma/2, \quad (157b)$$

Eq. (156) reduces to the following coupled equations

$$B_{XT} - BC_X - B_{YY} + C_Y + sB_X = 0, \quad C_{XT} + BB_X - C_{YY} - B_Y = 0, \quad (158)$$

provided $A = C_X$. For convenience latter, the independent variables X and T will be written in their lower cases, respectively. The system (158) has been shown [69] to possess

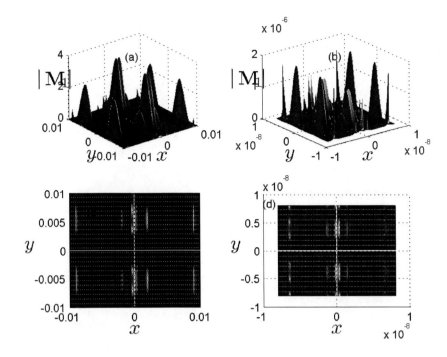

Figure 18. Nonlocal fractal excitations depicted at $t=0$ by the observable $|\mathbf{M}_1|$ which expression is given by Eq. (190). In this case, the parameters are selected as $a_0 = 1$, $a_1 = 1$, $a_2 = 1$, and $a_3 = 1/2$ such that: -For $p(x,t) = \Theta(x,t)$, $\lambda_1 = 1/4$, $\lambda_2 = 0$, $\theta_{01} = 0$, $k_1 = 1$, and $v_1 = 1$. -For $q(y,t) = \Theta(y,t)$, $\lambda_1 = 1/4$, $\lambda_2 = 0$, $\theta_{02} = 0$, $k_2 = 1$, and $v_2 = 0$. Panels (a) and (b) represent the pattern formations depicted in $3D$-perspective, and the the two others (c) and (d) are their corresponding densities represented within the square regions $[-1 \cdot 10^{-2}, 1 \cdot 10^{-2}] \times [-1 \cdot 10^{-2}, 1 \cdot 10^{-2}]$ and $[-8 \cdot 10^{-9}, 8 \cdot 10^{-9}] \times [-8 \cdot 10^{-9}, 8 \cdot 10^{-9}]$, respectively.

line solitons with background instability. We aim at extending the class of such solutions to another one in which fractal structures describing the fractal nature of ferromagnetic slabs in particular cases, can be unwrapped.

4.2. Painlevé-analysis

The P-singularity structure analysis of an analytical polynomial differential equation is carried out by seeking a generalized Laurent expansion [181] for the dependent variables

$$B = \sum_{j=0} B_j h^{j+\rho}, \quad C = \sum_{j=0} C_j h^{j+\delta}, \qquad (159)$$

in the neighborhood of the noncharacteristic singular manifold $h(x,y,t) = 0$, with nonvanishing derivatives. The quantities ρ and δ are arbitrary constants which should be negative-like integers in order to expect the system (158) to be integrable. The leading order behavior

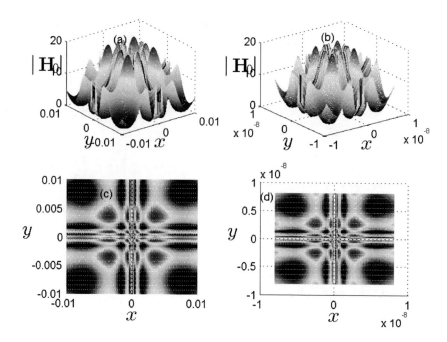

Figure 19. Fractal dromion excitations depicted at $t = 0$ by the observable $|\mathbf{H}_0|$ which expression is given by Eq. (190). In this case, the parameters are selected as $a_0 = 1$, $a_1 = 1$, $a_2 = 1$, $a_3 = 2$, $\theta_0 = 0$, and $k = 1$ such that: -For $p(x,t) = \Theta(x,t)$, $c = 1$, and, $v = 1$. -For $q(y,t) = \Theta(y,t)$, $c = 1$, and $v = 0$. Panels (a) and (b) represent the pattern formations depicted in $3D$-perspective, and the the two others (c) and (d) are their corresponding densities represented within the square regions $[-1 \cdot 10^{-2}, 1 \cdot 10^{-2}] \times [-1 \cdot 10^{-2}, 1 \cdot 10^{-2}]$ and $[-8 \cdot 10^{-9}, 8 \cdot 10^{-9}] \times [-8 \cdot 10^{-9}, 8 \cdot 10^{-9}]$, respectively.

of the solution is analyzed by assuming the following

$$B \sim B_0 h^\rho, \quad C \sim C_0 h^\delta. \tag{160}$$

Substituting Eq. (160) into (158) and balancing the most dominant terms yield

$$\rho = \delta = -1, \tag{161}$$

and

$$C_0 = -2(h_x h_t - h_y^2)/h_x, \quad B_0 = \pm \imath C_0, \tag{162}$$

where $\imath^2 = -1$.

Now, substituting the full Laurent expansion (159) into (158) and vanishing all the coefficients of the powers h^j, we obtain the recursion relations to determine the functions B_j and C_j as follows

$$\mathcal{M}_j \mathcal{V}_j = \mathcal{T}_j, \tag{163}$$

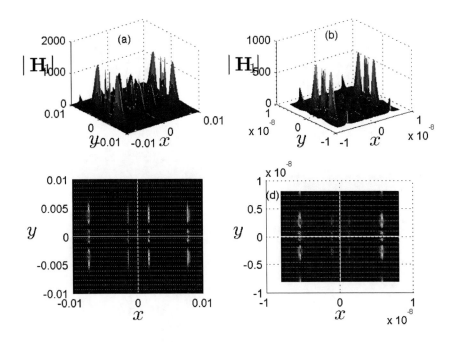

Figure 20. Fractal dromion excitations depicted at $t = 0$ by the observable $|\mathbf{H}_1|$ which expression is given by Eq. (190). In this case, the parameters are selected as $a_0 = 1$, $a_1 = 1$, $a_2 = 1$, $a_3 = 2$, $\theta_0 = 0$, and $k = 1$ such that: -For $p(x,t) = \Theta(x,t)$, $c = 1$, and, $v = 1$. - For $q(y,t) = \Theta(y,t)$, $c = 1$, and $v = 0$. Panels (a) and (b) represent the pattern formations depicted in $3D$-perspective, and the the two others (c) and (d) are their corresponding densities represented within the square regions $[-1 \cdot 10^{-2}, 1 \cdot 10^{-2}] \times [-1 \cdot 10^{-2}, 1 \cdot 10^{-2}]$ and $[-8 \cdot 10^{-9}, 8 \cdot 10^{-9}] \times [-8 \cdot 10^{-9}, 8 \cdot 10^{-9}]$, respectively.

where $\mathcal{M}_j$ is a square matrix expressed as

$$\mathcal{M}_j = \begin{pmatrix} (j-1)(j-2)(h_x h_t - h_y^2) + h_x C_0 & -(j-1)h_x B_0 \\ (j-2)h_x B_0 & (j-1)(j-2)(h_x h_t - h_y^2) \end{pmatrix}, \quad (164)$$

and $\mathcal{V}_j \equiv (B_j, C_j)^T$, $\mathcal{T}_j \equiv (\mathcal{P}_j, \mathcal{Q}_j)^T$ with

$$\begin{aligned}
\mathcal{P}_j &= B_{j-2,yy} - B_{j-2,xt} - C_{j-2,y} - sB_{j-2,x} \\
&+ (j-2)(2B_{j-1,y}h_y + B_{j-1}h_{yy} - B_{j-1,x}h_t - B_{j-1,t}h_x - B_{j-1}h_{xt}) \\
&- (j-2)C_{j-1}h_y - s(j-2)B_{j-1}h_x + \sum_{l=1}^{j} B_{j-l}C_{l-1,x} + \sum_{l=1}^{j-1}(l-1)B_{j-l}C_l h_x, \quad (165)
\end{aligned}$$

and

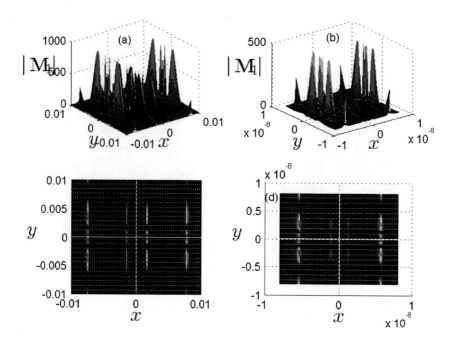

Figure 21. Fractal dromion excitations depicted at $t = 0$ by the observable $|\mathbf{M}_1|$ which expression is given by Eq. (190). In this case, the parameters are selected as $a_0 = 1$, $a_1 = 1$, $a_2 = 1$, $a_3 = 2$, $\theta_0 = 0$, and $k = 1$ such that: -For $p(x,t) = \Theta(x,t)$, $c = 1$, and, $v = 1$. - For $q(y,t) = \Theta(y,t)$, $c = 1$, and $v = 0$. Panels (a) and (b) represent the pattern formations depicted in $3D$-perspective, and the the two others (c) and (d) are their corresponding densities represented within the square regions $[-1 \cdot 10^{-2}, 1 \cdot 10^{-2}] \times [-1 \cdot 10^{-2}, 1 \cdot 10^{-2}]$ and $[-8 \cdot 10^{-9}, 8 \cdot 10^{-9}] \times [-8 \cdot 10^{-9}, 8 \cdot 10^{-9}]$, respectively.

$$\mathcal{Q}_j = C_{j-2,yy} - C_{j-2,xt} + B_{j-2,y}$$
$$+ (j-2)(2C_{j-1,y}h_y + C_{j-1}h_{yy} - C_{j-1,x}h_t - C_{j-1,t}h_x - C_{j-1}h_{xt})$$
$$+ (j-2)B_{j-1}h_y - \sum_{l=1}^{j}(B_{j-l}B_{l-1})_x/2 + (j-2)\sum_{l=1}^{j-1}B_{j-l}C_l h_x/2, \quad (166)$$

By requiring the determinant of the matrix $\mathcal{M}_j$ to be zero, the following resonance equation

$$(j+1)(j-1)(j-2)(j-4) = 0, \quad (167)$$

is obtained. Thus, resonances, that is, powers at which arbitrary functions are introduced into the Laurent expansion are found at

$$j = -1, 1, 2, 4. \quad (168)$$

The resonance at $j = -1$ corresponds to the arbitrariness of the singularity manifold $h(x, y, t)$. For the system to pass the P-analysis, three resonance conditions at $j = 1, 2$

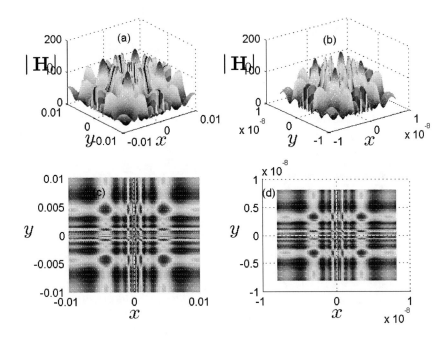

Figure 22. Fractal lump excitations depicted at $t = 0$ by the observable $|\mathbf{H}_0|$ which expression is given by Eq. (190). In this case, the parameters are selected as $a_0 = 1$, $a_1 = 1$, $a_2 = 1$, $a_3 = 2$, $\theta_0 = 0$, and $k = 1$ such that: -For $p(x,t) = \Theta(x,t)$, $v = 1$. -For $q(y,t) = \Theta(y,t)$, $v = 0$. Panels (a) and (b) represent the pattern formations depicted in 3D-perspective, and the the two others (c) and (d) are their corresponding densities represented within the square regions $[-1 \cdot 10^{-2}, 1 \cdot 10^{-2}] \times [-1 \cdot 10^{-2}, 1 \cdot 10^{-2}]$ and $[-8 \cdot 10^{-9}, 8 \cdot 10^{-9}] \times [-8 \cdot 10^{-9}, 8 \cdot 10^{-9}]$, respectively.

and 3 are required such that three arbitrary functions among B_j and C_j can be introduced into the general series expansion given by Eq. (159).

For $j = 1$, the following system

$$h_x C_0 B_1 = \mathcal{P}_1, \quad -h_x B_0 B_1 = \mathcal{Q}_1, \qquad (169)$$

is derived with

$$\mathcal{P}_1 = C_0 h_y + s B_0 h_x + B_0 C_{0,x} - (B_{0,y} h_y + B_0 h_{yy} - B_{0,x} h_t - B_{0,t} h_x - B_0 h_{xt}), \qquad (170)$$

and

$$\mathcal{Q}_1 = -(B_0 h_y + 2C_{0,y} h_y + C_0 h_{yy} - C_{0,x} h_t - C_{0,t} h_x - C_0 h_{xt} - B_0 B_{0x}). \qquad (171)$$

From the previous system, it appears that C_1 is arbitrary provided that the following equation

$$\mathcal{Q}_1 \pm \imath \mathcal{P}_1 = 0, \qquad (172)$$

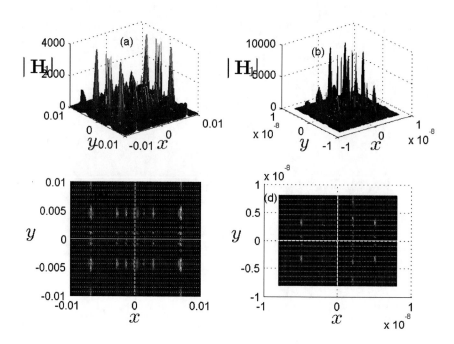

Figure 23. Fractal lump excitations depicted at $t = 0$ by the observable $|\mathbf{H}_1|$ which expression is given by Eq. (190). In this case, the parameters are selected as $a_0 = 1$, $a_1 = 1$, $a_2 = 1$, $a_3 = 2$, $\theta_0 = 0$, and $k = 1$ such that: -For $p(x,t) = \Theta(x,t)$, $v = 1$. -For $q(y,t) = \Theta(y,t)$, $v = 0$. Panels (a) and (b) represent the pattern formations depicted in 3D-perspective, and the the two others (c) and (d) are their corresponding densities represented within the square regions $[-1 \cdot 10^{-2}, 1 \cdot 10^{-2}] \times [-1 \cdot 10^{-2}, 1 \cdot 10^{-2}]$ and $[-8 \cdot 10^{-9}, 8 \cdot 10^{-9}] \times [-8 \cdot 10^{-9}, 8 \cdot 10^{-9}]$, respectively.

holds. However, direct calculation shows that

$$\mathcal{Q}_1 \pm i\mathcal{P}_1 = -sC_0 h_x, \tag{173}$$

which yields $s = 0$. Such result is actually interesting both from the viewpoint of existence of conservation laws of the system (158) and from the viewpoint of the stability of traveling pattern formations. In fact, in recent study [65, 66], Leblond and Manna have discussed the stability of line soliton solutions to the system (158) and have shown that the suppression of the background instability is straightforward by vanishing the damping term.

For $j = 2$, it comes

$$\mathcal{P}_2 = B_{0,yy} - B_{0,xt} - C_{0,y} - sB_{0,x} + B_1 C_{0,x} + B_0 C_{1,x}, \tag{174}$$

such that solving the system (163) provides

$$\mathcal{P}_2 = h_x C_0 (B_2 \pm iC_2), \tag{175}$$

showing that either B_2 or C_2 is arbitrary provided

$$C_{0,yy} - C_{0,xt} + B_{0,y} - (B_0 B_1)_x = 0, \tag{176}$$

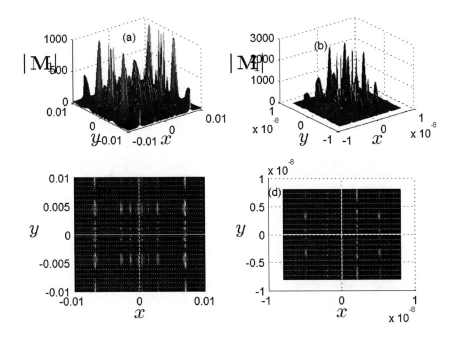

Figure 24. Fractal lump excitations depicted at $t = 0$ by the observable $|\mathbf{M}_1|$ which expression is given by Eq. (190). In this case, the parameters are selected as $a_0 = 1$, $a_1 = 1$, $a_2 = 1$, $a_3 = 2$, $\theta_0 = 0$, and $k = 1$ such that: -For $p(x,t) = \Theta(x,t)$, $v = 1$. -For $q(y,t) = \Theta(y,t)$, $v = 0$. Panels (a) and (b) represent the pattern formations depicted in $3D$-perspective, and the the two others (c) and (d) are their corresponding densities represented within the square regions $[-1 \cdot 10^{-2}, 1 \cdot 10^{-2}] \times [-1 \cdot 10^{-2}, 1 \cdot 10^{-2}]$ and $[-8 \cdot 10^{-9}, 8 \cdot 10^{-9}] \times [-8 \cdot 10^{-9}, 8 \cdot 10^{-9}]$, respectively.

hold. However, Eq. (175) actually get sense if the arbitrary function C_1 takes the following form

$$C_1 = \pm \imath B_1. \tag{177}$$

For $j = 3$, the system (163) leads to

$$B_3 = (\mathcal{Q}_3 \pm \imath \mathcal{P}_3/2)/h_x B_0, \quad C_3 = -\mathcal{P}_3/2h_x B_0, \tag{178}$$

which defines uniquely the functions B_3 and C_3 with

$$\begin{aligned}\mathcal{P}_3 = {}& 2B_{2,y}h_y + B_2 h_{yy} - B_{2,x}h_t - B_{2,t}h_x - B_2 h_{xt} - C_2 h_y \\ & + B_2 C_{0,x} + B_0 C_{2,x} + B_1 C_2 h_x + B_{1,yy} - B_{1,xt} - C_{1,y} + B_1 C_{1,x},\end{aligned} \tag{179}$$

and

$$\begin{aligned}\mathcal{Q}_3 = {}& 2C_{2,y}h_y + C_2 h_{yy} - C_{2,x}h_t - C_{2,t}h_x - C_2 h_{xt} + B_2 h_y \\ & + (B_2 B_0)_x - B_1 B_2 h_x + C_{1,yy} - C_{1,xt} + C_{1,y} - B_1 B_{1,x}.\end{aligned} \tag{180}$$

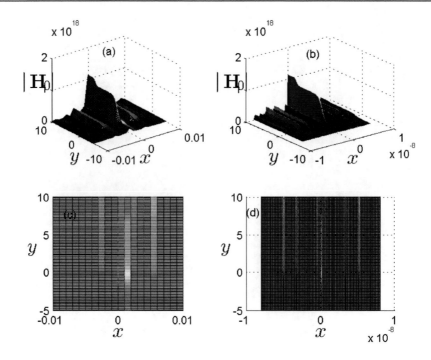

Figure 25. Stochastic fractal solitoff excitations depicted at $t = 0$ by the observable $|\mathbf{H}_0|$ which expression is given by Eq. (190). In this case, the parameters are selected as $a_0 = 1$, $a_1 = 1$, $a_2 = 1$, $a_3 = 2$, $\alpha = 3/2$, and $\beta = 3/2$ such that $p(x,t) = \Theta(x,t)$ and $q(y,t) = \tanh(y)/5 + \tanh(2y-15)/4$, with $\kappa = 2$, $M = 1$, $\eta_0 = 0$, $\eta_1 = 1/2$, $\eta_m = 0$ ($m \geq 2$), $\nu_1 = 1$, $k_1 = 4$, $v_1 = -1$, $\theta_{01} = -20$. Panels (a) and (b) represent the pattern formations depicted in $3D$-perspective, and the the two others (c) and (d) are their corresponding densities represented within the square regions $[-1 \cdot 10^{-2}, 1 \cdot 10^{-2}] \times [-5, 10]$ and $[-8 \cdot 10^{-9}, 8 \cdot 10^{-9}] \times [-5, 10]$, respectively.

Naturally, in such a case, there is no arbitrary functions as derived from the aforementioned resonance conditions.

Finally, solving the case $j = 4$, we obtain the system

$$2(h_x h_t - h_y^2)(3C_4 \pm \imath 2B_4) = \mathcal{Q}_4, \tag{181}$$

showing that either B_4 or C_4 is arbitrary provided to get $\mathcal{P}_4 = \pm \imath \mathcal{Q}_4$ as it can be checked straightforwardly from the following expressions

$$\mathcal{P}_4 = 2(2B_{3,y}h_y + B_3 h_{yy} - B_{3,x}h_t - B_{3,t}h_x - B_3 h_{xt} - C_3 h_y + B_1 C_3 h_x)$$
$$+ B_3 C_{0,x} + B_0 C_{3,x} + B_{2,yy} - B_{2,xt} - C_{2,y} + B_2 C_{1,x} + B_1 C_{2,x} + B_2 C_2 h_x, \tag{182}$$

and

$$\mathcal{Q}_4 = 2(2C_{3,y}h_y + C_3 h_{yy} - C_{3,x}h_t - C_{3,t}h_x - C_3 h_{xt} + B_3 h_y - B_1 B_3 h_x)$$
$$- (B_3 B_0)_x + C_{2,yy} - C_{2,xt} + B_{2,y} - (B_2 B_1)_x - B_2^2 h_x. \tag{183}$$

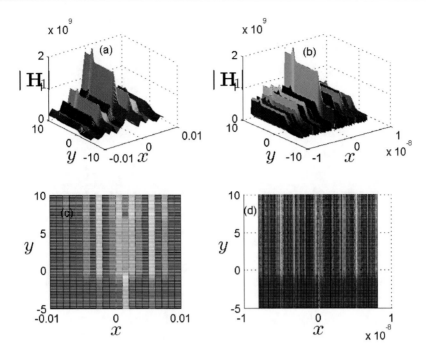

Figure 26. Stochastic fractal solitoff excitations depicted at $t = 0$ by the observable $|\mathbf{H}_1|$ which expression is given by Eq. (190). In this case, the parameters are selected as $a_0 = 1$, $a_1 = 1$, $a_2 = 1$, $a_3 = 2$, $\alpha = 3/2$, and $\beta = 3/2$ such that $p(x,t) = \Theta(x,t)$ and $q(y,t) = \tanh(y)/5 + \tanh(2y - 15)/4$, with $\kappa = 2$, $M = 1$, $\eta_0 = 0$, $\eta_1 = 1/2$, $\eta_m = 0$ ($m \geq 2$), $\nu_1 = 1$, $k_1 = 4$, $v_1 = -1$, $\theta_{01} = -20$. Panels (a) and (b) represent the pattern formations depicted in $3D$-perspective, and the the two others (c) and (d) are their corresponding densities represented within the square regions $[-1 \cdot 10^{-2}, 1 \cdot 10^{-2}] \times [-5, 10]$ and $[-8 \cdot 10^{-9}, 8 \cdot 10^{-9}] \times [-5, 10]$, respectively.

derived from Eqs. (165) and (166).

Our previous analysis shows that there exist sufficient number of arbitrary functions at the resonance values given by Eq. (167). One can proceed further to obtain the higher-order cœfficient functions for all $j \geq 5$ in terms of the previous cœfficients without the introduction of any movable critical singularity manifold into the Laurent expansion. Consequently, we conclude that the system investigated in this section actually passes the P-test. Throughout the achievement of this investigation, it has been derived some NLPDEM equations as compatibility systems under the setting $B_k = \pm \imath C_k$ ($k = 0, \cdots, 4$). The question logically arises whether the extension of this relation can hold even for $j \geq 5$ in general. Such issue constitutes a matter of current investigation which results will be reported elsewhere.

The P-test is actually useful in searching for the BT and hence Lax pairs, such that the spectral problem and also HB of the system can straightforwardly be written down. We aim at investigating deeply the HB of the aforementioned system in a further issue. Nonetheless, we derive the auto-BT of the system as follows.

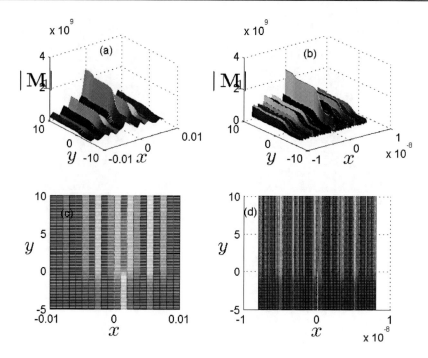

Figure 27. Stochastic fractal solitoff excitations depicted at $t = 0$ by the observable $|\mathbf{M}_1|$ which expression is given by Eq. (190). In this case, the parameters are selected as $a_0 = 1$, $a_1 = 1$, $a_2 = 1$, $a_3 = 2$, $\alpha = 3/2$, and $\beta = 3/2$ such that $p(x,t) = \Theta(x,t)$ and $q(y,t) = \tanh(y)/5 + \tanh(2y - 15)/4$, with $\kappa = 2$, $M = 1$, $\eta_0 = 0$, $\eta_1 = 1/2$, $\eta_m = 0$ ($m \geq 2$), $\nu_1 = 1$, $k_1 = 4$, $v_1 = -1$, $\theta_{01} = -20$. Panels (a) and (b) represent the pattern formations depicted in $3D$-perspective, and the the two others (c) and (d) are their corresponding densities represented within the square regions $[-1 \cdot 10^{-2}, 1 \cdot 10^{-2}] \times [-5, 10]$ and $[-8 \cdot 10^{-9}, 8 \cdot 10^{-9}] \times [-5, 10]$, respectively.

Setting

$$B_k = C_k = 0, \quad k \geq 2, \qquad (184)$$

Eq. (159) becomes

$$B = B_0/h + B_1, \quad C = C_0/h + C_1, \qquad (185)$$

where

$$B_{1,xt} - B_1 C_{1,x} - B_{1,yy} + C_{1,y} = 0, \quad C_{1,xt} + B_1 B_{1,x} - C_{1,yy} - B_{1,y} = 0, \qquad (186)$$

showing that $\{B_1, C_1\}$ is also a solution to the system (158). Thus, the truncated expansion given above is also a BT called for the case the auto-BT. As it has been obtained throughout this study, the quantities B_1 and C_1 which are arbitrary quantities will serve as a seed solution. Based upon the relation between these two observables and combining Eqs. (176) and (186) lead to differential equations satisfied by the noncharacteristic manifold $h(x, y, t)$.

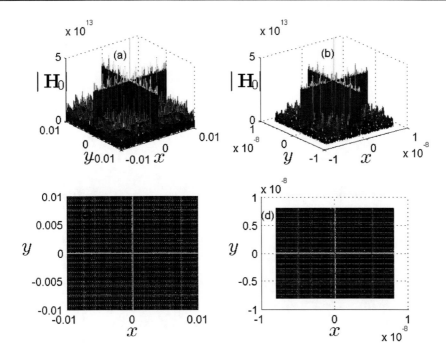

Figure 28. Stochastic fractal lump excitations depicted at $t = 0$ by the observable $|\mathbf{H}_0|$ which expression is given by Eq. (190). In this case, the parameters are selected as $a_0 = 1$, $a_1 = 1$, $a_2 = 1$, $a_3 = 2$, $\alpha = 3/2$, $\beta = 3/2$, and $\bar{N} = 2$ such that: -For $p(x,t) = \Theta(x,t)$, $k_1 = 1$, $v_1 = 1$, $\theta_{01} = 0$, $\varrho_0 = 0$, $\varrho_1 = 1$, $\varrho_2 = 0$, and $\mu_1 = 1000$. -For $q(y,t) = \Theta(y,t)$, $k_2 = 1$, $v_2 = 0$, $\theta_{02} = 0$, $\varrho_0 = 0$, $\varrho_1 = 0$, $\varrho_2 = 1$, and $\mu_2 = 1000$. Panels (a) and (b) represent the pattern formations depicted in 3D-perspective, and the the two others (c) and (d) are their corresponding densities represented within the square regions $[-1 \cdot 10^{-2}, 1 \cdot 10^{-2}] \times [-1 \cdot 10^{-2}, 1 \cdot 10^{-2}]$ and $[-8 \cdot 10^{-9}, 8 \cdot 10^{-9}] \times [-8 \cdot 10^{-9}, 8 \cdot 10^{-9}]$, respectively.

Solving such NLPDEM system shows that the manifold h is actually arbitrary and can be computed through arbitrary expressions as we shall see below.

From Eq. (185), we can introduce three arbitrary functions G, H and F such that

$$B \sim G/F, \quad C \sim H/F. \tag{187}$$

Substituting Eq. (187) into the system (158) leads to the following

$$F(D_x D_t G \cdot F - D_y^2 G \cdot F - D_y H \cdot F)$$
$$+ G(D_x D_t F \cdot F - D_x H \cdot F - D_y^2 F \cdot F) = 0, \tag{188a}$$
$$F(D_x D_t H \cdot F - D_y^2 G \cdot F - D_y G \cdot F)$$
$$+ H(D_x D_t F \cdot F - D_y^2 F \cdot F) + G D_x G \cdot F = 0, \tag{188b}$$

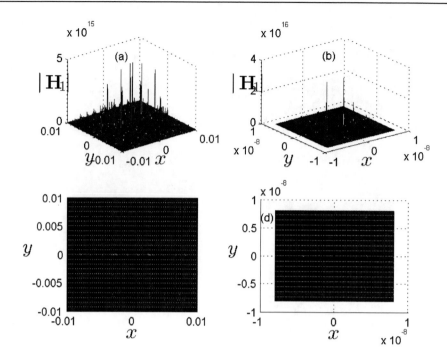

Figure 29. Stochastic fractal lump excitations depicted at $t = 0$ by the observable $|\mathbf{H}_1|$ which expression is given by Eq. (190). In this case, the parameters are selected as $a_0 = 1$, $a_1 = 1$, $a_2 = 1$, $a_3 = 2$, $\alpha = 3/2$, $\beta = 3/2$, and $\bar{N} = 2$ such that: -For $p(x,t) = \Theta(x,t)$, $k_1 = 1$, $v_1 = 1$, $\theta_{01} = 0$, $\varrho_0 = 0$, $\varrho_1 = 1$, $\varrho_2 = 0$, and $\mu_1 = 1000$. -For $q(y,t) = \Theta(y,t)$, $k_2 = 1$, $v_2 = 0$, $\theta_{02} = 0$, $\varrho_0 = 0$, $\varrho_1 = 0$, $\varrho_2 = 1$, and $\mu_2 = 1000$. Panels (a) and (b) represent the pattern formations depicted in $3D$-perspective, and the the two others (c) and (d) are their corresponding densities represented within the square regions $[-1 \cdot 10^{-2}, 1 \cdot 10^{-2}] \times [-1 \cdot 10^{-2}, 1 \cdot 10^{-2}]$ and $[-8 \cdot 10^{-9}, 8 \cdot 10^{-9}] \times [-8 \cdot 10^{-9}, 8 \cdot 10^{-9}]$, respectively.

which can be decoupled as

$$(D_x D_t - D_y^2 - \mu)G \cdot F - D_y H \cdot F = 0, \tag{189a}$$
$$(D_x D_t - D_y^2)H \cdot F - D_y G \cdot F + \lambda H^2 + \nu G^2 = 0, \tag{189b}$$
$$D_x G \cdot F - \nu GF = 0, \tag{189c}$$
$$(D_x - \lambda)H \cdot F - \mu F^2 = 0, \tag{189d}$$
$$(D_x D_t - D_y^2)F \cdot F - \lambda D_y HF = 0, \tag{189e}$$

with constants μ, ν and λ being arbitrary parameters. From the previous system, multibright and multidark soliton solutions can be constructed by expanding suitably the functions G, H and F as power series of an arbitrary perturbative parameter.

Following the previous results, we have shown that the system (158) passes the P-test and possesses some additional properties such as auto-BT and HB. Consequently, we can conclude that the system is completely integrable provided the system free from the back-

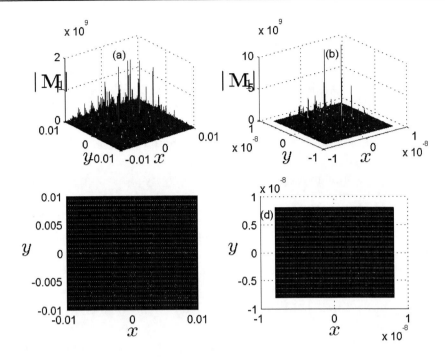

Figure 30. Stochastic fractal lump excitations depicted at $t = 0$ by the observable $|\mathbf{M}_1|$ which expression is given by Eq. (190). In this case, the parameters are selected as $a_0 = 1$, $a_1 = 1$, $a_2 = 1$, $a_3 = 2$, $\alpha = 3/2$, $\beta = 3/2$, and $\bar{N} = 2$ such that: -For $p(x,t) = \Theta(x,t)$, $k_1 = 1$, $v_1 = 1$, $\theta_{01} = 0$, $\varrho_0 = 0$, $\varrho_1 = 1$, $\varrho_2 = 0$, and $\mu_1 = 1000$. -For $q(y,t) = \Theta(y,t)$, $k_2 = 1$, $v_2 = 0$, $\theta_{02} = 0$, $\varrho_0 = 0$, $\varrho_1 = 0$, $\varrho_2 = 1$, and $\mu_2 = 1000$. Panels (a) and (b) represent the pattern formations depicted in $3D$-perspective, and the the two others (c) and (d) are their corresponding densities represented within the square regions $[-1 \cdot 10^{-2}, 1 \cdot 10^{-2}] \times [-1 \cdot 10^{-2}, 1 \cdot 10^{-2}]$ and $[-8 \cdot 10^{-9}, 8 \cdot 10^{-9}] \times [-8 \cdot 10^{-9}, 8 \cdot 10^{-9}]$, respectively.

ground instability. In the further step, we will make advantage of this interesting property to unwrap a quite deal of pattern formations with fractal structures of different kinds.

Let us pay attention to the dynamics of the observables $|\mathbf{H}_0/m|$, $|\mathbf{H}_1/m|$, and $|\mathbf{M}_1/m|$ expressed as follows.

$$|\mathbf{H}_0/m| = \sqrt{1 + 2A + 2A^2}, \quad |\mathbf{H}_1/m| = \sqrt{2(|C|^2 + |C_y|^2)}, \quad |\mathbf{M}_1/m| = 2|C|, \quad (190)$$

where quantities A and C are expressed above in terms of the arbitrary manifold h given by

$$h(x, y, t) = a_0 + a_1 p(x,t) + a_2 q(y,t) + a_3 p(x,t) q(y,t), \quad (191)$$

with constant a_i ($i = 0, \ldots, 3$) being arbitrary parameters while p and q stand for arbitrary functions of (x,t) and (y,t), respectively. Due to the arbitrariness of the manifold h, there arguably exists a wide set of solutions from which we convey our interest to structures with fractal properties. For convenience, we set the dimensionless magnetization to unity.

We search for appropriate parameters and arbitrary functions that can enter into the expression of the manifold h above, and we construct an interesting set of structures. As a matter of facts, in Fig. 16, we depict the variations of the $\mathbf{H}_0$-observable with space at $t = 0$. In a $3D$-representation, the features presented in panel $16(a)$ within the space region $[-1{\cdot}10^{-2}, 1{\cdot}10^{-2}]^2 \times \mathbf{H}_0$ and those depicted in $16(b)$ within region $[-8{\cdot}10^{-9}, 8{\cdot}10^{-9}]^2 \times \mathbf{H}_0$ are self-similar nonlocal. Such a similarity in the profiles is clearly shown in panels $16(c)$ and $16(d)$ standing for their density plots, respectively. The same observations are made when representing the features of the $\mathbf{H}_1$-observable both in planar and $3D$-perspectives. The nonlocal structures depicted in panels $17(a)$ and $17(b)$ both show a distribution of peaks in the xy-plane and are similarly represented in their density plots through panels $17(c)$ and $17(d)$, respectively. In the wake of the previous depictions, as presented in Fig. 18, the self-similarity is also observed in the dynamical behavior of the $\mathbf{M}_1$-observable which at initial time describes a nonlocal distribution of smooth peak within the same considered region as above.

Alongside the previous structures, by choosing appropriately lower-dimensional arbitrary dromion functions and suitable parameters as presented in the caption of Fig. 19, we obtain a typical pattern formation known as fractal dromion which is exponentially localized in a large scale of x and y. Nonetheless, in order to better present the self-similarity in structure of such fractal patterns, we enlarge a small region near the centre of the previous figure. For instance, by reducing the region of panel $19(a)$ $(x, y) \in [-1 \cdot 10^{-2}, 1 \cdot 10^{-2}]^2$ to $[-8 \cdot 10^{-9}, 8 \cdot 10^{-9}]^2$ of panel $19(b)$, we obtain a totally similar structure with density plots represented in panels $19(c)$ and $19(d)$, respectively. The process can continue further with lesser reducing orders of x- and y-independent variables. Such fractal dromion excitation is also expressive in Fig. 20 where features of $\mathbf{H}_1$-observables are depicted in both planar and $3D$-perspectives with sharp peaks. From these pictures represented in $3D$-perspective in panels $20(a)$ and $20(b)$, one can see how the regular distribution of the peaks is made transversally following parallel lines to the y-axis as depicted in panels $20(c)$ and $20(d)$. Such similarity in structure of the above pictures which also appears in panels $21(a)$ and $21(b)$ while plotting the dynamics at initial time of the $\mathbf{M}_1$-observable shows that the two previous observables actually possess the property of self-similarity. Compare to the previous nonlocal fractal patterns, it appears that the fractal dromions have relatively high amplitudes.

With the details presented in the caption of Fig. 22, we obtain a typical pattern formation which is algebraically localized in a large scale of x and y. Such kind of wave is known as fractal lump. In fact, near the centre, there are infinitely many peaks which are distributed in a fractal manner. In order to see the fractal structure of the previous pattern, we look at the structure more carefully. Panel $22(a)$ presents the $3D$-representation of the fractal features of $\mathbf{H}_0$-observable and its corresponding density plot at region $(x, y) \in [-1{\cdot}10^{-2}, 1{\cdot}10^{-2}]^2$ is depicted in panel $22(c)$. Reducing such a region to $[-8 \cdot 10^{-9}, 8 \cdot 10^{-9}]^2$ shows self-similar structures in panels $22(b)$ and $22(d)$. More detailed studies show the self-similar structure of the previous fractal pattern. For instance, if we reduce the region of $22(b)$ to $[-8 \cdot 10^{-15}, 8 \cdot 10^{-15}]^2, [-8 \cdot 10^{-22}, 8 \cdot 10^{-22}]^2, \ldots$, we retrieve a totally similar structure to that plotted in $22(a)$. Such a fractal lump excitation where there are an infinite distribution of many peaks near the centre is further investigated in Figs. 23 and 24 where the amplitudes are too higher compare to the previous ones. With the density plots shown in panels $23(c)$ and $23(d)$ for the $\mathbf{H}_1$-observable, and $24(c)$ and $24(d)$ for the $\mathbf{M}_1$-observable, one can clearly point out the self-similarity property of these observables.

In addition to the above self-similar regular fractal dromion and lump excitations, by using the lower-dimensional stochastic fractal functions, we construct some higher-dimensional stochastic fractal dromions and lump solutions. By trying the so-called Weierstrass function into the dromion or lump solutions leads to stochastic fractal dromions and lumps. Indeed, Fig. 25 presents a typical stochastic fractal solitoff pattern derived from a lower-dimensional stochastic arbitrary fractal dromion. By expanding the planar region of Fig. 25 to larger scale shows that the previous structure is actually localized. Investigating the self-similarity in detail, we reduce the region $(x, y) \in [-1 \cdot 10^{-2}, 1 \cdot 10^{-2}] \times [-5, 10]$ of panel $25(a)$ to $(x, y) \in [-8 \cdot 10^{-22}, 8 \cdot 10^{-22}] \times [-5, 10]$ as shown in panel $25(b)$. Their densities presented at the bottom in panels $25(c)$ and $25(d)$, respectively, clearly present such a similarity in structure. In the wake of such analysis, the depiction of the features of the $\mathbf{H}_1$-observables at time $t = 0$ is also presented in Fig. 26, and as it can be witnessed, the stochastic fractal nature of this structure is conserved even in the very small scale of representation. In such a configuration, panels $26(a)$ and $26(b)$ describe fractal solitoff excitations which amplitudes vary stochastically. Such kind of fractal solitoff is also investigated in Fig. 27 where the $3D$-representations $27(a)$ and $27(b)$ are presented in view of corroborating the self-similarity in structure as depicted in panels $27(c)$ and $27(d)$ for the $\mathbf{M}_1$-observable. These features would also be more expressive if the Weierstrass function was included in the expression of q. In Figs. 28, 29 and 30, with the selecting parameters and suitable choices of lower-dimensional arbitrary stochastic fractal lump functions as presented in the captions of these figures, we obtain higher-dimensional stochastic lump excitations. Through the panels $(28(a)$ and $28(b))$, $(29(a)$ and $29(b))$ and $(30(a)$ and $30(b))$ depicting the variations of $\mathbf{H}_0$-, $\mathbf{H}_1$-, and $\mathbf{M}_1$-observables, respectively, at $t = 0$, the self-similarity in structure of these observables shows how the peaks are distributed stochastically within the regions $(x, y) \in [-1 \cdot 10^{-2}, 1 \cdot 10^{-2}]^2$ and $(x, y) \in [-8 \cdot 10^{-22}, 8 \cdot 10^{-22}]^2$. In such a configuration, the stochastic fractal lump excitations appear to be the wave with greater amplitudes in comparison with the previous ones.

Now, from a physical point of view, in the wake of the previous results, it is actually remarkable to see that the bulk polaritons which propagate within a ferromagnet out of some arbitrary background instability can generate not only line solitons as previously pointed out by Leblond and Manna [65] in a recent investigation, but a wide class of pattern formations among which the fractal structures upon which we have paid attention throughout this section. In fact, such fractal patterns depicted above provide more information about the fractal nature of such model ferromagnet. While comparing the different amplitudes of the observables plotted above, it appears that the stochastic fractal patterns have greater amplitudes than the two other classes of fractal structures. Nonetheless, we remark with our results that the amplitude of the first-order component $\mathbf{H}_1$ of perturbation is greater than that of the magnetic field $\mathbf{H}_0$. In order to sustain the consistency of the previous multiscale scheme with the results obtained above, it would be worth considering a perturbative parameter ϵ very small which could compensate the high value of the stochastic patterns. The present work arguably casts on ones mind the idea of computing the appropriate value of parameter ϵ in account of the selected fractal structures to propagate in the ferromagnet as modeled above.

5. Higher-Dimensional Bose–Einstein Condensates within an Optical Trap: Spinors and their Fractal Background

5.1. Physical background

Let us consider bosons with hyperfine spin $f = 1$. This is the case of alkalis with nuclear spin $I = 3/2$ such as sodium 23, potassium 39 and rubidium 87. For alkali bosons with $f = 1$ such as rubidium 85 ($I = 5/2$) and cesium 135 ($I = 7/2$), the structures are richer and will be investigated in a separate issue. We begin by examining the confining potential produced by a laser with a linearly polarized electric field $\mathbf{E}$ in the x-space direction with frequency ω. Thus, an atom in the hyperfine state $/1, m\rangle$ where m is a magnetic quantum number sees the potential

$$\mathcal{U}_m = -\alpha_m \langle \mathbf{E}^2 \rangle / 2, \tag{192}$$

with

$$\alpha_m = \sum_l |eX_{lm}|^2 \omega_{lm} / (\omega_{lm}^2 - \omega), \quad \sum_l |X_{lm}^2| \omega_{lm} = \hbar Z / 2M, \tag{193}$$

and quantities α_m and M being the polarizability of the state $/1, m\rangle$ and the mass of atom, respectively. The quantity $\hbar\omega_{lm} > 0$ is the excitation energy from the ground state $/1, m >$ to excited state $/l\rangle$, eX_{lm} is the dipole matrix element between $/1, m\rangle$ and $/l\rangle$, and Z is the atomic number. Constant e refers to electric charge. The excitation energy ω_{lm} for different nS to nP transitions [182, 183] differ from each other only by the fine splitting typically smaller than the energy difference between these levels. Thus, for a far detuned laser, Eq. (193) can be approximated as

$$\alpha_m = - \left(\sum_l |eX_{lm}^2| \omega_{lm} \right) / (\omega_{PS}^2 - \omega^2) + o(\Delta_{fine}/\Omega), \tag{194}$$

where Δ_{fine} is the fine splitting and $\Omega \equiv \omega_{PS} - \omega$ is the detuning frequency. To the extent, the polarizability is independent of m.

In general, alkali atoms (bosons and fermions) have two hyperfine multiplets such that energy difference between the higher f_{high} and lower f_{low} cases are many orders larger than the frequencies of typical traps. The hyperfine states of atoms can be changed after the scattering because of the electron-spin dependency of the interaction between two alkali atoms. Nonetheless, owing to the fact that there is not enough energy to promote two f_{low}-atoms after the scattering to f_{high}-level, they will remain in the same multiplet after the process. Thus, in an optical trap, all atoms in the ground state will be in the lower multiplet, and the low energy dynamics of the system is therefore described by a pairwise interaction that is rotationally invariant in the hyperfine spin space and preserves the hyperspin of the individual atoms. The general form of this interaction is given by

$$\hat{V}(\mathbf{r}_1 - \mathbf{r}_2) = \delta(\mathbf{r}_1 - \mathbf{r}_2) \sum_{F=0}^{2f} g_F \mathcal{P}_F, \tag{195}$$

with $g_F = 4\pi\hbar^2 a_F/Ma_\perp^2(1 - Ca_F/a_\perp)$, $\mathcal{P}_F$ is the projection operator which projects the pair 1 and 2 into a total hyperfine spin F state, and a_F is the s-wave scattering length in the total spin F channel. Constant $C \simeq 1.46$ and $a_\perp$ is the size of the transverse ground state.

For bosons (or fermions), symmetry implies that only even (or odd) F terms appear in $\hat{V}$. For bosons with hyperfine spin $f = 1$, we have $\hat{V} = g_0\mathcal{P}_0 + g_2\mathcal{P}_2$. Likewise, the relation $\mathbf{F}_1 \cdot \mathbf{F}_2 = \sum_{F=0}^{2f} \lambda_F\mathcal{P}_F$, $\lambda_F \equiv F(F + 1)/2 - f(f + 1)$ becomes $\mathbf{F}_1 \cdot \mathbf{F}_2 = \mathcal{P}_2 - 2\mathcal{P}_0$ and the relation $\sum_{F=0}^{2f} \mathcal{P}_F = 1$ reads $\mathcal{P}_0 + \mathcal{P}_2 = 1$. Dropping the δ-function, $\hat{V} = c_0 + c_2\mathbf{F}_1 \cdot \mathbf{F}_2$ where $c_0 = (g_0 + 2g_2)/3$ and $c_2 = (g_2 - g_0)/3$. The Hamiltonian in the second quantized form is then

$$\mathcal{H} = \int \mathrm{d}\mathbf{r}\Big[\hbar^2\nabla\psi_a^\dagger \cdot \nabla\psi_a/2M + U\psi_a^\dagger\psi_a$$
$$+ c_0\psi_a^\dagger\psi_{a'}^\dagger\psi_{a'}\psi_a/2 + c_2\psi_a^\dagger\psi_{a'}^\dagger\mathbf{F}_{ab} \cdot \mathbf{F}_{a'b'}\psi_{b'}\psi_b/2\Big], \quad (196)$$

where $\psi_a(\mathbf{r})$ is the field annihilation operator for an atom in the hyperfine state $/1, a\rangle$ at point $\mathbf{r}$, $(a = 1, 0, -1)$, $\psi_a^\dagger(\mathbf{r})$ is its adjoint and U is the trapping potential. In order to discuss the ground state structure of the system, it is convenient to write the Bose condensate $\Psi_a(\mathbf{r}) \equiv \langle\psi_a(\mathbf{r})rangle$ as

$$\Phi_a(\mathbf{r}) = \sqrt{n(\mathbf{r})}\zeta_a(\mathbf{r}), \quad (197)$$

where $n(\mathbf{r})$ is the density and ζ_a is the normalized spinor. The ground state structure of $\Psi_a(\mathbf{r})$ is then determined by minimizing the energy with a fixed particle number, that is, $\delta\mathcal{K} = 0$,

$$\mathcal{K} = \int \mathrm{d}\mathbf{r}\left[\hbar^2(\nabla\sqrt{n})^2/2M + \hbar^2(\nabla\zeta)^2 n/2M - (\mu - U)n + n^2(c_0 + c_2\langle\mathbf{F}\rangle^2)/2\right], \quad (198)$$

with $\langle\mathbf{F}\rangle \equiv \zeta_a^\star\mathbf{F}_{ab}\zeta_b$ and μ being the chemical potential. Spinors are related to each other by gauge transformation $\exp(i\theta)$ and spin rotations $\mathcal{U}(\alpha, \beta, \tau) = \exp(-iF_z\alpha)\exp(-iF_z\beta)\exp(-iF_z\tau)$ are degenerate; α, β and τ being Euler angles. Two configurations appear:

- For the "polar" state in analogy with the polar state in superfluid helium 3, the constant c_2 is real-positive ($g_2 > g_0$) and the energy is minimized by $\langle\mathbf{F}\rangle = 0$ leading to

$$\zeta = \exp(i\theta)\mathcal{U}\begin{pmatrix} 0 \\ 1 \\ 0 \end{pmatrix}$$

$$= \exp(i\theta)\begin{pmatrix} -\exp(-i\alpha)\sin\beta/\sqrt{2} \\ \cos\beta \\ \exp(i\alpha)\sin\beta/\sqrt{2} \end{pmatrix}, \quad (199a)$$

$$n^0(\mathbf{r}) = [\mu - U(\mathbf{r}) - W(\mathbf{r})]/c_0. \quad (199b)$$

where $W(\mathbf{r}) = \hbar^2 \nabla^2 \sqrt{n^0}/2M\sqrt{n^0}$. The symmetry group of this state is therefore $U(1) \times S^2$ where $U(1)$ denotes the phase angle θ and S^2 is a surface of a unit sphere referring to all orientations (α, β) of the spin quantization axis.

- For the "Ferromagnetic" state where c_2 is negative $(g_2 < g_0)$, the energy is minimized by making $\langle \mathbf{F}^2 \rangle = 1$ leading to

$$\begin{aligned}
\zeta &= \exp(\imath\theta)\mathcal{U}\begin{pmatrix} 1 \\ 0 \\ 0 \end{pmatrix} \\
&= \exp[\imath(\theta - \tau)]\begin{pmatrix} \exp(-\imath\alpha)\cos^2(\beta/2) \\ \sin\beta/\sqrt{2} \\ \exp(\imath\alpha)\sin^2(\beta/2) \end{pmatrix}, \quad (200a) \\
n^0(\mathbf{r}) &= [\mu - U(\mathbf{r}) - W(\mathbf{r})]/g_2. \quad (200b)
\end{aligned}$$

The combination $(\theta - \tau)$ in Eq. (200) clearly displays a "spin-gauge" symmetry [184]. Owing to this symmetry, the distinct configurations of ζ are given by the full range of the Euler-angles. In such a configuration, the symmetry group is therefore $SO(3)$.

Following the values of the s-wave scattering lengths a_0 and a_2 for sodium 23 and rubidium 87 [185], the overlapping error bars of these lengths shed doubt about the genuine nature of their ground states. Nevertheless, if the inequalities suggested by current estimate [185] are true, it then appears that condensates of sodium 23 and rubidium 87 are polar state and ferromagnetic state, respectively.

Now, investigating the collective modes of trapped spinor BEC, the equation of motion in zero field is

$$\imath\hbar\psi_{m,t} = -(\hbar^2/2M)\nabla^2\psi_m + (U - \mu)\psi_m + c_0\psi_a^\dagger\psi_a\psi_m + c_2(\psi_a^\dagger\mathbf{F}_{ab}\psi_b)\cdot(\mathbf{F}\psi)_m, \quad (201)$$

where repeated subscripts $(a, b = -1, 0, 1)$ should be summed-up. Restricting our interests to $f = 1$ and using the spin-1 (3×3)-matrices, Eq. (201) is rewritten as

$$\begin{aligned}
\imath\hbar\Phi_{1,t} &= -(\hbar^2/2M)\Delta\Phi_1 + (c_0 + c_2)(|\Phi_1|^2 + |\Phi_0|^2)\Phi_1 \\
&\quad +(c_0 - c_2)|\Phi_{-1}|^2\Phi_1 + c_2\Phi_{-1}^\star\Phi_0^2 + (U - \mu)\Phi_1, \quad (202a) \\
\imath\hbar\Phi_{0,t} &= -(\hbar^2/2M)\Delta\Phi_0 + (c_0 + c_2)(|\Phi_1|^2 + |\Phi_{-1}|^2)\Phi_0 \\
&\quad +c_0|\Phi_0|^2\Phi_0 + 2c_2\Phi_0^\star\Phi_1\Phi_{-1} + (U - \mu)\Phi_0, \quad (202b) \\
\imath\hbar\Phi_{-1,t} &= -(\hbar^2/2M)\Delta\Phi_{-1} + (c_0 + c_2)(|\Phi_{-1}|^2 + |\Phi_0|^2)\Phi_{-1} \\
&\quad +(c_0 - c_2)|\Phi_1|^2\Phi_{-1} + c_2\Phi_1^\star\Phi_0^2 + (U - \mu)\Phi_{-1}, \quad (202c)
\end{aligned}$$

where symbol $\Delta \equiv \nabla^2$ stands for the Laplacian operator. With the transformation $t \to t\hbar$, $\mathbf{r} \to \hbar\mathbf{r}/\sqrt{2M}$, and $(\Phi_1, \Phi_0, \Phi_{-1}) \to (\Phi_1, \sqrt{2}\Phi_0, \Phi_{-1})$, Eq. (202) becomes

$$\begin{aligned}
\imath\Phi_{1,t} &= -\Delta\Phi_1 + (c_0 + c_2)(|\Phi_1|^2 + 2|\Phi_0|^2)\Phi_1 \\
&\quad +(c_0 - c_2)|\Phi_{-1}|^2\Phi_1 + 2c_2\Phi_{-1}^\star\Phi_0^2 + (U - \mu)\Phi_1, \quad (203a) \\
\imath\Phi_{0,t} &= -\Delta\Phi_0 + (c_0 + c_2)(|\Phi_1|^2 + |\Phi_{-1}|^2)\Phi_0 \\
&\quad +2c_0|\Phi_0|^2\Phi_0 + 2c_2\Phi_0^\star\Phi_1\Phi_{-1} + (U - \mu)\Phi_0, \quad (203b) \\
\imath\Phi_{-1,t} &= -\Delta\Phi_{-1} + (c_0 + c_2)(|\Phi_{-1}|^2 + 2|\Phi_0|^2)\Phi_{-1} \\
&\quad +(c_0 - c_2)|\Phi_1|^2\Phi_{-1} + 2c_2\Phi_1^\star\Phi_0^2 + (U - \mu)\Phi_{-1}. \quad (203c)
\end{aligned}$$

We refer to Eq. (203) as the three-component GP-type (2+1)-dimensional equations. The lower-dimensional counterparts of Eq. (203) has been solved by IST-method, and multi-component bright and dark soliton solutions have been reported for specific choices of c_0 and c_2 [186–189]. We aim at isolating all possible integrable models arising from Eq. (203) for arbitrary choices of c_0 and c_2 which can be tuned suitably through Feshbach resonance. For such a purpose, we perform a P-analysis to the above fairly generalized system.

5.2. Integrability properties of the spinor Gross–Pitaevskii system

Before performing the P-singularity structure analysis of Eq. (203), we use the following notations

$$\psi_1 = a, \quad \psi_1^\star = b, \quad \psi_{-1} = m, \quad \psi_{-1}^\star = n, \quad \psi_0 = p, \quad \psi_0^\star = q, \tag{204}$$

in such a way that Eq. (203) becomes

$$
\begin{aligned}
\imath a_t &= -\Delta a + (c_0 + c_2)(ab + 2pq)a \\
&\quad + (c_0 - c_2)mna + 2c_2np^2 + (U - \mu)a, &\tag{205a} \\
-\imath b_t &= -\Delta b + (c_0 + c_2)(ab + 2pq)b \\
&\quad + (c_0 - c_2)mnb + 2c_2mq^2 + (U - \mu)b, &\tag{205b} \\
\imath m_t &= -\Delta m + (c_0 + c_2)(mn + 2pq)m \\
&\quad + (c_0 - c_2)abm + 2c_2bp^2 + (U - \mu)m, &\tag{205c} \\
-\imath n_t &= -\Delta n + (c_0 + c_2)(mn + 2pq)n \\
&\quad + (c_0 - c_2)abn + 2c_2aq^2 + (U - \mu)n, &\tag{205d} \\
\imath p_t &= -\Delta p + (c_0 + c_2)(ab + mn)p \\
&\quad + 2c_0p^2q + 2c_2qam + (U - \mu)p, &\tag{205e} \\
-\imath q_t &= -\Delta q + (c_0 + c_2)(ab + mn)q \\
&\quad + 2c_0q^2p + 2c_2pbn + (U - \mu)q. &\tag{205f}
\end{aligned}
$$

We restrict our interests to bosons trapped within a time-independent potential. The case of an analytic time-dependent potential will be investigated in a separate issue. Thus, we use the following transformation

$$(a, b, m, n, p, q)$$
$$\rightarrow \Big(a \exp(\imath\theta t), b \exp(-\imath\theta t), m \exp(\imath\theta t), n \exp(-\imath\theta t), p \exp(\imath\theta t), q \exp(-\imath\theta t) \Big), \tag{206}$$

where $\theta = U - \mu$ is an arbitrary time-independent quantity.

Following the P-formalism due to Weiss, Tabor and Carnevale [89,90], we give expressions of the generalized Laurent expansion [190] for the dependent variables

$$a = \sum_{j=0} a_j w^{j+\tilde{\alpha}}, \quad b = \sum_{j=0} b_j w^{j+\tilde{\beta}}, \quad m = \sum_{j=0} m_j w^{j+\tilde{\gamma}}, \tag{207a}$$

$$n = \sum_{j=0} n_j w^{j+\tilde{\delta}}, \quad p = \sum_{j=0} p_j w^{j+\tilde{\epsilon}}, \quad q = \sum_{j=0} q_j w^{j+\tilde{\omega}}, \tag{207b}$$

in the neighborhood of the noncharacteristic singular manifold $w(x, y, t) = 0$ with nonvanishing derivatives. Naturally, it is required that the constants $\tilde{\alpha}$, $\tilde{\beta}$, $\tilde{\gamma}$, $\tilde{\delta}$, $\tilde{\epsilon}$, and $\tilde{\omega}$ being negative-definite integers such as to expect the previous reduced system passing the P-test. The first task to undertake is the search for such parameters following the leading order analysis of the solution in the following form

$$a \sim a_0 w^{\tilde{\alpha}}, \quad b \sim b_0 w^{\tilde{\beta}}, \quad m \sim m_0 w^{\tilde{\gamma}}, \quad n \sim n_0 w^{\tilde{\delta}}, \quad p \sim p_0 w^{\tilde{\epsilon}}, \quad q \sim q_0 w^{\tilde{\omega}}. \quad (208)$$

Substituting Eq. (208) into the reduced form of Eq. (205) and balancing the most dominant terms, at the leading order, one finds

$$\tilde{\alpha} = \tilde{\beta} = \tilde{\gamma} = \tilde{\delta} = \tilde{\epsilon} = \tilde{\omega} = -1, \quad (209)$$

with the following relations

$$p_0^2 = a_0 m_0, \quad q_0^2 = b_0 n_0, \quad |\nabla w|^2 = (c_0 + c_2) \left(\sqrt{a_0 b_0} + \sqrt{m_0 n_0} \right)^2 / 2, \quad (210)$$

From Eq. (210), it clearly appears that there are three equations for six unknowns revealing that three of them are arbitrary. Consequently, the value $j = 0$ refers to the power at which arbitrary functions can enter into the Laurent expansion. Such values of different powers are called resonances. Before searching for these powers, we substitute the full Laurent expansion given above into the reduced form of Eq. (205) such as to derive the following system

$$\mathcal{D}_j X_j^T = Y_j^T, \quad (211)$$

where upper case T refers to "matrix transpose" of $X_j = (X_j^l)$ and $Y_j = (Y_j^l)$ ($l = 1, \ldots, 6$), $X_j \equiv (a_j, b_j, m_j, n_j, p_j, q_j)$ and $Y_j \equiv (A_j, B_j, M_j, N_j, P_j, Q_j)$ with

$$
\begin{aligned}
A_j = {} & i[a_{j-2,t} + (j-2)a_{j-1}w_t] + \Delta a_{j-2} + (j-2)(2\nabla a_{j-1} \cdot \nabla w + a_{j-1}\Delta w) \\
& -r\left[a_0 \sum_{l=1}^{j-1}(a_{j-l}b_l + 2p_{j-l}q_l) + \sum_{k=1}^{j-1}\sum_{l=0}^{k}(a_{j-k}a_{k-l}b_l + 2a_{j-k}p_{k-l}q_l) \right] \\
& - sa_0 \sum_{l=1}^{j-1} m_{j-l}n_l - s\sum_{k=1}^{j-1}\sum_{l=0}^{k} a_{j-k}m_{k-l}n_l - 2c_2 p_0 \sum_{l=1}^{j-1} p_{j-l}n_l \\
& - 2c_2 \sum_{k=1}^{j-1}\sum_{l=0}^{k} p_{j-k}p_{k-l}n_l, \quad (212)
\end{aligned}
$$

$$
\begin{aligned}
B_j = {} & -i[b_{j-2,t} + (j-2)b_{j-1}w_t] + \Delta b_{j-2} + (j-2)(2\nabla b_{j-1} \cdot \nabla w + b_{j-1}\Delta w) \\
& -r\left[b_0 \sum_{l=1}^{j-1}(b_{j-l}a_l + 2q_{j-l}p_l) + \sum_{k=1}^{j-1}\sum_{l=0}^{k}(b_{j-k}b_{k-l}a_l + 2b_{j-k}q_{k-l}p_l) \right] \\
& - sb_0 \sum_{l=1}^{j-1} n_{j-l}m_l - s\sum_{k=1}^{j-1}\sum_{l=0}^{k} b_{j-k}n_{k-l}m_l - 2c_2 q_0 \sum_{l=1}^{j-1} q_{j-l}m_l \\
& - 2c_2 \sum_{k=1}^{j-1}\sum_{l=0}^{k} q_{j-k}q_{k-l}m_l, \quad (213)
\end{aligned}
$$

$$M_j = A_j \Big|_{\substack{a_j \leftrightarrow m_j \\ b_j \leftrightarrow n_j}} \ , \qquad N_j = B_j \Big|_{\substack{a_j \leftrightarrow m_j \\ b_j \leftrightarrow n_j}} \ , \tag{214}$$

$$
\begin{aligned}
P_j \ = \ & \imath[p_{j-2,t} + (j-2)p_{j-1}w_t] + \Delta p_{j-2} + (j-2)(2\nabla p_{j-1} \cdot \nabla w + p_{j-1}\Delta w) \\
& - r\left[\sum_{l=1}^{j-1}(a_0 b_{j-l} + n_0 m_{j-l})p_l + \sum_{k=1}^{j-1}\sum_{l=0}^{k}(a_{j-k}b_{k-l} + n_{j-k}m_{k-l})p_l\right] \\
& - 2\sum_{l=1}^{j-1}(c_0 p_0 p_{j-l} + c_2 a_0 m_{j-l})q_l \\
& - 2\sum_{k=1}^{j-1}\sum_{l=0}^{k}(c_0 p_{j-k}p_{k-l} + c_2 a_{j-k}m_{k-l})q_l, \tag{215}
\end{aligned}
$$

and

$$
\begin{aligned}
Q_j \ = \ & -\imath[q_{j-2,t} + (j-2)q_{j-1}w_t] + \Delta q_{j-2} + (j-2)(2\nabla q_{j-1} \cdot \nabla w + q_{j-1}\Delta w) \\
& - r\left[\sum_{l=1}^{j-1}(b_0 a_{j-l} + m_0 n_{j-l})q_l + \sum_{k=1}^{j-1}\sum_{l=0}^{k}(b_{j-k}a_{k-l} + m_{j-k}n_{k-l})q_l\right] \\
& - 2\sum_{l=1}^{j-1}(c_0 q_0 q_{j-l} + c_2 b_0 n_{j-l})p_l \\
& - 2\sum_{k=1}^{j-1}\sum_{l=0}^{k}(c_0 q_{j-k}q_{k-l} + c_2 b_{j-k}n_{k-l})p_l, \tag{216}
\end{aligned}
$$

in which we have set $r = c_0 + c_2$ and $s = c_0 - c_2$. The (6×6)-matrix $\mathcal{D}_j$ is expressed as

$$\mathcal{D}_j = \begin{pmatrix} Q_{1j} & ra_0^2 & sn_0 a_0 & rm_0 a_0 & P_{5j} & 2rp_0 a_0 \\ rb_0^2 & Q_{1j} & rb_0 n_0 & sm_0 b_0 & 2rq_0 b_0 & P_{7j} \\ sb_0 m_0 & ra_0 m_0 & Q_{2j} & rm_0^2 & P_{6j} & 2rp_0 m_0 \\ rb_0 n_0 & sa_0 n_0 & rn_0^2 & Q_{2j} & 2rq_0 n_0 & P_{8j} \\ P_{1j} & ra_0 p_0 & P_{3j} & rm_0 p_0 & Q_{3j} & 2ra_0 m_0 \\ rb_0 q_0 & P_{2j} & rn_0 q_0 & P_{4j} & 2rb_0 n_0 & Q_{3j} \end{pmatrix}, \tag{217}$$

where

$$
\begin{aligned}
Q_{1j} \ &= \ -(j-1)(j-2)|\nabla w|^2 + 2r(a_0 b_0 + p_0 q_0) + sm_0 n_0, \tag{218a} \\
Q_{2j} \ &= \ -(j-1)(j-2)|\nabla w|^2 + 2r(m_0 n_0 + p_0 q_0) + sa_0 b_0, \tag{218b} \\
Q_{3j} \ &= \ -(j-1)(j-2)|\nabla w|^2 + r(a_0 b_0 + m_0 n_0) + 2(r+s)p_0 q_0, \tag{218c} \\
P_{1j} \ &= \ rb_0 p_0 + (r-s)q_0 m_0, \quad P_{2j} = ra_0 q_0 + (r-s)p_0 n_0, \tag{218d} \\
P_{3j} \ &= \ rn_0 p_0 + (r-s)a_0 q_0, \quad P_{4j} = rm_0 q_0 + (r-s)p_0 b_0, \tag{218e} \\
P_{5j} \ &= \ 2ra_0 q_0 + 2(r-s)n_0 p_0, \quad P_{6j} = 2rq_0 m_0 + 2(r-s)b_0 p_0, \tag{218f} \\
P_{7j} \ &= \ 2rp_0 b_0 + 2(r-s)m_0 q_0, \quad P_{8j} = 2rp_0 n_0 + 2(r-s)a_0 q_0. \tag{218g}
\end{aligned}
$$

The resonance equation is therefore derived as

$$j^3(j+1)(j-3)^3(j-4)[2(r-s)-3jr+j^2r]^2 = 0, \qquad (219)$$

which yields

$$j \in \{-1, 0, 3, 4, \tilde{N}_1, \tilde{N}_2\}, \qquad (220)$$

provided

$$\tilde{N}_{1,2} = \left\{ 3 \pm \sqrt{r(r+8s)}/r \right\}/2. \qquad (221)$$

It is suitable that $\tilde{N}_{1,2}$ being integers. Thus, setting $\tilde{N}_1 = m$ with $m \in \mathbb{N}$, it comes $\tilde{N}_1 = 3 - m$ such that:

- For $m \in \{0, 3\}$, $c_2 = 0$. The reduced form of the system (205) simplifies to a set of three-coupled higher-dimensional NLS equations with resonances $j = -1, 0, 0, 0, 0, 3, 3, 3, 3, 4$. Although integrability and P-analysis of the lower-dimensional counterpart of this system has already been studied in detail [191], the higher-dimensional case remains worth investigating. However, in a recent work, the present authors [92] have shown that the two-coupled case is integrable and possesses miscellaneous pattern formations.

- For $m \in \mathbb{N}\backslash\{0, 3\}$, $c_0 = -[m(m-3)+4]c_2/m(m-3)$. Particularly, we notice that for $m \in \{1, 2\}$, we get $c_0 = c_2 = c$, c being an arbitrary parameter, and the resonances read $j = -1, 0, 0, 0, 1, 1, 2, 2, 3, 3, 3, 4$. In a previous study, Kanna et al. [189] investigated the lower-dimensional counterpart of this system and showed that the system is integrable. The completion of such analysis to higher-dimensional systems is straightforward while unearthing more other properties of these systems alongside the construction of various kinds of excitations.

  Naturally, for $m \geq 4$, there are more negative resonances implying a lesser number of arbitrary functions to enter into the Laurent expansion. Such a case should be investigated with much care. In the present work, we restrict our interests to $m \in \{1, 2\}$ and investigate its structure based upon the viewpoint of the existence of sufficient arbitrary functions to enter into the Laurent expansion.

Thus, for $j = -1$, the noncharacteristic manifold w is arbitrary.

For $j = 0$, as previously obtained in Eq. (210) while solving the leading-order problem of the system, there are three arbitrary functions as expected from the resonance equation.

For $j = 1$, it is suitable to write Eq. (211) in the form

$$\tilde{D}_1 \tilde{X}_1^T = \tilde{Y}_1^T, \qquad (222)$$

where $\tilde{X}_1^j \equiv X_1^j/X_0^j$ and $\tilde{Y}_1^j \equiv Y_1^j/X_0^j$, $(j = 1, \ldots, 6)$ with

$$Y_1^j = (-1)^j \imath w_t - (2\nabla \ln|X_0^j| \cdot \nabla w + \Delta w). \qquad (223)$$

The matrix $\tilde{D}_1$ is given by

$$
\tilde{D}_1 = \begin{pmatrix}
4cl_1 & 2cX_0^1X_0^2 & 0 & 2cX_0^3X_0^4 & 4cl_2 & 4cX_0^5X_0^6 \\
2cX_0^1X_0^2 & 4cl_1 & 2cX_0^3X_0^4 & 0 & 4cX_0^5X_0^6 & 4cl_2 \\
0 & 2cX_0^1X_0^2 & 4cl_2 & 2cX_0^3X_0^4 & 4cl_1 & 4cX_0^5X_0^6 \\
2cX_0^1X_0^2 & 0 & 2cX_0^3X_0^4 & 4cl_2 & 4cX_0^5X_0^6 & 4cl_1 \\
2cl_1 & 2cX_0^1X_0^2 & 2cl_2 & 2cX_0^3X_0^4 & 2c(l_1+l_2) & 4cX_0^5X_0^6 \\
2cX_0^1X_0^2 & 2cl_1 & 2cX_0^3X_0^4 & 2cl_2 & 4cX_0^5X_0^6 & 2c(l_1+l_2)
\end{pmatrix}, \quad (224)
$$

with

$$
l_1 = X_0^1X_0^2 + X_0^5X_0^6, \quad l_2 = X_0^3X_0^4 + X_0^5X_0^6. \tag{225}
$$

Solving Eqs. (222)–(225), we find that the following system

$$
\tilde{Y}_1^1 + \tilde{Y}_1^3 = 2\tilde{Y}_1^5, \quad \tilde{Y}_1^2 + \tilde{Y}_1^4 = 2\tilde{Y}_1^6, \tag{226}
$$

actually holds. This straightforwardly implies that among the six functions $\{X_1^j\}_{j=1,\ldots,6}$, there are two of them which are arbitrary. Such result is actually consistent with the double resonance at $j = 1$.

For $j = 2$, Eq. (211) can be written as

$$
\tilde{D}_1 \tilde{X}_2^T = \tilde{Y}_2^T, \tag{227}
$$

where $\tilde{X}_2^j \equiv X_2^j/X_0^j$ and $\tilde{Y}_2^j \equiv Y_2^j/X_0^j$ $(j = 1,\ldots,6)$ with

$$
\begin{aligned}
\tilde{Y}_2^1 = {}& \imath X_{0,t}^1/X_0^1 + \Delta X_0^1/X_0^1 - 2c\left[(X_1^5)^2 X_0^4 + (X_1^1)^2 X_0^2\right]/X_0^1 \\
& - 4c\left(X_1^1X_1^5X_0^6 + X_1^1X_0^5X_1^6 + X_1^1X_1^2X_0^1 + X_1^1X_1^6X_0^5 + X_0^5X_1^1X_1^1\right)/X_0^1, \quad (228)
\end{aligned}
$$

$$
\begin{aligned}
\tilde{Y}_2^2 = {}& -\imath X_{0,t}^2/X_0^2 + \Delta X_0^2/X_0^2 - 2c\left[(X_1^2)^2 X_0^1 + (X_1^6)^2 X_0^3\right]/X_0^2 \\
& - 4c\left(X_1^2X_0^5X_1^1 + X_1^2X_1^1X_0^5 + X_1^1X_1^2X_0^2 + X_1^5X_1^6X_0^2 + X_0^6X_1^2X_1^1\right)/X_0^2, \quad (229)
\end{aligned}
$$

$$
\tilde{Y}_2^3 = \tilde{Y}_2^1\left|\begin{matrix} X_0^1 \leftrightarrow X_0^3, X_1^1 \leftrightarrow X_1^3 \\ X_0^2 \leftrightarrow X_0^4, X_1^2 \leftrightarrow X_1^4 \end{matrix}\right., \quad
\tilde{Y}_2^4 = \tilde{Y}_2^2\left|\begin{matrix} X_0^1 \leftrightarrow X_0^3, X_1^1 \leftrightarrow X_1^3 \\ X_0^2 \leftrightarrow X_0^4, X_1^2 \leftrightarrow X_1^4 \end{matrix}\right., \quad (230)
$$

$$
\begin{aligned}
\tilde{Y}_2^5 = {}& \imath X_{0,t}^5/X_0^5 + \Delta X_0^5/X_0^5 - 2c\Big[(X_1^5)^2 X_0^6 + 2X_1^5X_1^6X_0^5 + X_1^1X_1^6X_0^3 \\
& + X_1^3X_1^4X_0^5 + X_1^1X_1^2X_0^5 + X_1^1X_1^5X_0^2 + X_1^2X_1^5X_0^1 + X_1^4X_1^5X_0^3 \\
& \qquad\qquad\qquad X_1^1X_1^3X_0^6 + X_1^3X_1^6X_0^1 + X_1^3X_1^5X_0^4\Big]/X_0^5, \quad (231)
\end{aligned}
$$

and

$$\tilde{Y}_2^6 = -\imath X_{0,t}^6/X_0^6 + \Delta X_0^6/X_0^6 - 2c\Big[(X_1^6)^2 X_0^5 + 2X_1^6 X_1^5 X_0^6 + X_1^2 X_1^5 X_0^4$$
$$+ X_1^3 X_1^4 X_0^6 + X_1^1 X_1^2 X_0^6 + X_1^2 X_1^6 X_0^1 + X_1^1 X_1^6 X_0^2 + X_1^3 X_1^6 X_0^4$$
$$X_1^2 X_1^4 X_0^5 + X_1^4 X_1^5 X_0^2 + X_1^4 X_1^6 X_0^3\Big]/X_0^6. \quad (232)$$

Solving Eqs. (227)–(232), after some tedious calculations of $\{\tilde{Y}_2^j\}$ ($j = 1,\ldots,6$) while replacing $\{X_0^j\}$ and $\{X_1^j\}$ ($j = 1,\ldots,6$) into Eqs. (228)–(232) by their expressions, respectively, we found that the system

$$\tilde{Y}_2^1 + \tilde{Y}_2^3 = 2\tilde{Y}_2^5, \quad \tilde{Y}_2^2 + \tilde{Y}_2^4 = 2\tilde{Y}_2^6, \quad (233)$$

is consistent with the requirements. Such result shows that there exist two arbitrary functions among the six set of dependent variables $\{X_2^j\}$ ($j = 1,\ldots,6$). Such consistency agrees with the multiplicity degree two of the resonance $j = 2$.

For $j = 3$, Eq. (211) is simplified to

$$(\tilde{D}_1 - 2|\nabla w|^2 I)\tilde{X}_3^T = \tilde{Y}_3^T, \quad (234)$$

where letter I stands for (6×6)-identity matrix, $\tilde{X}_3^j \equiv X_3^j/X_0^j$ and $\tilde{Y}_3^j \equiv Y_3^j/X_0^j$, ($j = 1,\ldots,6$) with

$$\tilde{Y}_3^1 = \imath(X_{1,t}^1 + X_2^1 w_t)/X_0^1 + (\Delta X_1^1 + 2\nabla X_2^1 \cdot \nabla w + X_2^1 \Delta w)/X_0^1$$
$$- 2c\left[(X_1^1)^2 X_1^2 + (X_1^5)^2 X_1^4\right]/X_0^1 - 4c\Big(X_2^1 X_1^5 X_0^6 + X_2^1 X_1^1 X_0^2 + X_2^1 X_1^2 X_1^1$$
$$+ X_2^5 X_1^6 X_0^1 + X_2^4 X_1^5 X_0^5 + X_2^5 X_1^4 X_0^5 + X_2^6 X_1^1 X_0^5 + X_2^5 X_1^1 X_0^6$$
$$X_2^2 X_1^1 X_0^1 + X_1^1 X_1^5 X_1^6 + X_2^1 X_1^6 X_0^5 + X_2^6 X_1^1 X_0^1 + X_2^5 X_1^5 X_0^4\Big)/X_0^1, \quad (235)$$

$$\tilde{Y}_3^2 = -\imath(X_{1,t}^2 + X_2^2 w_t)/X_0^2 + (\Delta X_1^2 + 2\nabla X_2^2 \cdot \nabla w + X_2^2 \Delta w)/X_0^2$$
$$- 2c\left[(X_1^2)^2 X_1^1 + (X_1^6)^2 X_1^3\right]/X_0^2 - 4c\Big(X_2^2 X_1^6 X_0^5 + X_2^2 X_1^2 X_0^1 + X_2^2 X_1^1 X_0^2$$
$$+ X_2^6 X_1^5 X_0^2 + X_2^3 X_1^6 X_0^6 + X_2^6 X_1^3 X_0^5 + X_2^5 X_1^2 X_0^6 + X_2^6 X_1^2 X_0^5$$
$$X_2^1 X_1^2 X_0^2 + X_1^2 X_1^5 X_1^6 + X_2^2 X_1^1 X_0^6 + X_2^5 X_1^2 X_0^4 + X_2^6 X_1^6 X_0^3\Big)/X_0^2, \quad (236)$$

$$\tilde{Y}_3^3 = \tilde{Y}_3^1 \Big|_{\substack{X_l^1 \leftrightarrow X_l^3 \\ X_l^2 \leftrightarrow X_l^4, \, l=0,1,2}} \quad , \quad \tilde{Y}_3^4 = \tilde{Y}_3^2 \Big|_{\substack{X_l^1 \leftrightarrow X_l^3 \\ X_l^2 \leftrightarrow X_l^4, \, l=0,1,2}} \quad , \quad (237)$$

$$\tilde{Y}_3^5 = \imath(X_{1,t}^5 + X_2^5 w_t)/X_0^5 + (\Delta X_1^5 + 2\nabla X_2^5 \cdot \nabla w + X_2^5 \Delta w)/X_0^5$$
$$- 4c(X_2^5 X_1^6 X_0^5 + X_2^6 X_1^5 X_0^5 + X_2^5 X_1^5 X_0^6)/X_0^5$$
$$- 2c\Big[X_2^3 X_1^5 X_0^4 + X_1^3 X_1^4 X_1^5 + X_1^1 X_1^2 X_1^5 + X_2^1 X_1^2 X_0^5 + X_2^4 X_1^3 X_0^5$$
$$+ X_2^3 X_1^4 X_0^5 + X_2^2 X_1^1 X_0^5 + X_2^5 X_1^2 X_0^1 + X_2^5 X_1^3 X_0^4 + X_2^4 X_1^5 X_0^3$$
$$+ X_2^5 X_1^4 X_0^3 + X_2^2 X_1^5 X_0^1 + X_2^5 X_1^1 X_0^2 + X_1^1 X_1^5 X_0^2 + X_2^6 X_1^3 X_0^1$$
$$+ X_2^3 X_1^6 X_0^1 + X_2^3 X_1^1 X_0^6 + X_1^3 X_1^1 X_1^6 + X_2^1 X_1^3 X_0^6 + X_2^1 X_1^6 X_0^3$$
$$+ X_2^6 X_1^1 X_0^3 + (X_1^5)^2 X_1^6\Big]/X_0^5, \quad (238)$$

and

$$\tilde{Y}_3^6 = -\imath(X_{1,t}^6 + X_2^6 w_t)/X_0^6 + (\Delta X_1^6 + 2\nabla X_2^6 \cdot \nabla w + X_2^6 \Delta w)/X_0^6$$
$$- 4c(X_2^6 X_1^5 X_0^6 + X_2^5 X_1^6 X_0^6 + X_2^6 X_1^6 X_0^5)/X_0^6$$
$$- 2c\Big[X_2^4 X_1^6 X_0^3 + X_1^3 X_1^4 X_1^6 + X_1^1 X_1^2 X_1^6 + X_2^2 X_1^1 X_0^6 + X_1^4 X_2^3 X_0^6$$
$$+ X_2^4 X_1^3 X_0^6 + X_2^1 X_1^2 X_0^6 + X_2^6 X_1^1 X_0^2 + X_2^6 X_1^4 X_0^3 + X_2^3 X_1^6 X_0^4$$
$$+ X_2^6 X_1^3 X_0^4 + X_2^1 X_1^6 X_0^2 + X_2^6 X_1^2 X_0^1 + X_2^2 X_1^6 X_0^1 + X_2^5 X_1^4 X_0^2$$
$$+ X_2^4 X_1^5 X_0^2 + X_2^4 X_1^2 X_0^5 + X_1^2 X_1^4 X_1^5 + X_2^2 X_1^4 X_0^5 + X_2^2 X_1^5 X_0^4$$
$$+ X_2^5 X_1^2 X_0^4 + (X_1^6)^2 X_1^5\Big]/X_0^6. \quad (239)$$

Now, we solve the system of Eqs. (234)–(239) following the substitution of $\{X_l^j\}$ ($l = 0, 1, 2;\ j = 1, \ldots, 6$) into Eqs. (235)–(239) by their different expressions derived from Eqs. (210), (222) and (227). Then, after tedious calculations, it is found that the following compatibility equations

$$\tilde{Y}_3^1 + l_2 \tilde{Y}_3^3/l_1 = (1 - l_2/l_1)\tilde{Y}_3^5, \quad (240a)$$
$$\tilde{Y}_3^2 + l_2 \tilde{Y}_3^4/l_1 = (1 - l_2/l_1)\tilde{Y}_3^6, \quad (240b)$$
$$\tilde{Y}_3^3 - \tilde{Y}_3^4 = -(1 + l_1/X_0^5 X_0^6)(\tilde{Y}_3^5 - \tilde{Y}_3^6), \quad (240c)$$

hold. These three equations reveal that three arbitrary functions among the set $\{X_3^j\}$ ($j = 1, \ldots, 6$) should enter into the Laurent series at this order of expansion. Such a result is actually in consistency with the multiplicity degree three of the resonance $j = 3$.

Finally, we examine the case $j = 4$. In this situation, for some convenience, we write Eq. (211) as follows

$$(\tilde{D}_1 - 6|\nabla w|^2 I)\tilde{X}_4^T = \tilde{Y}_4^T, \quad (241)$$

where $\tilde{X}_4^j \equiv X_4^j/X_0^j$ and $\tilde{Y}_4^j \equiv Y_4^j/X_0^j$, $(j = 1, \ldots, 6)$ with

$$\tilde{Y}_4^1 = \imath(X_{2,t}^1 + 2X_3^1 w_t)/X_0^1 + (\Delta X_2^1 + 4\nabla X_3^1 \cdot \nabla w + 2X_3^1 \Delta w)/X_0^1 \quad (242)$$

$$-2c\Big[(X_2^1)^2 X_0^2 + (X_1^1)^2 X_2^2 + (X_0^5)^2 X_3^4 + (X_2^5)^2 X_0^4 + (X_1^5)^2 X_2^4 + X_3^4 X_1^5 X_0^5\Big]/X_0^1 \quad (243)$$

$$-4c\Big(X_3^1 X_1^2 X_0^1 + X_2^1 X_2^2 X_0^1 + X_3^2 X_1^1 X_0^1 + X_3^1 X_1^1 X_0^2 + X_2^1 X_1^1 X_1^2 + X_3^5 X_1^4 X_0^5 \quad (244)$$

$$+X_2^5 X_2^4 X_0^5 + X_3^5 X_1^5 X_0^4 + X_2^5 X_1^5 X_1^4 + X_3^5 X_1^6 X_0^1 + X_2^5 X_2^6 X_0^1 + X_3^6 X_1^5 X_0^1 \quad (245)$$

$$+X_3^1 X_1^5 X_0^6 + X_3^1 X_1^6 X_0^5 + X_2^1 X_2^5 X_0^6 + X_2^1 X_1^5 X_1^6 + X_2^1 X_2^6 X_0^5 + X_3^5 X_1^1 X_0^6 \quad (246)$$

$$+X_2^5 X_1^6 X_1^1 + X_2^6 X_1^5 X_1^1 + X_3^1 X_1^1 X_0^5\Big)/X_0^1 \quad (247)$$

$$\tilde{Y}_4^2 = -\imath(X_{2,t}^2 + 2X_3^2 w_t)/X_0^2 + (\Delta X_2^2 + 4\nabla X_3^2 \cdot \nabla w + 2X_3^2 \Delta w)/X_0^2$$

$$- 2c\Big[(X_2^2)^2 X_0^1 + (X_1^2)^2 X_2^1 + (X_0^6)^2 X_3^3 + (X_2^6)^2 X_0^3 + (X_1^6)^2 X_2^3 + X_3^3 X_1^6 X_0^6\Big]/X_0^2$$

$$- 4c\Big(X_3^2 X_1^1 X_0^2 + X_2^2 X_2^1 X_0^2 + X_3^1 X_1^2 X_0^2 + X_3^2 X_1^2 X_0^1 + X_2^2 X_1^2 X_1^1 + X_3^6 X_1^3 X_0^6$$

$$+ X_2^6 X_2^3 X_0^6 + X_3^6 X_1^6 X_0^3 + X_2^6 X_1^6 X_1^3 + X_3^6 X_1^5 X_0^2 + X_2^6 X_2^5 X_0^2 + X_3^5 X_1^6 X_0^2$$

$$+ X_3^2 X_1^6 X_0^5 + X_3^2 X_1^5 X_0^6 + X_2^2 X_2^6 X_0^5 + X_2^2 X_1^5 X_1^6 + X_2^2 X_2^5 X_0^6 + X_3^6 X_1^2 X_0^5$$

$$+ X_2^6 X_1^5 X_1^2 + X_2^5 X_1^6 X_1^2 + X_3^2 X_1^2 X_0^6\Big)/X_0^2, \quad (248)$$

$$\tilde{Y}_4^3 = \tilde{Y}_4^1 \bigg|_{\substack{X_l^1 \leftrightarrow X_l^3 \\ X_l^2 \leftrightarrow X_l^4,\ l = 0, 1, 2, 3}} \quad , \quad \tilde{Y}_4^4 = \tilde{Y}_4^2 \bigg|_{\substack{X_l^1 \leftrightarrow X_l^3 \\ X_l^2 \leftrightarrow X_l^4,\ l = 0, 1, 2, 3}} \quad , \quad (249)$$

$$\tilde{Y}_4^5 = \imath(X_{2,t}^5 + 2X_3^5 w_t)/X_0^5 + (\Delta X_2^5 + 4\nabla X_3^5 \cdot \nabla w + 2X_3^5 \Delta w)/X_0^5$$

$$- 4c\Big(X_3^5 X_1^6 X_0^5 + X_2^5 X_2^6 X_0^5 + X_3^6 X_1^5 X_0^5 + X_3^5 X_1^5 X_0^6 + X_2^5 X_1^5 X_1^6\Big)/X_0^5$$

$$- 2c\Big[X_3^2 X_1^5 X_0^1 + X_3^3 X_1^5 X_0^4 + X_2^2 X_2^5 X_0^1 + X_2^3 X_2^5 X_0^4 + X_3^5 X_1^2 X_0^1 + X_3^5 X_1^3 X_0^4$$

$$+ X_3^1 X_1^2 X_0^5 + X_3^4 X_1^3 X_0^5 + X_3^1 X_1^5 X_0^2 + X_3^4 X_1^5 X_0^3 + X_2^1 X_2^2 X_0^5 + X_2^4 X_2^3 X_0^5$$

$$+ X_2^1 X_1^2 X_1^5 + X_2^4 X_1^3 X_1^5 + X_2^1 X_1^5 X_0^2 + X_2^4 X_2^5 X_0^3 + X_3^2 X_1^1 X_0^5 + X_3^3 X_1^4 X_0^5$$

$$+ X_2^2 X_1^1 X_1^5 + X_2^3 X_1^1 X_1^5 + X_2^5 X_1^1 X_1^2 + X_2^5 X_1^4 X_1^3 + X_3^5 X_1^1 X_0^2 + X_3^5 X_1^4 X_0^3$$

$$+ X_3^3 X_1^6 X_0^1 + X_2^3 X_2^6 X_0^1 + X_3^6 X_1^3 X_0^1 + X_3^1 X_1^3 X_0^6 + X_3^1 X_1^6 X_0^3 + X_2^1 X_2^3 X_0^6$$

$$+ X_2^1 X_1^3 X_1^6 + X_2^1 X_2^6 X_0^3 + X_3^3 X_1^1 X_0^6 + X_2^3 X_1^1 X_1^6 + X_2^6 X_1^1 X_1^3 + X_3^6 X_1^1 X_0^3$$

$$+ (X_2^5)^2 X_0^6 + (X_1^5)^2 X_2^6\Big]/X_0^5, \quad (250)$$

and

$$\tilde{Y}_4^6 = -\imath(X_{2,t}^6 + 2X_3^6 w_t)/X_0^6 + (\Delta X_2^6 + 4\nabla X_3^6 \cdot \nabla w + 2X_3^6 \Delta w)/X_0^6$$
$$- 4c\Big(X_3^6 X_1^5 X_0^6 + X_2^6 X_2^5 X_0^6 + X_3^5 X_1^6 X_0^6 + X_3^6 X_1^6 X_0^5 + X_2^6 X_1^6 X_1^5\Big)/X_0^6$$
$$- 2c\Big[X_3^1 X_1^6 X_0^2 + X_3^4 X_1^6 X_0^3 + X_2^1 X_2^6 X_0^2 + X_2^4 X_2^6 X_0^3 + X_3^6 X_1^1 X_0^2 + X_3^6 X_1^4 X_0^3$$
$$+ X_3^2 X_1^1 X_0^6 + X_3^3 X_1^4 X_0^6 + X_3^2 X_1^6 X_0^1 + X_3^3 X_1^6 X_0^4 + X_2^1 X_2^2 X_0^6 + X_2^3 X_2^4 X_0^6$$
$$+ X_2^2 X_1^1 X_1^6 + X_2^3 X_1^4 X_1^6 + X_2^2 X_2^6 X_0^1 + X_2^3 X_2^6 X_0^4 + X_3^1 X_1^2 X_0^6 + X_3^4 X_1^3 X_0^6$$
$$+ X_2^1 X_1^2 X_1^6 + X_2^4 X_1^6 X_1^3 + X_2^6 X_1^2 X_1^1 + X_2^6 X_1^3 X_1^4 + X_3^6 X_1^2 X_0^1 + X_3^6 X_1^3 X_0^4$$
$$+ X_3^4 X_1^5 X_0^2 + X_2^4 X_2^5 X_0^2 + X_3^5 X_1^4 X_0^2 + X_3^2 X_1^4 X_0^5 + X_3^2 X_1^5 X_0^4 + X_2^2 X_2^4 X_0^5$$
$$+ X_2^2 X_1^4 X_1^5 + X_2^2 X_2^5 X_0^4 + X_3^4 X_1^2 X_0^5 + X_2^4 X_1^2 X_1^5 + X_2^5 X_1^2 X_1^4 + X_3^5 X_1^2 X_0^4$$
$$+ (X_2^6)^2 X_0^5 + (X_1^6)^2 X_2^5\Big]/X_0^6, \quad (251)$$

Solving the system of Eqs. (241)–(251) after the the substitution of $\{X_l^j\}$ ($l = 0, 1, 2, 3$; $j = 1, \ldots, 6$) into Eqs. (242)–(251) by their different expressions derived from Eqs. (210), (222), (227) and (234), by means of a MAPLE software, after tedious calculations, it is shown that the compatibility equation

$$2c(l_1 + l_2)\Big\{l_1\gamma_1 + X_0^3 X_0^4 \gamma_2 + \alpha_3(X_0^5 X_0^6 - l_1 - l_2)$$
$$+ c[\alpha_5 + X_0^5 X_0^6(\alpha_4 - \alpha_5)] + \tilde{Y}_4^6/2\Big\} + (l_1 - l_2)[\alpha_1 + \alpha_3 c(l_2 - l_1)] = 0, \quad (252)$$

is actually consistent, provided

$$\alpha_1 = (\tilde{Y}_4^1 - \tilde{Y}_4^3)/2, \quad \alpha_2 = (\tilde{Y}_4^2 - \tilde{Y}_4^4)/2, \quad (253a)$$
$$\alpha_3 = -(\tilde{Y}_4^1 + \tilde{Y}_4^3 - \tilde{Y}_4^5)/6c(l_1 + l_2), \quad (253b)$$
$$\alpha_4 = -(\tilde{Y}_4^2 + \tilde{Y}_4^4 - \tilde{Y}_4^6)/6c(l_1 + l_2), \quad (253c)$$
$$\alpha_5 = (\tilde{Y}_4^5 - \tilde{Y}_4^6)/2, \quad (253d)$$

and

$$\gamma_1 = \Big\{(\alpha_2 - \alpha_1)(l_1 + l_2)$$
$$- c[\alpha_5 + X_0^5 X_0^6(\alpha_4 - \alpha_3)](2l_1 + 3l_2)\Big\}/4c(l_1 + l_2)^2, \quad (254a)$$
$$\gamma_2 = \Big\{(\alpha_1 - \alpha_2)(l_1 + l_2)$$
$$- c[\alpha_5 + X_0^5 X_0^6(\alpha_4 - \alpha_3)](3l_1 + 2l_2)\Big\}/4c(l_1 + l_2)^2. \quad (254b)$$

Thus, at this level of truncation, there exists only one arbitrary function among the set $\{X_4^j\}$ ($j = 1, \ldots, 6$) to enter into the series. This result corroborates the one obtained from the resonance equation for $j = 4$ about its multiplicity degree one.

From the aforementioned results which stem from the analysis of arbitrary functions to enter into the Laurent expansion, we have found about nine arbitrary quantities as predicted by the resonance values. It is possible to proceed further to obtain higher-order cœfficient functions for all $j \geq 5$ without the introduction of any movable critical singular manifold into the Laurent expansion. We can conclude that the system under interest passes the P-test only for the case $c_0 = c_2 = c$ and is expected to be integrable. The case $c_0 = 0$ is still an open investigation. The complete integrability will be established if some additional properties such as BT, HB, Miura transformation, just to name a few, are derived.

Thus, let us set

$$\{X_l^k\} \equiv 0, \quad k = 1, \ldots, 6, \quad l \geq 2. \tag{255}$$

Then, the full Laurent expansion is truncated to

$$X^1 = X_0^1/w + X_1^1, \quad X^2 = X_0^2/w + X_1^2, \quad X^3 = X_0^3/w + X_1^3, \tag{256}$$
$$X^4 = X_0^4/w + X_1^4, \quad X^5 = X_0^5/w + X_1^5, \quad X^6 = X_0^6/w + X_1^6, \tag{257}$$

provided

$$\tilde{Y}_2^k \equiv 0, \quad k = 1, \ldots, 6, \tag{258}$$

and $\{X_1^k\}$ ($k = 1, \ldots, 6$) satisfy to the reduced form of Eq. (205) for $c_0 = c_2 = c$. Consequently, Eq. (256) stands to be an auto-BT of the above system. In order to find all possible solutions to the aforementioned equations, it is useful to regard the system $\{X_1^k\}$ as seed solutions. For a simple class of solutions, it is straightforward to choose the following case

$$\{X_1^k\} \equiv 0, \quad k = 1, \ldots, 6, \tag{259}$$

which obviously leads to

$$[(-1)^{k+1} i \partial_t + \Delta] X_0^k \equiv 0, \quad k = 1, \ldots, 6, \tag{260}$$

standing for the diffusion-type system.

Moreover, let us set

$$X^1 = G^1/F, \quad X^2 = G^2/F, \quad X^3 = G^3/F, \tag{261}$$
$$X^4 = G^4/F, \quad X^5 = G^5/F, \quad X^6 = G^6/F. \tag{262}$$

Substituting Eq. (261) into (205) yields

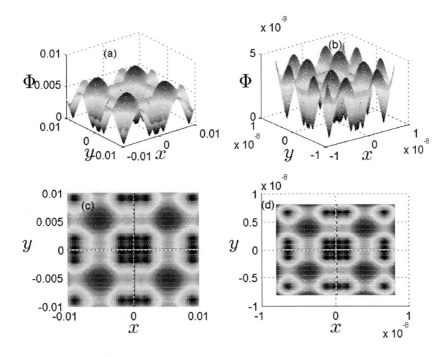

Figure 31. Nonlocal fractal excitations depicted at $t = 0$ by the observable Φ which expression is given by Eq. (273). In this case, the parameters are selected as $a^0 = 1$, $a^1 = 1$, $a^2 = 1$, $a^3 = 2$ such that: -For $f^1(x,t) = \Theta(x,t)$, $\lambda_1 = 1/4$, $\lambda_2 = 0$, $\theta_{01} = 0$, $k_1 = 1$, and $v_1 = 1$. -For $f^2(y,t) = \Theta(y,t)$, $\lambda_1 = 1/4$, $\lambda_2 = 0$, $\theta_{02} = 0$, $k_2 = 1$, and $v_2 = 0$. Panels (a) and (b) represent the pattern formations depicted in 3D-perspective, and the the two others (c) and (d) are their corresponding densities represented within the square regions $[-1 \cdot 10^{-2}, 1 \cdot 10^{-2}] \times [-1 \cdot 10^{-2}, 1 \cdot 10^{-2}]$ and $[-8 \cdot 10^{-9}, 8 \cdot 10^{-9}] \times [-8 \cdot 10^{-9}, 8 \cdot 10^{-9}]$, respectively.

$$F\left[\imath D_t G^1 \cdot F + (D_x^2 + D_y^2)G^1 \cdot F\right]$$
$$+G^1\left[(D_x^2 + D_y^2)F \cdot F - 2c(G^1G^2 + 2G^5G^6)\right] - 2cG^4(G^5)^2 = 0, \quad (263a)$$
$$F\left[-\imath D_t G^2 \cdot F + (D_x^2 + D_y^2)G^2 \cdot F\right]$$
$$+G^2\left[(D_x^2 + D_y^2)F \cdot F - 2c(G^1G^2 + 2G^5G^6)\right] - 2cG^3(G^6)^2 = 0, \quad (263b)$$
$$F\left[\imath D_t G^3 \cdot F + (D_x^2 + D_y^2)G^3 \cdot F\right]$$
$$+G^3\left[(D_x^2 + D_y^2)F \cdot F - 2c(G^3G^4 + 2G^5G^6)\right] - 2cG^2(G^5)^2 = 0, \quad (263c)$$
$$F\left[-\imath D_t G^4 \cdot F + (D_x^2 + D_y^2)G^4 \cdot F\right]$$
$$+G^4\left[(D_x^2 + D_y^2)F \cdot F - 2c(G^3G^4 + 2G^5G^6)\right] - 2cG^1(G^6)^2 = 0, \quad (263d)$$
$$F\left[\imath D_t G^5 \cdot F + (D_x^2 + D_y^2)G^5 \cdot F\right]$$
$$+G^5\left[(D_x^2+D_y^2)F \cdot F - 2c(G^1G^2 + G^3G^4 + G^5G^6)\right] - 2cG^1G^3G^6 = 0, \quad (263e)$$
$$F\left[-\imath D_t G^6 \cdot F + (D_x^2 + D_y^2)G^6 \cdot F\right]$$
$$+G^6\left[(D_x^2+D_y^2)F \cdot F - 2c(G^1G^2 + G^3G^4 + G^5G^6)\right] - 2cG^2G^4G^5 = 0, \quad (263f)$$

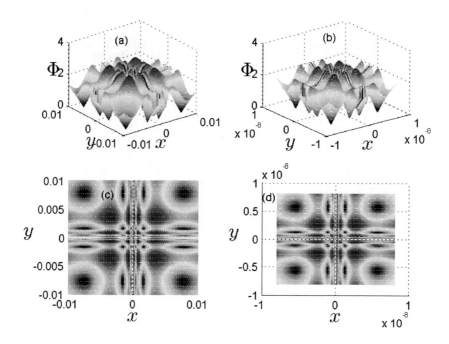

Figure 32. Fractal dromion depicted at $t = 0$ by the observable Φ which expression is given by Eq. (273). In this case, the parameters are selected as $a^0 = 1$, $a^1 = 1$, $a^2 = 1$, $a^3 = 2$, $\theta_0 = 0$, and $k = 1$ such that: -For $f^1(x,t) = \Theta(x,t)$, $c = 1$, and, $v = 1$. -For $f^2(y,t) = \Theta(y,t)$, $c = 1$, and $v = 0$. Panels (a) and (b) represent the pattern formations depicted in 3D-perspective, and the the two others (c) and (d) are their corresponding densities represented within the square regions $[-1 \cdot 10^{-2}, 1 \cdot 10^{-2}] \times [-1 \cdot 10^{-2}, 1 \cdot 10^{-2}]$ and $[-8 \cdot 10^{-9}, 8 \cdot 10^{-9}] \times [-8 \cdot 10^{-9}, 8 \cdot 10^{-9}]$, respectively.

which can be decoupled as

$$[(-1)^{j+1} \imath D_t + D_x^2 + D_y^2 - \mu^j] G^j \cdot F = 0, \quad j = 1, \ldots, 6, \quad (264a)$$
$$(D_x^2 + D_y^2 + \mu^1) F \cdot F - 2c(G^1 G^2 + 2G^5 G^6 + G^4 (G^5)^2 / G^1) = 0, \quad (264b)$$
$$(D_x^2 + D_y^2 + \mu^2) F \cdot F - 2c(G^1 G^2 + 2G^5 G^6 + G^3 (G^6)^2 / G^2) = 0, \quad (264c)$$
$$(D_x^2 + D_y^2 + \mu^3) F \cdot F - 2c(G^3 G^4 + 2G^5 G^6 + G^2 (G^5)^2 / G^3) = 0, \quad (264d)$$
$$(D_x^2 + D_y^2 + \mu^4) F \cdot F - 2c(G^3 G^4 + 2G^5 G^6 + G^1 (G^6)^2 / G^4) = 0, \quad (264e)$$
$$(D_x^2 + D_y^2 + \mu^5) F \cdot F - 2c(G^1 G^2 + G^3 G^4 + G^5 G^6 + G^1 G^3 G^6 / G^5) = 0, \quad (264f)$$
$$(D_x^2 + D_y^2 + \mu^6) F \cdot F - 2c(G^1 G^2 + G^3 G^4 + G^5 G^6 + G^2 G^4 G^5 / G^6) = 0, \quad (264g)$$

where constant μ^j ($j = 1, \ldots, 6$) are arbitrary parameter. Symbols D_x, D_y and D_t refer to Hirota's operators [85,151,167,168]. From Eq. (264), it is possible to construct multibright soliton solutions by expanding the function G^k ($k = 1, \ldots, 6$) and F in power series of an arbitrary perturbative parameter. We reserve such an issue to a further analysis in a separate commitment.

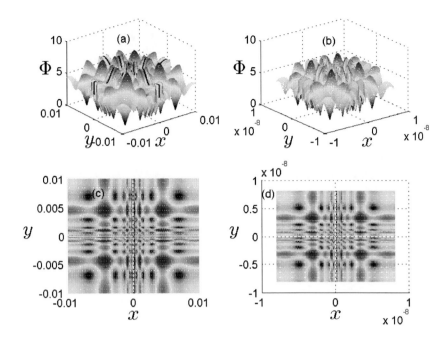

Figure 33. Fractal lump excitations depicted at $t = 0$ by the observable Φ which expression is given by Eq. (273). In this case, the parameters are selected as $a^0 = 1$, $a^1 = 1$, $a^2 = 1$, $a^3 = 2$, $\theta_0 = 0$, and $k = 1$ such that: -For $f^1(x,t) = \Theta(x,t)$, $v = 1$. -For $f^2(y,t) = \Theta(y,t)$, $v = 0$. Panels (a) and (b) represent the pattern formations depicted in $3D$-perspective, and the the two others (c) and (d) are their corresponding densities represented within the square regions $[-1 \cdot 10^{-2}, 1 \cdot 10^{-2}] \times [-1 \cdot 10^{-2}, 1 \cdot 10^{-2}]$ and $[-8 \cdot 10^{-9}, 8 \cdot 10^{-9}] \times [-8 \cdot 10^{-9}, 8 \cdot 10^{-9}]$, respectively.

In the wake of the aforementioned results, we can conclude that the system under which we conveyed all our attention is completely integrable. We now exploit this property in view of discussing the dynamics of a typical class of solutions constitutive of "fractal" pattern formations.

Indeed, solving Eq. (260) by means of the Fourier analysis yields the following solution

$$X_0^j(x, y, t) = \int\int d\lambda d\rho \hat{X}_0^j(\lambda, \rho, t) \exp\left\{ - \imath \left[\lambda x + \rho y + (-1)^{j+1}(\lambda^2 + \rho^2)\right] \right\}, \quad (265)$$

where the arbitrary quantity $\hat{X}_0^j$ is the Fourier transform of X_0^j.

Besides, Eq. (259) also implies that $\tilde{Y}_1^j \equiv 0$ such as

$$(-1)^j \imath w_t - (2\nabla \ln |X_0^j| \cdot \nabla w + \Delta w) = 0, \quad (266)$$

which can be solved by using the characteristic method as follows

$$X_0^j = \mathcal{G}_0^j\left(x - \int^x dy w_x/w_y\right) \exp\left[\int^x dx(\Delta w + (-1)^{j+1}\imath w_t)/2w_x\right], \quad (267)$$

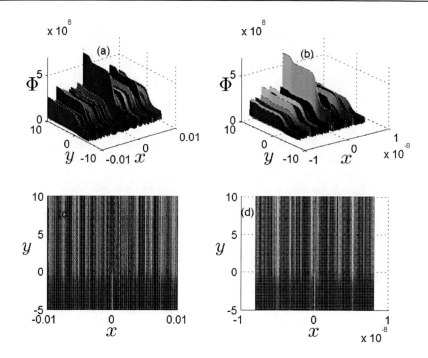

Figure 34. Stochastic fractal solitoff excitations depicted at $t = 0$ by the observable Φ which expression is given by Eq. (273). In this case, the parameters are selected as $a^0 = 1$, $a^1 = 1$, $a^2 = 1$, $a^3 = 2$, $\alpha = 3/2$, and $\beta = 3/2$ such that $f^1(x,t) = \Theta(x,t)$ and $f^2(y,t) = \tanh(y)/5 + \tanh(2y - 15)/4$, with $\kappa = 2$, $M = 1$, $\eta_0 = 0$, $\eta_1 = 1/2$, $\eta_m = 0$ $(m \geq 2)$, $\nu_1 = 1$, $k_1 = 4$, $v_1 = -1$, $\theta_{01} = -20$. Panels (a) and (b) represent the pattern formations depicted in $3D$-perspective, and the the two others (c) and (d) are their corresponding densities represented within the square regions $[-1\cdot 10^{-2}, 1\cdot 10^{-2}] \times [-5, 10]$ and $[-8\cdot 10^{-9}, 8\cdot 10^{-9}] \times [-5, 10]$, respectively.

where $\mathcal{G}$ is an arbitrary function-array which can be expressed in terms of $\hat{X}_0^j$ or rather satisfies to a NLPDEM system while making direct substitution of Eq. (266) into (260).

From the results obtained above, it is appears that the quantity $\left[\sqrt{X_0^1 X_0^2} + \sqrt{X_0^3 X_0^4}\right]$ is arbitrary, and hence, provides to the manifold w arbitrary values. Conversely, choosing arbitrary expressions of the manifold w, one can find a set of various kinds of such quantity. A simple and interesting way to compute such arbitrary quantities is to write the manifold as a system of x-and y-independent functions as follows

$$w(x,y,t) = a^0(t) + a^1(t)f^1(x,t) + a^2(t)f^2(y,t) + a^3(t)f^1(x,t)f^2(y,t), \quad (268)$$

where quantities a^l $(l = 0, 1, 2, 3)$ are arbitrary parameters, and f^1 and f^2 stand for arbitrary functions of (x,t) and (y,t), respectively. Besides, it is shown that the observables Φ_1, Φ_0

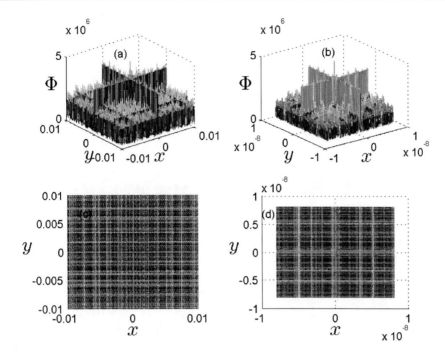

Figure 35. Stochastic fractal lump excitations depicted at $t = 0$ by the observable Φ which expression is given by Eq. (273). In this case, the parameters are selected as $a^0 = 1$, $a^1 = 1$, $a^2 = 1$, $a^3 = 2$, $\alpha = 3/2$, $\beta = 3/2$, and $\bar{N} = 2$ such that: -For $f^1 = \Theta(x,t)$, $k_1 = 1$, $v_1 = 1$, $\theta_{01} = 0$, $\varrho_0 = 0$, $\varrho_1 = 1$, $\varrho_2 = 0$, and $\mu_1 = 1000$. -For $f^2(y,t) = \Theta(y,t)$, $k_2 = 1$, $v_2 = 0$, $\theta_{02} = 0$, $\varrho_0 = 0$, $\varrho_1 = 0$, $\varrho_2 = 1$, and $\mu_2 = 1000$. Panels (a) and (b) represent the pattern formations depicted in $3D$-perspective, and the the two others (c) and (d) are their corresponding densities represented within the square regions $[-1 \cdot 10^{-2}, 1 \cdot 10^{-2}] \times [-1 \cdot 10^{-2}, 1 \cdot 10^{-2}]$ and $[-8 \cdot 10^{-9}, 8 \cdot 10^{-9}] \times [-8 \cdot 10^{-9}, 8 \cdot 10^{-9}]$, respectively.

and Φ_{-1} are related through the following system

$$|\Phi_1|^2 = \left[|\nabla w|/w + \sqrt{|\nabla w|^2/w^2 - 4c|\Phi_0|^2}\right]^2 / 4c, \qquad (269a)$$

$$|\Phi_{-1}|^2 = \left[|\nabla w|/w - \sqrt{|\nabla w|^2/w^2 - 4c|\Phi_0|^2}\right]^2 / 4c, \qquad (269b)$$

provided

$$|\Phi_0|^2 \leq |\nabla w|^2 / 4cw^2. \qquad (270)$$

In the next, we aim at discussing different kinds of fractal patterns depicted by the previous densities. A straightforward attempt is to express $|\Phi_0|$ as follows

$$|\Phi_0|^2 = |\nabla w|^2 / 4\sigma c w^2, \qquad (271)$$

where $\sigma \geq 1$. In fact, the motivation to consider such an expression stems from the fact that the densities $|\Phi_1|^2$ and $|\Phi_{-1}|^2$ are related to each other through the following system

$$(|\Phi_1| + |\Phi_{-1}|)^2 = |\nabla w|^2/4cw^2, \tag{272}$$

from which one can straightforwardly and arguably reckon $(|\Phi_1|^2, |\Phi_{-1}|^2) \propto (|\nabla w|^2/4cw^2)$ to constitute a family of pattern formations of condensates. Thus, let us set

$$\Phi \equiv \sqrt{c}(|\Phi_1| + |\Phi_{-1}|), \tag{273}$$

with $c > 0$ corresponding to the "polar" state of the spinor BEC.

While choosing appropriate parameters and arbitrary functions which enter into the expression of the manifold w above arguably yields an interesting set of pattern formations. For instance, in Fig. 31, the Φ-observable at $t = 0$ varies self-similarly. Indeed, in a $3D$-representation, the features presented in panel $31(a)$ within the space region $[-1 \cdot 10^{-2}, 1 \cdot 10^{-2}]^2 \times \Phi$ and those depicted in $31(b)$ within region $[-8 \cdot 10^{-9}, 8 \cdot 10^{-9}]^2 \times \Phi$ are structurally identical. Such a similarity in profile is clearly shown in panels $31(c)$ and $31(d)$ representing their density plots, respectively. By pursuing further with different scales of x- and y-independent space variables to the reducing orders, we obtain similar pictures. The dynamics of the aforementioned pattern formations can be studied by depicting their profiles at different evolving times and also by studying the scattering among such structures.

Moreover, selecting appropriately lower-dimensional arbitrary dromion functions and suitable parameters as presented in the caption of Fig. 32, we obtain a typical pattern formation which is exponentially localized in a large scale of x and y. As it is observed in such pictures, this pattern is a dromion-like fractal structure. Nonetheless, in order to better present the self-similarity in structure of such fractal patterns, we reduce the region of panel $32(a)$ $(x, y) \in [-1 \cdot 10^{-2}, 1 \cdot 10^{-2}]^2$ to $[-8 \cdot 10^{-9}, 8 \cdot 10^{-9}]^2$ of panel $32(b)$, and we obtain a totally similar structure with density plots represented in panels $32(c)$ and $32(d)$, respectively. The process can continue further with lesser reducing orders of x and y.

With the details presented in the caption of Fig. 33, we obtain a typical pattern formation known as fractal lump excitation, which is algebraically localized in a large scale of x and y. In fact, near the centre, as it can be obviously seen, there are infinitely many peaks which are distributed in a fractal manner. In order to see the fractal structure of the previous pattern, we look at the structure more deeply. Panel $33(a)$ presents the $3D$-representation of the fractal features of Φ-observable and its corresponding density plot at region $(x, y) \in [-1 \cdot 10^{-2}, 1 \cdot 10^{-2}]^2$ is depicted in panel $33(c)$. Reducing such a region to $[-8 \cdot 10^{-9}, 8 \cdot 10^{-9}]^2$ shows self-similar structures in panels $33(b)$ and $33(d)$.

In addition to the previous traveling waves, Fig. 34 presents a typical stochastic fractal solitoff pattern derived from a lower-dimensional stochastic arbitrary fractal dromion. In fact, in large scale, the previous structure is actually localized. Investigating the self-similarity in detail, we reduce the region $(x, y) \in [-1 \cdot 10^{-2}, 1 \cdot 10^{-2}] \times [-5, 10]$ of panel $34(a)$ to $(x, y) \in [-8 \cdot 10^{-22}, 8 \cdot 10^{-22}] \times [-5, 10]$ as shown in panel $34(b)$. Their densities presented at the bottom in panels $34(c)$ and $34(d)$, respectively, clearly show such a similarity in structure. Also, in Fig. 35, with the selecting parameters and suitable choices of lower-dimensional arbitrary stochastic fractal lump functions as presented in the caption of this figure, we obtain higher-dimensional stochastic lump excitations. Through the panels $35(a)$ and $35(b)$ depicting the variations of the Φ-observable at $t = 0$, the self-similarity in structure of this observable shows how the peaks are distributed stochastically. As a matter of remark, one can see that these stochastic fractal structures stand for traveling waves having greater amplitudes compare to the previous ones.

It is interesting to understand how dynamically the "polar" state spinor BEC behaves when the time elapses during the process of optical trapping. Such an investigation is carried out by means of the particle density of each species in the mixture. In the present case, although the Φ-observable appears as a sum of two particle densities related to two different quantum states, its expression can provide further information on the dynamics of each quantum state investigated in this work. Indeed, the results obtained previously show that by trapping optically a mixture of three species BEC having similar hyperfine spin $f = 1$ state and identical masses, the condensates interact mutually within a nonlinear process in such a way that their particle densities describe nonlocal pattern formations, localized structures, but also stochastic fractal waves. Each pattern arguably provides more idea about the genuine structure of the considered media. For instance, the existence of fractal patterns constructed above sheds more light on the probably fragmented internal geometric shape of the system in such a way that it is sufficient to study its dynamics within a reduced-sized copy of its whole to get more idea about its dynamical behavior. Further surveys towards this issue will constitute a matter of forthcoming interests.

Conclusion

Throughout this chapter, we have conveyed all our attention to the fractal structure of four higher-dimensional dynamical systems, namely, the coupled nonlinear extension of the reaction-diffusion equation modeling the development of highly complex organisms based upon nonlinear interactions between common genes, the two-coupled nonlinear Schrödinger equation arising in the description of dynamics of miscellaneous BEC mixtures confined within a time-independent anisotropic parabolic trap potential mapped onto the higher-dimensional time-gated Manakov system up to a first-order of accuracy, the dynamics of bulk polaritons in ferromagnetic slab through the single-oscillation two-dimensional soliton system, and the three-coupled Gross–Pitaevskii type nonlinear equations arising in the context of spinor BEC of atomic hyperfine spin $f = 1$ species. We have thus provided a classification of such fractal structures according to the type of pattern formations generated from the computation. Following the physical background of each system investigated in a straightforward manner, the first task that has been undertaken is the survey of the integrability properties of each system.

Indeed, in the first section of the chapter, we have investigated the P-property of the (2+1)-dimensional coupled NLERD system and proven that it is completely integrable as revealed earlier also by Zhai et al. [26], following the 'prolongation structure'-analysis originally due to Wahlquist and Estabrook [25]. We have then judiciously made use of the truncated P-analysis to derive Hirota's bilinearization [85, 151] of the system prior to a construction of some soliton solutions. The powerfulness of the WTC-method stems from the fact that it does not merely provide more information about the integrability properties of the investigated system, but is helpful in searching for and constructing miscellaneous pattern formations. Indeed, recently, based upon the arbitrariness of some basic functions, we have constructed a wide class of localized and periodic single-and multivalued excitations to the (2+1)-dimensional coupled NLERD system [91]. We pointed out the interaction between half-straight line solitons with different kinds of features. By selecting appropriate functions, we constructed many other structures such as typical kinks, breathers but also some periodic waves, just to name a few. We also presented a systematic way of assessing the soliton structure of a solitary wave solution to the (2+1)-dimensional coupled NLERD system. Sometimes, when two waves interact, they do not only repel or pass through each other. They can also coalesce to form a single moving wave. This is a fusion phenomenon. Besides, when a single solitary wave moves, a peculiar phenomenon sometimes appears where the initial wave splits into two, three or more other similar waves, but with different amplitudes. This kind of process is termed as fission. We showed that the (2+1)-dimensional coupled NLERD system actually possesses these kinds of soliton phenomena [91]. As illustration, we also showed [91] that an initial exponentially decaying ring soliton can split into two baby-solitons repelling each other further. Besides, two exponentially decaying ring solitons with different amplitudes can fuse together to form a single moving structure. Now, in this chapter, we have enriched the class of such structures by providing new ones known as fractal pattern formations. What is interesting here is that it is possible to construct fractal patterns from the associated aforementioned structures such as fractal dromion by introducing some fractal functions into their expressions. Physically, the different pictures depicted previously merely provide more information on that the nonlinear interactions between common genes of an organism satisfying to a reaction-diffusion equation can produce intriguing and fascinating pattern formations, particularly underlying in the understanding of the dynamical properties of some highly natural complex organisms such as zebra stripes, reticulated dragonflies, and butterfly stripes, just to name a few. However, further attention should be payed to the scattering behavior of such structures for a better understanding of the appearance of some other kinds of patterns.

Paying particular attention to two-species mixtures and looking forward to deriving a panel of miscellaneous excitations to the previous equations, we have systematically analyzed the Painlevé-properties of the (2+1)-dimensional CNLS equations, and shown that they are completely integrable according to the particular case $\varrho = -1$ and $\alpha = 1$ of self-defocusing Kerr-nonlinearity [50, 192, 193]. In this analysis, the associated BT and their connection with Hirota's bilinear formalism [85,151] have been presented in such a way that even N-soliton solutions can properly be constructed. In order to check that the aforementioned system is completely integrable, we have constructed its general Lax-representation following the ZS-scheme. It is important to note that there are actually no formal relations between the components of vector NLS solutions. However, in some particular cases, the CNLS system can be reduced to a one-component NLS equation. Although it can appear that such a one-component system is not integrable in general, one cannot conclude straightforwardly that the CNLS system is nonintegrable (see [194] and references therein). Besides, the higher-dimensional systems has a peculiarity in that it enriches the different classes of solutions of lower-dimensional counterparts with new ones no matter whether these lower systems are integrable or not. This is due to the fact that the independent variables spanning the higher-dimensional manifold in which the system is embedded are not related to any equation. However, using a similarity reduction of the system to a lower-dimensional one, the obtained evolution equation is only a part of all possible reductions of the former system. Thus, the integrability properties of the reduced system cannot infer whether the higher-dimensional system is integrable or not. Also, using the arbitrariness of the expansion cœfficients of the solutions to the (2+1)-dimensional CNLS equations, expressed in the form of series, we recently paved the ways for construction of a wide range of localized and periodic excitations among which the singlevalued and multivalued waves such as exponentially decaying ring solitons, "saddle"-like ring solitons, peakons, bubbles, foldons, solitons with compact support, "breather"-like solitons, half-straight-line solitons, and worms, just to name a few (see [92] and references therein). We also studied the scattering among such structures, and we unveiled fusion and fission phenomena occurring during the propagation of such waves [92]. By including conveniently some fractal functions into the expressions of the previous waves, we can generate a panel of fractal pattern formations with physical significance. In the present work, we have pointed out that the composite matter-wave pattern $\mathcal{W}$ in which square form is regarded as the total energy density of the two-species condensates, in a fractal manner describes the scattering among two-dimensional dark-bright matter-waves, but also some induced instabilities in the BEC, such as the thermal instability combined to the diffusivity of the atoms. During the matter-wave propagation, it can appear some peculiarities within the system such as singularities in the propagation distance related to the fractal nature of the matter. In such situations, the fractal patterns can be representative of such a state of the matter.

In the wake of the previous analysis, we have been concerned with the deep survey of the singularity structure of the aforementioned single-oscillation two-dimensional soliton system which has actually passed the P-test provided to suppress the background instability in the system. Following such a study, the associated BT and its connection with Hirota's bilinear formalism [85, 151] have been presented such that we can even construct N-soliton solutions to the system and study its scattering properties in detail. With such properties, we have shown that the aforementioned system is completely integrable. Taking advantage of the existence of arbitrary functions to enter into the Laurent series of each solution to the single-oscillation two-dimensional soliton system, we have generated a set of pattern formations with fractal background. Although there is still more to do with the different types of localized and periodic solutions to the aforementioned system, based upon a physical point of view, it is actually worth interesting to see that the bulk polaritons which propagate within a ferromagnet out of some arbitrary background instability can generate not only line solitons as previously pointed out by Leblond and Manna [65] in a recent investigation, but a wide class of pattern formations among which the fractal structures. In fact, such fractal patterns depicted above provide more information about the fractal nature of such model ferromagnet. By comparing the different amplitudes of the observables plotted above, it appears that the stochastic fractal patterns have greater amplitudes than the others. Also, noticing that the amplitude of the first-order component $\mathbf{H}_1$ of perturbation is greater than that of the magnetic field $\mathbf{H}_0$, in order to make multiscale scheme consistent with the results obtained above, it would be worth considering a perturbative parameter ϵ very small which could compensate the high value of the stochastic patterns. Such analysis has arguably cast on ones mind the idea of computing the appropriate value of parameter ϵ in account of the selected fractal structures to propagate in a the previous ferromagnet system.

Still performing the WTC-approach to the study of the probably existence of conserved quantities of the spinor BEC, we have studied the integrability properties of this system and shown that it actually passes the P-test conditionally. The complete integrability has been unearthed from the unwrapping of the associated BT and Hirota's bilinearization [85, 151] prior to the construction of soliton solutions. In the wake of this singularity structure analysis, we have generated another type of fractal structures to the three-coupled Gross–Pitaevskii type nonlinear equations above. Though there is still more to do with the different types of localized and periodic solutions to the aforementioned system, restricting our interests to the fractal patterns, their physical meanings are actually hidden in the particle density of each species of the mixture. Indeed, the results obtained previously show that by trapping optically a mixture of three-species BEC having similar hyperfine spin $f = 1$ state and identical masses, the condensates interact mutually within a nonlinear process in such a way that their densities describe nonlocal pattern formations, localized structures, but also stochastic fractal waves, each pattern providing more idea about the genuine structure of the considered media. For instance, the existence of fractal patterns constructed above sheds more light on the probably fragmented internal geometric shape of the system in such a way that it is sufficient to study its dynamics within a reduced-size copy of its whole.

Naturally, as mentioned previously, self-similarity is not the sole criterion for an object to be termed fractal. Examples of self-similar objects that are not fractals include the logarithmic spiral and straight lines, which contain copies of themselves at increasingly small scales. These do not qualify, since they have the same Hausdorff dimension as topological dimension. Though we have excessively discussed the self-similarity of the aforementioned physical systems, it is obvious that the question of what on earth their fractal structures are defined, should cast on our mind. Nonetheless, self-similarity is a necessary condition for a system to possess fractal structure. As far as we are concerned, for the sake of shortening this work for a fluent reading while avoiding cumbersome calculus, we leave the further complementary properties on the earth of forthcoming surveys. More structurally, it is essential to mention that the fractal patterns constructed in this work roughly belong to a set of classification according to the self-similarity. Indeed, it should be noted that there are three kinds of self-similarity:

- *Exact self-similarity*. Known as the strongest type of self-similarity, this item endows the fractal with rough identical structures at different scales. Fractals defined by iterated function systems often display exact self-similarity. For example, the Sierpinski triangle and Koch snowflake [101] exhibit exact self-similarity.

- *Quasi-self-similarity*. This constitutes the looser form of self-similarity. The fractal appears approximately identical at different scales. Quasi-self-similar fractals contain small copies of the whole fractal in distorted and degenerate forms. Fractal defined by recurrence relations are usually quasi-self-similar but not exactly self-similar. As illustrations, the Mandelbrot set [99] is quasi-self-similar, the satellites are approximations of the entire set, but not exact copies.

- *Statistical self-similarity*. Regarded as the weakest type of self-similarity, this item endows the fractal with preservation of numerical or statistical measures across scales. In fact, most reasonable definitions of "fractal" trivially imply some form of statistical self-similarity. For instance, random fractals are statistically self-similar, but neither exactly nor quasi-self-similar.

It is actually worth studying fractal structure of systems because they are easily found in the nature while displaying self-similar structures over an extended, but finite scale range. As a matter of examples, there are clouds, snowflakes, mountain ranges, river networks, cauliflower or broccoli, and system of blood vessels and pulmonary vessels. Applications are straightforward. As mentioned previously, random fractals can be used to describe many highly irregular real-world objects. Also, other applications can be found in classification of histopathology slides in medicine, generation of music, seismology, computer and video game designs, technical analysis of price series, signal and image compression, enzymology, and soil mechanics, among many others [101].

Acknowledgments

The authors would like to express their sincere thanks to Nova Science Publishers for their financial support in publication.

References

[1] M. Hausser, N. Spruston, and G. J. Stuart, Science **290**, 739 (2000).

[2] A. M. Lacasta, I. R. Cantalapiedra, C. E. Auguet, A. Peñaranda, and L. R. Piscina, Phys. Rev. E **59**, 7036 (1999).

[3] T. A. Witten and L. M. Sander, Phys. Rev. Lett. **47**, 1400 (1981).

[4] M. Mimura, H. Sakaguchi, and M. Matsushita, Physica A **282**, 283 (2000).

[5] E. B. Jacob, O. Shochet, A. Tenenbaum, I. Cohen, A. Czirok, and T. Vicsek, Nature (London) **368**, 46 (1994).

[6] H. Levine and E. B. Jacob, Phys. Biol. **1**, 14 (2004).

[7] G. B. Ermentrout and L. E. Keshet, J. Theor. Biol. **160**, 97 (1993).

[8] S. C. Trewick, T. F. Henshaw, R. P. Hausinger, T. Lindahl, and B. Sedgwick, Nature (London) **419**, 174 (2002).

[9] S. W. Lockless and R. Ranganathan, Science **286**, 295 (1999).

[10] P. A. Lindgard and H. Bohr, Phys. Rev. Lett. **77**, 779 (1996).

[11] M. B. Goodman, G. G. Ernstrom, D. S. Chelur, R. O'Hagan, C. A. Yao, and M. Chalfie, Nature (London) **415**, 1039 (2002).

[12] B. L. MacInnis and R. B. Campenot, Science **295**, 1536 (2002).

[13] K. Konno and M. Wadati, Prog. Theor. Phys. **52**, 1652 (1975).

[14] P. J. Olver, *Applications of Lie Groups to Differential Equations* (Springer, New-York, 1993).

[15] V. B. Matveev and M. A. Salle, *Darboux Transformations and Solitons* (Springer, Berlin, 1991).

[16] G. W. Bluman and S. Kumei, *Symmetries and Differential Equations* (Springer, New-York, 1989).

[17] H. C. Hu and Q. P. Liu, Chaos Solitons Fractals **17**, 921 (2003).

[18] Z. J. Lian, L. L. Chen and S. Y. Lou, Chin. Phys. **14**, 1486 (2005).

[19] R. M. Miura, C. S. Gardner, and M. D. Kruskal, J. Math. Phys. **9**, 1204 (1968).

[20] P. Lax, Comm. Pure. App. Math. **21**, 467 (1968).

[21] A. C. Scott, F. Y. F. Chu, and D. W. McLaughlin, Proc. IEEE **61**, 1443 (1973).

[22] C. S. Gardner, J. M. Greene, M. D. Kruskal, and R. M. Miura, Phys. Rev. Lett. **19**, 1095 (1967).

[23] G. L. Jr. Lamb, Rev. Mod. Phys. **43**, 99 (1971).

[24] M. J. Ablowitz, D. J. Kaup, A. C. Newell, and H. Segur, Phys. Rev. Lett. **31**, 125 (1973).

[25] H. D. Wahlquist and F. B. Estabrook, J. Math. Phys. **16**, 1 (1975).

[26] Y. Zhai, S. Albeverio, W. Z. Zhao, and K. Wu, J. Phys. A: Math. Gen. **39**, 2117 (2006).

[27] R. Myrzakulov, G. N. Nugmanova, and R. N. Syzdykova, J. Phys. A: Math. Gen. **31**, 9535 (1998).

[28] X. J. Duan, M. Deng, W. Z. Zhao, and K. Wu, J. Phys. A: Math. Theor. **40**, 3831 (2007).

[29] S. N. Bose, Z. Phys. **26**, 178 (1924).

[30] A. Einstein, Sitzber Kgl. Preuss. Akad. Wiss. **261**, (1924).

[31] C. J. Myatt, E. A. Burt, R. W. Ghrist, E. A. Cornell, and C. E. Wieman, Phys. Rev. Lett. **78**, 586 (1977).

[32] D. M. Stamper-Kurn, M. R. Andrews, A. P. Chikkatur, S. Inouye, H. J. Meisner, J. Stenger, and W. Ketterle, Phys. Rev. Lett. **80**, 2027 (1998); D. S. Hall, M. R. Matthews, J. R. Ensher, C. E. Wieman, and E. A. Cornell, Phys. Rev. Lett. **81**, 1539 (1998).

[33] M. Modugno, F. Dalfovo, C. Fort, P. Maddaloni, and F. Minardi, Phys. Rev. A **62**, 063607 (2000).

[34] N. A. Kostov, V. Z. Enol'skii, V. S. Gerdjikov, V. V. Konotop and M. Salerno, Phys. Rev. E **70**, 056617 (2004).

Dynamics of Miscellaneous Fractal Structures ...

[35] K. Kasamatsu, M. Tsubota, and M. Ueda, Phys. Rev. A **69**, 043621 (2004).

[36] G. Huang, X. Li, and J. Szeftel, Phys. Rev. A **69**, 065601 (2004).

[37] G. Mudugno, M. Modugno, F. Riboli, G. Roati, and M. Inguscio, Phys. Rev. Lett. **89**, 190404 (2002).

[38] F. Riboli and M. Modugno, Phys. Rev. A **65**, 063614 (2002).

[39] T. Busch and J. R. Anglin, Phys. Rev. Lett. **87**, 010401 (2001).

[40] B. Deconnick, J. N. Kutz, M. S. Patterson, and B. W. Warner, J. Phys. A **36**, 5431 (2003).

[41] S. V. Manakov, Sov. Phys. JETP **38**, 248 (1974).

[42] D. Schumayer and B. Apagyi, J. Phys. A **34**, 4969 (2001).

[43] V. M. Perez-Garcia, H. Michinel, and H. Herrero, Phys. Rev. A **57**, 3837 (1998); H. Michinel, V. M. Perez-Garcia, and R. de la Fuente, Phys. Rev. A **60**, 1513 (1999); L. D. Carr, M. A. Leung, and W. P. Reinhardt, J. Phys. B **33**, 3983 (2000).

[44] L. Khaykovich, F. Schreck, G. Ferrari, T. Bourdel, J. Cubizolles, L. D. Carr, Y. Castin, and C. Salomon, Science **296**, 1290 (2002).

[45] K. E. Strecker, G. B. Patridge, A. G. Truscott, and R. G. Hulet, Nature (London) **417**, 150 (2002).

[46] M. Soljacic, K. Steiglitz, S. M. Sears, M. Segev, M. H. Jakubowski, and R. Squier, Phys. Rev. Lett. **90**, 254102 (2003).

[47] D. S. Petrov, M. Holzmann, and G. V. Shylapnikov, Phys. Rev. Lett. **84**, 2551 (2000).

[48] D. S. Petrov, G. V. Shlyapnikov, and J. T. M. Walraven, Phys. Rev. Lett. **85**, 3745 (2000).

[49] H. Q. Zhang, X. H. Meng, T. Xu, L. L. Li, and B. Tian, Phys. Scripta **75**, 537 (2007).

[50] Y. S. Kivshar and G. P. Agrawal, *Optical Solitons for Fibers to Photonic Crystals* (Academic, San Diego, 2003).

[51] R. W. Boyd, *Nonlinear Optics*(Academic, San Diego, 1992).

[52] F. Dalfovo, S. Giorgini, L. P. Pitaevskii, and S. Stringari, Rev. Mod. Phys. **71**, 463 (1999).

[53] P. E. Gross, J. Math. Phys. **4**, 19 (1963).

[54] L. P. Pitaevskii, Sov. Phys.-JETP **13**, 451 (1961).

[55] A. K. Zvezdin and A. F. Popkov, Sov. Phys.-JETP **57**, 350 (1983).

[56] F. G. Bass, N. N. Nasonov, and O. V. Naumenko, Sov. Phys. Tech. Phys. **33**, 742 (1988).

[57] H. Leblond and M. Manna, Phys. Rev. E **50**, 2275 (1994).

[58] P. Degasperis, R. Marcelli, and G. Miccoli, Phys. Rev. Lett. **59**, 481 (1987).

[59] B. A. Kalinikos, N. G. Kovshikov, and A. N. Slavin, Phys. Rev. B **42**, 8658 (1990).

[60] A. N. Slavin and I. V. Rojdestvenski, IEEE Trans. Magn. **30**, 37 (1994).

[61] H. Leblond, J. Phys. A: Math. Gen. **28**, 3763 (1995).

[62] H. Leblond, J. Phys. A: Math. Gen. **32**, 7907 (1999).

[63] H. Leblond, J. Phys. A: Math. Gen. **35**, 10149 (2002).

[64] R. A. Kraenkel, M. A. Manna, and V. Merle, Phys. Rev. E **61**, 976 (2000).

[65] H. Leblond and M. Manna, J. Phys. A: Math. Gen. **41**, 185201 (2008).

[66] H. Leblond and M. Manna, Phys. Rev. Lett. **99**, 064102 (2007).

[67] E. Infeld and G. Rowlands, *Nonlinear Waves, Solitons and Chaos* (Cambridge Univ. Press, Cambridge, 1990).

[68] H. D. Ablowitz and H. Segur, *Solitons and the Inverse Scattering Transform* (PA: SIAM, Philadelphia, 1981).

[69] M. Manna and H. Leblond, J. Phys. A: Math. Gen. **39**, 10437 (2006).

[70] P. G. Kevrekidis, D. J. Frantzeskakis, and R. C. Gonzalez, *Emergent Nonlinear Phenomena in Bose-Einstein Condensates* (Springer, Berlin, 2008).

[71] T. L. Ho, Phys. Rev. Lett. **81**, 742 (1998).

[72] C. K. Law, H. Pu, and N. P. Bigelow, Phys. Rev. Lett. **81**, 5257 (1998).

[73] T. Ohmi and K. J. Machida, J. Phys. Soc. Jpn. **67**, 1822 (1998).

[74] P. Meystre, *Atom Optics* (Springer-Verlag, New-York, 2001).

[75] J. Stenger et al., Nature (London) **396**, 345 (1998).

[76] D. M. Stamper et al., Phys. Rev. Lett. **80**, 2027 (1998).

[77] J. Miesner et al., Phys. Rev. Lett. **82**, 2228 (1999).

[78] H. Pu et al., Phys. Rev. A **60**, 1463 (1999).

[79] H. Schmaljohann et al., Phys. Rev. Lett. **92**, 040402 (2004).

[80] M. S. Chang et al., Phys. Rev. Lett. **92**, 140403 (2004).

[81] J. Ieda, T. Miyakawa, and M. Wadati, Phys. Rev. Lett. **93**, 194102 (2004).

[82] K. E. Strecker et al., Nature (London) **417**, 150 (2002).

[83] C. S. Garner, J. M. Greene, M. D. Kruskal, R. M. Miura, Phys. Rev. Lett. **19**, 1095 (1095).

[84] C. H. Gu, Lett. Math. Phys. **26**, 199 (1992).

[85] R. Hirota, Phys. Rev. Lett. **27**, 1192 (1971).

[86] S. Y. Lou, J. Phys. A: Math. Gen. **32**, 4521 (1999).

[87] S. Y. Lou, Z. Naturf. **53**, 251 (1998).

[88] T. B. Bouetou, B. Gambo, V. K. Kuetche, and T. C. Kofane, Acta Appl. Math. **110**, 945 (2010).

[89] J. Weiss, M. Tabor, and G. Carnevale, J. Math. Phys. **24**, 522 (1983).

[90] J. Weiss, M. Tabor, and G. Carnevale, J. Math. Phys. **25**, 13 (1984)

[91] V. K. Kuetche, T. B. Bouetou, and T. C. Kofane, Phys. Rev. E **79**, 056605 (2009).

[92] V. K. Kuetche, T. B. Bouetou, T. C. Kofane, A. B. Moubissi, and K. Porsezian, Phys. Rev. A **82**, 053619 (2010).

[93] X. Y. Tang, S. Y. Lou, and Y. Zhang, Phys. Rev. E **66**, 046601 (2002).

[94] A. S. Fokas and P. M. Santini, Physica D **44**, 99 (1990).

[95] P. M. Santini, Physica D **41**, 26 (1990).

[96] A. S. Fokas and P. M. Santini, Phys. Rev. Lett. **63**, 1329 (1989).

[97] B. G. Konopelchenko and V. G. Dubrovsky, Stud. Appl. Math. **90**, 189 (1993).

[98] V. G. Dubrovsky and B. G. Konopelchenko, Inverse Probl. **9**, 391 (1993).

[99] B. Mandelbrot, *The Fractal Geometry of Nature* (W. H. Freeman and Co., New-York, 1982).

[100] J. Briggs, *Fractals: The Patterns of Chaos* (Thames and Hudson, London, 1992).

[101] K. J. Falconer, *Fractal Geometry: Mathematical Foundations and Applications* (Wiley and Sons, New Jersey, 1982).

[102] K. J. Falconer, *The Geometry of Fractal Sets* (Cambridge University Press, Cambridge 1985).

[103] T. Vicsek, *Fractal Growth Phenomena* (World Scientific, Singapore, 1989).

[104] W. D. McComb, *The Physics of Fluid Turbulence* (Oxford University Press, USA, 1992).

[105] J. L. Linsky, Space Sci. Rev. **130**, 367 (2007).

[106] T. Michely and J. Krug, *Islands, Mounds, and Atoms: Patterns and Processes in Crystal Growth Far From Equilibrium* (Springer, Berlin, 2003).

[107] G. I. Stegeman and M. Segev, Science **286**, 1518 (1999).

[108] Y. S. Kivshar and B. A. Malomed, Rev. Mod. Phys. **61**, 765 (1989).

[109] A. G. Abanov and P. B. Wiegmann, Phys. Rev. Lett. **86**, 1319 (2001).

[110] J. P. Gollub and M. C. Cross, Nature **404**, 710 (2000).

[111] R. A. Jalabert and H. M. Patawski, Phys. Rev. Lett. **86**, 2490 (2001).

[112] M. V. Hecke and M. Howard, Phys. Rev. Lett. **86**, 2018 (2001).

[113] G. Hu, Y. Zhang, H. A. Cerdeira, and S. G. Chen, Phys. Rev. Lett. **85**, 3377 (2000).

[114] X. Y. Tang, C. L. Chen, and S. Y. Lou, J. Phys. A: Math. Gen. **35**, L293 (2002).

[115] S. Y. Lou, Phys. Lett. A **276**, 94 (2000).

[116] S. Y. Lou, X. Y. Tang, X. M. Qian, C. L. Chen, J. Lin, and S. L. Zhang, Mod. Phys. Lett. B **16**, 1075 (2002).

[117] X. Y. Tang and S. Y. Lou, Chaos Solitons Fractals **14**, 1451 (2002).

[118] C. L. Zheng, Chin. J. Phys. **41**, 442 (2003).

[119] J. Lin and F. M. Wu, Chaos Solitons Fractals **19**, 189 (2004).

[120] A. Maccari, Phys. Lett. A **336**, 117 (2005).

[121] A. Maccari, Chaos Solitons Fractals **27**, 363 (2006).

[122] J. F. Ye and C. L. Zheng, Chin. J. Phys. **45**, 1 (2007).

[123] A. Maccari, J. Math. Phys. **49**, 022702 (2008).

[124] A. Maccari, Chaos Solitons Fractals **43**, 86 (2010).

[125] P. Z. Huan, M. S. Hua, and F. J. Ping, Chin. Phys. B **19**, 100301 (2010).

[126] C. L. Bai, H. Zhao, and X. Y. Wang, Nonlinearity **19**, 1697 (2006).

[127] K. Nakayama, J. Phys. Soc. Jpn. **67**, 3031 (1998).

[128] A. J. Koch and H. Meinhardt, Rev. Mod. Phys. **66**, 1481 (1994).

[129] A. M. Turing, Philos. Trans. R. Soc. London B **237**, 37 (1952).

[130] A. Gierer and H. Meinhardt, Kybernetic **12**, 30 (1972).

[131] L. A. Segel and J. L. Jackson, J. Theor. Biol. **37**, 545 (1972).

[132] D. H. Hubel, T. N. Wiesel, and S. LeVay, Philos. Trans. R. Soc. London B **278**, 377 (1977).

[133] R. Lefever, J. Chem. Phys. **49**, 4977 (1968).

[134] G. Cocho, R. P. Pascual, J. L. Rius, and F. Soto, J. Theor. Biol. **125**, 437 (1987).

[135] H. F. Nijhout, J. Exp. Zool. **206**, 119 (1978).

[136] L. Wolpert, J. Theor. Biol. **25**, 1 (1969).

[137] H. Meinhardt, J. Theor. Biol. **74**, 307 (1978).

[138] M. A. Kuziora and W. McGinnis, Mech. Dev. **33**, 83 (1990).

[139] E. Serfling, Trend Genetics **5**, 131 (1989).

[140] M. A. Lewis and J. D. Murray, J. Math. Biol. **31**, 25 (1992).

[141] D. E. Bentil and J. D. Murray, Physica D **63**, 161 (1993).

[142] L. F. Jaffe, Philos. Trans. R. Soc. London B **295**, 553 (1981).

[143] C. D. Stern, BioEssays **4**, 180 (1986).

[144] A. Babloyantz, J. Theor. Biol. **68**, 551 (1977).

[145] H. Meinhardt, Nature **376**, 722 (1995).

[146] R. Kapral, Physica D **86**, 149 (1995).

[147] C. Varea, J. L. Aragon, and R. A. Barrio, Phys. Rev. E **56**, 1250 (1997).

[148] M. Beccaria and G. Soliani, Physica A **260**, 301 (1998).

[149] L. Martina, O. K. Pashaev, and G. Soliani, Class. Quantum Gravity **14**, 3179 (1997).

[150] E. Celeghini, M. Rasetti, G. Vitiello, Ann. Phys. **215**, 156 (1992).

[151] R. Hirota, *Direct Methods in Soliton Theory* in Soliton, (Springer, Berlin, 1980).

[152] X. Y. Tang and S. Y. Lou, J. Math. Phys. **44**, 4000 (2003).

[153] X. Y. Tang and S. Y. Lou, Comm. Theor. Phys. **40**, 62 (2003).

[154] X. Y. Tang, J. Lin, X. M. Qian, and S. Y. Lou, Int. J. Mod. Phys. B **17**, 4343 (2003).

[155] C. L. Bai and H. Zhao, J. Phys. A **38**, 4375 (2005).

[156] C. L. Bai and H. Zhao, J. Phys. A **39**, 3283 (2006).

[157] C. L. Bai and H. Zhao, Chin. J. Phys. **44**, 94 (2006).

[158] F. Dalfovo and S. Stringari, Phys. Rev. A **53**, 2477 (1996).

[159] G. P. Agrawal, *Nonlinear Fiber Otics* 2nd ed. (Academic, San Diego, 1995).

[160] V. V. Konotop and M. Salerno, Phys. Rev. A **65**, 021602 (2002).

[161] B. B. Baizakov, V. V. Konotop, and M. Salerno, J. Phys. B **35**, 5105 (2002).

[162] L. D. Landau and E. M. Lifshitz, *Quantum Mechanics: Nonrelativistic Theory* (Pergamon Press, New-York, 1977).

[163] A. S. Desyatnikov, D. E. Pelinovsky, and J. Yang, J. Math. Sci. **151**, 3091 (2008).

[164] K. Bongs, S. Burger, S. Dettmer, D. Hellweg, J. Arlt, W. Ertmer, and K. Sengtock, Phys. Rev. A **63**, 031602 (2001).

[165] N. Syassen, D. Bauer, M. Lettner, D. D. T. Volz, J. J. Garca-Ripoli, I. J. Cirac, G. Rempe, and S. Drr, Science **320**, 1329 (2008); J. Anglin, Phys. Rev. Lett. **79**, 6 (1997); J. Ruostekoski and D. F. Walls, Phys. Rev. A **58**, R50 (1998); A. Vardi and J. R. Anglin, Phys. Rev. Lett. **86**, 568 (2001); A. V. Ponomarev, J. M. Nero, A. R. Kolovsky, and A. Buchleitner, Phys. Rev. Lett. **96**, 050404 (2006); W. Wang, L. B. Fu, and X. X. Yi, Phys. Rev. A **75**, 045601 (2007).

[166] X. G. He, D. Zhao, L. Li, and H.G. Luo, Phys. Rev. E **79**, 056610 (2009).

[167] R. Hirota and J. Satsuma, J. Phys. Soc. Jpn. **40**, 611 (1980).

[168] R. Hirota, *Direct Methods in Soliton Theory* (Cambridge University Press, Cambridge, 2004).

[169] J. D. Gibbon, Radmore, M. Tabor, and D. Wood, Stud. Appl. Math. **72**, 39 (1985).

[170] R. Sahadevan, K. M. Tamizhmani, and M. Lakshmanan, J. Phys. A: Math. Gen. **19**, 1783 (1986).

[171] A. B. Shabat, Sov. Math. Dokl. **14**, 1266 (1973).

[172] V. E. Zakharov and A. B. Shabat, Funct. Anal. Appl. **8**, 226 (1974).

[173] P. G. Drazin and R. S. Johnson, *Solitons: an Introduction* (Cambridge University Press, Cambridge, 1989).

[174] J. Babarro, M. J. Paz-Alonso, H. Michinel, J. R. Salguiero, and D. N. Olivieri, Phys. Rev. A **71**, 043608 (2005).

[175] L. Landau and E. Lifschitz, Phys. Z. Sowjet. **8**, 153 (1935).

[176] M. A. Manna and V. Merle, Phys. Rev. E **57**, 6206 (1998).

[177] M. A. Manna, J. Phys. A: Math. Gen. **34**, 4475 (2001).

[178] A. N. Slavin and I. V. Rojdestvenski, IEEE Trans. Magn. **30**, 37 (1994).

[179] H. Leblond, J. Phys. A: Math. Gen. **29**, 4623 (1996).

[180] R. C. Lecraw, E. G. Spencer, and C. S. Porter, Phys. Rev. **110**, 1311 (1958).

[181] M. Daniel, M. D. Kruskal, M. Lakshmanan, and K. Nakamura, J. Math. Phys. **33**, 771 (1992).

[182] I. I. Sobelman, *Atomic Spectra and Radiative Transitions* (Springer-Verlag, Berlin, 1979).

[183] W. L. Wiese, M. W. Smith, and B. M. Glennon, *Atomic Transition Probabilities* (Natl. Burt. Stand (US), Washington, 1966).

[184] T. L. Ho and V. B. Shenoy, Phys. Rev. Lett. **77**, 2595 (1996); ibid. **77**, 3276 (1996); ibid. **81**, 742 (1998).

[185] J. Burke, J. Bohn, and C. Greene (Private communication).

[186] J. Ieda, T. Miyakawa, and M. Wadati, J. Phys. Soc. Jpn. **73**, 2996 (2004).

[187] M. Uchiyama, J. Ieda, and M. Wadati, J. Phys. Soc. Jpn. **75**, 064002 (2006).

[188] T. Kurosaki and M. Wadati, J. Phys. Soc. Jpn. **76**, 084002 (2007).

[189] T. Kanna, K. Sakkaravarthi, C. S. Kumar, M. Lakshmanan, and M. Wadati, J. Math. Phys. **50**, 113520 (2009).

[190] M. Daniel, M. D. Kruskal, M. Lakshmanan, and K. Nakamura, J. Math. Phys. **33**, 771 (1992).

[191] R. Radhakrishnan, R. Sahadevan, and M. Lakshmanan, Chaos Solitons Fractals **5**, 2315 (1995).

[192] N. B. Abraham and W. J. Firth, J. Opt. Soc. Am. B **7**, 951 (1990).

[193] P. D. Maker and R. W. Terhume, Phys. Rev. A **137**, 801 (1965).

[194] V. E. Zakharov and E. I. Schulman, Physica D **4**, 270 (1982).

In: Classification and Application of Fractals
Editor: William L. Hagen

ISBN978-1-61209-967-5
© 2012 Nova Science Publishers, Inc

Chapter 8

MULTIFRACTALS: CONCEPTS AND APPLICATIONS

Ashok Razdan
Astrophysical Sciences Division, Bhabha Atomic Research Centre
Trombay, Mumbai

1 Fractals and Multifractals

Dimensions can be Ecludian d_E, Topological d_T or Fractal d_F. Euclidean dimension d_E derives it definition from co-ordinate system. Topological dimension d_T derives it definition from flexibility of a given form which can be changed to another form without change in value of d_T. Fractal dimension derives its definition from "¥Û ½øÆß which means that a part of an object when magnified, resembles ¨ " © ±¥Ú In general $d_T \leq d_F \leq d_E$. Some objects coexist both as standard as well as fractal. For regular objects $d_F = d_T$, they are called dimensionally concordant sets. For irregular objects $d_F > D_T$ and such sets are called dimensionally discordant sets. Mandelbrot [1] realised that many natural boundaries would not satisfy self-similar property completely. He suggested that fractal needs to redefined in terms of "statistically self similar" behaviour. Statistically self similar means that the magnified local structure is not completely similar to the whole structure. On the average (i.e. statistically) the part of the profile looks similar to the whole profile. Fractals are of two kinds, (a) deterministic (b) random. For deterministic case, a given rule of construction does not change in the entire stage of their construction e.g. koch curves, cantor sets. However, in the case of random fractals a random sequence of formation in the construction gets repeated randomly. This pattern thus formed is generally disordered. One of the examples of random fractals is [2] diffusion limited aggression (DLA).

Multifractals come into picture when different regions of a given object say DLA structures have different fractals dimensions in different regions. Thus for a single complex pattern more than one fractal dimension is needed and hence a need for frame work of generalized dimensions. According to Feder [2] multifractal measures are related to the study of distribution of various quantities on a geometric support. Multifractals describe the statistical properties of the measure in terms of the generalized dimensions or singularity spectrum.

1.1 Hausdroff–Bescovitch dimensions

Mandelbrot, the father of modern fractals, defines a fractal as a set for which Hausdroff–Bescovitch dimension D_H exceeds topological dimension D_T. For any measurement we can define a measuring unit, say $m(l) = al^d$, then the total measure is

$$M_d = a \sum m(l), \tag{1}$$

where 'a' is a constant which is equal to 1 for lines, cubes, squares etc. In the limit of l tending to zero M_d is finite or infinite or zero. In d dimensions if l^d is the unit of measurement and there are $N(l)$ units in the whole measurement, equation (1) can be written as

$$M_d = aN(l)(l)^d. \tag{2}$$

Hausdroff–Bescovitch dimension d_H of a set is the critical dimension for which M_d changes from zero to infinity.

In the case of one scaled Cantor set, one begins with a unit interval [0,1].In the next stage this unit interval is replaced with two new intervals of length (l) $\frac{1}{3}$. Repeat this procedure till nth stage where unit interval is divided into 3^n equal intervals and we need 2^n such pieces to cover the Cantor set. Thus for $N(l)=2^n$ and $l=3^{-n}$

$$M_d = \lim_{l \to 0} 2^n 3^{-dn}. \tag{3}$$

This measure M_d diverges or approaches zero unless d=D=ln(2)/ln(3)=0.6309.

Consider space S in d [3] dimensions and let us cover this space with varying box size l_i^d, where $i = 1, 2, \ldots, N$, N is the total number of boxes. The D-dimensional Hausdroff measure of this set can be written as

$$M_d = \lim_{l \to 0} \sum_j N(l_j) l_i^D. \tag{4}$$

For some unique value of D (say D_f) , M_d will be infinity if $D < D_f$ or zero if $D > D_f$. D_f is the Hausdroff–Bescovitch dimension.

1.2 Generalized dimensions

Fractals are self-similar objects which look same on many different scales of observations and are defined in terms of Hausdroff–Bescovitch dimensions. However, fractal dimensions characterize the geometric support of a structure but can not provide any information about a possible distribution or a probability that may be part of a given structure. This problem has been solved by defining an infinite set of dimensions known as generalised dimensions which are achieved by dividing the object under study into pieces, each piece is labeled by an index $i = 0, 1, 2, \ldots, N$. If we associate a probability p_i with each piece of size l_i than partition function [4] for finite l_i can be written as

$$\Gamma(q, \tau) = \lim_{l \to 0} \Gamma(q, \tau, l), \tag{5}$$

where

$$\Gamma(q, \tau, l) = \sum_{i=1}^n \frac{p_i^q}{l_i^\tau}. \tag{6}$$

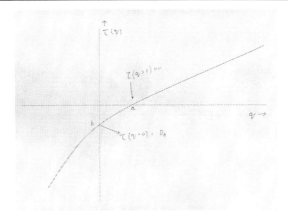

Figure 1: Typical ($\tau(q)$) vs q behaviour of a multifractal.

For unique function $\tau(q)$ it has been shown that

$$\Gamma(q,\tau) = \infty \tag{7}$$

for $\tau < \tau(q)$

$$\Gamma(q,\tau) = 0 \tag{8}$$

for $\tau > \tau(q)$.

This permits to define generalised dimension D_q

$$(q-1)D_q = \tau(q). \tag{9}$$

Here q is a parameter which can take all values between $-\infty$ to ∞. This formalism is called as multi-fractal formalism which characterizes both the geometry of a given structure and the probability measure associated with it.

For $q = 0$, D_0 gives fractal dimension. For $q = 1$, D_1 is the information dimension which encodes the entropy scaling and for $q = 2$, D_2 is the correlation dimension which measures scaling of two point density correlation. In the figure 1, general dependence of $\tau(q)$ on q is shown for a multifractal. Point 'a' in the figure 1 corresponds to value $\tau(q=1)=0$. Point 'b' in this figure corresponds to $\tau(q=0)=D_0$. For multifractals we have

$$D_\infty = \lim_{l \to 0} \frac{ln(MaxP_i)}{ln(l)}, \tag{10}$$

$$D_{-\infty} = \lim_{l \to 0} \frac{ln(MinP_i)}{ln(l)}. \tag{11}$$

Apart from D_0, D_1, D_2 there are infinite set of other exponents from which information can be obtained by constructing an equivalent picture of the system in terms of scaling indices 'α' for the probability measure defined on a support of fractal dimension f(α). In the figure '2' the general shape of the q dependence on D_q is displayed. Point 'a' in the figure '2' corresponds to value of $D_{-\infty}$, point 'b' corresponds to value of D_0 and point 'c' corresponds to D_∞.

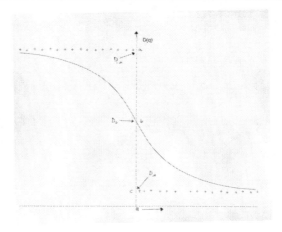

Figure 2: Typical dependence of generalized dimension D(q) on q.

2 Singularity Picture

The measure which is characterized by continuous spectrum of fractal dimension is called multifractal. This situation may be characterized by a singularity exponent α which is known as Lipschitz Holder exponent.

The probability distribution [5] is characterized by its moments

$$z(q) = \sum_p n(p) p^q. \qquad (12)$$

From critical phenomena theory it is known that probability distributions can be related to exponents

$$Z(q) \propto L^{-\tau(q)}, \qquad (13)$$

where L is the size of the system. In critical phenomena τ_q has two exponents. But for examples like DLA, it is found that τ_q is continuous curve having infinite hierarchy of independent exponents which are related by the Legendre transform to fractal dimension $f(\alpha)$.

$$Z(q) \propto L^{f-\alpha q} \qquad (14)$$

An equivalent picture of the system in terms of scaling indices 'α' can be developed for the probability measure defined on a support of fractal dimension f(α). This is achieved by defining probability measure p_i in terms of α.

$$p_i = l_i^{\alpha(q)}, \qquad (15)$$

$$\alpha(q) \geq 0. \qquad (16)$$

It has been shown that

$$D_q = \frac{q\alpha(q) - f(\alpha(q))}{q-1}, \qquad (17)$$

$$\alpha(q) = \frac{d}{dq}[(q-1)D_q], \qquad (18)$$

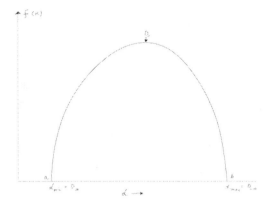

Figure 3: Typical f(α) vs α behaviour of a multifractal.

$$\tau(q) = f(\alpha(q) - q\alpha(q)), \qquad (19)$$

f(α(q)) is the fractal dimension of the set. The generalized dimensions characterize the non uniformity of the measure whereas singularity picture [6] describes mulitifractals in terms of interwoven set of variable singularity strengths α, whose Hausdroff dimension is f($\alpha(q)$). In figure '3' a general shape of a multifractal spectrum is shown. This figure shows α dependence of the multifractal function $f(\alpha)$. The point 'a' in figure '3' corresponds to α_{min} which is also equal to D_∞ and the point 'b' in the figure '3' corresponds to α_{max} which is also equal to $D_{-\infty}$.

Box counting method is one of the most common methods used in multifractals. In this method boxes of different sizes are used to cover the measure and count the number of particles in each box. The minimum size of the box is equal to smallest particle in the measure. The probability to find a particle in a box is number of particles in the box divided by total number of particles in the whole measure from which partition function is calculated. In the following we discuss this method in detail.

2.1 Multifractal analysis procedure

Multifractal properties of a complex structure like DLA can calculated from its digital image in the following steps.

(1) Digitize a given image into numerical values such that data is obtained in a square grid of M x M values. As a case of illustration, we consider 16x16 square grid. i.e $16 \times 16 = 256$ data values.

(2) Identify measure for which multifractal nature is to be investigated. For example in our case of illustration let us consider measure of multifractal moment $G_q(M)$ which is defined as

$$G_q(M) = \sum_{j=1}^{M} \left(\frac{k_j}{N}\right)^q, \qquad (20)$$

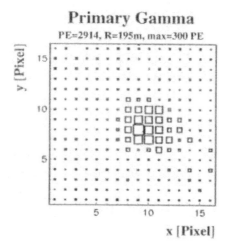

Figure 4: Simulated Cherenkov image for γ-ray initiated showers.

where k_j is the numerical value in j-cell and N is total numerical value of whole image. Again q will vary from $-\infty$ to ∞. M is discussed below.

(3) Divide square grid into equal parts ($\frac{M}{2} \times \frac{M}{2}$) and calculate measure of each part independently. Now add all measures. In the case of our example divide 16×16 grid into M=4 equal parts. we will have 4 square grids of 8×8 values. Calculate $G_q(M)$ of all four 8×8 grids for different values of q. We will have $G_q(1), G_q(2), G_q(3)$ and $G_q(4)$ for each value of q. Now calculate total fractal moment $G_q(1-4) = G_q(1)+G_q(2)+G_q(3)+G_q(4)$.

(4) Divide square grid further into equal parts ($\frac{M}{4} \times \frac{M}{4}$) and calculate measure of each part independently. Now add all measures. In the case of our example divide 16×16 grid into $M = 16$ equal parts. We will have 16 square grids of 4×4 values. Calculate $G_q(M)$ of all sixteen 4×4 grids for different values of q. We will have $G_q(1), G_q(2), G_q(3), \ldots, G_q(16)$ for each value of q. Now calculate total fractal moment $G_q(1-16) = G_q(1) + G_q(2) + \cdots + G_q(16)$. If $G_q(1-16)$ is greater than $G_q(1-4)$ for a fractal.

(5) For fifth step repeat this process of dividing square grid into further values and calculating measure for each part independently. For our example of 16 x 16 grid, we have to repeat this studies for M=64 and 256 and obtain values of $G_q(1-64)$ and $G_q(1-256)$. respectively.

(6)Thus for our example we have M=4,16,64 and 256 and corresponding values of $G_q(1-4), G_q(1-16), G_q(1-64)$ and $G_q(1-256)$ respectively. For a fractal we observe that $G_q(1-4) < G_q(1-16) < G_q(1-64) < G_q(1-256)$ i.e G_q shows power law relation with M

$$G_q \propto M^{\tau_q}. \tag{21}$$

The exponent τ_q is determined from G_q from

$$\tau_q = \frac{1}{ln(2)} \frac{dln(G_q)}{d\nu}, \tag{22}$$

Multifractals 253

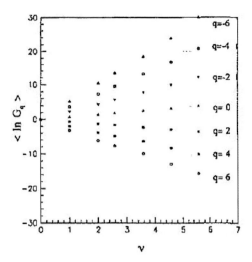

Figure 5: G(q) dependence on q for a γ-ray Cherenkov image.

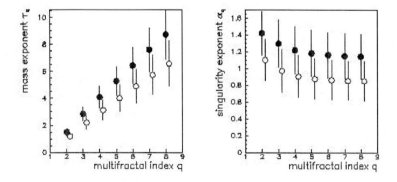

Figure 6: $\tau(q)$ and $\alpha(q)$ of γ-ray and proton initiated Cherenkov images.

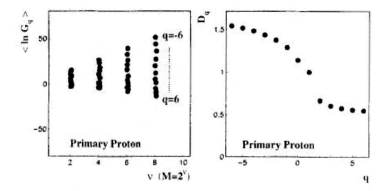

Figure 7: G_q and D_q dependence on q for proton Cherenkov images.

where $M=2^\nu$. For a multifractal structure, there exists a linear relationship between natural logarithm of G_q and scale length ν and the slope of this line, τ_q is related to generalized dimension, D_q by

$$D_q = \frac{\tau_q}{(q-1)}, \qquad (23)$$

where $q \neq 1$.

In order to illustrate this method we consider a case of applying above discussed procedure to a simulated Cherenkov image [7], [8]. This method has been applied in detail to simulated Cherenkov images initiated by γ-rays, protons, neons and Iron nuclei. In the figure '4' is shown a typical gamma initiated simulated Cherenkov image obtained in 16 x 16 pixel camera used in γ-ray astronomy observations. The box size represents number of photoelectrons per pixel. The total number of photoelectrons in the whole image along with maximum number of photoelectrons (in one pixel) are indicated along with distance R from the shower axis.

Simulated image is divided into $M = 2^\nu$ where $\nu = 2, 4, 6, 8$ is the scale. The multifractal moments are given by equation (20) where k_j in the present case, is the number of photoelectrons in the kth cell and N is the total number of photoelectrons in whole image.

Thus by applying the above discussed procedure to the simulated image we obtain figure '5' which shows scaling behaviour of G_q versus q for γ-ray initiated shower. In figure '6' $\tau(q)$ and $\alpha(q)$ dependence on q is shown for γ-rays and protons. In figure '7' G_q scaling and D_q versus q dependence for protons is shown. The multifractal nature of Cherenkov images was experimentally confirmed by HEGRA collaboration [9].

3 Direct Method of Measuring Multifractal Measures

The singularity spectrum $f(\alpha)$ is smooth function of α and generalized dimension D_q is the smooth function of q. $f(\alpha)$ and D_q are related by a Legendre transformation. A method to directly calculate $f(\alpha)$ (with out D_q) from the measure was suggested by Chhabra and Jensen [10]. This method is based on using a relationship between entropy and Hausdroff

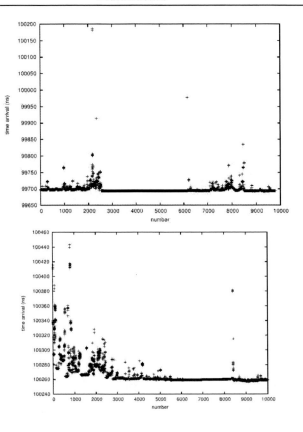

Figure 8: Simulated Cherenkov arrival time for cosmic rays.

dimension. The entropy is related to the probabilities P_i of multiplicative process

$$S = \sum_i P_i ln(P_i). \qquad (24)$$

Billingsley's theorem relates entropy to Hausdroff's dimension of the measure

$$d_h(M) = -\lim_{l \to 0} \frac{1}{ln(N_i)} \sum_{i=1}^{N} P_i(lnP_i). \qquad (25)$$

In this approach suggested by Chhabra et. al. [10], [11] whole experimental /simulation measure is covered with boxes of size l and probability $P_i(l)$ is computed. From this probability construct a one parameter family of normalized measure $\mu(q)$

$$\mu_i(q,l) = \frac{[P_i(l)]^q}{\sum_j [P_j(l)]^q}. \qquad (26)$$

The Hausdroff dimension of this measure is given as

$$f(q) = \lim_{l \to 0} \frac{\sum_i \mu_i(q,l) ln(\mu_i(q,l))}{ln(l)}. \qquad (27)$$

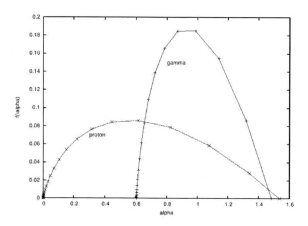

Figure 9: f(α) vs α behaviour using direct method.

The corresponding singularity strength α is given as

$$\alpha(q) = \lim_{l \to 0} \frac{\sum_i P_i(q,l) ln(P_i(q,l))}{ln(l)}. \qquad (28)$$

As in the case of generalized dimensions, parameter q here also works as a microscope to explore different regions of singular measure. For $q > 1$, $\mu(q)$ amplifies the singular regions of the measure. For $q < 1$, $\mu(q)$ accentuates the lesser regions of singular measure. For q=1, the measure $\mu(1)$ replicates the original measure.

Simulated Cherenkov arrival time data for γ-rays and protons are shown for energies of 1 TeV and 2 TeV respectively [12] in figure 8 (upper figure for γ and lower figure for protons) respectively. It is observed that there are lot of fluctuations from shower to shower basis. To do multifractal analysis we choose only those γ-ray and proton showers where number of photons generated were roughly same. In Cherenkov arrival times of γ-rays and protons, probability $P_i(l)$ was calculated for various box sizes. By calculating $\mu_i(q,l)$ in each box Hausdroff dimension and singularity strength was computed. It is clear from figure 9 that for both γ-ray and proton initiated showers we obtain multifractal spectrum. $f(\alpha(q))$ is the fractal dimension of the set.

In this studies it is shown that temporal character of extensive air showers (EAS) is multifractal in nature. It is clear that spatial character of EAS is also multifractal (figures 4–7) in nature. So it can be concluded that EAS is a multifractal process both in spatial and temporal manifestation. It is important to observe form figure 9 that γ-ray component and proton component of the cosmic rays can be segregated by using multifractal spectrum.

In the above studies simulations were carried out using CORSIKA (version 5.6211) [13] along with EGS4, VENUS, GHEISHA codes for Cherenkov option. Simulated data is generated for TACTIC telescope [14] like configuration, each element of the size 4m X 4m. Simulated data corresponds to Mt. Abu altitude (1300 m) and appropriate magnetic field. Cherenkov arrival time data corresponds to wavelength band of 300–450 nm.

4 Method of Histogram Scaling

This is another method to study multifractal [11] property. Let $p_i(l)$ be the total measure and r_i is its logarithmic value,than

$$r_i = log_{10}(p_i(l)), \tag{29}$$

r_i will vary between r_{max} and r_{min} for different values of l. The slope of r_{max} for various values of l will give α_{max} and slope of r_{min} for various values of l will give α_{min}. The values between α_{min} and α_{max} are obtained by dividing the interval into equal parts and the slope of r_i with l will give all α values.

$f(\alpha)$ values are obtained by counting number N of boxes in an interval say Δr. The slope of the plot between $N \, \Delta r^{\frac{1}{2}}$ and l is used to calculate $f(\alpha)$. However, in this approach there is scope of errors. For large scaling range, errors are less and for small scaling range errors are large.

4.1 Projective Covering method

Fractals have been used to study morphology of fracture surfaces like rocks etc. Projective covering method has been employed to study fracture surfaces. To measure fracture on surfaces experimentally laser profilometer is used [15]. For projective covering method probability has been defined by

$$p_i(l) = \frac{A_i(l)}{\sum_j A_j(l)} = \frac{A_i(l)}{A_T(l)}, \tag{30}$$

$A_i(l)$ is the area of i-th cell, l is scale size and $A_T(l)$ is the total area of the fracture surface. In reference [15] application of this method to study fracture is given in detail.

5 Thermodynamics and Multifractals

The partition function Z for multifractals

$$Z = \sum_i P_i^q(l) \sim l^{-\tau} = \sum_i exp(qln(P_i)). \tag{31}$$

In thermodynamics, partition function is defined as

$$Z = \sum_i exp(-\beta E_i) = exp(-nF(\beta)). \tag{32}$$

Again equation (32) can be also written as

$$Z = exp(-\tau ln(l)). \tag{33}$$

Following the discussion in reference [5], if we make comparisons between partition functions in thermodynamics and multifractals, following identifications can be made

$$q = \beta \tag{34}$$

corresponding to inverse of temperature

$$\tau = F(\beta) \tag{35}$$

corresponding to free energy

$$f = s \tag{36}$$

corresponding to entropy

$$\alpha = U \tag{37}$$

corresponding to the internal energy $f(\alpha)$ versus α curve corresponds to entropy versus internal energy curve and

$$m_i(q, l) = \frac{(p_i(l))^q}{\sum_k (p_i(l)^q} \tag{38}$$

corresponds to Boltzmann weights of canonical ensemble

$$m_i(q, l) = \frac{exp(-\beta E_i(l))}{\sum_k \beta E_k(l)}. \tag{39}$$

5.1   Multifractals in Cantor Process

Cantor process has unit length at $n = 0$. Second stage is obtained by dividing this unit length into three unequal parts in the ratio $l_1 : y : l_2$. Let p_1 and p_2 be probabilities corresponding to l_1 and l_2. Length y is discarded at each stage. Let us repeat this process of division of lengths (n times) in the same ratio [4] and discarding length y at all stages. Three cases emerge.

For a cantor process at next stage if each piece of object is further divided into $j = 0, 1, 2, \ldots, N$ sequence than we have

$$\Gamma(q, \tau, l) = (\Gamma(q, \tau, l))^2, \tag{40}$$

$\tau(q)$ is obtained from

$$\Gamma(q, \tau, l) = 1 \tag{41}$$

Case I

This is the case of equal division in length and equal probability. In this case no part is discarded i.e y=0. Using equations (40) and (41)

$$\sum_{i=1}^{n} \frac{p_i^q}{l_i^\tau} = 2\left(\frac{p^q}{l^\tau}\right) = 1. \tag{42}$$

Letting $l_1 = l_2 = \frac{1}{2}$ and $p_1 = p_2 = p = \frac{1}{2}$, where $p_1 + p_2 = 1$. Thus we have

$$2\left(\frac{\frac{1}{2}^q}{\frac{1}{2}^\tau}\right) = 1 \tag{43}$$

from which we get $\tau = (q - 1)$ and $D_q = 1$. This is a non fractal situation.

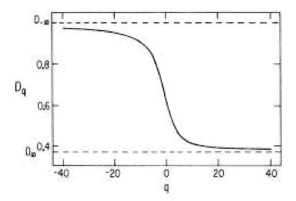

Figure 10: D_q vs q behaviour for two scale Cantor set. Reprinted with permission from AIP , ref T. C. Hasley et. al. Phys. Rev. A 33, 1141 (1986).

Case II

This is the case of equal division in length and equal probability but at all stages a length y is chopped off and thrown away. Using equations (40) and (41) and letting $l_1 = l_2 = \frac{l}{3}$ and $p_1 = p_2 = p = \frac{1}{2}$ where $p_1 + p_2 = 1$. we have

$$2\left(\frac{\frac{1}{3}^q}{\frac{1}{2}^\tau}\right) = 1 \tag{44}$$

from which

$$\tau = (q-1)\frac{ln(2)}{ln(3)} \tag{45}$$

and

$$D_q = \frac{ln(2)}{ln(3)} \tag{46}$$

from which it has been easily shown that

$$\tau(q) = q\alpha - f. \tag{47}$$

This situation corresponds to a monofractal.

Case III

For the case $l_1 \neq l_2$ and $p_1 \neq p_2$, using equations (40) and (41), we have

$$\frac{p_1^q}{l_1^\tau} + \frac{p_2^q}{l_2^\tau} = 1. \tag{48}$$

At nth stage of cantor process

$$\left(\frac{p_1^q}{l_1^\tau} + \frac{p_2^q}{l_2^\tau}\right)^n = 1. \tag{49}$$

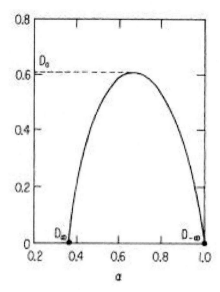

Figure 11: Multifractal spectrum for two scale Cantor set. Reprinted with permission from AIP, ref T. C. Hasley et. al. Phys. Rev. A 33, 1141 (1986).

Above equation can be solved for $\tau = \tau_q$ from which $D_q(q-1) = \tau_q$ can be calculated. This is the case of a multifractal. Figure 10 shows D_q dependence on q for a Cantor set with $l_1 = 0.25$, $l_2 = 0.4$ with $p_1 = 0.6$ and $p_2 = 0.4$. The corresponding multifractal spectrum is shown in figure 11 with $D_{-\infty} = 1.0$ and $D_\infty = 0.3684$

General case

At nth stage of multiplication, the length is $l_1^m l_2^m$ where $m = 0, 1, \ldots, n$. The probability at nth stage is $p_1^m p_2^m$. For the case $l_2 > l_1$, we have

$$\tau = \frac{ln(\frac{n}{m-1}) + q \ln\left(\frac{p_1}{p_2}\right)}{\ln\left(\frac{l_1}{l_2}\right)} \tag{50}$$

with the constrain

$$q(ln(p_1)ln(l_2) - ln(p_2)ln(l_2)) = ln l_2 ln\left(\frac{m}{n}\right) - ln(l_1)ln\left(1 - \frac{m}{n}\right). \tag{51}$$

This constrain arise out of minimisation. For the case $l_1 = l_2 = l$, constraint and $\tau(q)$ can be written as

$$q ln\left(\frac{p_1}{p_2}\right) = ln\left(\frac{m}{n}\right) - ln\left(1 - \frac{m}{n}\right) \tag{52}$$

and

$$\tau_q = \frac{ln(p_1^q + p_2^q)}{ln(l_1)}. \tag{53}$$

For a general case it has been shown that α and f(α) can be written as

$$\alpha = \frac{u ln p_1 + (1-u) ln p_2}{u ln l_1 + (1-u) ln l_2} \tag{54}$$

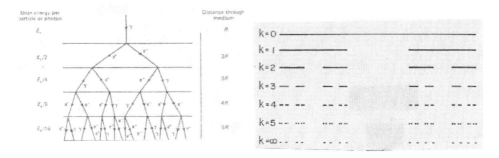

Figure 12: Comparison between electromagnetic shower and cantor set.

and
$$f(\alpha) = \frac{u ln u + (1-u) ln(1-u)}{u ln l_1 + (1-u) ln l_2} \qquad (55)$$

where u $=\frac{m}{n}$.

Thus for a cantor process with unequal length and unequal probability we have a multifractal structure. If we plot $f - \alpha$ curve, it shows a maximum. This maximum value corresponds to D_0. As $q \to \pm\infty$ the slope of $f - \alpha$ curve tends to $\pm\infty$. Minimum value of α corresponds to D_∞ and maximum value corresponds to $D_{-\infty}$.

6 A Possible Practical Example of Cantor Process

A ultra relativistic γ-ray enters atmosphere from the top and interacts with air molecule to produce electron-photon cascade. The radiation $length(x)$ for pair production and bremsstrahlung process is equal in UHE/VHE region. The charged particle -photon cascade in the atmosphere is sustained alternately by electron (e^-)-positron (e^+) pair production and bremsstrahlung process till the average energy per particle reaches critical value E_c, below which energy loss process is mainly dominated by ionisation.

This process can be visualised as follows. Figure 12 displays similarity between extensive air showers (left figure) with cantor process (right figure). A γ-ray of energy E_0 after traveling distance x (on average) produces electron-positron pair each having energy $\frac{E_0}{2}$. Since energy is getting divided into two equal parts, we can attribute a resolution of energy E $=2^{-1}$. In the next radiation length both electron and positron lose half of their energy (on average) and each radiates one photon. Thus in this radiation length there are two photons and two charged particles (e^{-1} and e^{+1}) each having energy $\frac{E_0}{4}$ which can be attributed to energy resolution E= 2^{-2}. At this stage fraction of photons ($p_1 = \frac{1}{2}$) is same as fraction of charged particles ($p_2 = \frac{1}{2}$). In third radiation length there are two photons and six charged particles each having energy $\frac{E_0}{8}$. This stage can be attributed to the energy resolution E=2^{-3}. In this radiation length the fraction of photons is $p_1 = \frac{1}{3}$ and fraction of charged particles (electron +positrons) is $p_2 = \frac{2}{3}$. The fourth radiation length corresponds to energy resolution E=2^{-4} as total of 16 photons and charged particles are produced each having energy $\frac{E_0}{16}$. However, there are 10 charged particles (5 positrons + 5 electrons) and 6 photons. Here again the fraction of photons is $p_1 = \frac{1}{3}$ and fractions of charged particles is $p_2 = \frac{2}{3}$.

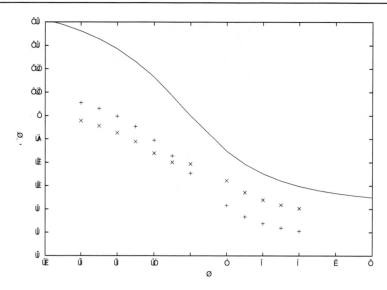

Figure 13: D_q vs q behaviour for $E=\frac{1}{2}$, $p_1 = \frac{1}{3}$ and $p_2 = \frac{2}{3}$ corresponding to continuous curve. (+) corresponds to protons and (x) corresponds to γ-rays.

At a distance of nx, the total number of charged particles and photons is 2^n, each having average energy $\frac{E_0}{2^n}$ and on an average shower consists of fraction of $\frac{2}{3}$ charged particles and $\frac{1}{3}$ photons even at nth stage. This corresponds to energy resolution of $E = 2^{-n}$. At nth stage each charged particle or photon can be labeled sequentially with $i = 0, 1, 2, \ldots$. The probability or fraction of particles and photons can be written as $p_i = p_1^k p_2^{n-k}$. The partition function for finite energy can be written as

$$\Gamma(q, \tau, E) = \sum_{i=1}^{n} \frac{N_i p_i^q}{E_i^{\tau}}, \quad (56)$$

where N_i is the number of charged particles and photons each with energy E_i. In the limit of $E \to 0$, the most dominant contribution to this partition function will survive when $\tau = \tau(q)$.

$$\Gamma_{max}(q, \tau, E) \geq \sum_{i=1}^{n} \frac{N_i p_i^q}{E_i^{\tau}}, \quad (57)$$

For the simplicity of calculations, we assume that incoming energy E_0 is equal to unit energy. For this assumption energy of each charged particle or photon at nth stage will be 2^{-n} instead of $\frac{E_0}{2^n}$. $\tau(q)$ is the solution of the equation

$$(p_1^q E^{-\tau(q)} + p_2^q E^{-\tau(q)})^n = 1. \quad (58)$$

Above equation can be easily solved to get $\tau(q)$. The number of charged particles in the nth radiation length are 2^n, each having average energy of $\frac{E}{2^n}$. It is important to note here that the measure of photons and charged particles fluctuates in each radiation length, but on the average, the shower consists of $\frac{2}{3}$ positrons and electrons and $\frac{1}{3}$ photons. We will

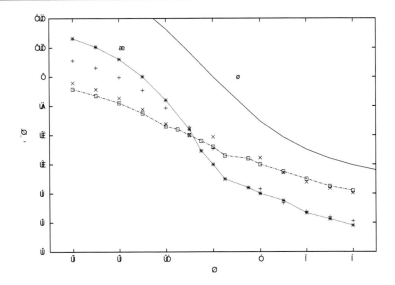

Figure 14: D_q vs q behaviour for (a) $E_1 = E_2 = \frac{1}{2}$, $p_1 = \frac{1}{3}$ and $p_2 = \frac{2}{3}$, (b)$E_1 = \frac{10}{35}$, $E_2 = \frac{25}{50}$, $p_1 = \frac{2}{5}$ and $p_2 = \frac{3}{5}$, (c)$E_1 = \frac{13}{35}$, $E_2 = \frac{22}{50}$, $p_1 = \frac{2}{5}$ and $p_2 = \frac{3}{5}$, (+) to simulated images of protons and (×) corresponds to simulated images of γ-rays.

take average values of E_1, E_2, p_1 and p_2 to calculate various multifractal parameters. In the present case energy E = $\frac{1}{2}$, $p_1 = \frac{1}{3}$, $p_2 = \frac{2}{3}$. For this case D_q can be written as

$$D_q = \frac{1}{q-1} \frac{ln(p_1{}^q + p_2{}^q)}{lnE} . \qquad (59)$$

Figure 13 depicts D_q vs q results. D_q values shown in the figure for γ-rays (×) and protons (+) are the average values of 100 Cherenkov images. The continuous curve corresponds to E=$\frac{1}{2}$, $p_1 = \frac{1}{3}$ and $p_2 = \frac{2}{3}$.

It is clear that the simulated and the calculated curves do not match with each other as shown in figure 13. If in our energy cantor process (as discussed in case III)we chop off y=23% from protons and y=18% from γ-rays at all stages of constructions, we obtain results in which the simulated results match perfectly with the calculated results. Thus, this example is a perfect example of cantor process. Results are shown in figure 14. However it should be noted that in the present case energies $E_1 \neq E_2$ and probabilities $p_1 \neq p_2$ and also there is chopping of energy "y" at all stages of progression.

Multifractals and Self Affinity

Brownian motion is self-affine by nature. A transformation that scales time and distance by different factors is called affine and behaviour that reproduces itself under affine transformation is called self-affine. The concept of multifractals has been applied to self affine [17] systems also. For a class of iteratively constructed self affine functions there exists an infinite hierarchy of exponents H_q such that

$$C_q(x) \propto x^{qH_q} \qquad (60)$$

where $C_q(x)$ is the qth-order height-height correlation function defined as

$$C_q(x) = \frac{1}{N} \sum_i^N |H(x_i) - H(x_i + x)|^q, \tag{61}$$

where N is the number of points. H_q describes the scaling of the qth order height-height correlation function.

7 Holder Exponents

For a smooth time series fractal dimension is one. For times series where dimension is between 1 and 1.5 we have a persistent times series. For fractal dimension equal to 1.5, nature of time series is random walk and for fractal dimension between 1.5 and 2, we have a anti-persistent time series. In many real time series, we see sudden changes, discontinuities, transients and sometimes singularities. In such time series singularity strength is characterized by the Holder exponent. If there exists a time series f(t) such that

$$|f(t + h) - f(t)| < Constant \quad h^\alpha. \tag{62}$$

In this case α depends on t and $\alpha(t)$ is known as a local Holder exponent. Multifractal time series can be characterized by local Holder exponent $\alpha(t)$ given that $f(t, \Delta t) \sim (\Delta t)^{\alpha(t)}$. For classical Brownian motion $\alpha(t) = 1$. Small value of Holder exponent corresponds to irregularity in signal and larger value corresponds to regular signal.

7.1 Global hurst exponent

Box counting method is a standard method used for multifractal studies but this method is mainly preferred for problems of spatial nature. To study multifractal nature of time series involves calculating qth order height-height correlation function or qth order structure function [18] of a normalized time series $y(t_i)$

$$c_q(\tau) = \langle |y(t_{i+r}) - y(t_i)|^q \rangle, \tag{63}$$

where only non-zero terms are considered in the average taken over all pairs (t_{i+r}, t_i) such that

$$\tau = |t_{i+r} - t_i| \tag{64}$$

and

$$c_q(\tau) \sim \tau^{\eta(q)}, \tag{65}$$

where $q \geq 0$ is the order of the moment and $\eta(q)$ is the scale invariant structure function exponent. For q=1, $\eta(1) = H$, which is the Hurst exponent. In general $\eta(q)$ is given by

$$\eta(q) = qH - \frac{C_1}{\alpha - 1} (q_\alpha - q), \tag{66}$$

where $C_1 \leq$ d is an intermittency parameter, d is the dimension of space (here $d = 1$) and α varies between 0 and 2. α is the Levy index. A multifractal process is characterized by a non linear behaviour of $\eta(q)$ (for multiplicative cascades)where as for processes which are additive in nature $\eta(q)$ is linear or bi-linear. For Brownian motion (bm) $\eta(q) = \frac{q}{2}$ and for fractional Brownian motion (fbm) $\eta(q) = qH$. Thus for a purely bm or fbm, $\eta(q)$ is linear, whereas for multifractal nature $\eta(q)$ is non-linear. Figure 16 is an example of multifractal nature because it shows non-linear behaviour of $qH(q)$ and $K(q)$ with q for liquid water path data (shown in figure 15).

7.2   Intermittency

Intermittency of a time series can be defined as

$$E(r, l) = \frac{r^{-1} \sum_{i=l}^{l+r-1} |y(t_{i+r} - y(t_i)|}{\langle |y(t_{i+r}) - y(t_i)| \rangle} \tag{67}$$

with $i = 0, \ldots, n - r$ and $r = 1, \ldots, n = 2^m$. A function $X_q(\tau)$ is defined such that

$$X_q(\tau) = \langle E(r, l)^q \rangle \sim \tau^{-K(q)} \tag{68}$$

for $q \geq 0$. The generalized dimension D_q and K(q) are related by

$$D_q = 1 - \frac{K(q)}{q - 1}. \tag{69}$$

The non-linear relationship between $qH(q)$ and q or between $K(q)$ and q is reflection of multifractal behaviour or multiscaling. For the case $qH(q) = 1$, that value of q is called as threshold value q_t. The exponent H_1 characterizes roughness and exponent C_1 where $C_1 = 1 - D_1$ characterizes sparseness of data.

D_1 is the information dimension and C_1 is the measure of intermittency in the signal

$$C_1 = \left. \frac{dH_q}{dq} \right|_{q=1}. \tag{70}$$

Time profile of liquid water content of various clouds is very important input for geophysical studies. The time profile of samples was collected by aircraft during various missions using microwave radiometer. The aim of the studies was to associate non-stationarity and intermittency of the time profile with geophysical features. In figure 16 multifractal features of time profile are depicted by non linear behaviour of $qH(q)$ and (q). Liquid water path is the vertical integral of liquid water content. The non-stationarity and intermittency data are shown as $(H1, C1)$ plot which is a multifractal plane (figure 17).

7.3   Lacunarity

Roughly speaking, lacunarity is related to degree of empty spaces or holes in a fractal set. Lacunarity is low for a fractal which has few holes and lacunarity is high for a fractal which has large gaps or holes or spaces. Gefen et.al. [21] defines lacunarity as the deviation of a fractal from translational invariance. In order to explain the importance of lacunarity we will heavily depend on reference Plotnik et. al. [22], which uses gliding box method earlier

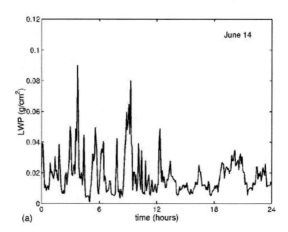

Figure 15: Liquid water path measured on a typical day. Reprinted with permission from AIP, K. Ivanova et. al. Phys. Rev. E 59 (1999) 2278.

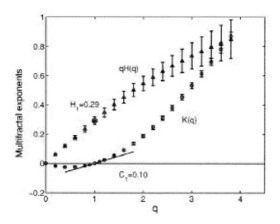

Figure 16: Multifractal behaviour of liquid water path. Reprinted with permission from AIP, K. Ivanova et. al. Phys. Rev. E 59 (1999) 2778.

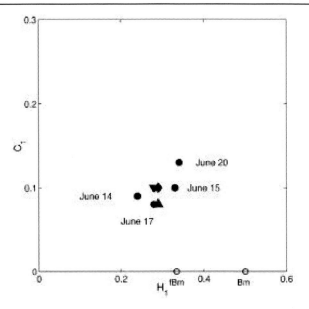

Figure 17: (H1,C1) plot of liquid water data. Reprinted with permission from AIP, K. Ivanova et. al. Phys. Rev. E 59 (1999) 2778.

developed by Allain and Cloitre [23]. Figure 18 displays five sets of one dimensional data each having different translational invariance. There are 44 occupied sites out of 256 possibilities in each case. Now move a box of size s and count number of occupied sites within this box. The number of occupied sites within this box will give box mass m. Box scanning over whole set will provide frequency distribution $N(m, x)$. The probability distribution $p(m, x)$ is obtained by $\frac{N(m,x)}{N(x)}$, where $N(x)$ is the number of boxes. The first and second moments are given as

$$z(1) = \sum mP(m, x), \qquad (71)$$

$$z(2) = \sum m^2 p(m, x). \qquad (72)$$

The lacunarity is given as

$$\lambda(x) = \frac{z(2)}{z(1)^2}. \qquad (73)$$

Lacunarity can be calculated for $x = 1, 2, \ldots, n$. Lacunarity curves for one dimensional data (figure 19) have breaks in slopes which correspond to scales characterizing that set.

From the figures it is clear that lacunarity goes to infinity if number of occupied sites goes to zero. Lacunarity also decreases with the increase of size of box. For fractals, lacunarity curve is straight line because fractals appear same at all scales.

Lacunarity concept can also be used to distinguish fractal from multifractal. Figure 20 displays binomial multiplicative of 11 generations for different values of p; (A) corresponds to $p = 0.3$, (B) corresponds to $p = 0.1$ and (C) $p = 0.1$ for generations $1 - 6$ and $p = 0.3$ generation $7 - 11$. The lacunarity analysis of information in figure 20 is shown in the figure 21.

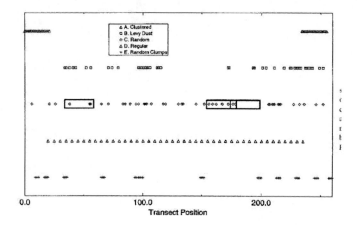

Figure 18: This figure depicts five different spatial distributions of one dimensional data. The boxes are gliding boxes of length 9. Reprinted with permission from AIP, ref: Plotnick et. al. Phys. Rev. E 53 (1996) 5461.

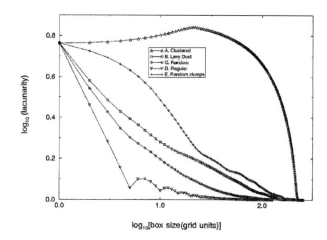

Figure 19: Lacunarity analysis for above five types of data. Reprinted with permission from AIP, ref: Plotnick et. al. Phys. Rev. E 53 (1996) 5461.

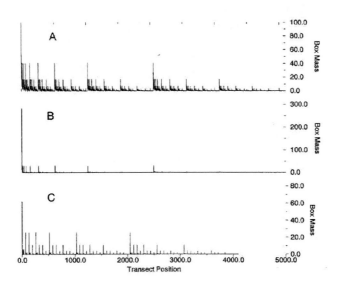

Figure 20: Figure A and Figure B correspond to 11 generations of binomial multiplicative process with $p = 0.3$ and $p = 0.1$ respectively. Figure C corresponds to $p = 0.1$ for first $1 - 6$ generations and $p = 0.3$ from $7 - 11$ generations. Reprinted with permission from AIP, ref: Plotnick et. al. Phys. Rev. E 53 (1996) 5461.

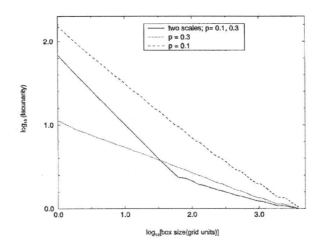

Figure 21: Lacunarity analysis of previous figure. Reprinted with permission from AIP, ref: Plotnick et. al. Phys. Rev. E 53 (1996) 5461.

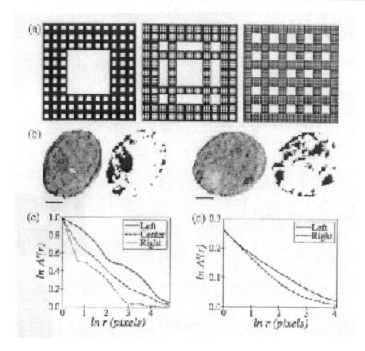

Figure 22: (a)Various stages of Sierpinksi carpets (b) Two breast epitheliel cells and (c) lacunarity curves for figures a and b. Reprinted with permsission from AIP, ref: A. J. Einstein et. al. Phys. Rev. Lett. 80 (1998) 397.

In figure 22 is displayed lacunarity studies of (a) two stages of Sierpinksi carpets and (b) two breast epithelial nuclei and thresholded at first quartile of the intensity histogram. Siepinksi carpets have Hausdroff dimension of 1.90. Lacunarity studies in the case of cells is used to quantify morphological differences [24] such as chromatin clumping and nuclei.

8 Some Examples of Multifractal Nature

Cosmic rays come from outer space and enter our atmosphere from top. Cosmic ray primarily consist of protons and some high z nuclei. Less than 1 % of cosmic ray contain γ-rays. Cosmic rays lose their energy and direction information because of interstellar magnetic fields. Only γ-ray part of cosmic ray maintain their direction and energy information. So it is very important to separate γ-ray content of cosmic from charged component. Multifractal analysis of cosmic ray showers has been used to segregate γ-ray initiated cascades from protons or high z nuclei initiated cascades. Figure 23 shows $f(\alpha)$ versus α distribution for simulated EAS [25] showers of γ-ray, proton and iron nuclei.

Multifractal analysis of Cherenkov images (extensive air shower products) has been used to segregate γ-ray initiated image from protons or high z nuclei initiated image. Figure 24 shows $f(\alpha)$ versus α distribution [7] for simulated Cherenkov images of γ-ray, proton, neon and iron nuclei initiated showers.

Neurosurgery is needed for treatment of Parkinson's disease and it is very important

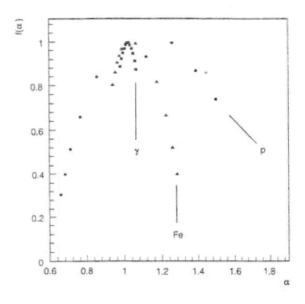

Figure 23: $f(\alpha)$ v/s α behaviour for extensive air showers of γ-ray, proton and iron nuclei. Reprinted with Permission from IOP Ltd. ref: Kempa et. al. J. of phys. G 24 (1998) 1039.

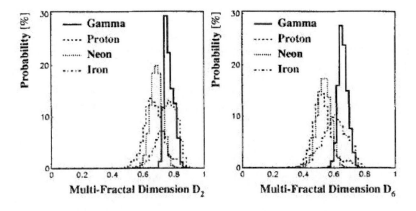

Figure 24: Distribution of generalized dimensions of simulated Cherenkov images of γ-ray, proton, neon and iron nuclei.

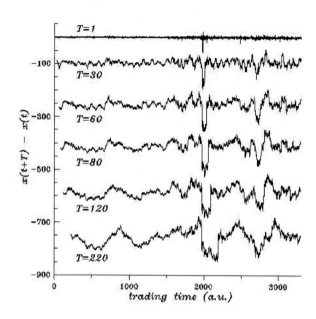

Figure 25: Time fluctuations of (S and P 500) index for the period 1982-92 v/s trading time lags. Reprinted with permission from IOP Ltd. ref: E. Canessa, J. of Phys. A: 33 (2000) 3637.

to understand deep brain stimulation. The Globus Pallidus (GP) is a complex structure in basal ganglia. The GP can be divided into two parts (a) Globus Pallidus externa (GPe) and Globu Pallidus interna (GPi). For neurosurgery in Parkinson's disease it is important to distinguish GPe from GPi. It is interesting to observe that multifractal formalism is very powerful method as compared to conventional approaches [26] to distinguish between GPe and GPi.

Fractals / multifractals have been extensively used to study stock market [18] indices, crashes, volatility etc. It has also found applications in currency exchange market dynamics [19]. Concepts of multifractal analysis were applied to study crashes in S and P index from 1980–1992 [27]. Figure 25 displays time fluctuations of S and P index for different trading time lags. It has been shown [27] that at the time of crashes a 'shoulder' appears in the right side of main peak of the 'analogous' specific heat obtained from multifractal formalism. This feature has been associated with the various impacts which a stock market dynamics may be subjected. Figure 26 displays very interesting behaviour of D_q dependence on q for different time lags.

To distinguish between a DNA sequence and a random sequence and to distinguish between different DNA sequences, multifractal approach has been applied. Subintervals in one dimensional space were used to represent substrings and histogram were obtained to represent complete genome. This histogram was called measure representation and used to study classification and evolution of different organisms. This was further extended [28 and references therein] to provide complete characterization of DNA sequence by obtaining

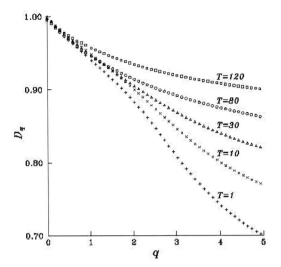

Figure 26: D_q dependence with q for different time lags for S and P 500 data. Reprinted with permission from ref: E. Canessa, J. of Phys. A 33 (2000) 3637 arXiV/0004170v1.

probability distribution of the measure uniquely determined by the exponent $K(q)$ of the multifractal analysis (see figure 27).

The multifractal analysis of databases of heart beats of healthy persons and heart patients has been conducted[29]. The data for healthy subjects corresponded to both day time and night time series records. The heart patients were suffering from congestive heart failure condition. It was observed that the partition function $Z_q(a)$ scales as power law with scaling exponent 'a' for different values of q for healthy subjects as well as for patients. But $\tau(q)$ versus q shows a clear distinction between healthy subjects and patients at least for positive values of q. From multifractal spectrum it is clear that healthy subjects show multifractal spectrum over a wide range as compared to narrow range for heart patients. The case of heart patients shows loss of multifractality.

One of the assumptions in cosmology is that the matter in universe is homogeneously distributed. Multifractal analysis was done to CFA galaxy catalog data. The multifractal [30] behaviour indicates absence of homogeneity for length scales probed by CFA catalogue. The distribution of the luminous objects in the universe may be result of the multifractal cascade process. It has been shown that the this cascade process was "smoothed" [31] in a subspace of dimension $\propto 0.45$.

It has been observed that wavelength distribution of emission lines of energized atoms displays [32] multifractal properties. The emission spectrum is divided into various segments. In each segment number of lines $n_i(r)$ is counted and the partition function is defined as

$$Z(q,r) = \sum_{i=1}^{N(r)} \left(\frac{n_i(r)}{N}\right)^q. \qquad (74)$$

Using multifractal approach it has been possible to compare spectral line series of different

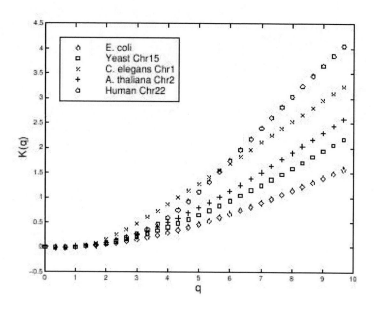

Figure 27: K(q) dependence on q of chromosome 22 of Homo Sapiens, chromosome 2 of A.Thaliana ,chromosome 1 o c ele elegans , chromosome 15 of S.cerevisiae and E.Coli. Reprinted with permission from IOP Ltd., Anh et. al. J. of Phys. A 34 (2001) 7127.

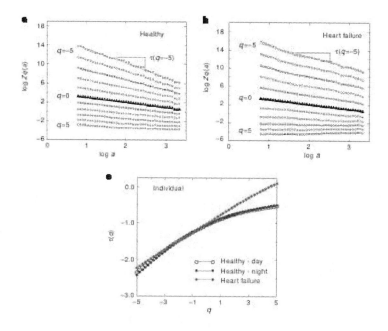

Figure 28: Partition function scaling of healthy and of heart failure and corresponding $\tau(q)$ dependence, Reprinted by permission form Macmillan Publishers Ltd: Nature 399 (1999) 461, copyright (1999).

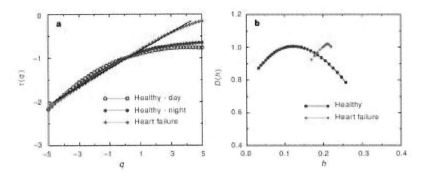

Figure 29: Generalized dimensions of for healthy and heart failure, Reprinted by permission from Macmillan Publishers Ltd: Nature 399 (1999) 461 copyright (1999).

atoms. Figure 30 displays Balmer and Pschen series of hydrogen atom (upper figure), scaling behaviour of the partition function and generalized dimension D_q dependence on q (middle figure) and corresponding multifractal spectrum (lower/last figure).

The temporal fluctuations of geo-electrical signals are related to tectonic plate dynamics [33]. The complexity of seismic activity is connected to the nature of geo-electrical signals. It has been reported that the multifractal behaviour becomes prominent prior to seismic occurrences.

Wavelets

Wavelet analysis [34,35] is being used increasingly to study given structures in different scales. Wavelets can detect both the location and a scale of a structure. Wavelets are parameterized both by scale $a > 0$ (dilation parameter) and a translation parameter 'b' $(-\infty < b < \infty)$ such that

$$\Psi_{a,b} = \frac{\psi(x-b)}{a}. \tag{75}$$

The wavelet domain of one dimensional function Ψ is rather two dimensional in nature; one dimension corresponds to scale and other to translation. The continuous wavelet transform for one dimensions is defined as

$$W(a,b) = \int dx f(x) \frac{\psi^*(|x-b|)}{a}, \tag{76}$$

where a is the scale. Here $f(x)$ is one dimensional function and ψ^* ($*$ is complex conjugate) is the analysing wavelet or also known as mother wavelet. Mexican hat wavelet is very commonly used wavelet can be written as

$$\psi \frac{(|x-b|)}{a} = \frac{1}{(2\pi)^{0.5}a} \left(2 - \frac{|x-b|^2}{a^2}\right) exp\left(-\frac{|x-b|^2}{2a^2}\right). \tag{77}$$

Mexican wavelet is an isotropic wavelet having minimum number of oscillations. An important property of wavelet ψ to be used as analysing wave,it must have zero mean value

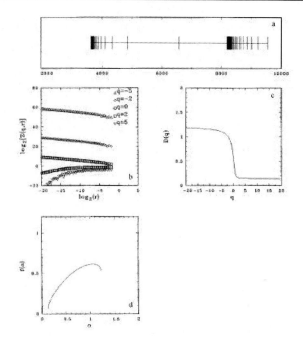

Figure 30: Balmer and Paschen series for hydrogen atom, partition function, generalized dimension dependence on q and corresponding multifractal spectrum. Reprinted with permission from APS, ref Saiz et. al. Phys. Rev. E 54 (1996) 2431.

i.e. $\int \psi(x)\,dx=0$. ψ is also required to be orthogonal to some lower order polynomials i.e.

$$\int_{-\infty}^{\infty} x^m \psi(x)\,dx = 0, \tag{78}$$

where $0 \leq m \leq n$. Here n is the upper limit related to the order of wavelet.

The scaling operation either compresses a signal or dilates it.

9 Partition Function of Wavelets

To calculate multifractal measure box counting method is common approach. The partition function for multifractals is given as

$$Z_q(l) = \sum \langle p^q(x,l) \rangle \sim l^{\tau(q)}, \quad l \to 1, \tag{79}$$

where $x = 0, l, 2l, 4l, \ldots$ Legendre transform $f(\alpha)$ can be calculated from $\tau(q)$. As discussed in the section on multifractals f(α) shows non linear dependence on q for multifractal measure. Similarly partition function [36] has been defined for wavelets

$$W_q(l) = \sum \langle |p(x,l) - p(x+l,l)|^q \rangle \sim l^{\beta(q)}, \quad l \to 1, \tag{80}$$

where $x = 0, l, 2l, 4l, \ldots$. The scaling exponent $\beta(q)$ plays similar role in partition function of wavelets as $\tau(q)$ plays in fractals. Wavelet partition function is actually scaling property of box probabilities in different scales and hence is very useful to study fluctuations.

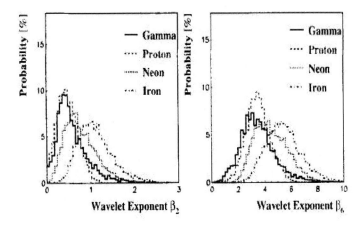

Figure 31: Distribution of Wavelet moments of Cherenkov images of cosmic rays.

10 Application of Partition Function of Wavelets

Let us now apply wavelets [7], [8] to a Cherenkov image of TACTIC like telescope. As discussed in this chapter Cherenkov images (Figures 4–7) display multifractal character. In the context [7], [8] of Cherenkov images, wavelet moment can be written as

$$W_q(M) = \sum_{j=1}^{M} \frac{|k_{j+1} - k_j|^q}{N}, \qquad (81)$$

where k_j is the number of photoelectrons in the jth cell in a particular scale and k_{j+1} is the number of photoelectrons in the jth cell of a consecutive scale. The wavelet moments have been obtained by dividing the Cherenkov image into $M = 4, 16, 64, 256$ equally sized parts and counting the number of photoelectrons in each part. The difference of probability in each scale gives the wavelet moment and shows proportional behaviour

$$W_q \propto M^{\beta_q}. \qquad (82)$$

Cherenkov images are produced when Cherenkov radiation produced by incoming cosmic rays, is collected by γ-ray astronomy telescope. Most of the cosmic rays consist of protons but γ-rays and other high Z nuclei are also present in lesser ratio. The challenge is to segregate Cherenkov images which are formed due to various components of cosmic rays. By calculating β_q for each Cherenkov images, it was observed that wavelet exponents like β_2 and β_6 are useful parameters to segregate Cherenkov images of different origins as shown in the figure 31.

11 Fractals and Wavelets

$f(\alpha)$ spectrum displays all singularities but it does not give any spatial location of these singularities. This problem has been solved by using wavelets. Wavelet transforms have

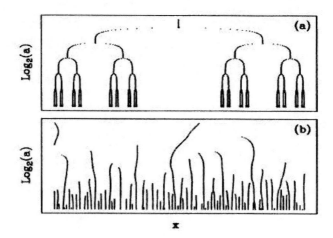

Figure 32: Wavelet scaling of devil's stair case. Reprinted with permission from AIP, ref Muzy et. al. Phys. Rev. Lett. 67 (1991) 3515.

been used to study scaling properties of fractals [37]. For a fractal measure $dm(x)$, the wavelet transform is defined as

$$W(a,b) = \frac{1}{a^n} \sum \psi\left(\frac{x-b}{a}\right) dm(x), \qquad (83)$$

where a^n is the scaling factor. Wavelet transform analysis has been applied to standard cantor process for the cases of $p_1 = p_2 = \frac{1}{2}$ and $p_1 = \frac{3}{4}$, $p_2 = \frac{1}{4}$. The results are discussed in the reference [37]. From figures of reference [37] it is clear that singularities appear in wavelet transform as oscillatory behaviour and can be identified along 'b' axis. The difference between equal probability and unequal probability of the cantor sets is clear.

Again wavelet approach has been used to generalize multifractal formalism. Multifractal approach to intermittency involves extracting spectrum of Holder exponent using scaling properties. Muzy et. al. [38] used wavelet transform to define partition function. From the scaling properties of this partition function singularity spectrum is determined. The partition function in the limit $a \to 0$, is defined as

$$Z(a,q) = \sum_{x_i(a)_i} |W(a, x_i(a))|^q = a^{\tau(q)}. \qquad (84)$$

The approach is based on taking only local maxima of wavelet transform at a given scale. Wavelet transfer modulus maxima contains all the information for the hierarchical distribution of singularities in the signal. Figure 32 displays positions of modulus maxima of wavelet for a devil's staircase.

The whole singularity spectrum has been obtained from [38] wavelet transform. The wavelet transform of a signal f(x) is given as

$$T_g(a, x_0) = \frac{1}{a} \int_{\infty}^{\infty} f(x) g(\frac{x - x_0}{a}) dx \qquad (85)$$

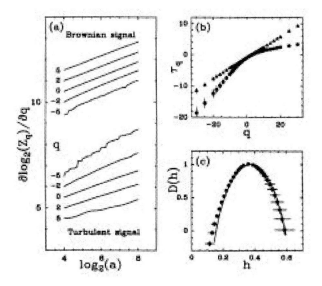

Figure 33: Wavelet transform of a Brownian motion and turbulence velocity and their singularity spectrum. Reprinted with permission from AIP, ref: Muzy et. al. Phys. Rev. Lett. 67 (1991) 3515.

for a > 0 where 'a' is scale parameter. The partition function in the limit of $a \to 0$ is given as

$$z(a,q) = \sum |T_g(a, x_i(a))|^q \propto a^{\tau(q)}. \tag{86}$$

The summation in the above equation is over the wavelet transform modulus maxima $x_i(a)_i$. By this approach partition function gets incorporated with the multiplicative structure. The singularity spectrum D(h) is given as

$$h(q) = \lim_{a \to 0} \frac{1}{lna} \sum T^+(q; q, x_i(a)) ln|T_g(q; a, x_i(a))| \tag{87}$$

and

$$D(h(q)) = \lim_{a \to 0} \frac{1}{lna} \sum T^+(q; q, x_i(a)) ln T_g(q; a, x_i(a)) \tag{88}$$

where

$$T_g^+(q; a, x_i) = \frac{|T_g(a, x_i)|^q}{\sum_{x_i} |T_g(a, x_i)|^q}. \tag{89}$$

Thus from log-log plots of above equations, Holder exponent 'h' and corresponding '$D(h)$' can be obtained. This approach has been applied to fully turbulence data as shown in the figure 33. Part 'a' of the figure shows wavelet transform of a Brownian motion and turbulent velocity signals

This method has been further applied [39] to study fully developed tubulence data and DNA sequences.

12 Detrended Fluctuation Analysis

Detrended Fluctuation Analysis (DFA) [40] has been applied to many studies in physics related problems. A time series $x(i)$, where $i = 1, 2, \ldots, N$ where N is the length of signal. Mean value of $\bar{x}$ is defined as

$$\bar{x} = \frac{1}{N} \sum_{i=1}^{N} x(i) \tag{90}$$

from which we calculate $y(i)$

$$y(i) = \sum_{i=1}^{i} (x(i) - \bar{x}). \tag{91}$$

The profile $y(i)$ is divided into boxes of equal length and to data of each box a polynomial $Z_n(i)$ is fitted. Thus we have DFA-1, DFA-2, DFA-3 ... corresponding to the order of fitting polynomial like 1, 2, 3 ..... The detrended fluctuation function $F^2(n)$ is obtained as

$$F^2(n) = \frac{1}{N} |y(i) - z_n(i)|^2. \tag{92}$$

All these above mentioned steps are repeated for variable box sizes to obtain a distribution of $F(n)$ over large scale. An average value of $F(n)$ so obtained follows a power law

$$F(n) \sim n^{\alpha}. \tag{93}$$

The value of α quantifies the degree of correlation in the time series. For $\alpha = \frac{1}{2}$, the signal is uncorrelated. For $\alpha > \frac{1}{2}$ signal is correlated and for $\alpha < \frac{1}{2}$ signal is anti-correlated. DFA method has become very popular to study mono fractal scaling properties. This method has been extended to study multifractal [41] properties also. For this purpose average is taken over all segments to obtain qth order fluctuation function

$$F_q(n) = \left[\frac{1}{2N} \sum_{i=1}^{2N} [F^2(i, n)]^{\frac{q}{2}} \right]^{\frac{1}{q}}, \tag{94}$$

where q can take any real value. For $q = 2$, this method reduces to standard DFA approach. By plotting log-log of F_q with n for different values of q, multifractal scaling behaviour can be obtained. F_q follows a power law

$$F_q(n) = n^{h(q)}, \tag{95}$$

where $h(q)$ is related to $\tau(q)$,

$$\tau(q) = qh(q) - 1. \tag{96}$$

Zips'law

Zipf's method first applied to text words consists in sorting out given words and calculating frequency of their occurrence. If R is rank associated with each word and R=1 is assigned for most frequent word then frequency and rank are related by a power law

$$f \sim R^{-\varsigma}, \tag{97}$$

where ζ is exponent obtained from log-log plot between f and R. Power law appears because of hierarchical structure in words of various languages. Zipf's law has been applied to other type of problems [42] like meteorology, random walks, econophysics etc by converting a given time series into words on an alphabet of k characters and calculating ζ on time series of constant size (m) of words. ζ is related to Hurst exponent H by

$$\zeta = |2H - 1|. \tag{98}$$

13   Moving Averages Method and Hurst Exponent

Moving average measures the trend present in time interval T. For a time series $x(t)$, where $t = 0, 1, 2, \ldots, T_1, \ldots, T_2, \ldots$. Moving average $X_m(t)$ [43] can be written as

$$X_m(t) = \frac{1}{T_1} \sum_{i=0}^{T-1} x(t - i), \tag{99}$$

$X_m(t)$ gives the average for time T_1. If X_1 and X_2 are moving average of a self affine signal for short period of time and long period of time respectively then if X_1 will cross X_2 from above event is called death cross and if X_1 will cross X_2 from below event is called Gold cross. Presence of crossing points between X_1 and X_2 reflects power law behaviour of signal and density of these crosses measures power law correlation in the signal. Density of these crosses scales as power of $(H - 1)$ where H is Hurst exponent.

Standard deviation about moving average can be written as

$$\sigma_{dma} = \sqrt{\frac{1}{T_{max} - T} \sum_{t=T}^{T_{max}} (x(t) - X_m(t))^2}, \tag{100}$$

where $X_m(t)$ is the moving average in the window T. σ_{dma} shows power law behaviour with Hurst exponent

$$\sigma_{dma} \propto T^H. \tag{101}$$

Moving average method has been extended for multifractals also [44]. This method involves finding cumulative sums $y(t)$ of a time series and calculating moving average function $z(t)$ in a moving window. The signal is detrended by removing the moving average function. The new series in segmented from which root mean square function $F^2(n)$ is obtained. The qth order fluctuation obtained has a power law relationship with scale size n

$$F_q(n) \propto n^{h(q)}, \tag{102}$$

where exponent $h(q)$ is related to

$$\tau(q) = qH(q) - D_f, \tag{103}$$

where D_f is the fractal dimension of the geometric support of the multifractal measure.

Recently a new method has been proposed which is called as detrended cross correlation [45] analysis. This method is based on detrended co-variance.

14   Retracing Multifractals

Multifractals are remnants of a dynamical process which in general is a multiplicative process. In figure 4, a simulated Cherenkov is displayed. A multifractal procedure has been discussed to go from figure 4 (image) to figure 7 (generalized dimensions D_q). Now the question - is it possible to get some information about figure 4 from figure 7 ?. Earlier Feigenbaum–Jensen–Procaccia (FJP) recognized this problem in chaos theory. This led them to develop a method [46] which connects multifractal measures with underlying dynamics using thermodynamic approach. In general, multifractal measures are done using box counting method. In any measure if the size of the box is fixed and probability in that box varies, such a multiplicative process is known as P model. If the size of the box is allowed to vary such that probability in each box is fixed then such a multiplicative process is known as L model. Measure in which both box size and probability are allowed to vary according to some criteria, such a process is called as LP model [47].

In the following we explore the possibility of using FJP method to study electromagnetic (EM) showers. For this purpose we use information of D_q versus q curve of a simulated Cherenkov image i.e figure 7. For a cascade like extensive air showers (EAS), the underlying dynamics means that there are energy splits and probability variations of charged particles and photons. So the idea is, to retrieve some information about energy splits or probabilities or both, from the Cherenkov image using FJP method. Since we are using P-model [47] approach, we will be able to obtain information about probabilities only because in P-model there is assumption of equal split of energies.

15   FJP Method

In this section we discuss physical outline of FJP method. The detailed mathematical approach is given in the references [46]. The core of FJP method is a transfer matrix. The elements of this transfer matrix are scaling functions of the dynamical process. The scaling functions describe the contraction factors of each interval along each branch. The scaling functions are obtained from the partition function. In general transfer matrix is ∞ x∞ matrix. However, for practical applications mostly 2x2 or sometimes 3x3 matrix is used.

For L-model [46], [47] the ratio of partition functions for two successive refinements is

$$\frac{\Gamma^{n+1}(\tau)}{\Gamma^n(\tau)} = \frac{\sum_{i=1}^{N_n+1}(L_i^{n+1})^q}{\sum_{i=1}^{N_n}(L_i^n)^q} = a^q \tag{104}$$

where 'a' is a number. The scaling function σ_L for L Model is

$$\sigma_L(\epsilon_{n+1}\cdots\epsilon_0) = \frac{L(\epsilon_{n+1},\ldots,\epsilon_0)}{L(\epsilon_n,\ldots,\epsilon_0)}\delta_{\epsilon_n,\epsilon_n'}\cdots\delta_{\epsilon_1,\epsilon_1'} \tag{105}$$

$L(\epsilon_n,\ldots,\epsilon_0)$ is the length of the interval belonging to the tree and δ is Kronecker delta function.

For any tree structure, each parent produces number of off springs. At each level of refinement the number of off springs are increasing. For any two successive refinement levels, ratio R of the partition functions can be obtained. It has been found that this ratio R=

$\lambda(\tau)$ is the leading eigen value of the transfer matrix. The characteristic equations of this transfer matrix is given as

$$\lambda^2(\tau) - \lambda(\tau)Tr(T) + DET(T) = 0 \tag{106}$$

where Tr and DET are trace and determinant of a matrix T respectively. By solving equation (106), information about underlying dynamics can be obtained.

Multiplicative processes in extensive air showers can be visualized in three ways. In the first case, at each level of refinement there is unequal split in energy but equal probability (L-model). In the second case, at each level of refinement there is equal split in energy with unequal probability. (P-model). In the third case, there is unequal split in energy involving unequal probability (LP-model). Most of the problems have been solved using L or P model. Solutions of LP-model have been found to be unstable.

Transfer matrix for EM showers:

To study EM showers we consider P-model approach. In P-model, re-arrangement of probabilities in the cascade results in a multifractal measure. P-models are preferred over other models when there is no information or data available about the underlying dynamics. The concept of multifractal measures was first conceived in turbulence [48] by using P-model.

For P-model [47] the ratio of partition functions for two successive refinements is

$$\frac{\Gamma^{n+1}(q)}{\Gamma^n(q)} = \frac{\sum\limits_{i=1}^{N_n+1} (P_i^{n+1})^q}{\sum\limits_{i=1}^{N_n} (P_i^n)^q} , = R^{-\tau} \tag{107}$$

where P_i^n is the probability in the ith-box for nth level of refinement.

The scaling function σ_p is

$$\sigma_p(\epsilon_{n+1} \cdots \epsilon_0) = \frac{P(\epsilon_{n+1}, \ldots, \epsilon_0)}{P(\epsilon_n, \ldots, \epsilon_0)} \delta_{\epsilon_n,\epsilon_n'} \cdots \delta_{\epsilon_1,\epsilon_1'} . \tag{108}$$

$P(\epsilon_n, \ldots, \epsilon_0) = P_i^n$, where ϵ_i gives the location of probability on the path of the tree and δ is Kronecker delta function.

The elements of 2x2 transfer matrix for EM showers are $\sigma_p(00)$, $\sigma_p(01)$, $\sigma_p(10)$ and $\sigma_p(11)$. This is the case of one step memory process. The binary digits 0 and 1 correspond to the left (charged particle) and right (photon) offspring of the parent. In the next level of refinement there are two digits (00,01,10,11), the first digit denoting the offspring being left or right and the second digit corresponds to parent being left or right.

For a given Cherenkov image, $\sigma_p(00)$, $\sigma_p(01)$, $\sigma_p(10)$ and $\sigma_p(11)$ are unknown. Transfer matrix T for a P-model can be written as

$$\begin{pmatrix} \sigma_p(00) & \sigma_p(01) \\ \sigma_p(10) & \sigma_p(11) \end{pmatrix}$$

with the condition

$$\sigma_p(00) + \sigma_p(10) = 1, \tag{109}$$

$$\sigma_p(10) + \sigma_p(11) = 1, \tag{110}$$

σ_p's correspond to the probability of charged particles and photons with $\sigma_p(00) \neq \sigma_p(01)$ and $\sigma_p(10) \neq \sigma_p(11)$, meaning unequal probabilities for charged particles and photons of the same parent. The characteristic equation of the transfer matrix is

$$a^{2q} - [\sigma_p^{-\tau}(00) + \sigma_p^{-\tau}(11)]a^q + [\sigma_p^{-\tau}(00)\sigma_p^{-\tau}(11) - \sigma_p^{-\tau}(01)\sigma_p^{-\tau}(10)] = 0. \tag{111}$$

For a given Cherenkov image whose D_q versus q behaviour is known, $D_{-\infty}$, $D_{+\infty}$ can be calculated. For a P-model

$$D_\infty = \frac{[log(\sigma_p(00))]}{[log(R^{-1})]}, \tag{112}$$

$$D_{-\infty} = \frac{[log(\sigma_p(11))]}{[log(R^{-1})]}, \tag{113}$$

and equation (111) for q=0, can be written as

$$1 - [\sigma_p^{-\tau}(00) + \sigma_p^{-\tau}(11)] + [\sigma_p^{-\tau}(00)\sigma_p^{-\tau}(11) - \sigma_p^{-\tau}(01)\sigma_p^{-\tau}(10)] = 0. \tag{114}$$

For a typical γ-ray initiated simulated Cherenkov image corresponding to 50 TeV [7] energy, $D_{-\infty} = 1.5$, $D_\infty = 0.6$ and $D_0 = 1.0$. Using equations (112) and (113), we obtain the value of $\sigma_p(00) = 0.66$ and $\sigma_p(11) = 0.34$ and from equation (114), we get the value of $\sigma_p(01)\sigma_p(10)$. Using $\sigma_p(00)$, $\sigma_p(11)$ and $\sigma_p(01)\sigma_p(10)$, equation(111), can be solved for different values of q to obtain $\tau(q)$. The resulting $\tau(q)$ versus q values can be compared with simulated or experimental data. Thus using equations (109) and (110), we have $\sigma_p(00) = \sigma_p(10) = 0.66$ and $\sigma_p(11) = \sigma_p(01) = 0.34$.

Feigenbaum et.al. [46] called multifractal measures as "static objects". Fractals/ multifractal measures are remnants of a complex underlying dynamics. The connection between the dynamics and the resulting generalized dimensions obtained by Feigenbaum et al was indeed a breakthrough. Chabbra et. al. [47] investigated FJP method in detail and found that D_q versus q results, obtained using L-model, P-model or LP-model may not always be unique. However, Chabbra et. al. [47] also concluded that that FJP method will give accurate D_q versus q results if (a) there is proper and independent choice of ratio 'R' (b) there may be some independent clue for choosing L-model or P-model or LP-model. Thus these two conditions become important when D_q versus q is calculated for comparative studies with experimental or simulated data.

Feigenbaum et. al.[46] applied FJP method to chaos theory. Chabara et. al. [47] investigated FJP method and applied it to the study of energy dissipation in turbulence. Batumin and Sergeev [49] applied FJP method to the study of intermittency in hadron collisions.

In EM showers it is well known that EAS tree structure is of binary nature. A γ-ray produces $e^\pm$ pair which initiates charged particle / photon cascade. So at all levels of refinement there are only two possibilities and ratio $R = 2$ will not change. Again in γ-ray initiated showers, it is also well known that there is small loss of energy. For hadron initiated showers we cannot use P-model because at each level of interaction $R > 2$.

The resulting values of two probabilities $P_1 = 0.66$ and $P_2 = 0.34$ are obtained from transfer matrix method of Cherenkov images [50]. These values are unique because using L-model we get equal probabilities and LP model is inherently unstable. These values of probabilities are very close to the results obtained from Heitler's model. Heitler's model gives a simplified picture of EM showers. However, despite its simplicity it predicts some important features of EM showers which include (a) the proportionality between total number of particles and energy (b) the relationship between shower maxima with energy. Recently Heitler's [51] model has been extended to explain important features of hadron showers.

Concluding Remarks:

In this chapter we have reviewed multifractals in a simplified manner so that concepts can be understood by non experts and beginners also. Following a very simple approach, care has been taken that everything important about multifractals is covered in this chapter. Examples from cosmic ray physics, γ ray astronomy, stock markets to biology have been discussed so that new readers can think of applying multifractals to such problems where they have not been applied till now.

Acknowledgments

I am thankful to Dr. S. Kailash, Director Physics Group, Bhabha Atomic Research Centre, for his support and encouragement. I am thankful to Sh. R.Koul, Head ApSd for his support. I am thankful to my family and friends for their support. I am thankful to the Institute of Physics (IOP Ltd), American Physical Society (APS) and Macmillan Publishers for their permission to reproduce some of their figures.

References

[1] B. B. Mandelbrot. The Fractal Geometry of Nature (Freeman , San Francisco 1980).

[2] J. Feder. Fractals, Plenum Press New York, 1988.

[3] R. E. Amritkar. Indian J. Physics 66 A (1992) 429.

[4] T. C. Hasley, M. H. Jensen, L. P. Kadanoff, I.Procaccia and B. I. Shraiman. Physical Review A 33 (1986) 1141.

[5] H. E. Stanley and P. Meakin. Nature 33 (1988) 405.

[6] L. Turkevich and H. Scher. Phys. Rev. Lett. 55 (1985) 1024.

[7] A. Haungs, A. Razdan, C. L. Bhat, R. C. Rannot and H. Rebel. Astroparticle Physics 12 (1999) 145.

[8] A. Razdan, A. Haungs, H. Rebel and C. L. Bhat, Astroparticle Physics 17 (2002) 497.

[9] B. M. Schafer, W. Hofmann, H. Lampeitl and M. Hemberger. arXiv.org:astro-ph/0101318v1,2001.

[10] A. Chabbra and R. V. Jensen. Phys. Rev. Lett. 62 (1989) 1327.

[11] A. S. Chabbra, R. V. Jensen and K. R. Sreenivasan, Physical Review A 40 (1989) 4593.

[12] A. Razdan, ICRC 2005, Pune Proceedings.

[13] D. Heck, J. Knapp, J. N. Capdevielle, G. Schatz, T. Thow, FZKA Report 6019, Forschungs Zentrum, Karlsruhe (1998).

[14] C. L. Bhat et.al. In: T. Kifune, Editor, Towards a major atmospheric Cherenkov detector III, Universal Academic Press; 1994 page 207.

[15] H. Xie, J. Wang and E. Stein. Phys. Lett. A 242 (1998) 41.

[16] A. Razdan. Chaos, Solitons and Fractals 42 (2009) 2735.

[17] A. Barabasi, T. Vicsek. Phys. Rev. A 44 (1991) 2730.

[18] M. Ausloos and K. Ivanova. arXiv:cond-mat/0108394.

[19] M. Ausloos and K. Ivanova. Eur. Phys. J. B 8 (1999) 665.

[20] K. Ivanova and T. Ackerman. Phys. Rev. E 59 (1999) 2778.

[21] Y. Gefen, Y. Meir and A. Aharony. Phys. Rev. Lett. 50 (1983) 145.

[22] R. E. Plotnick, R. H. Gardner, W. W. Hargrove, K. Prestegaard, M. Permutter. Phys. Rev. E 53 (1996) 5461.

[23] C. Allain and M. Cloitre. Phys. Rev. A. 44 (1991) 3552.

[24] A. J. Einstein and J. Gil. Phys. Rev. Lett. 80 (1998) 397.

[25] J. Kempa and M. Sarmorski. Journal of Physics G: Nucl. Part. Phys. 24 (1998) 1039.

[26] J. Zheng, J. Gao, J. C. Sanchez, J. C. Principe and M. S. Okun; Phys. Lett. A 344 (2005) 253.

[27] E. Canessa; Journal of Physics A: Math. Gen. 33 (2000) 3637.

[28] V. Anh, K. Lau and Z. Yu; Journal of Physics A: Math. Gen. 34 (2001) 7127.

[29] P. Ch. Ivanov, L. A. N. Amarel, A. L. Goldberger, S. Halvina, M. G. Rosenblum, Z. R. Struzik and H. E. Stanley. Nature 399 (1999) 461.

[30] P. H. Coleman and L. Pietronero. Phys. Rep. 213 (1992) 311.

[31] P. Garrido, S. Lovejoy and D. Schertzer. Physica A 225 (1996) 294.

[32] A. Saiz and V. J. Martinez. Phys. Rev. E 54 (1996) 2431.

[33] L. Telesca, G. Colangelo and V. Lapenna. Natural Hazard and Earth System Sciences 5 (2005) 673.

[34] I. Daubechies. Commun. Pure Appl. Math. 41 (1988) 909.

[35] Y. Meyer, Wavelet and Operators, Cambridge University Press, New York, 1992.

[36] J. W. Kantelhardt, H. E. Roman and M. Greiner. Physica A 220 (1995) 219.

[37] A. Arneodo, G. Grasseau, M. Holschneider. Phys. Rev. Lett. 61 (1988) 2281.

[38] J. F. Muzy, E. Bracy and A. Arneodo. Phys. Rev. Lett. 67 (1991) 3515.

[39] A. Arneodo, B. Audit, E. Bracy, S. Manneville, J. F. Muzy and S. G. Roux. Physica A 254 (1998) 24.

[40] C.-K. Peng, S. U. Beldyrev, S. Halvin, M. Simons, H. E. Stanley and A. L. Goldberger. Phys. Rev. E 49 (1994) 1685.

[41] J. W. Kantelhardt, S. A. Zschiegner, E. K. Bunde, S. Havlin, A.Bunde and H. E. Stanley. Physica A 316 (2002) 87.

[42] M. Ausloos, K. Ivanova; Physica A 270 (1999) 526.

[43] N. Vandewalle and M. Ausloos. Phys. Rev. E 58 (1998) 6832.

[44] G. Gu and W. Zhou. Phys. Rev. E 82 (2010) 011136.

[45] B. Podobnik and H. E. Stanley. Phys. Rev. Lett. 100 (2008) 084102.

[46] M. J. Feigenbaum, M. H. Jensen and Itamar Procaccia. Physical Review Letters 57 (1503) 1986.

[47] A. S. Chabbra, R. V. Jensen and K. R. Sreenivasan. Physical Review A 40 (1989) 4593.

[48] B. B. Mandelbrot. Journ. Fluid Mech. 62 (1974) 331.

[49] A. V. Batunin and S. M. Sergeev, Physics Letters B 327 (1994) 293.

[50] A. Razdan. Chaos, Solitons and Fractals. 42 (2009) 253.

[51] J. Matthews. Astroparticle Physics, 22 (2005) 387.

In: Classification and Application of Fractals
Editor: William L. Hagen

ISBN978-1-61209-967-5
© 2012 Nova Science Publishers, Inc.

Chapter 9

SUBGRID MODELING OF STEADY-STATE FLOW PROCESSES IN FRACTAL MEDIUM

O. N. Soboleva and E. P. Kurochkina

[1] Institute of Computational Mathematics and Mathematical, Novosibirsk, Russia
[2] Institute of of Thermophysics, Novosibirsk, Russia

ABSTRACT

The effective coefficients in the equations of steady-state flow processes are calculated for a multiscale medium by using a subgrid modeling approach. The physical parameters in the medium are mathematically represented by a Kolmogorov multiplicative continuous cascades with a log-normal or log-stable probability distributions. The scale of the solution domain is assumed to be large as compared with the scale of heterogeneities of the medium. The theoretical results obtained in the chapter are compared with the results of a direct 3D numerical simulation and the results of the conventional perturbation theory.

PACS 05.45-a, 52.35.Mw, 96.50.Fm.

Keywords: Effective coefficients; Subgrid modeling; Multiscale random parameters, Multiplicative cascades.

1. Introduction

Heterogeneity of a medium essentially affects processes of a wave propagation, passage of current, heat transfer, flow of fluid, and so on [1]. The physical processes in heterogenous media are described by mathematical models. The equations described the physical processes can be accurate models for real phenomena, but they require great computational costs if fluctuations of all the scales are taken into account. The large-scale medium heterogeneities, such as layers, intercalations, are taken into account in these models with the help of some boundary conditions (see, for example, [2], [3]). As a rule, the small-scale fluctuations of parameters are taken into account by some effective coefficients, i.e.

---

*E-mail address: olga@nmsf.sscc.ru

some simplified models with computationally resolvable scales are sought. The solution to governing equations in these models must be approximate, for example, to the ensemble-averaged solution of the initial governing equations. This is a major subject of physical and engineering science that is encountered under various names, e.g. homogenization, coarse graining and subgrid modeling. The spatial distributions of small-scale heterogeneities may not be exactly known. It is customary to assume that these parameters are random fields characterized by joint probability distribution functions. Then the effective coefficients someway or other depend on all the parameters of the random field, for example, all the statistical moments.

There exists a simple method to construct the effective coefficients based on perturbation theory. In this method, the solution to the initial equations is sought in the form of a power series expansion of a small parameter. The means of physical values (for example, velocity, flow rate. electric and magnetic field strength vectors and the current density) can be found by averaging each term of the series over a given statistic of physical parameters. The effective physical parameters can be constructed using these mean values. In practice it is difficult to measure the higher orders statistical moments of the geophysical parameters. At best, only mean values and correlation functions are known. In addition the direct use of the higher orders statistical moments can only impair result [4]. Therefore only the first terms of the series of perturbation theory are used. A detailed description of this approach, based on the low-order approximations of perturbation theory is presented in [5], [6], [7]. As a rule this approach is called the conventional perturbation theory.

There exist some robust methods of finding effective coefficients. The good approximation of a mean hydraulic conductivity in the steady filtration theory gives an effective hydraulic conductivity calculated by the well-known Landau–Lifshitz–Matheron formula [8], [9]:

$$\sigma\left(\mathbf{x}\right)_{ef} = \sigma_G \exp\left(\frac{d-2}{2d} D_{\ln\sigma}\right), \tag{1}$$

where d is the space dimension, $D_{\ln\sigma}$ is the variance of the conductivity logarithm, $\sigma_G = \exp\langle\ln\sigma\rangle$. This formula gives a high accuracy in estimation of the mean filtration velocity at $D_{\ln\sigma} < 2$. An attempt to improve formula (1) was made in [10]. The authors used a self-similar approximation theory developed in [11] and [12]. Some effective formulas of hydraulic conductivity are constructed for $D_{\ln\sigma} \gg 1$. The authors tentatively concluded that among these formulas only formulas based on expansion in terms of $1/d$ up to the third order provide a solution with good accuracy.

The self-similar approximation theory and the above mentioned method of subgrid modeling can be applied to a "scale regular" medium. The field measurements of e many geophysical parameters and experimental data have shown that the irregularity of the electric conductivity, permeability, porosity, density, etc. increases as the scale of measurement decreases [1], [13]–[15]. The latter property is indicative of that the statistical parameters have correlation length increasing with the scale of measurements. Systematic increase of the irregularity with scale has led several researchers to apply fractal concepts [1], [16]. Therefore many natural media are considered "scale regular" in the sense that they can be described by fractals and multiplicative cascades.

Kolmogorov [17] and Yaglom [18] were the first to induce the concept of a multi-

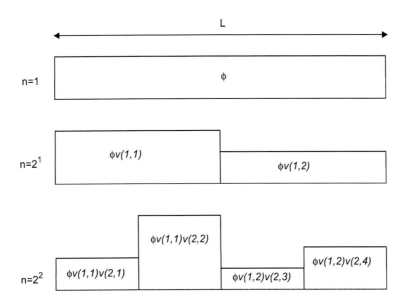

Figure 1. A multiplicative cascade. The ensemble average of ϕ is conserved at any level.

plicative cascade in the problems of fully developed turbulence. Figure 1 illustrates the construction of a three-level multiplicative cascade in a one-dimensional space.

Consider the field ϕ at step 0 uniformly distributed over the length L, divide L in n segments ($n = 2$ in Figure 1, but any other integer value could be used), generate n positive random variables $v(i,j)$ ($i = 1$, $j = 1, n$), and multiply by each one of them in a corresponding segment selected randomly. Repeat at level 2 the procedure conducted at level 1 using n^2 weights. Hence the field $\phi(i,j)$ is random for $i > 1$, and the scale ratio $\lambda = 2^i$ increases with each cascade step i. The weights have to be selected such that their average is equal to 1 at each cascade level to conserve the average value of the field at each level. When the cascade is iterated indefinitely (λ tends to infinity), the resulting field ϕ is a statistical multifractal (as opposed to a deterministic multifractal)). However, from a practical point of view, it suffices that the scale ratio becomes sufficiently large $i > 7$ to obtain an essentially multifractal field (see, for example, [19]). If we require conservation of the ensemble average of the field at each cascade level $\langle V \rangle = 1$ (where angle brackets denote ensemble average), then we obtain a conservative multifractal field. Let ϕ_0 be equal to 1, for simplicity. After i cascade steps, $\phi(i,j)$ is given by the formula

$$\phi(i,j) = v(1) \times v(2) \times \cdots \times v(i), \tag{2}$$

where the spatial dependence of the weights at steps less than i are omitted for simplicity. Taking the logarithm on both sides of (2) results in

$$\log \phi(i,j) = \log v(1) + \log v(2) + \cdots + \log v(i). \tag{3}$$

If weights v are independent random values with the same distribution and a finite variance, $\log \phi(i,j)$ has a normal distribution according to the central limit theorem. Thus $\phi(i,j)$ has a lognormal distribution. Mandelbrot argued that the lognormal model can only be obtained "under extreme and unlikely conditions" [20]. But this statement has been disproved in [21].

A generalization of the lognormal model was introduced in [22] and independently in [23]. It includes situations where the variance of v is infinite. In such a case, the generalized central limit theorem results in $\log \phi(i, j)$ following a Levy-stable probability distribution [24].

In the present chapter, the physical parameters are approximated by a 3D multiplicative continuous cascade with lognormal distribution and a Levy-stable probability distribution. We obtain effective coefficients to estimate the first and the second statistical moments of physical fields, for example, pressure, filtration velocity, electric or magnetic field strengths and current density.

2. The Physical Parameters Model

Let the field of electrical conductivity be known. This means that the field is measured on a small scale l_0 at each point $\mathbf{x}$, $\sigma(\mathbf{x})_{l_0} = \sigma(\mathbf{x})$. To pass to a coarser scale grid, it is not sufficient to smooth the field $\sigma(\mathbf{x})_{l_0}$ on a scale l, $l > l_0$. The field thus smoothed is not a physical parameter that can describe, the physical process , governed by equations, on the scales (l, L) where L is the maximum scale of heterogeneities. This is due to the fact that the fluctuations of physical parameter on the scale interval (l_0, l) correlate with the fluctuations of physical value, for example, the electric field strength induced by the electric conductivity correlates with the electric conductivity. To find a physical parameter that can describe the physical process on the scales (l, L), one has to repeat measurements on a sample of the scale l. This is not always possible. To obtain effective parameters that can describe the physical process on the scales (l, L) governing equations will be used.

Following [17] consider a dimensionless field ψ, which is equal to the ratio of two fields obtained by smoothing the field $\sigma(\mathbf{x})_{l_0}$ on two different scales l', l. Let $\sigma(\mathbf{x})_l$ denote the parameter $\sigma(\mathbf{x})_{l_0}$ smoothed on the scale l:

$$\psi(\mathbf{x}, l, l') = \sigma_{l'}(\mathbf{x})/\sigma_l(\mathbf{x}, l), \quad l' < l.$$

The field $\psi(\mathbf{x}, l, l')$ is smoother function than $\sigma(\mathbf{x})_l$ but it has too many arguments. To find a simpler field uniquely connected with $\psi(\mathbf{x}, l, l')$ and possessing all its properties, we use the following apparent equality

$$\psi(\mathbf{x}, l, l'') = \psi(\mathbf{x}, l, l')\psi(\mathbf{x}, l', l''). \tag{4}$$

For scale l', l'' close to l', we expand the function $\psi(\mathbf{x}, l, l'')$ in series in l', l'' at the point l':

$$\psi(\mathbf{x}, l, l'') = \psi(\mathbf{x}, l, l') + \frac{\partial \psi(\mathbf{x}, l, l')}{\partial l'} (l'' - l') + \cdots \tag{5}$$

In the same way we expand the function $\psi(\mathbf{x}, l', l'')$ into the series in l', l'' at the point l':

$$\psi(\mathbf{x}, l', l'') = \psi(\mathbf{x}, l', l') + \frac{\partial \psi(\mathbf{x}, l', l'')}{\partial l''}\bigg|_{l''=l'} (l'' - l') + \cdots =$$

$$= 1 + \frac{\partial \psi(\mathbf{x}, l', l'y)}{l'\partial y}\bigg|_{y=1} (l'' - l') + \cdots , \tag{6}$$

Subgrid Modeling of Steady-State Flow Processes in Fractal Medium 293

where $y = l''/l'$. Substituting (5) and (6) in (4) and discarding terms of second-order in $l'' - l'$, obtain

$$\frac{\partial \psi(\mathbf{x}, l, l')}{\partial l'} = \frac{1}{l'} \psi(\mathbf{x}, l, l') \, \varphi(\mathbf{x}, l'), \tag{7}$$

where $\varphi(\mathbf{x}, l') = \partial \psi(\mathbf{x}, l', l'y)/l'\partial y \,|_{y=1}$. Equation (7) yields the relation

$$\varphi(\mathbf{x}, l) = \frac{\partial \ln \varepsilon(\mathbf{x}, l)}{\partial \ln l} . \tag{8}$$

The solution of equation (8) is

$$\sigma_{l_0}(\mathbf{x}) = \sigma_0 \exp \left(- \int_{l_0}^{L} \varphi(\mathbf{x}, l_1) \frac{dl_1}{l_1} \right), \tag{9}$$

where σ_0 is a constant. The field φ determines statistical properties of the physical parameter σ_{l_0}. This approach is described in detail in [25]. According to the limit theorem for sums of independent random variables [24] if the variance of $\varphi(\mathbf{x}, l)$ is finite, the integral in (9) tends to a field with a normal distribution as the ratio L/l_0 increases. If the variance of $\varphi(\mathbf{x}, l)$ is infinite and there exists a nondegenerate limit of the integral in (9), the integral tends to a field with a stable distribution. Let a correlation function of φ is equal to:

$$\Phi(\mathbf{x}, \mathbf{y}, l, l')\delta \left(\ln l - \ln l' \right) = \langle \varphi(\mathbf{x}, l)\varphi(\mathbf{y}, l') \rangle - \langle \varphi(\mathbf{x}, l) \rangle \langle \varphi(\mathbf{y}, l') \rangle . \tag{10}$$

It follows from (10) that the fluctuations of $\varphi(\mathbf{x}, l)$ on different scales do not correlate. This assumption is standard in the scaling models [17]. This is due to the fact that the statistical dependence is small if the scales of fluctuations are different. To derive subgrid formulas to calculate effective coefficients this assumption may be ignored. However, this assumption is important for the numerical simulation of the field σ. If the field $\varphi(\mathbf{x}, l)$ is isotropic with a statistically homogeneous correlation function, then

$$\Phi(\mathbf{x}, \mathbf{y}, l, l') = \Phi(|\mathbf{x} - \mathbf{y}|, l, l'). \tag{11}$$

For a scale-invariant medium the following relation holds for any positive K

$$\Phi(|\mathbf{x} - \mathbf{y}|, l, l') = \Phi(K|\mathbf{x} - \mathbf{y}|, Kl, Kl').$$

Choosing $K = 1/l$ we obtain

$$\Phi \left((\mathbf{x} - \mathbf{y})^2, l, l' \right) = \Phi \left(\frac{(\mathbf{x} - \mathbf{y})^2}{l^2}, \frac{l'}{l} \right). \tag{12}$$

It follows from this equation that the function Φ depends on the two arguments only.

2.1. Log-normal model

If a random field φ has the Gaussian distribution, the number of its independent correlation functions reduces to the first two functions. In the theory of probability, random Gaussian quantities are considered as the simplest objects. The parameter $\sigma_{l_0}(\mathbf{x})$ has a log-normal distribution. Thus, the scale-invariant, log-normal model is the simplest one in the class of scale-symmetric models. For conservative cascade (9), the following equality should be satisfied in this log-normal model for any l:

$$\langle \sigma_l(\mathbf{x}) \rangle = \sigma_0. \tag{13}$$

For such fields as a porosity field, condition (13) follows from their physical sense. This condition is also valid for the fields of permeability and conductivity, if smoothing over large volumes is equivalent to the statistical averaging in according to the ergodic hypothesis. Using (9) and (13), we obtain

$$\left\langle \exp\left(-\int_{l_0}^{L} \varphi(\mathbf{x}, l_1) \frac{dl_1}{l_1} \right) \right\rangle = 1. \tag{14}$$

According to [18], the following formula is valid for an arbitrary non-random function $\theta(l)$ and the Gaussian field $f(l)$:

$$\left\langle \exp\left(-i\int_{l}^{L} \theta(l_1) f(l_1) dl_1 \right) \right\rangle =$$

$$= \exp\left(-i\int_{l}^{L} \theta(l_1) \langle f(l_1) \rangle dl_1 - \frac{1}{2}\int_{l}^{L} dl_1 \int_{l}^{L} dl_1 \theta(l_1) \theta(l_2) \langle f(l_1) f(l_2) \rangle_c \right), \tag{15}$$

where the subscript c denotes central statistical moment. Choosing $f(l) = \varphi(\mathbf{x}, l)$ and $\theta(l) = -i/l_,$, we obtain from (14) and (15) the equation

$$\int_{l_0}^{L} \frac{dl_1}{l_1} \int_{l_0}^{L} \frac{dl_2}{l_2} \langle \varphi(\mathbf{x}, l_1) \varphi(\mathbf{x}, l_2) \rangle_c - 2\int_{l_0}^{L} \langle \varphi(\mathbf{x}, l_1) \rangle \frac{dl_1}{l_1} = 0. \tag{16}$$

If $\varphi(\mathbf{x}, l)$ is the homogeneity and scale-invariance, $\langle \varphi(\mathbf{x}, l) \rangle$ is a constant. In a particular case of non-correlated fluctuations of the field φ of different scales, the model is simplified. We choose the correlation function in the form

$$\Phi\left(\frac{(\mathbf{x} - \mathbf{y})^2}{l^2}, \frac{l'}{l} \right) = \Phi_0 \exp\left\{ -\frac{(\mathbf{x} - \mathbf{y})^2}{l^2} \right\} \delta\left(\ln l - \ln l' \right). \tag{17}$$

Then for $\mathbf{x} = \mathbf{y}$, equation (16) yields

$$\Phi_0 = 2\langle \varphi \rangle. \tag{18}$$

If relation (13) is not used, then the constant Φ_0 may depend on the scale and a cascade is not conservative. Let us consider the correlation function σ at the points $\mathbf{x}$ and $\mathbf{x} + \mathbf{r}$. It follows from (9) that

$$\langle \sigma_{l_0}(\mathbf{x}) \sigma_{l_0}(\mathbf{x} + \mathbf{r},) \rangle = \left\langle \sigma_0^2 \exp\left[-\left(\int_{l_0}^{L} (\varphi(\mathbf{x}, l_1) + \varphi(\mathbf{x} + \mathbf{r}, l_1)) \frac{dl_1}{l_1} \right) \right] \right\rangle.$$

Subgrid Modeling of Steady-State Flow Processes in Fractal Medium 295

Choosing $f(l) = \varphi(\mathbf{x}, l) + \varphi(\mathbf{x} + \mathbf{r}, l)$ and $\theta(l) = -i/l$ in (15), we obtain

$$\langle \sigma_{l_0}(\mathbf{x})\sigma_{l_0}(\mathbf{x} + \mathbf{r}) \rangle =$$

$$= \sigma_0^2 \exp\left(-2\langle\varphi\rangle \ln\frac{L}{l_0} + \int_{l_0}^{L}\frac{dl_1}{l_1}\int_{l_0}^{L}\frac{dl_2}{l_2}\langle\varphi(\mathbf{x}, l_1)\varphi(\mathbf{x} + \mathbf{r}, l_2)\rangle_c\right) =$$

$$= \sigma_0^2 \exp\left(-2\langle\varphi\rangle \ln\frac{L}{l_0} + \int_{l_0}^{L}\frac{dl_1}{l_1}\int_{l_0}^{L}\Phi\left(\frac{r^2}{l_1^2}, \frac{l_1}{l_2}\right)\frac{dl_2}{l_2}\right). \quad (19)$$

If the correlation function has the form of (17), then

$$\langle \sigma_{l_0}(\mathbf{x})\sigma_{l_0}(\mathbf{x} + \mathbf{r}) \rangle = \sigma_0^2 \exp\left\{-2\langle\varphi\rangle \ln\frac{L}{l_0} + \Phi_0\int_{l_0}^{L}\exp\left(-\frac{r^2}{l_1}\right)\frac{dl_1}{l_1}\right\}. \quad (20)$$

At $l_0 < r < L$ the integral in (20) is equal to

$$\int_{l_0}^{L}\exp\left(-\frac{r^2}{l_1}\right)\frac{dl_1}{l_1} = -\frac{1}{2}Ei\left(\frac{r^2}{L^2}\right) + \frac{1}{2}Ei\left(\frac{r^2}{l_0^2}\right). \quad (21)$$

At $x > 1$ the integral exponential function $Ei(x)$ is small and $Ei(x) \sim \gamma + \ln x$ at $x < 1$. Then

$$\langle \sigma_{l_0}(\mathbf{x})\sigma_{l_0}(\mathbf{x} + \mathbf{r}) \rangle \approx C\left(\frac{r}{L}\right)^{-\Phi_0}. \quad (22)$$

Formula (22) is analogous to the formula for the correlation function of the energy-dissipation rate for the isotropic turbulence [18]. The constant in (22) is defined by the expression $C = \sigma_0^2(\frac{L}{l_0})^{-2\langle\varphi\rangle}e^{-\Phi_0\gamma}$, where $\gamma = 0.57722$ is the Euler constant and l_0 is the minimum scale. For $r \gg L$, we have $\langle \sigma_{l_0}(\mathbf{x})\sigma_{l_0}(\mathbf{x} + \mathbf{r}) \rangle \to \sigma_0^2$. This case is not interesting and is not further considered. In the case of conformal symmetry, similar estimates were obtained in [26] without using the assumption about the absence of correlation in terms of $\ln l$.

2.2. Log-stable model

A strong irregularity and intermittency in the behavior of the physical fields have led some researchers to investigation of random fields with a log-stable distribution [22], [23], [27]. In [28] the authors, using experimental data for boreholes, have obtained distributions of permeability fields and some statistical characteristics and showed that the permeability fields can be simulated by the fields with a log-stable distribution.

All stable distributions are described by characteristic functions that depend on four parameters: α, β, μ, and λ [29]. The parameter α is such that $0 < \alpha \leq 2$, where the situation $\alpha = 2$ corresponds to the Gaussian distribution. The statistical moments of order m are not defined for $m \geq \alpha$, with the exception of the Gaussian case where all statistical moments are defined. Thus the variance is infinite for $\alpha < 2$ and the mean is infinite for

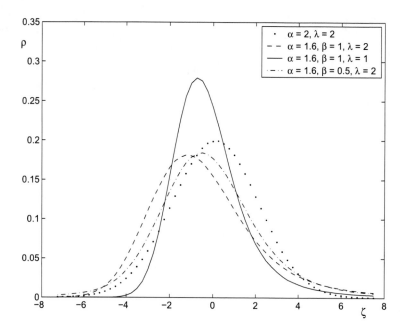

Figure 2. Probability density functions of stable distributions at $\mu = 0$; $\alpha = 2$ corresponds to the Gaussian case.

$\alpha < 1$. The parameter μ is a real number and is a centering term. This parameter is equal to the mean of the distribution only when $\alpha > 1$. and $\beta = 0$. The parameter $\beta \in [-1, 1]$ is known as the skewness parameter: $\beta = -1$ gives an extreme negative antisymmetric distribution, where the probability of producing negatively large random variables is high, the converse occurs for $\beta = 1$, and $\beta = 0$ results in a symmetric distribution with respect to μ. The parameter λ is known as the scale parameter; it is equal to half the variance for $\alpha = 2$ and plays a similar role when $\alpha < 2$ (i.e., it is a measure of the width of the distribution).

Figure 2 gives the one-dimensional probability densities of the stable laws for $\mu = 0$. The field $\varphi(\mathbf{x}, l)$ with a stable distribution law is modeled using the approach described in [29]. At the points $(\mathbf{x}_j, l)$, the field φ is expressed in terms of the sum of random independent variables which have stable distributions having the same parameters α, β, $\mu = 0$ and $\lambda = 1$ (the form A is considered [30]):

$$\varphi(\mathbf{x}_j, l) = \left(\frac{\Phi_0(l)}{2(\Delta \tau \ln 2)^{\alpha-1}} \right)^{\frac{1}{\alpha}} a_{\mathbf{j}\mathbf{i}}^l \zeta_{\mathbf{i}}^l + \langle \varphi \rangle, \qquad (23)$$

Here $l = 2^\tau$ and $\Delta \tau$ is the τ grid-size; $a_{\mathbf{j}\mathbf{i}}^l$ denotes $a_{j_x,j_y,j_z;i_x,i_y,i_z}^l$; the coefficients $a_{\mathbf{j}\mathbf{i}}^l$ have a support of size l^3 and depend only on the modulus of the difference of the indices $a_{\mathbf{j}\mathbf{i}}^l \equiv a^l(|\mathbf{i} - \mathbf{j}|)$; therefore, the subscript j can be omitted below. For all l, the condition $\sum_{k_x} \sum_{k_y} \sum_{k_z} \left(a_{k_x k_y k_z}^l \right)^\alpha = 1$ is satisfied. For $1 \leq \alpha \leq 2$, the thus constructed field φ is stable, homogeneous, and isotropic in the spatial variables [29]. If the coefficients $a_{\mathbf{j}\mathbf{i}}^l$ satisfy the

Subgrid Modeling of Steady-State Flow Processes in Fractal Medium 297

condition $a_{ji}^l \equiv a^l \left(\frac{|\mathbf{i}-\mathbf{j}|}{l} \right)$ and if the constants $\Phi_0(l)$ and $\langle \varphi \rangle$ are the same values for all l, the field φ is invariant under scale transformations. The mean value of the field φ exists and is equal to $\langle \varphi \rangle$; as regards the second moments, for $\alpha \neq 2$, they are infinite. So. it is not possible to apply a correlation analysis that is performed, for example, in [25]. Nevertheless, for the extreme point $\beta = 1$ the second moments for the field σ exist despite the absence of variance of the field φ. This case is of interest because it was confirmed by field measurements in geophysics [28]. In the above described model the field σ has the form

$$\sigma_{l_0}(\mathbf{x}) = \sigma_0 \exp \left[- \left(\ln 2 \sum_{\hat{l}_0}^{\hat{L}} \varphi(\mathbf{x}, \tau_l) \delta \tau \right) \right], \tag{24}$$

where $L = 2^{\hat{L}\Delta\tau}$ and $l_0 = 2^{\hat{l}_0 \Delta\tau}$. The integral in formula (9) is replaced by a sum.

Consider the mean value of the field σ. The mean value of the random variable $e^{-b\zeta}$ is given by the formula [29]

$$\langle \exp(-b\zeta) \rangle = \exp \left[\lambda \mu b - \lambda b^\alpha / \cos(\pi\alpha/2) \right] \tag{25}$$

Using formula (25) and taking into account that $\mu = 0, \sigma = 1, \beta = 1, 1 < \alpha < 2$ for ζ_k we obtain

$$\langle \sigma_{l_0}(\mathbf{x}) \rangle = \sigma_0 \left\langle \exp \left[-\Delta\tau \ln 2 \left(\sum_{\hat{l}=\hat{l}_0}^{\hat{L}} \sum_k \left(\frac{\Phi_0(\tau_{\hat{l}})}{2(\Delta\tau \ln 2)^{\alpha-1}} \right)^{\frac{1}{\alpha}} a_{\mathbf{k}}^l \zeta_{\mathbf{k}} + \langle \varphi \rangle \right) \right] \right\rangle =$$

$$= \prod_{l=\hat{l}_0}^{\hat{L}} \left\langle \exp \left[-\Delta\tau \ln 2 \left(\sum_k \left(\frac{\Phi_0(\tau_{\hat{l}})}{2(\Delta\tau \ln 2)^{\alpha-1}} \right)^{\frac{1}{\alpha}} a_{\mathbf{k}}^l \zeta_{\mathbf{k}} + \langle \varphi \rangle \right) \right] \right\rangle =$$

$$= \prod_{l=\hat{l}_0}^{\hat{L}} \exp \left[\frac{\Phi_0(\tau_{\hat{l}}) \Delta\tau \ln 2}{2} \left[\cos \left(\frac{\pi\alpha}{2} \right) \right]^{-1} \sum_k \left(a_{\mathbf{k}}^l \right)^\alpha - \langle \varphi \rangle \Delta\tau ln2 \right] =$$

$$= \sigma_0 \exp \left[-\Delta\tau \ln 2 \left(\frac{1}{2 \cos \left(\frac{\pi\alpha}{2} \right)} \sum_{\hat{l}=\hat{l}_0}^{\hat{L}} \Phi_0(\tau_{\hat{l}}) + \langle \varphi \rangle \right) \right]. \tag{26}$$

If the medium is scale-invariant, Φ_0 does not depend on l and the mean value of the field $\sigma_{l_0}(\mathbf{x})$ given by the formula

$$\langle \sigma_{l_0}(\mathbf{x}) \rangle = \sigma_0 \exp \left[- \left(\frac{\Phi_0}{2 \cos \left(\frac{\pi\alpha}{2} \right)} + \langle \varphi \rangle \right) (\ln L - \ln l_0) \right] =$$

$$= \sigma_0 \left(\frac{L}{l_0} \right)^{-\left(\frac{\Phi_0}{2 \cos \left(\frac{\pi\alpha}{2} \right)} + \langle \varphi \rangle \right)}. \tag{27}$$

Consider the second single-point moment for $\sigma_{l_0}(\mathbf{x})$. Using formula (25) we obtain

$$\langle \sigma_{l_0}^2(\mathbf{x}) \rangle = \sigma_0^2 \left\langle \exp\left[-\sum_{\widehat{l_0}}^{\widehat{L}} 2\varphi(\mathbf{x}, \tau_l) \Delta\tau \ln 2 \right] \right\rangle =$$

$$= \sigma_0^2 \exp\left[-\Delta\tau \ln 2 \left(\frac{2^{(\alpha-1)}}{\cos\left(\frac{\pi\alpha}{2}\right)} \sum_{\widehat{l}=\widehat{l_0}}^{\widehat{L}} \Phi_0(\tau_{\widehat{l}}) + 2\langle\varphi\rangle \right) \right]. \qquad (28)$$

If $\langle \sigma_l(\mathbf{x}) \rangle = \sigma_0$ for each l, then the parameters Φ_0 and $\langle\varphi\rangle$ satisfy the equality

$$\frac{\Phi_0}{2\cos\left(\frac{\pi\alpha}{2}\right)} = -\langle\varphi\rangle. \qquad (29)$$

Let us consider the correlation function $\langle \sigma_{l_0}(\mathbf{x})\sigma_{l_0}(\mathbf{x}+\mathbf{r}) \rangle$:

$$\langle \sigma_{l_0}(\mathbf{x})\sigma_{l_0}(\mathbf{x}+\mathbf{r}) \rangle =$$

$$= \sigma_0^2 \left\langle \exp\left[-\sqrt[\alpha]{\frac{1}{2}\Delta\tau \ln 2} \sum_{\widehat{l_0}}^{\widehat{L}} \sqrt[\alpha]{\Phi_0(\widehat{l})} \sum_k \left(a_\mathbf{k}^l \zeta_\mathbf{k}^l + a_\mathbf{k}^l \zeta_{\mathbf{k}+\mathbf{k}_r}^l \right) - 2\langle\varphi\rangle \right] \right\rangle. \qquad (30)$$

The sum in the exponent in (30) is divided with respect to the scale $\widehat{l}$ into groups $\widehat{l_0} \leq \widehat{l} \ll \widehat{l_r}$, and $\widehat{l_r} < \widehat{l} < \widehat{L}$, where $r = 2^{\widehat{l_r}\Delta\tau}$. The estimation of the first group gives the constant C. For the second group, formula (25) is used. In this case, it is taken into account that for $\zeta_\mathbf{k}$, the parameters in form A [30] are equal to $\mu = 0$, $\lambda = 1$, $\beta = 1$, $1 < \alpha \leq 2$. The mean value of $\sigma_{l_0}(\mathbf{x})$ is calculated by formula (26). Finally, we obtain the estimation

$$\langle \sigma_{l_0}(\mathbf{x})\sigma_{l_0}(\mathbf{x}+\mathbf{r}) \rangle \simeq C \exp\left[-\Delta\tau \ln 2 \left(\frac{2^{\alpha-1}}{\cos\left(\frac{\pi}{2}\alpha\right)} \sum_{\widehat{l_r}}^{\widehat{L}} \Phi_0(\widehat{l}) + 2\langle\varphi\rangle \right) \right]. \qquad (31)$$

For a scale-invariant medium, for the second group we obtain the following estimation

$$\langle \sigma_{l_0}(\mathbf{x})\sigma_{l_0}(\mathbf{x}+\mathbf{r}) \rangle \simeq C \exp\left[-\left(\frac{2^{\alpha-1}\Phi_0}{\cos\left(\frac{\pi}{2}\alpha\right)} + 2\langle\varphi\rangle \right) (\ln L - \ln r) \right] \simeq$$

$$\simeq C \left(\frac{L}{r} \right)^{-\left(\frac{2^{\alpha-1}\Phi_0}{\cos\left(\frac{\pi}{2}\alpha\right)} + 2\langle\varphi\rangle \right)}. \qquad (32)$$

The constant C is not universal, and the exponent in (32) for a fractal medium is universal and, according to [28] can be determined.

3. Subgrid Modeling of Transfer Processes in Isotropic Multi-Scale Media

Let a local flow $\mathbf{v}$ and a field $\mathbf{h}$ be linked by the system of relations

$$\mathbf{v}(\mathbf{x}) = \sigma(\mathbf{x})\mathbf{h}(\mathbf{x}), \quad div\,\mathbf{v}(\mathbf{x}) = 0, \quad \mathbf{h}(\mathbf{x}) = -\nabla U(\mathbf{x}). \qquad (33)$$

Subgrid Modeling of Steady-State Flow Processes in Fractal Medium 299

In this problem a direct-current flow through a heterogeneous medium, the vector $\mathbf{v}$ is the electric current density vector, the field $\mathbf{h}$ defined by the potential $U(\mathbf{x})$ is the electric field, and the conductivity $\sigma(\mathbf{x})$ is a random field of the electrical conductivity of a medium. In the problem of filtration of a single-phase liquid in a heterogeneous medium at small Reynolds numbers, the vector $\mathbf{v}$ is the filtration velocity vector, vector $\mathbf{h}$ is the field defined by the pressure gradient $\mathbf{h}(\mathbf{x}) = -\nabla p(\mathbf{x})$; the conductivity $\sigma(\mathbf{x})$ is equal to the ratio of permeability to viscosity. The pressure and velocity are related by the Darcy equation. We assume that on the boundary Γ of the region V in which equation (33) is solved, some boundary conditions are specified. The dimensions of the region V are larger than the heterogeneity scale.

The conductivity function $\sigma(\mathbf{x}) = \sigma(\mathbf{x})_{l_0}$ is divided into two components with respect to the scale l. The large-scale (ongrid) component $\sigma(\mathbf{x}, l)$ is obtained by statistical averaging over all $\varphi(\mathbf{x}, l_1)$ with $, l_0 < l_1 < l, l - l_0 = dl$, where dl is small. The small-scale (subgrid) component is equal to $\sigma'(\mathbf{x}) = \sigma(\mathbf{x}) - \sigma(\mathbf{x}, l)$:

$$\sigma(\mathbf{x}, l) = \sigma_0 \exp\left[-\int\limits_l^L \varphi(\mathbf{x}, l_1)\frac{dl_1}{l_1}\right] \left\langle \exp\left[-\int\limits_{l_0}^l \varphi(\mathbf{x}, l_1)\frac{dl_1}{l_1}\right]\right\rangle,$$

$$\sigma'(\mathbf{x}) = \sigma(\mathbf{x}, l) \left[\frac{\exp\left[-\int\limits_{l_0}^l \varphi(\mathbf{x}, l_1)\frac{dl_1}{l_1}\right]}{\left\langle \exp\left[-\int\limits_{l_0}^l \varphi(\mathbf{x}, l_1)\frac{dl_1}{l_1}\right]\right\rangle} - 1\right]. \tag{34}$$

The large-scale (ongrid) component of the potential $\mathbf{U}(\mathbf{x}, l)$ is obtained by averaging the solutions to system (33), in which the large-scale component of conductivity $\sigma(\mathbf{x}, l)$ is fixed and the small component $\sigma'(\mathbf{x})$ is a random variable. The subgrid component of the potential is equal to $U'(\mathbf{x}) = U(\mathbf{x}) - U(\mathbf{x}, l)$. Substituting the relation for $U'(\mathbf{x})$ and $\sigma(\mathbf{x})$ into system (33) and averaging over small-scale components, we have

$$\nabla\left[\sigma(\mathbf{x}, l)\nabla U(\mathbf{x}, l) + \langle\sigma'(\mathbf{x})\nabla U'(\mathbf{x})\rangle\right] = 0. \tag{35}$$

The subgrid term $\langle\sigma'\mathbf{U}'\rangle$ in equation (35) is unknown. This term cannot be neglected without some preliminary estimation, since the correlation between the conductivity and the gradient of the potential field may be significant. The form of this term in (35) determines a subgrid model. The subgrid term is estimated using perturbation theory. Subtracting system (35) from system (33) and taking into account only the first order terms, we obtain the subgrid equations:

$$\Delta U'(\mathbf{x}) = -\frac{1}{\sigma(\mathbf{x}, l)}\nabla\sigma'(\mathbf{x})\nabla U(\mathbf{x}, l). \tag{36}$$

The variable $\nabla U(\mathbf{x}, l)$ on the right-hand side of (36) is assumed to be known. Solving system (36) we have

$$U'(\mathbf{x}) = \frac{1}{4\pi\sigma(\mathbf{x}, l)}\left(\int\limits_V \frac{1}{r}\nabla'_j\sigma'(\mathbf{x}')\,d\mathbf{x}'\right)\nabla_j U(\mathbf{x}, l), \tag{37}$$

where $r = |\mathbf{x} - \mathbf{x}'|$ and the summation of repeated indices is implied. Since a small change in the scale of σ produces considerable fluctuations in the field (which is typical of fractal fields), the field $\sigma(\mathbf{x}, l)$ and its derivatives are believed to change slower than σ' and its derivatives. Similar assumptions are made for $U(\mathbf{x}, l)$. Therefore $U(\mathbf{x}, l)$, $\sigma(\mathbf{x}, l)$ and their derivatives are factored outside the integral sign in (37). Using (37) the subgrid term can be written as

$$
\langle \sigma'(\mathbf{x}) \nabla_i U'(\mathbf{x}) \rangle = \frac{1}{4\pi\sigma(\mathbf{x}, l)} \left\langle \sigma'(\mathbf{x}) \int_V \nabla_i \frac{1}{r} \nabla'_j \sigma'(\mathbf{x}') \, d\mathbf{x}' \right\rangle \nabla_j U(\mathbf{x}, l) =
$$

$$
= -\frac{1}{4\pi\sigma(\mathbf{x}, l)} \int_V \nabla'_i \frac{1}{r} \nabla'_j \langle \sigma'(\mathbf{x}) \sigma'(\mathbf{x}') \rangle \, d\mathbf{x}' \, \nabla_j U(\mathbf{x}, l), \qquad (38)
$$

where the equality $\nabla_i(1/r) = -\nabla'_i(1/r)$ is used. If the field of conductivity is isotropic and statistically homogeneous with *log-normal distribution of probability* , from (34) for small $dl = l - l_0$, we have

$$
\sigma(\mathbf{x}, l) \simeq \left[1 - \langle \varphi \rangle \frac{dl}{l} + \frac{1}{2} \Phi_0(l) \frac{dl}{l} \right] \sigma_l(\mathbf{x}). \qquad (39)
$$

$$
\langle \sigma'(\mathbf{x}) \sigma'(\mathbf{x}') \rangle =
$$

$$
\sigma(\mathbf{x}, l)^2 \left\langle \exp\left(-\int_{l_0}^{l} (\varphi(\mathbf{x}, l_1) + \varphi(\mathbf{x}', l_1) - 2\langle \varphi \rangle + \Phi_0(l_1)) \frac{dl_1}{l_1} \right) - 1 \right\rangle \approx
$$

$$
\approx \sigma(\mathbf{x}, l)^2 \, \Phi\left(|\mathbf{x} - \mathbf{x}'|, l \right) \frac{dl}{l}, \qquad (40)
$$

where $\Phi_0(l) = \Phi(0, l)$. Substitution of (40) into (38) yields

$$
\langle \sigma'(\mathbf{x}) \nabla_i U'(\mathbf{x}) \rangle = -\frac{1}{4\pi} \int_V \nabla'_i \frac{1}{r} \nabla'_j \Phi\left(|\mathbf{x} - \mathbf{x}'|, l \right) d\mathbf{x}' \sigma(\mathbf{x}, l) \nabla_j U(\mathbf{x}, l) \frac{dl}{l}. \qquad (41)
$$

The integral over V in (41) can be changed by an integral with infinite limits, since the correlation function Φ is small if $L \ll L_0$, where L_0 is minimum size of V. This change gives a sensible error only in a narrow region of the correlation radius size near the boundary. In formula (41), the Cartesian coordinates are changed for spherical coordinates. Integrating $n_j n_m$, where $n_m = x_m/r$, over the complete solid angle we arrive at the formula $\int n_j n_m d\vartheta = \frac{4\pi}{3} \delta_{jm}$. Using this formula and integrating (41) by parts we obtain

$$
\langle \sigma' \nabla_i U' \rangle = -\frac{1}{3} \Phi_0 \sigma(\mathbf{x}, l) \nabla_i U(\mathbf{x}, l) \frac{dl}{l}. \qquad (42)
$$

Substituting (42) into (35), we have

$$
\nabla_i \sigma_{0l} \exp\left[-\int_{l}^{L} \varphi(\mathbf{x}, l_1) \frac{dl_1}{l_1} \right] \nabla_i U(\mathbf{x}, l) = 0,
$$

$$
\sigma_{0l} = \sigma_0 \left[\left(1 - \frac{1}{3} \Phi_0 \frac{dl}{l} \right) \left(1 + \frac{\Phi_0}{2} \frac{dl}{l} - \langle \varphi \rangle \frac{dl}{l} \right) \right]. \qquad (43)
$$

It follows from (43) that the new coefficient σ_{l0} is

$$\sigma_{0l} = \left[1 + \left(\frac{\Phi_0\,(l)}{6} - \langle\varphi\rangle\right)\frac{dl}{l}\right]\sigma_0, \tag{44}$$

with second order of accuracy. As $dl \to 0$ we obtain the equation

$$\frac{d\ln\sigma_{0l}}{d\ln l} = \frac{\Phi_0\,(l)}{6} - \langle\varphi(l)\rangle. \tag{45}$$

For a scale-invariant medium, the effective coefficient has the following simple form

$$\sigma_{0l} = \sigma_{0L}\left(\frac{l}{L}\right)^{\frac{\Phi_0}{6} - \langle\varphi\rangle}. \tag{46}$$

At $l = l_0$ the formula for the effective conductivity in (46) is the Landau-Lifshitz-Matheron formula (1):

$$\begin{aligned}\left(\frac{l}{L}\right)^{\langle\varphi\rangle - \Phi_0/6} &= \exp\left[(\Phi_0/6 - \langle\varphi\rangle)\,(\ln L - \ln l_0)\right], \\ &= \sigma_G\exp\left(\frac{d-2}{2d}D_{ln\sigma}\right).\end{aligned}$$

Here $D_{\ln\sigma} = \Phi_0\,(\ln L - \ln l_0)$ and $\sigma_G = \exp\left[-\langle\varphi\rangle\,(\ln L - \ln l_0)\right]$.

If field of conductivity is isotropic and statistically homogeneous with *log-stable distribution of probability* from (27) and (34) we have

$$\sigma(\mathbf{x}, l) = \sigma_0\exp\left[-\Delta\tau\ln 2\left(\sum_{\widehat{l}_1=\widehat{l}}^{\widehat{L}}\varphi\left(\mathbf{x}, \tau_{\widehat{l}_1}\right) + \sum_{\widehat{l}_1=\widehat{l}_0}^{\widehat{l}}\frac{\Phi_0(\tau_{\widehat{l}_1})}{2\cos\left(\frac{\pi}{2}\alpha\right)} + \langle\varphi\rangle\right)\right]. \tag{47}$$

$$\sigma(\mathbf{x}, l) = \sigma_l(\mathbf{x})\left(1 - \Delta\tau\ln 2\left(\frac{\Phi_0(\tau_{\widehat{l}_1})}{2\cos\left(\frac{\pi}{2}\alpha\right)} + \langle\varphi\rangle\right)\right). \tag{48}$$

$$\sigma'(\mathbf{x}) = \sigma(\mathbf{x}, l)\left\{\exp\left[\Delta\tau\ln 2\sum_{\widehat{l}_1=\widehat{l}_0}^{\widehat{l}}\left(-\varphi(\mathbf{x}, \tau_{\widehat{l}_1}) + \frac{\Phi_0\left(\tau_{\widehat{l}_1}\right)}{2\cos\left(\frac{\pi}{2}\alpha\right)} + \langle\varphi\rangle\right)\right] - 1\right\}. \tag{49}$$

Then

$$\langle\sigma'(\mathbf{x})\sigma'(\mathbf{x})\rangle \approx \sigma(\mathbf{x}, l)^2\Delta\tau\ln 2\left[\cos\left(\frac{\pi}{2}\alpha\right)\right]^{-1}\left(1 - 2^{\alpha-1}\right)\Phi_0\left(\tau_{\widehat{l}}\right). \tag{50}$$

Using formulas (38) and (50), we arrive at the following formula for the subgrid term

$$\langle\sigma'(\mathbf{x})\nabla U'(\mathbf{x})\rangle \approx -\frac{\Delta\tau\ln 2}{3}\left[\cos\left(\frac{\pi}{2}\alpha\right)\right]^{-1}\left(1 - 2^{\alpha-1}\right)\Phi_0\left(\tau_{\widehat{l}}\right)\sigma(\mathbf{x}, l)\nabla U(\mathbf{x}, l). \tag{51}$$

Substituting (51) into (35), using (48) and taking into account only first order smallness in $\Delta\tau$, we obtain that the new coefficient σ_{l0} is

$$\sigma_{0l} = \sigma_0\left[1 - \Delta\tau\ln 2\left(\Phi_0\left(\tau_{\widehat{l}_1}\right)\frac{2\left(1 - 2^{\alpha-1}\right) + 3}{6\cos\left(\frac{\pi}{2}\alpha\right)} + \langle\varphi\rangle\right)\right]. \tag{52}$$

Replacing the variable τ by l and letting $dl \to 0$ we obtain the equation

$$\frac{d \ln \sigma_{0l}}{d \ln l} = -\Phi_0(l) \frac{2\left(1 - 2^{\alpha-1}\right) + 3}{6 \cos\left(\frac{\pi}{2}\alpha\right)} - \langle \varphi \rangle. \tag{53}$$

For a scale-invariant medium, the effective coefficient has the following simple form:

$$\sigma_{0l} = \sigma_{0L} \left(\frac{l}{L}\right)^{\Phi_0\left(2\left(1 - 2^{\alpha-1}\right) + 3\right)/(6\cos(\alpha\pi/2)) - \langle\varphi\rangle}. \tag{54}$$

3.1. Estimation of the second statistical moments

The estimation of the second statistical moments is described in detail in [31]. For the components of the covariance tensor of the field $\mathbf{h}$, the ongrid equation has the form

$$\langle h_n(\mathbf{x}) h_j(\mathbf{x}) \rangle = \nabla_n U(\mathbf{x}, l) \nabla_j U(\mathbf{x}, l) + \langle \nabla_n U'(\mathbf{x}) \nabla_j U'(\mathbf{x}) \rangle. \tag{55}$$

Let us estimate the second term in (55) for the isotropic case. We denote $r_1 = |\mathbf{x} - \mathbf{x}'|$ and $r_2 = |\mathbf{x} - \mathbf{x}''|$. Using formula (36) we obtain components of the covariance tensor $\nabla U'$ at a log-normal distribution $\sigma(\mathbf{x}, l)$:

$$\langle \nabla_n U'(\mathbf{x}) \nabla_j U'(\mathbf{x}) \rangle = \frac{1}{16\pi^2 \sigma(\mathbf{x}, l)^2} \times$$

$$\times \left\langle \int_V \int_V \nabla_n \frac{1}{r_1} \nabla'_m \sigma'(\mathbf{x}') \nabla_j \frac{1}{r_2} \nabla''_k \sigma'(\mathbf{x}'') \, d\mathbf{x}'' d\mathbf{x}' \nabla_m U(\mathbf{x}, l) \nabla_k U(\mathbf{x}, l) \right\rangle =$$

$$= \frac{1}{16\pi^2} \int_V \int_V \nabla'_n \frac{1}{r_1} \nabla''_j \frac{1}{r_2} \nabla'_m \nabla''_k \Phi\left(|\mathbf{x}' - \mathbf{x}''|, l\right) d\mathbf{x}'' d\mathbf{x}' \frac{dl}{l} \nabla_m U(\mathbf{x}, l) \nabla_k U(\mathbf{x}, l), \tag{56}$$

Changing the integrals over V in (56) by integrals with infinite limits and denoting $\mathbf{y} = \mathbf{x}' - \mathbf{x}''$, $\mathbf{z} = \mathbf{x} - \mathbf{x}'$, we have

$$\langle \nabla_n U'(\mathbf{x}) \nabla_j U'(\mathbf{x}) \rangle = \frac{1}{16\pi^2} \times$$

$$\times \int_{-\infty}^{\infty} \int_{-\infty}^{\infty} \frac{\partial}{\partial z_n} \frac{1}{z} \frac{\partial}{\partial z_j} \frac{1}{|\mathbf{z} - \mathbf{y}|} \frac{\partial}{\partial y_m} \frac{\partial}{\partial y_k} \Phi(y, l) \, dz dy \frac{dl}{l} \nabla_m U(\mathbf{x}, l) \nabla_k U(\mathbf{x}, l), \tag{57}$$

where $z = |\mathbf{z}|$ and $y = |\mathbf{y}|$. The convolution of the Green function and its partial derivatives in (57) are integrated with the Fourier transform. The Fourier transform formulas for a spherical symmetric function [32] are

$$F^{-1}\left(F\left(\frac{\partial}{\partial z_n}\frac{1}{z}\right) F\left(\frac{\partial}{\partial z_j}\frac{1}{|\mathbf{z} - \mathbf{y}|}\right)\right) = -2\pi \frac{\partial^2}{\partial y_n \partial y_j} y. \tag{58}$$

Substituting (90) into (57), passing to spherical coordinates and using the formula

$$\int_0^{2\pi} \int_0^{\pi} n_\alpha n_\beta n_\gamma n_\nu \sin\theta d\chi d\theta = 4\pi \left(\delta_{\alpha\beta}\delta_{\gamma\nu} + \delta_{\alpha\gamma}\delta_{\beta\nu} + \delta_{\alpha\nu}\delta_{\beta\gamma}\right)/15,$$

Subgrid Modeling of Steady-State Flow Processes in Fractal Medium 303

we obtain

$$\langle \nabla_n U'(\mathbf{x}) \nabla_j U'(\mathbf{x}) \rangle = \frac{\Phi_0}{15} (l) (\delta_{nj}\delta_{km} + \delta_{nk}\delta_{jm} + \delta_{nm}\delta_{kj}) \frac{dl}{l} \nabla_m U(\mathbf{x}, l) \nabla_k U(\mathbf{x}, l). \quad (59)$$

It follows from (59) that at $n = j$ the components of the covariance tensor satisfy the relations

$$\left\langle h_1 (\mathbf{x})^2 - h_2 (\mathbf{x})^2 \right\rangle \simeq \left(1 + \frac{2}{15}\Phi_0 (l) \frac{dl}{l}\right) \left(h_1 (\mathbf{x}, l)^2 - h_2 (\mathbf{x}, l)^2\right),$$

$$\left\langle h_3 (\mathbf{x})^2 - h_2 (\mathbf{x})^2 \right\rangle \simeq \left(1 + \frac{2}{15}\Phi_0 (l) \frac{dl}{l}\right) \left(h_3 (\mathbf{x}, l)^2 - h_2 (\mathbf{x}, l)^2\right),$$

$$\left\langle h_1 (\mathbf{x})^2 + h_2 (\mathbf{x})^2 + h_3 (\mathbf{x})^2 \right\rangle \simeq \quad (60)$$

$$\simeq \left(1 + \frac{1}{3}\Phi_0 (l) \frac{dl}{l}\right) \left(h_1 (\mathbf{x}, l)^2 + h_2 (\mathbf{x}, l)^2 + h_3 (\mathbf{x}, l)^2\right).$$

By analogy with (43) we can write:

$$\sigma_{0l}^{(1)} = \sigma_0^2 \left(1 + \frac{2}{15}\Phi_0 (l) \frac{dl}{l}\right),$$

$$\sigma_{0l}^{(2)} = \sigma_0^2 \left(1 + \frac{1}{3}\Phi_0 (l) \frac{dl}{l}\right),$$

with second order of accuracy. As $dl \to 0$ we obtain the equations

$$\frac{d \ln \sigma_{0l}^{(1)}}{d \ln l} = \frac{2}{15} \Phi_0 (l), \quad \sigma_{0l}^{(1)}|_{l=L} = 1,$$

$$\frac{d \ln \sigma_{0l}^{(2)}}{d \ln l} = \frac{1}{3} \Phi_0 (l), \quad \sigma_{0l}^{(2)}|_{l=L} = 1. \quad (61)$$

Solving system (60) with the effective coefficients $\sigma_{0l}^{(1)}$ and $\sigma_{0l}^{(2)}$, we find the components $\left\langle h_i (\mathbf{x})^2 \right\rangle$.

In the scale-invariant medium, the effective system takes the form

$$\begin{bmatrix} \left\langle h_1 (\mathbf{x})^2 \right\rangle_{ef} \\ \left\langle h_2 (\mathbf{x})^2 \right\rangle_{ef} \\ \left\langle h_3 (\mathbf{x})^2 \right\rangle_{ef} \end{bmatrix} = \frac{1}{3} (l/L)^{\Phi_0/3} \mathcal{A} \begin{bmatrix} h_1 (\mathbf{x}, l)^2 \\ h_2 (\mathbf{x}, l)^2 \\ h_3 (\mathbf{x}, l)^2 \end{bmatrix}, \quad (62)$$

where the elements of the matrix $\mathcal{A}$ are equal to: $a_{ii} = 2(\frac{l}{L})^{-\frac{\Phi_0}{5}} + 1, i = 1, 2, 3$ and $a_{ij} = -(\frac{l}{L})^{-\frac{\Phi_0}{5}} + 1, i \neq j$. It follows from formulas (55) and (59) that for $n \neq j$:

$$\langle h_n (\mathbf{x}) h_j (\mathbf{x}) \rangle \sim \left(1 + \frac{2}{15}\Phi_0 \frac{dl}{l}\right) h_n (\mathbf{x}, l) h_j (\mathbf{x}, l). \quad (63)$$

Hence for estimating these components of the correlation tensor we can use the first equation from (61).

Similar estimations can be obtained for components of the correlation tensor of $\mathbf{v}(\mathbf{x}) = -\sigma(\mathbf{x})\nabla U(\mathbf{x})$. For the components of this correlation tensor the ongrid equation has the form

$$
\begin{aligned}
\langle v_i(\mathbf{x})\, v_k(\mathbf{x})\rangle &= \sigma(\mathbf{x}, l)^2\, \nabla_i U(\mathbf{x}, l)\, \nabla_k U(\mathbf{x}, l) + \sigma(\mathbf{x}, l)^2\, \langle \nabla_k U'(\mathbf{x})\, \nabla_i U'(\mathbf{x})\rangle + \\
&+ \left\langle \sigma'(\mathbf{x})^2\right\rangle \nabla_k U(\mathbf{x}, l)\, \nabla_i U(\mathbf{x}, l) + 2\left\langle \sigma'(\mathbf{x})\, \nabla_k U'(\mathbf{x})\right\rangle \sigma(\mathbf{x}, l)\, U_i(\mathbf{x}, l) + \\
&+ 2\left\langle \sigma'(\mathbf{x})\, \nabla_i U'(\mathbf{x})\right\rangle \sigma(\mathbf{x}, l)\, U_k(\mathbf{x}, l) + \left\langle \sigma'(\mathbf{x})^2\, \nabla_i U'(\mathbf{x})\, \nabla_k U'(\mathbf{x})\right\rangle + \\
&+ 2\sigma(\mathbf{x}, l)\left\langle \sigma'(\mathbf{x})\, \nabla_k U'(\mathbf{x})\, \nabla_i U'(\mathbf{x})\right\rangle + \\
&+ \left\langle \sigma'(\mathbf{x})^2\, \nabla_k U'(\mathbf{x})\right\rangle \nabla_i U(\mathbf{x}, l) + \left\langle \sigma'(\mathbf{x})^2\, \nabla_i U'(\mathbf{x})\right\rangle \nabla_k U(\mathbf{x}, l). \quad (64)
\end{aligned}
$$

The estimation of the second and the third terms in (64) follows from formulas (59) and (40) respectively. The fourth and the fifth terms are estimated by formula (42). All the other terms are second order in dl/l and can be omitted. For $i = k$ from (64) the components can be written as

$$
\left\langle v_1(\mathbf{x})^2 - v_2(\mathbf{x})^2\right\rangle \approx
$$

$$
\approx \left(1 - 2\langle\varphi\rangle\frac{dl}{l} + \frac{4}{5}\Phi_0\frac{dl}{l}\right)\sigma_l(\mathbf{x})^2\left[\nabla_1 U(\mathbf{x}, l)^2 - \nabla_2 U(\mathbf{x}, l)^2\right],
$$

$$
\left\langle v_3(\mathbf{x})^2 - v_2(\mathbf{x})^2\right\rangle \approx
$$

$$
\approx \left(1 - 2\langle\varphi\rangle\frac{dl}{l} + \frac{4}{5}\Phi_0\frac{dl}{l}\right)\sigma_l(\mathbf{x})^2\left[\nabla_3 U(\mathbf{x}, l)^2 - \nabla_2 U(\mathbf{x}, l)^2\right], \quad (65)
$$

$$
\left\langle v_1(\mathbf{x})^2 + v_2(\mathbf{x})^2 + v_3(\mathbf{x})^2\right\rangle \approx \left(1 - 2\langle\varphi\rangle\frac{dl}{l} + \Phi_0\frac{dl}{l}\right) \times
$$

$$
\times\sigma_l(\mathbf{x})^2\left[\nabla_1 U(\mathbf{x}, l)^2 + \nabla_2 U(\mathbf{x}, l)^2 + \nabla_3 U(\mathbf{x}, l)^2\right].
$$

As $dl \to 0$, we obtain the following equations for the effective coefficients

$$
\frac{d\ln\sigma_{0l}^{(3)}}{d\ln l} = \left(\frac{4}{5}\Phi_0(l) - 2\langle\varphi\rangle\right),
$$

$$
\frac{d\ln\sigma_{0l}^{(4)}}{d\ln l} = \left(\Phi_0(l) - 2\langle\varphi\rangle\right). \quad (66)
$$

To estimate $\left\langle v_i(\mathbf{x})^2\right\rangle$ one can solve linear system (65) with effective coefficients (66) instead of the scale-difference relations. In a scale-invariant medium, the effective system for variance of the components v_i takes the form

$$
\begin{bmatrix} \left\langle v_1(\mathbf{x})^2\right\rangle - \langle v_1(\mathbf{x})\rangle^2 \\ \left\langle v_2(\mathbf{x})^2\right\rangle - \langle v_2(\mathbf{x})\rangle^2 \\ \left\langle v_3(\mathbf{x})^2\right\rangle - \langle v_3(\mathbf{x})\rangle^2 \end{bmatrix}_{ef} = \Omega_1\mathcal{A}\begin{bmatrix} v_1(\mathbf{x}, l)^2 \\ v_2(\mathbf{x}, l)^2 \\ v_3(\mathbf{x}, l)^2 \end{bmatrix} - \Omega_2\begin{bmatrix} v_1(\mathbf{x}, l)^2 \\ v_2(\mathbf{x}, l)^2 \\ v_3(\mathbf{x}, L)^2 \end{bmatrix},
$$

Subgrid Modeling of Steady-State Flow Processes in Fractal Medium 305

where $\Omega_1 = (l/L)^{\Phi_0 - 2\langle\varphi\rangle}/3$ and $\Omega_2 = (/L)^{2\Phi_0/3 - 2\langle\varphi\rangle}$.

For the components of tensor at $i \neq k$ we have the relation

$$\langle v_i(\mathbf{x}) v_k(\mathbf{x})\rangle \approx \left(1 + \frac{4}{5}\Phi_0(l)\frac{dl}{l} - 2\langle\varphi\rangle\frac{dl}{l}\right)\sigma_l^2\nabla_i U(\mathbf{x}, l)\nabla_k U(\mathbf{x}, l). \tag{67}$$

From this it follows that we can use the first equation from (66) to estimate $\langle v_i(\mathbf{x}) v_k(\mathbf{x})\rangle$.

3.2. Numerical testing

To verify the above formulas, as in the isotropic case, using dimensionless variables we numerically solve the problem (33) for a unit cube with a unit jump of the potential for $\sigma_0 = 1$ with boundary conditions: on the cube sides $y = 0$ and $y = L_0$, the constant potential is specified: $U(x, y, z)|_{y=0} = U_1$ and $U(x, y, z)|_{y=L_0} = U_2$ $(U_1 > U_2)$. The potential on the other faces of the cube is given as a linear function of y: $U = U_1 + (U_1 - U_2)y/L_0$. In this case, the largest component of the local flow is directed along the axis y, and the mean values of the components v_x and v_z are equal to zero. A $256 \times 256 \times 256$ grid is used for the space variables. The conductivity field in the cube is simulated by formula (9). The integral in (9) is approximated by a finite-difference formula, in which it is convenient to pass to a logarithm to base 2:

$$\sigma(\mathbf{x})_{l_0} = \exp\left[-\ln 2 \int_{\log_2 l_0}^{\log_2 L} \varphi(\mathbf{x}, \tau)\, d\tau \approx 2^{-\sum_{k=-8}^{0}\varphi(\mathbf{x}, \tau_k)\Delta\tau}\right], \tag{68}$$

where $\langle\sigma_{l_0}(\mathbf{x})\rangle = 1$, $l = 2^\tau$, $\Delta\tau$ is the τ grid size. In our calculations, $\Delta\tau$ is taken to be one. For the field φ the following correlation function is used

$$\langle\varphi(\mathbf{x}, \tau_i)\varphi(\mathbf{y}, \tau_j)\rangle - \langle\varphi(\mathbf{x}, \tau_i)\rangle\langle\varphi(\mathbf{y}, \tau_j)\rangle = (\Phi_0/\ln 2)\exp\left[-(\mathbf{x} - \mathbf{y})^2/2^{2\tau_i}\right], \tag{69}$$

where $\Phi_0 = 2\langle\varphi\rangle$ is a constant taken from experimental data. In [1] approximate values for Φ_0 are given for some natural media. We use $\Phi_0 = 0.3$. The field φ is generated independently for each τ_i, since the field φ is assumed to be statistically independent on the various scales. The number of terms in (68) is chosen so that the ensemble average can be replaced by the average in space on the largest fluctuation scale. The smallest fluctuations scale is chosen in such a way as to approximate (33) by a difference scheme with good accuracy on all scales. The field in the exponent of (68) is generated as the sum of three scales, $\tau_k = -6, -5, -4$. The minimum scale is $l_0 = 1/64$, the maximum scale is $L = 1/16$. The correlation matrix (69) can be represented in the form of a direct product of four matrices with lower dimension. This enables us to apply the algorithm "along rows and columns" [33] to the numerical simulation of a Gaussian field. Figure 3 shows the change of conductivity with increasing the number of the summand in (68) at a cross-section $z = 1/2$.

To numerically solve system (33) an iterative method combined with the fast Fourier transform and second-order sweep method is used [34].

In according to the procedure of deriving the subgrid formulas, we have to solve numerically the complete problem and perform the probability averaging over small-scale

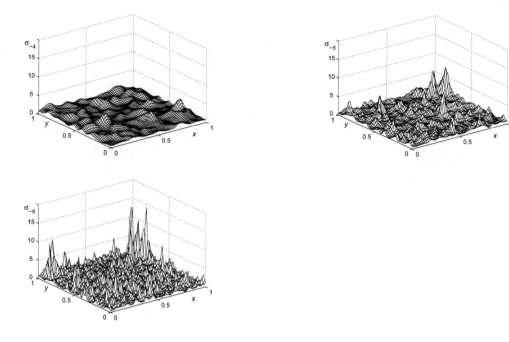

Figure 3. Variation of conductivity for $\sigma(\mathbf{x})_{-4} = 2^{-\varphi(\mathbf{x},\tau_4)\Delta\tau}$, $\sigma(\mathbf{x})_{-5} = 2^{-\sum_{k=-4}^{-5}\varphi(\mathbf{x},\tau_k)\Delta\tau}$, $\sigma(\mathbf{x})_{-6} = 2^{-\sum_{k=-4}^{-6}\varphi(\mathbf{x},\tau_k)\Delta\tau}$.

fluctuations to verify these formulas. As a result, we obtain subgrid terms, which can be compared with the theoretical expressions. the probability average requires multiple solution of the initial problem with a fixed large-scale component of conductivity but a random subgrid component. However a different procedure based on the power dependence of mean physical fields is used. The field v_y is calculated on the scales (l_0, L) then this field is averaged over space. Then this field is additionally averaged over the Gibbs ensemble if it is required. For the mean $\langle v_y \rangle$ thus obtained is verified formula

$$\left\langle \exp\left(-\int_l^L \varphi(\mathbf{x}, l_1) \frac{dl_1}{l_1}\right) \nabla y U \right\rangle \approx \left(\frac{l}{L}\right)^{\frac{\Phi_0}{6} - \langle\varphi\rangle} (U_1 - U_2). \tag{70}$$

The conventional perturbation theory gives the estimation

$$\left\langle \exp\left(-\int_l^L \varphi(\mathbf{x}, l_1) \frac{dl_1}{l_1}\right) \nabla U \right\rangle \approx \left(1 - \frac{1}{3}\exp(\Phi_0 K \ln 2)\right)(U_1 - U_2), \tag{71}$$

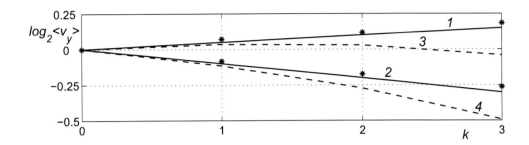

Figure 4. Logarithm of mean v_y at $\Phi_0 = 0.3$ obtained by: 1,2 - effective equation at $\langle \varphi \rangle = 0$ and $\langle \varphi \rangle = 0.15$ respectively, 3,4 - conventional perturbation theory at $\langle \varphi \rangle = 0$ and $\langle \varphi \rangle = 0.15$, stars - numerical method.

where K is number of scales in (68). For the values $\langle v_x^2 \rangle$, $\langle v_y^2 \rangle$ and $\langle v_z^2 \rangle$ are verified formulas

$$\left\langle \exp\left(-2\int_l^L \varphi(\mathbf{x},l_1)\frac{dl_1}{l_1}\right) \left((\nabla_x U(\mathbf{x}))^2 - (\nabla_y U(\mathbf{x}))^2\right)\right\rangle \approx$$

$$\approx \left(\frac{l_0}{L}\right)^{-\frac{4}{5}\Phi_0 + 2\langle\varphi\rangle} [U_1 - U_2]^2,$$

$$\left\langle \exp\left(-2\int_l^L \varphi(\mathbf{x},l_1)\frac{dl_1}{l_1}\right) \left((\nabla_z U(\mathbf{x}))^2 - (\nabla_y U(\mathbf{x}))^2\right)\right\rangle \approx \qquad (72)$$

$$\approx \left(\frac{l_0}{L}\right)^{-\frac{4}{5}\Phi_0 + 2\langle\varphi\rangle} [U_1 - U_2]^2,$$

$$\left\langle \exp\left(-2\int_l^L \varphi(\mathbf{x},l_1)\frac{dl_1}{l_1}\right) \left((\nabla_x U(\mathbf{x}))^2 + (\nabla_y U(\mathbf{x}))^2 + (\nabla_z U(\mathbf{x}))^2\right)\right\rangle \approx$$

$$\approx \left(\frac{l_0}{L}\right)^{-\Phi_0 + 2\langle\varphi\rangle} [U_1 - U_2]^2.$$

The conventional perturbation theory gives the estimation

$$\langle v_x^2(\mathbf{x})\rangle - \langle v_x(\mathbf{x})\rangle^2 = \frac{1}{15}\sigma_0^2 \exp(\Phi_0 K \ln 2)(U_1 - U_2)^2,$$
$$\langle v_y^2(\mathbf{x})\rangle - \langle v_y(\mathbf{x})\rangle^2 = \frac{8}{15}\sigma_0^2 \exp(\Phi_0 K \ln 2)(U_1 - U_2)^2, \qquad (73)$$
$$\langle v_z^2(\mathbf{x})\rangle - \langle v_z(\mathbf{x})\rangle^2 = \frac{1}{15}\sigma_0^2 \exp(\Phi_0 K \ln 2)(U_1 - U_2)^2.$$

Figure 4 shows the logarithm of mean v_y obtained by the described above numerical method compared with the effective fields obtained with the help of (46) and by the conventional perturbation theory.

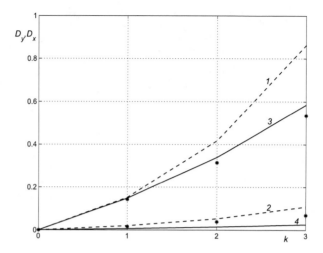

Figure 5. D_x and D_y denote the variance in x and y coordinate at $\Phi_0 = 0.3$ and $\langle \varphi \rangle = 0.15$ obtained by: 1,2 - conventional perturbation theory, 3,4 - effective equation, stars - numerical method.

Figure 5 shows the variance of v_x and v_y at $\langle \varphi(\mathbf{x}) \rangle = 0.15$ (conservative cascade), D_x and D_y denote the variance in x and y coordinates respectively.

If the conductivity has the log-stable distribution, then a random field $\varphi(\mathbf{x}, \tau)$ is generated by formula (23). The coefficients a_{ji}^l are taken in the form

$$a_{ji}^l = \left(\frac{\sqrt{\alpha}}{l\sqrt{\pi}}\right)^{3/\alpha} \exp\left(-\frac{(\mathbf{x_j} - \mathbf{x_i})^2}{l^2}\right). \qquad (74)$$

The independent random values ζ_j^l in (23) are simulated by the algorithm and the program proposed in [35] at $\alpha = 1.6$, $\beta = 1$ and $\mu = 0$. The field $\varphi(\mathbf{x}, \tau_k)$ is generated separately for each k. For approximated calculations, it is possible to use a certain limited number of k. In our case, they are three: $\tau_k = -4, -5$, and -6. The scale of the largest conductivity fluctuations is chosen such that ensemble average can be replaced by space average, and the scale of the smallest conductivity fluctuations is chosen such that the finite differences yields a good approximation. Figure 6, 7 show isolines of the scale-invariant conductivity for the three scales at a cross-section $z = 1/2$ calculated by formula (23) for $\alpha = 2$ (which corresponds t a log-normal conductivity model) and for $\alpha = 1.6$, $\beta = 1$ and $\mu = 0$ (which corresponds to a log-stable model) with $\Phi_0 = 0.3$ and $\langle \varphi \rangle = 0$. Figures 8, 9 show $\log_2 \langle v_y(\mathbf{x}) \rangle$ obtained by the described above numerical method compared with value obtained by the effective equation for the scale-invariant and not scale-invariant conductivity models.

Subgrid Modeling of Steady-State Flow Processes in Fractal Medium 309

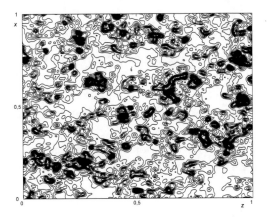

Figure 6. Gaussian model.

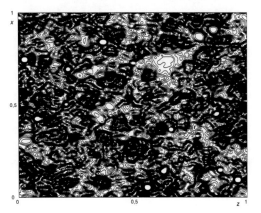

Figure 7. Log-stable model.

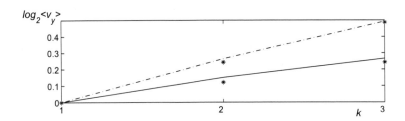

Figure 8. Solid curve – theoretical result at $\Phi_0 = 0.3$; dash-and-dotted curve – theoretical result at $\Phi_0 = 0.6$; stars – numerical results

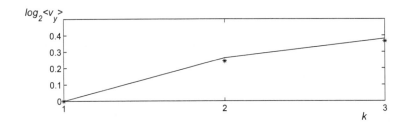

Figure 9. Solid curve – theoretical result at $\Phi_0(1) = 0.3$, $\Phi_0(2) = 0.6$, $\Phi_0(3) = 0.3$; stars – numerical results

4. Effective Coefficients of Quasi-Steady Maxwell's Equations with Multiscale Isotropic Random Conductivity

Electromagnetic logging is an effective tool for studying of a medium structure. The aim of this method is to estimate the medium conductivity as precisely as possible. In the present chapter the electric conductivity is approximated by a multiplicative continuous cascade. We obtain effective coefficients to estimate the first and the second statistical moments of electric or magnetic field strengths and current density in the quasi-steady Maxwell's equations [36]. According to [8], the quasi-steady Maxwell's equations for monochromatic fields $\widetilde{\mathbf{E}}(\mathbf{x}, t) = Re\left(\mathbf{E}(\mathbf{x}) e^{-i\omega t}\right)$, $\widetilde{\mathbf{H}}(\mathbf{x}, t) = Re\left(\mathbf{H}(\mathbf{x}) e^{-i\omega t}\right)$ in the absence of extraneous currents can be written as

$$
\begin{aligned}
rot\,\mathbf{H}(\mathbf{x}) &= \sigma(\mathbf{x})\,\mathbf{E}(\mathbf{x}), \\
rot\,\mathbf{E} &= i\omega\mu\mathbf{H},
\end{aligned}
\tag{75}
$$

where $\mathbf{E}$ and $\mathbf{H}$ are the vectors of electric and magnetic field strengths, respectively; μ is the magnetic permeability; $\sigma(\mathbf{x})$ is the electric conductivity; ω is the cyclic frequency; and $\mathbf{x}$ is the vector of spatial coordinates. The magnetic permeability is assumed to be equal to the magnetic permeability of vacuum. We also assume that the electric conductivity is constant outside a finite volume V with a smooth surface S. At the surface S, the tangent components of electric and magnetic field strengths are continuous.

The electric conductivity function $\sigma(\mathbf{x}) = \sigma(\mathbf{x})_{l_0}$ is divided into two components with respect to the scale l: large-scale (ongrid) component $\sigma(\mathbf{x}, l)$ and small-scale (subgrid) component $\sigma'(\mathbf{x})$. These components described above by formulas (34). The large-scale (ongrid) components of electric and magnetic field strengths $\mathbf{E}(\mathbf{x}, l)$, $\mathbf{H}(\mathbf{x}, l)$ are obtained by averaging the solutions to system (75), in which the large-scale component of conductivity $\sigma(\mathbf{x}, l)$ is fixed and the small component $\sigma'(\mathbf{x})$ is a random variable. The subgrid components of the electric and magnetic field strengths are equal to $\mathbf{H}'(\mathbf{x}) = \mathbf{H}(\mathbf{x}) - \mathbf{H}(\mathbf{x}, l)$, $\mathbf{E}'(\mathbf{x}) = \mathbf{E}(\mathbf{x}) - \mathbf{E}(\mathbf{x}, l)$. Substituting the relations for $\mathbf{E}(\mathbf{x})$, $\mathbf{H}(\mathbf{x})$ and $\sigma(\mathbf{x})$ into system (75) and averaging over small-scale components, we have

$$
\begin{aligned}
rot\,\mathbf{H}(\mathbf{x}, l) &= \sigma(\mathbf{x}, l)\mathbf{E}(\mathbf{x}, l) + \langle\sigma'\mathbf{E}'\rangle, \\
rot\,\mathbf{E}(\mathbf{x}, l) &= i\omega\mu\mathbf{H}(\mathbf{x}, l).
\end{aligned}
\tag{76}
$$

The subgrid term $\langle\sigma'\mathbf{E}'\rangle$ in system (76) is unknown. This term cannot be neglected without some preliminary estimation, since the correlation between the electric conductivity and the electric field strength may be significant. The form of this term in (76) determines the subgrid model. The subgrid term is estimated using perturbation theory. Subtracting system (76) from system (75) and taking into account only the first order terms, we obtain the subgrid equations:

$$
\begin{aligned}
rot\,\mathbf{H}' &= \sigma(\mathbf{x}, l)\,\mathbf{E}' + \sigma'\mathbf{E}(\mathbf{x}, l), \\
rot\,\mathbf{E}' &= i\omega\mu\mathbf{H}'.
\end{aligned}
\tag{77}
$$

The variable $\mathbf{E}(\mathbf{x}, l)$ on the right-hand side of (77) is assumed to be known. Solving system

Subgrid Modeling of Steady-State Flow Processes in Fractal Medium   311

(77) for the components of the electric field strength we have [37]

$$E'_\alpha(\mathbf{x}) = \frac{1}{4\pi}\, i\omega\mu \int_V \frac{e^{ikr}}{r} \sigma'(\mathbf{x}')\, E_\alpha(\mathbf{x}', l)\, d\mathbf{x}' +$$

$$+ \frac{1}{4\pi} \frac{\partial}{\partial x_\alpha} \frac{1}{\sigma(\mathbf{x}, l)}\, div \int_V \frac{e^{ikr}}{r} \sigma'(\mathbf{x}')\, \mathbf{E}(\mathbf{x}', l)\, d\mathbf{x}', \quad (78)$$

where $r = |\mathbf{x} - \mathbf{x}'|$, $k^2 = i\omega\mu\sigma(\mathbf{x}, l)$, $k = (1+i)\sqrt{\omega\mu\sigma(\mathbf{x}, l)/2}$. We take the square root such that $Re\, k > 0$, $Im\, k > 0$. Using (78) the subgrid term can be written as

$$\langle \sigma' E'_\alpha \rangle = \frac{1}{4\pi} i\omega\mu \int_V \frac{e^{ikr}}{r} \langle \sigma'(\mathbf{x})\sigma'(\mathbf{x}') \rangle\, E_\alpha(\mathbf{x}', l)\, d\mathbf{x}' +$$

$$+ \frac{1}{4\pi} \left\langle \sigma'(\mathbf{x}) \frac{\partial}{\partial x_\alpha} \frac{1}{\sigma(\mathbf{x}, l)}\, div \int_V \frac{e^{ikr}}{r} \sigma'(\mathbf{x}')\, \mathbf{E}(\mathbf{x}', l)\, d\mathbf{x}' \right\rangle. \quad (79)$$

Since a small change in the scale of σ produces considerable fluctuations in the field (which is typical of fractal fields), the field $\sigma(\mathbf{x}, l)$ and its derivatives are believed to change slower than σ' and its derivatives. Similar assumptions are made for $\mathbf{E}(\mathbf{x}, l)$ and $\mathbf{H}(\mathbf{x}, l)$. Therefore $\mathbf{E}(\mathbf{x}, l)$, $\sigma(\mathbf{x}, l)$ and their derivatives can be factored outside the integral sign in (79). Integrating by parts (79) we have

$$\langle \sigma' E'_\alpha \rangle = \frac{1}{4\pi} i\omega\mu \int_V \frac{e^{ikr}}{r} \langle \sigma'(\mathbf{x})\sigma'(\mathbf{x}') \rangle\, d\mathbf{x}'\, E_\alpha(\mathbf{x}, l) +$$

$$+ \frac{1}{4\pi\sigma(\mathbf{x}, l)} \int_V \frac{\partial}{\partial x'_\alpha} \frac{\partial}{\partial x'_\beta} \left(\frac{e^{ikr}}{r} \right) \langle \sigma'(\mathbf{x})\sigma'(\mathbf{x}') \rangle\, d\mathbf{x}'\, E_\beta(\mathbf{x}, l). \quad (80)$$

Here the summation of repeated indices is implied. Substituting formula $\langle \sigma'(\mathbf{x})\sigma'(\mathbf{x}') \rangle = \Phi(r, l)\sigma^2(\mathbf{x}, l)\frac{dl}{l}$ into (80) yields

$$\langle \sigma' E'_\alpha \rangle = \frac{1}{4\pi} i\omega\mu\sigma(\mathbf{x}, l) \int_V \frac{e^{ikr}}{r} \Phi(r, l)\, d\mathbf{x}'\sigma(\mathbf{x}, l)\, E_\alpha(\mathbf{x}, l)\frac{dl}{l} +$$

$$+ \frac{1}{4\pi} \int_V \frac{\partial}{\partial x'_\alpha} \frac{\partial}{\partial x'_\beta} \left(\frac{e^{ikr}}{r} \right) \Phi(r, l)\, d\mathbf{x}'\sigma(\mathbf{x}, l)\, E_\beta(\mathbf{x}, l)\frac{dl}{l}. \quad (81)$$

The integral over V in (81) can be changed by an integral with infinite limits, since the correlation function Φ is small if $L \ll L_0$, where L_0 is minimum size of V. This change gives a sensible error only in a narrow region of the correlation radius size near the boundary. In formula (81), Cartesian coordinates are changed for spherical coordinates. Integrating $n_j n_m$, where $n_m = x_m/r$, over the complete solid angle we arrive at the formula $\int n_j n_m d\vartheta = \frac{4\pi}{3}\delta_{jm}$. Using this formula and integrating (81) by parts we obtain

$$\langle \sigma' E'_\alpha \rangle = \left(-\frac{1}{3}\Phi_0 + \frac{2}{3} k^2 \int_0^\infty r e^{ikr} \Phi(r, l)\, dr \right) \sigma(\mathbf{x}, l)\, E_\alpha(\mathbf{x}, l)\frac{dl}{l}. \quad (82)$$

If $\omega\mu L^2\sigma(\mathbf{x}, l) \ll 1$, the integral in (82) is small [7]. This inequality is not restrictive for the problems of electromagnetic logging. In these problems monochromatic sources are often used to generate electromagnetic fields. Therefore this inequality can be made valid for almost the entire frequency interval. Hence, the integral in (82) can be neglected. Substituting (82) into (76), we have

$$rot\,\mathbf{H}\,(\mathbf{x}, l) = \sigma_{l0}\exp\left[-\int_l^L \varphi(\mathbf{x}, l_1)\frac{dl_1}{l_1}\right]\mathbf{E}\,(\mathbf{x}, l),$$

$$rot\,\mathbf{E}\,(\mathbf{x}, l) = i\omega\mu\mathbf{H}\,(\mathbf{x}, l),$$

$$\sigma_{l0} = \left(1 - \frac{\Phi_0}{3}\frac{dl}{l}\right)\left[1 + \left(\frac{\Phi_0}{2} - \langle\varphi\rangle\right)\frac{dl}{l}\right]\sigma_0.$$

(83)

It follows from (83) that the new coefficient σ_{l0} is

$$\sigma_{l0} = \sigma_0 + \left(\frac{\Phi_0}{6} - \langle\varphi\rangle\right)\sigma_0\frac{dl}{l},$$

with second order of accuracy. As $dl \to 0$ we obtain the equation

$$\frac{d\ln\sigma_{0l}}{d\ln l} = \frac{1}{6}\Phi_0\,(l) - \langle\varphi\rangle.$$

(84)

For a scale-invariant medium, effective equations have the following simple form

$$rot\,\mathbf{H}\,(\mathbf{x}, l) = \left(\frac{l}{L}\right)^{\langle\varphi\rangle - \Phi_0/6}\sigma_l\,(\mathbf{x})\,\mathbf{E}\,(\mathbf{x}, l),$$

$$rot\,\mathbf{E}\,(\mathbf{x}, l) = i\omega\mu\mathbf{H}\,(\mathbf{x}, l).$$

(85)

4.1. Estimation of the second statistical moments of current density and electric field strength

For the components of the first covariance tensor [7], the ongrid equation has the form

$$\left\langle E_\alpha\,(\mathbf{x})\,\overline{E}_\beta\,(\mathbf{x})\right\rangle = E_\alpha\,(\mathbf{x}, l)\,\overline{E}_\beta\,(\mathbf{x}, l) + \left\langle E'_\alpha\,(\mathbf{x})\,\overline{E'_\beta}\,(\mathbf{x})\right\rangle.$$

(86)

Here the overscribed bar denotes complex conjugation. Changing the integrals over V in (78) by integrals with infinite limits, we have

$$\left\langle E'_\alpha\,(\mathbf{x})\,\overline{E'_\beta}\,(\mathbf{x})\right\rangle \approx$$

$$\approx \frac{1}{16\pi^2}\omega^2\mu^2\sigma^2\,(\mathbf{x}, l)\frac{dl}{l}\iint\frac{e^{ikr'}}{r'}\frac{\overline{e^{ikr''}}}{r''}\Phi\left(\left|\mathbf{x}' - \mathbf{x}''\right|, l\right)d\mathbf{x}'d\mathbf{x}''E_\alpha\,(\mathbf{x}, l)\,\overline{E}_\beta\,(\mathbf{x}, l) +$$

$$+ \frac{1}{16\pi^2}\frac{dl}{l}\iint\frac{\partial}{\partial x'_\alpha}\frac{\partial}{\partial x'_\gamma}\left(\frac{e^{ikr'}}{r'}\right)\frac{\partial}{\partial x''_\beta}\frac{\partial}{\partial x''_\nu}\left(\frac{\overline{e^{ikr''}}}{r''}\right)\Phi\left(\left|\mathbf{x}'' - \mathbf{x}'\right|, l\right)d\mathbf{x}'d\mathbf{x}''\times$$

$$\times E_\gamma\,(\mathbf{x}, l)\,\overline{E}_\nu\,(\mathbf{x}, l),$$

(87)

Subgrid Modeling of Steady-State Flow Processes in Fractal Medium   313

where $\mathbf{r}'' = \mathbf{x} - \mathbf{x}''$ and $\mathbf{r}' = \mathbf{x} - \mathbf{x}'$. Replacing $\mathbf{x}'' - \mathbf{x}'$ with $\mathbf{y}$ in (87) we arrive at

$$I_1 = \frac{k_1^4}{4\pi^2} \frac{dl}{l} \iint \frac{e^{(-k_1+ik_1)r'}}{r'} \frac{e^{(-k_1-ik_1)|r'-\mathbf{y}|}}{|r'-\mathbf{y}|} \Phi\left(|\mathbf{y}|,l\right) dr' d\mathbf{y} \, E_\alpha\left(\mathbf{x},l\right) \overline{E}_\beta\left(\mathbf{x},l\right), \quad (88)$$

$$I_2 = \frac{1}{16\pi^2} \frac{dl}{l} \iint \frac{\partial}{\partial x_\alpha'} \frac{\partial}{\partial x_\gamma'} \left(\frac{e^{-k_1+ik_1 r'}}{r'}\right) \left(\frac{e^{(-k_1-ik_1)|r'-\mathbf{y}|}}{|r'-\mathbf{y}|}\right) \frac{\partial}{\partial y_\beta} \frac{\partial}{\partial y_\nu} \Phi\left(|\mathbf{y}|,l\right) dx' d\mathbf{y} \times$$
$$\times E_\gamma\left(\mathbf{x},l\right) \overline{E}_\nu\left(\mathbf{x},l\right), \quad (89)$$

where $k_1 = k(i-1)/2$. The convolution of the Green function and its partial derivatives in (88), (89) is integrated with the Fourier transform. The Fourier transform formulas for a spherical symmetric function [32] are

$$F^{-1}\left(F\left(\frac{e^{(-k_1+ik_1)r'}}{r'}\right) F\left(\frac{e^{(-k_1-ik_1)r'}}{r'}\right)\right) =$$

$$= \frac{8}{r'}\pi^2 \int_0^\infty \frac{\xi}{k_1^4 + 4\pi^4\xi^4} \sin\left(2\pi\xi r'\right) dr' = 2\frac{\pi}{r'k_1^2} \exp\left(-k_1 r'\right) \sin k_1 r'. \quad (90)$$

Substituting (90) into (88) and (89), passing to spherical coordinates and using the formula

$$\int_0^{2\pi}\int_0^\pi n_\alpha n_\beta n_\gamma n_\nu \sin\theta d\chi d\theta = 4\pi\left(\delta_{\alpha\beta}\delta_{\gamma\nu} + \delta_{\alpha\gamma}\delta_{\beta\nu} + \delta_{\alpha\nu}\delta_{\beta\gamma}\right)/15,$$

we obtain

$$I_1 = 2k_1^2 \frac{dl}{l} \int_0^\infty r \exp\left(-k_1 r\right) \sin k_1 r \Phi\left(r,l\right) dr \, E_\alpha\left(\mathbf{x},l\right) \overline{E}_\beta\left(\mathbf{x},l\right), \quad (91)$$

$$I_2 = \frac{1}{30k_1^2} \int_0^\infty \left[\left(\frac{1}{r^2} + \frac{k_1}{r}\right) \sin rk_1 - \frac{k_1}{r} \cos rk_1\right] e^{-rk_1} \times$$
$$\times \left(r^2\Phi'''\left(r,l\right) + 2r\Phi''\left(r,l\right) - 2\Phi'\left(r,l\right)\right) dr \times$$
$$\times \left(\delta_{\alpha\beta}\delta_{\gamma\nu} + \delta_{\alpha\gamma}\delta_{\beta\nu} + \delta_{\alpha\nu}\delta_{\beta\gamma}\right) E_\gamma\left(\mathbf{x},l\right) \overline{E}_\nu\left(\mathbf{x},l\right) \frac{dl}{l}. \quad (92)$$

Integrating (92) by parts, for the components of the first covariance tensor, we have

$$\left\langle E_\alpha\left(\mathbf{x}\right) \overline{E}_\beta\left(\mathbf{x}\right)\right\rangle \approx$$
$$\approx E_\alpha\left(\mathbf{x},l\right) \overline{E}_\beta\left(\mathbf{x},l\right) + 2k_1^2 \int re^{-rk_1} \Phi\left(r,l\right) \sin rk_1 dr \, E_\alpha\left(\mathbf{x},l\right) \overline{E}_\beta\left(\mathbf{x},l\right) \frac{dl}{l} +$$
$$+ \left[\frac{1}{15}\Phi_0\left(l\right) - \frac{2}{15}k_1^2 \int re^{-rk_1} \Phi\left(r,l\right) \sin rk_1 dr\right] \times$$
$$\times \left(\delta_{\alpha\beta}\delta_{\gamma\nu} + \delta_{\alpha\gamma}\delta_{\beta\nu} + \delta_{\alpha\nu}\delta_{\beta\gamma}\right) E_\gamma\left(\mathbf{x},l\right) \overline{E}_\nu\left(\mathbf{x},l\right) \frac{dl}{l}. \quad (93)$$

A similar formula can be obtained for the components of the second covariance tensor of the electric field strength:

$$\langle E_\alpha (\mathbf{x}) E_\beta (\mathbf{x}) \rangle \approx$$

$$\approx E_\alpha (\mathbf{x}, l) E_\beta (\mathbf{x}, l) - \frac{k_1^2}{3} \int r e^{-ikr} (-kr + 4i) \Phi (r, l) \, dr \, E_\alpha (\mathbf{x}, l)^2 \frac{dl}{l}$$

$$+ \frac{1}{15} \left[\Phi_0 (l) + k_1^2 \int r e^{-ikr} (kr + 4i) \Phi (r, l) \, dr \right] \times$$

$$\times (\delta_{\alpha\beta}\delta_{\gamma\nu} + \delta_{\alpha\gamma}\delta_{\beta\nu} + \delta_{\alpha\nu}\delta_{\beta\gamma}) E_\gamma (\mathbf{x}, l) E_\nu (\mathbf{x}, l)^2 \frac{dl}{l}. \quad (94)$$

If $\omega\mu L^2 \sigma(\mathbf{x}, l) \ll 1$, the integrals in (93), (94) can be neglected [7]. It follows from (94) that at $\alpha = \beta$ the components of the second covariance tensor satisfy the relations

$$\left\langle E_1 (\mathbf{x})^2 \right\rangle - \left\langle E_2 (\mathbf{x})^2 \right\rangle = \left(1 + \frac{2}{15} \Phi_0 (l) \frac{dl}{l} \right) \left(E_1 (\mathbf{x}, l)^2 - E_2 (\mathbf{x}, l)^2 \right),$$

$$\left\langle E_1 (\mathbf{x})^2 \right\rangle - \left\langle E_3 (\mathbf{x})^2 \right\rangle = \left(1 + \frac{2}{15} \Phi_0 (l) \frac{dl}{l} \right) \left(E_1^2 (\mathbf{x}, l) - E_3 (\mathbf{x}, l)^2 \right),$$

$$\left\langle E_3 (\mathbf{x})^2 \right\rangle + \left\langle E_2 (\mathbf{x})^2 \right\rangle + \left\langle E_3 (\mathbf{x})^2 \right\rangle =$$

$$= \left(1 + \frac{1}{3} \Phi_0 (l) \frac{dl}{l} \right) \left(E_1 (\mathbf{x}, l)^2 + E_2 (\mathbf{x}, l)^2 + E_3 (\mathbf{x}, l)^2 \right). \quad (95)$$

By analogy with (83) we can write:

$$\widehat{\sigma}_{1l} = \widehat{\sigma}_{1l_0} + \frac{2}{15} \Phi_0 (l) \frac{dl}{l},$$

$$\widehat{\sigma}_{2l} = \widehat{\sigma}_{2l_0} + \frac{1}{3} \Phi_0 (l) \frac{dl}{l}. \quad (96)$$

As $dl \to 0$, we obtain the following equations for the effective coefficients $\widehat{\sigma}_1$ and $\widehat{\sigma}_2$:

$$\frac{d \ln \widehat{\sigma}_{1l}}{d \ln l} = \frac{2}{15} \Phi_0 (l), \quad \widehat{\sigma}_1|_{l_0} = 1,$$

$$\frac{d \ln \widehat{\sigma}_{2l}}{d \ln l} = \frac{1}{3} \Phi_0 (l), \quad \widehat{\sigma}_2|_{l_0} = 1. \quad (97)$$

Solving system (95) with the effective coefficients $\widehat{\sigma}_{1l}$ and $\widehat{\sigma}_{2l}$, we find the components $\left\langle E_\alpha (\mathbf{x})^2 \right\rangle$. The components of the first covariance tensor $\left\langle E_\alpha (\mathbf{x}) \overline{E}_\alpha (\mathbf{x}) \right\rangle$ satisfy exactly the same system as (95) but with effective coefficients. The covariance tensor components for the real and the imaginary parts of the electric field strength can be written as

$$\left\langle (ReE_\alpha (\mathbf{x}))^2 \right\rangle = \frac{1}{2} Re \left[\left\langle E_\alpha (\mathbf{x}) \overline{E}_\alpha (\mathbf{x}) \right\rangle + \left\langle E_\alpha (\mathbf{x})^2 \right\rangle \right],$$

$$\left\langle (ImE_\alpha (\mathbf{x}))^2 \right\rangle = \frac{1}{2} Re \left[\left\langle E_\alpha (\mathbf{x}) \overline{E}_\alpha (\mathbf{x}) \right\rangle - \left\langle E_\alpha (\mathbf{x})^2 \right\rangle \right], \quad (98)$$

$$\langle ReE_\alpha (\mathbf{x}) \, Im \, E_\alpha (\mathbf{x}) \rangle = \frac{1}{2} Im \left[\left\langle E_\alpha (\mathbf{x})^2 \right\rangle - \left\langle E_\alpha (\mathbf{x}) \overline{E}_\alpha (\mathbf{x}) \right\rangle \right].$$

In the scale-invariant medium, the effective system takes the form

$$
\begin{bmatrix} \left\langle (ReE_1\,(\mathbf{x}))^2 \right\rangle \\[4pt] \left\langle (ReE_2\,(\mathbf{x}))^2 \right\rangle \\[4pt] \left\langle (ReE_3\,(\mathbf{x}))^2 \right\rangle \end{bmatrix} = \frac{1}{3} \left(\frac{l}{L} \right)^{-\frac{\Phi_0}{3}} \mathcal{A} \begin{bmatrix} (ReE_1\,(\mathbf{x},l))^2 \\[4pt] (ReE_2\,(\mathbf{x},l))^2 \\[4pt] (ReE_3\,(\mathbf{x},l))^2 \end{bmatrix}, \tag{99}
$$

where the elements of symmetrical matrix $\mathcal{A}$ are equal to: $a_{ii} = 2(\frac{l}{L})^{-\frac{\Phi_0}{5}} + 1, i = 1,2,3$ and $a_{ij} = -(\frac{l}{L})^{-\frac{\Phi_0}{5}} + 1, i \neq j$. The components $\left\langle (Im\,E_\alpha(\mathbf{x}))^2 \right\rangle$ and $\left\langle ReE_\alpha(\mathbf{x})ImE_\alpha(\mathbf{x}) \right\rangle$ satisfy exactly the same relations as (99). If $\alpha \neq \beta$, we have

$$
\left\langle ReE_\alpha\,(\mathbf{x})\,ReE_\beta\,(\mathbf{x}) \right\rangle \approx \left(\frac{l}{L} \right)^{\frac{2\Phi_0}{15}} ReE_\alpha\,(\mathbf{x},l)\,ReE_\beta\,(\mathbf{x},l). \tag{100}
$$

The components $\left\langle ImE_\alpha\,(\mathbf{x})\,ImE_\beta\,(\mathbf{x}) \right\rangle$ satisfy the same formula as (100).

Similar estimations can be obtained for the components of current density tensor. If $\alpha = \beta$, the components can be written as

$$
\left\langle \sigma\,(\mathbf{x})^2\,\overline{E}_\alpha\,(\mathbf{x})\,E_\alpha\,(\mathbf{x}) \right\rangle \approx \sigma\,(\mathbf{x},l)^2\,\overline{E}_\alpha\,(\mathbf{x},l)\,E_\alpha\,(\mathbf{x},l) +
$$

$$
+ \sigma\,(\mathbf{x},l)^2 \left\langle E'_\alpha \overline{E}'_\alpha \right\rangle + \left\langle \sigma'^2 \right\rangle \overline{E}_\alpha\,(\mathbf{x},l)\,E_\alpha\,(\mathbf{x},l) +
$$

$$
+ 2\sigma\,(\mathbf{x},l)\,\overline{E}_\alpha\,(\mathbf{x},l) \left\langle \sigma'E'_\alpha \right\rangle + 2\sigma\,(\mathbf{x},l)\,E_\alpha\,(\mathbf{x},l) \left\langle \sigma' \overline{E}'_\alpha \right\rangle. \tag{101}
$$

If $\omega\mu L^2 \sigma(\mathbf{x},l) \ll 1$ then it follows from formulas (82) and (93) that

$$
\left\langle \sigma\,(\mathbf{x})^2\,\overline{E}_\alpha\,(\mathbf{x})\,E_\alpha\,(\mathbf{x}) \right\rangle \approx
$$

$$
\approx \sigma\,(\mathbf{x},l)^2\,\overline{E}_\alpha\,(\mathbf{x},l)\,E_\alpha\,(\mathbf{x},l) - \frac{1}{3}\,\Phi_0\,(l)\,\sigma\,(\mathbf{x},l)^2\,\overline{E}_\alpha\,(\mathbf{x},l)\,E_\alpha\,(\mathbf{x},l)\,\frac{dl}{l} +
$$

$$
+ \frac{1}{15}\,\Phi_0\,(l)\,(\delta_{\beta\beta} + 2\delta_{\alpha\beta})\,\sigma\,(\mathbf{x},l)^2\,E_\beta\,(\mathbf{x},l)\,\overline{E}_\beta\,(\mathbf{x},l)\,\frac{dl}{l}.
$$

Then, using (39), we obtain

$$
\left\langle \sigma\,(\mathbf{x})^2\,\overline{E}_1\,(\mathbf{x})\,E_1\,(\mathbf{x}) \right\rangle - \left\langle \sigma\,(\mathbf{x})^2\,\overline{E}_2\,(\mathbf{x})\,E_2\,(\mathbf{x}) \right\rangle =
$$

$$
= \left[1 + \frac{4}{5}\,\Phi_0\,(l)\,\frac{dl}{l} - 2 \left\langle \varphi \right\rangle \frac{dl}{l} \right] \sigma_l\,(\mathbf{x})^2 \times \left[\overline{E}_1\,(\mathbf{x},l)\,E_1\,(\mathbf{x},l) - \overline{E}_2\,(\mathbf{x},l)\,E_2\,(\mathbf{x},l) \right],
$$

$$
\left\langle \sigma\,(\mathbf{x})^2\,\overline{E}_1\,(\mathbf{x})\,E_1\,(\mathbf{x}) \right\rangle - \left\langle \sigma\,(\mathbf{x})^2\,\overline{E}_3\,(\mathbf{x})\,E_3\,(\mathbf{x}) \right\rangle =
$$

$$
= \left[1 + \frac{4}{5}\,\Phi_0\,(l)\,\frac{dl}{l} - 2 \left\langle \varphi \right\rangle \frac{dl}{l} \right] \sigma_l\,(\mathbf{x})^2 \times \left[\overline{E}_1\,(\mathbf{x},l)\,E_1\,(\mathbf{x},l) - \overline{E}_3\,(\mathbf{x},l)\,E_3\,(\mathbf{x},l) \right],
$$

$$
\left\langle \sigma\,(\mathbf{x})^2 \sum_{\alpha=1}^{3} \overline{E}_\alpha\,(\mathbf{x})\,E_\alpha\,(\mathbf{x}) \right\rangle =
$$

$$
= \left[1 + (\Phi_0\,(l) - 2 \left\langle \varphi \right\rangle)\,\frac{dl}{l} \right] \sigma_l\,(\mathbf{x})^2 \sum_{\alpha=1}^{3} \overline{E}_\alpha\,(\mathbf{x},l)\,E_\alpha\,(\mathbf{x},l).
$$

316 O. N. Soboleva and E. P. Kurochkina

As $dl \to 0$, we obtain the following equations for the effective coefficients $\widehat{\sigma}_3$ and $\widehat{\sigma}_4$:

$$\frac{d\ln\widehat{\sigma}_3}{d\ln l} = \frac{4}{5}\Phi_0(l) - 2\langle\varphi\rangle, \quad \widehat{\sigma}_3|_{l_0} = 1,$$
$$\frac{d\ln\widehat{\sigma}_4}{d\ln l} = \Phi_0(l) - 2\langle\varphi\rangle, \quad \widehat{\sigma}_4|_{l_0} = 1. \tag{102}$$

To estimate the components of the second covariance tensor, the equation of effective coefficients (102) can also be used, since the formulas for the first and second tensors differ in the integral terms only . In a scale-invariant medium, the effective system takes the form

$$\begin{bmatrix} \left\langle (\sigma(\mathbf{x})\,ReE_1(\mathbf{x}))^2 \right\rangle \\ \left\langle (\sigma(\mathbf{x})\,ReE_2(\mathbf{x}))^2 \right\rangle \\ \left\langle (\sigma(\mathbf{x})\,ReE_3(\mathbf{x}))^2 \right\rangle \end{bmatrix} = \Omega\mathcal{A} \begin{bmatrix} (\sigma_l(\mathbf{x})\,ReE_1(\mathbf{x},l))^2 \\ (\sigma_l(\mathbf{x})\,ReE_2(\mathbf{x},l))^2 \\ (\sigma_l(\mathbf{x})\,ReE_3(\mathbf{x},l))^2 \end{bmatrix}, \tag{103}$$

where

$$\Omega = \frac{1}{3}(l/L)^{-\Phi_0(l)+2\langle\varphi\rangle}.$$

The components $\left\langle (\sigma(\mathbf{x})\,Im\,E_\alpha(\mathbf{x}))^2 \right\rangle$, $\left\langle \sigma(\mathbf{x})^2\,Re\,E_\alpha(\mathbf{x})\,ImE_\alpha(\mathbf{x}) \right\rangle$ also satisfy system (103).

5. Numerical Testing

In order to verify the formulas obtained above the following numerical problem is solved. Equations (75) are solved in a cube with edge L_0. An alternating magnetic field with cyclic frequency ω acts on a conducting medium. This field satisfies the condition $\mathbf{H} = (0, H_y(z), 0)$. On the boundary with vacuum, $z = 0$, the component $H_y(z)$ is assumed to be constant, $H_y(0) = H_0$. The following dimensionless variables are used: $\mathbf{x} = \widehat{\mathbf{x}}/L_0$, $\sigma = \widehat{\sigma}/\sigma_0$, $\sigma_0 = \langle\widehat{\sigma}\rangle$, $\mathbf{H} = \widehat{\mathbf{H}}/H_0$, $\mathbf{E} = (L_0\sigma_0/k_1 H_0)\widehat{\mathbf{E}}$, $k_1 = L_0\sqrt{\mu\omega\sigma_0}$. Thus, the problem is solved in a unit cube with $\sigma_0 = 1$, $H_0 = 1$. Equations (75) in dimensionless form are written as

$$rot\,\mathbf{H}(\mathbf{x}) = k_1\sigma(\mathbf{x})\,\mathbf{E}(\mathbf{x}),$$
$$rot\,\mathbf{E}(\mathbf{x}) = ik_1\mathbf{H}(\mathbf{x}). \tag{104}$$

Outside the cube and on the boundaries, the magnetic and electric fields strengths are specified by the formulas

$$H_y = \exp\left(-k_1 z/\sqrt{2}\right)\exp\left(ik_1 z/\sqrt{2}\right), \quad H_x = H_z = 0,$$
$$E_x = \exp\left(-k_1 z/\sqrt{2}\right)\exp\left(ik_1 z/\sqrt{2} - \pi/4\right), \tag{105}$$
$$E_y = E_z = 0.$$

The conductivity field in the cube is simulated by formula (68). A $256 \times 256 \times 256$ grid is used for the space variables. For the field φ the correlation function is defined by formula (69). We use $\Phi_0 = 0.3$ and $\langle\varphi\rangle = \Phi_0/2$. The field in the exponent of (68) is generated as

Subgrid Modeling of Steady-State Flow Processes in Fractal Medium   317

the sum of three scales, $\tau_i = -6, -5, -4$. The minimum scale is $l_0 = 1/64$, the maximum scale is $L = 1/16$.

To numerically solve system (104) a method based on a finite-difference scheme proposed in [38] and a decomposition method from [39] are used. We use a square 3D grid $\mathbf{x} = (x_m, y_n, z_j)$ with a constant spacing. The cube is divided into two subdomains, P and R. The subdomain P contains the points with even sums of indices (three even numbers or one even number and two odd ones). The subdomain R contains the remaining points. Denote the functions defined on P and R by upper indices P and R respectively. Then both subdomains P and R are divided into four independent subdomains in which equations (104) are solved independently. The derivatives with respect to x are approximated by

$$
\begin{aligned}
\left(f_x^P\right) &= \frac{f_{m+1,n,j}^R - f_{m-1,n,j}^R}{h}, \\
\left(f_x^R\right) &= \frac{f_{m+1,n,j}^P - f_{m-1,n,j}^P}{h}.
\end{aligned}
\tag{106}
$$

The derivatives with respect to y and z are approximated in a similar way. For each grid point, twelve values are calculated: $\mathbf{H}$ and $\mathbf{E}$ vectors, each having three real and three imaginary components. The problem is solved eight times on a $128 \times 128 \times 128$ grid. To solve the linear systems obtained by discretization in the grid subdomains, an iterative method for a self-adjoint operator with an arbitrary spectrum, SYMMLQ, is used. The algorithm is described in [40].

The theoretical formulas obtained in the previous section can be verified by solving the above initial problem and performing probabilistic averaging over the small-scale fluctuations. The solution thus obtained can be compared with the solution of effective equations (85). However a different procedure is used in the present paper. The mean characteristics of the current density and the electric and magnetic field strengths are calculated on the scales (l_0, L). At each z, these fields are averaged over the planes (x, y). Then these fields are additionally averaged over the Gibbs ensemble. Equations (104) are solved eighty times. The fields thus obtained are compared with the solution to effective equations (85) and fields obtained by conventional perturbation theory with a Born expansion describing only single scattered fields [7]. For the primary wave (105) denoted by $\mathbf{E}_{00}$, $\mathbf{H}_{00}$ in the Born expansion we have

$$
E_{x_k}(\mathbf{x}) = E_{x_k 00}(\mathbf{x}) + \frac{1}{4\pi} i\omega\mu \int_V \frac{e^{ikr}}{r} \sigma'(\mathbf{x}') E_{x_k 00}(\mathbf{x}')\, d\mathbf{x}' +
$$

$$
+ \frac{1}{4\pi\sigma_0} \frac{\partial}{\partial x_k} div \int_V \frac{e^{ikr}}{r} \sigma'(\mathbf{x}') \mathbf{E}_{00}(\mathbf{x}')\, d\mathbf{x}',
\tag{107}
$$

$$
\mathbf{H}(\mathbf{x}) = \mathbf{H}_{00}(\mathbf{x}) + rot \frac{1}{4\pi} i\omega\mu \int_V \frac{e^{ikr}}{r} \sigma'(\mathbf{x}') \mathbf{E}_{00}(\mathbf{x}')\, d\mathbf{x}'.
$$

The mean electric and magnetic field strengths are equal to $\mathbf{E}_{00}$ and $\mathbf{H}_{00}$, since $\langle\sigma'\rangle = 0$. The random field σ' varies faster than $\mathbf{E}_{00}$ and $\mathbf{H}_{00}$. Therefore, in formulas (107) these functions can be factored outside the integral sign. The current density is given by

$$\langle \sigma E_x \rangle = \sigma_0 E_{x00} + \frac{1}{4\pi} i\omega\mu \int_V \frac{e^{ikr}}{r} \Omega(r)\, d\mathbf{x}' E_{x00}(\mathbf{x}) +$$

$$+ \frac{1}{4\pi\sigma_0} \frac{\partial}{\partial x} \frac{\partial}{\partial x} \int_V \frac{e^{ikr}}{r} \Omega(r)\, d\mathbf{x}' E_{x00}(\mathbf{x}) +$$

$$+ \frac{1}{4\pi\sigma_0} \frac{\partial}{\partial x} \frac{\partial}{\partial y} \int_V \frac{e^{ikr}}{r} \Omega(r)\, d\mathbf{x}' E_{y00}(\mathbf{x}) +$$

$$+ \frac{1}{4\pi\sigma_0} \frac{\partial}{\partial x} \frac{\partial}{\partial z} \int_V \frac{e^{ikr}}{r} \Omega(r)\, d\mathbf{x}' E_{z00}(\mathbf{x}) =$$

$$= \left(-\frac{\sigma_0}{3} \Omega_0 + \frac{2}{3} i\omega\mu\sigma_0 \int r e^{ikr} \Omega(r)\, dr \right) E_{x00}(\mathbf{x}), \qquad (108)$$

where

$$\Omega_0 = \langle \sigma'(\mathbf{x}) \sigma'(\mathbf{x}) \rangle = \exp(\Phi_0 K \ln 2) - 1 = 2^{\Phi_0 K} - 1, \quad r = |\mathbf{x} - \mathbf{x}'|,$$

and K is the number of scales in formula (68). It follows from formula (105) that the mean values of the other components of the current density are equal to zero. The integral in (108) can also be neglected at $\omega\mu L^2 \sigma(\mathbf{x}, l) \ll 1$. For the components of the second statistical moment of the current density, the Born expansion gives

$$\langle E_x^2 \rangle - E_{x00}^2 = \frac{1}{5} \sigma_0^2 \Omega(0) E_{x00}^2,$$

$$\langle E_y^2 \rangle - E_{y00}^2 = \frac{1}{15} \sigma_0^2 \Omega(0) E_{x00}^2, \qquad (109)$$

$$\langle E_z^2 \rangle - E_{z00}^2 = \frac{1}{15} \sigma_0^2 \Omega(0) E_{x00}^2$$

$$\langle \sigma(\mathbf{x}) E_x^2 \rangle - \sigma_0^2 E_{x00}^2 = \frac{8}{15} \sigma_0^2 \Omega(0) E_{x00}^2,$$

$$\langle \sigma(\mathbf{x}) E_y^2 \rangle - \sigma_0^2 E_{y00}^2 = \frac{1}{15} \sigma_0^2 \Omega(0) E_{x00}^2, \qquad (110)$$

$$\langle \sigma(\mathbf{x}) E_z^2 \rangle - \sigma_0^2 E_{z00}^2 = \frac{1}{15} \sigma_0^2 \Omega(0) E_{x00}^2.$$

In the calculations we use: $\Phi_0 = 0.3$, $\varphi = 0.15$, $k_1 = 6\sqrt{2}$, i.e., $L_0 = 6h_{skin}$, where h_{skin} is the thickness of the skin-layer at $\sigma = 1$.

Figures 10-12 present the mean fields obtained by the numerical method described above compared with the effective fields obtained by equations (85) and by the conventional perturbation theory.

Figures 13-15 present the variance of fields obtained by the numerical method compared with the fields obtained by the effective system of equations and the variances of fields obtained by the conventional perturbation theory.

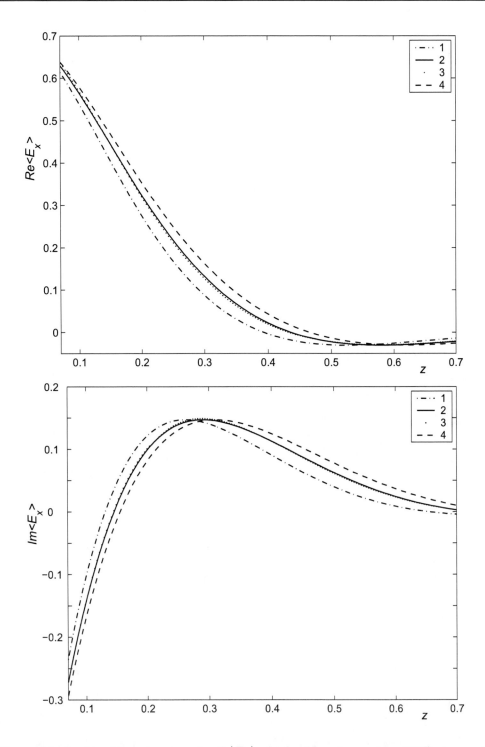

Figure 10. Real and imaginary parts of $\langle E_x \rangle$ obtained by: 1 – system (104) at $\sigma = 1$; 2 – effective system; 3 – numerical method; 4 – Born expansion.

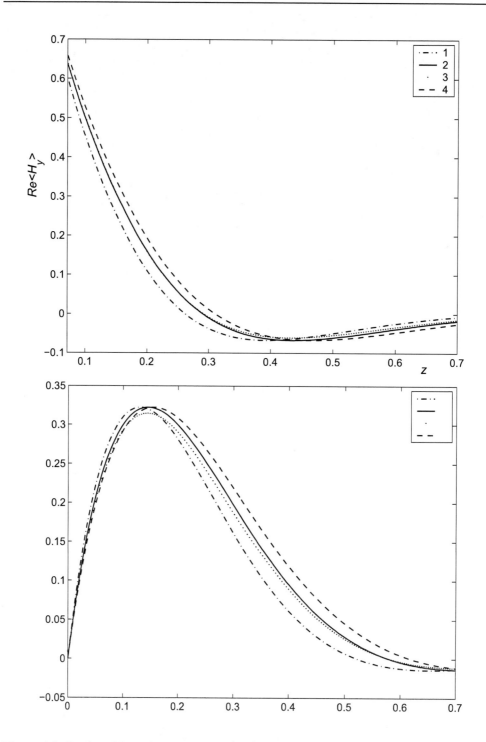

Figure 11. Real and imaginary parts of $\langle H_y \rangle$ obtained by: 1 – system (104) at $\sigma = 1$; 2 – effective system; 3 – numerical method; 4 – Born expansion.

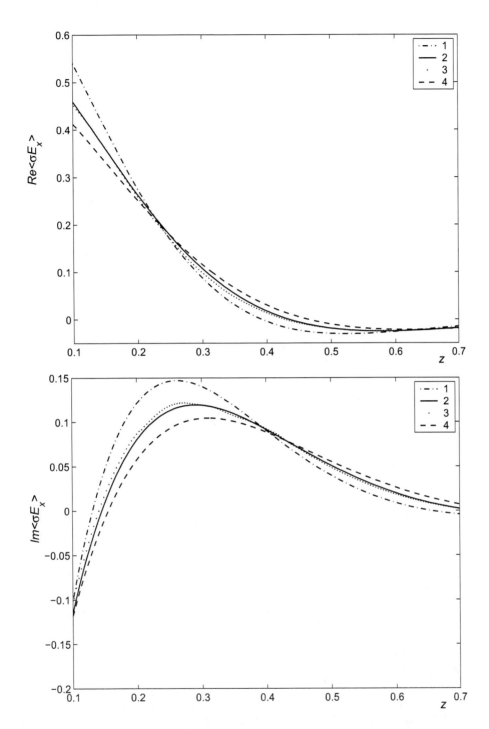

Figure 12. Real and imaginary parts of $\langle \sigma E_x \rangle$ obtained by: 1 – system (104) at $\sigma = 1$; 2 – effective system; 3 – numerical method; 4 – Born expansion.

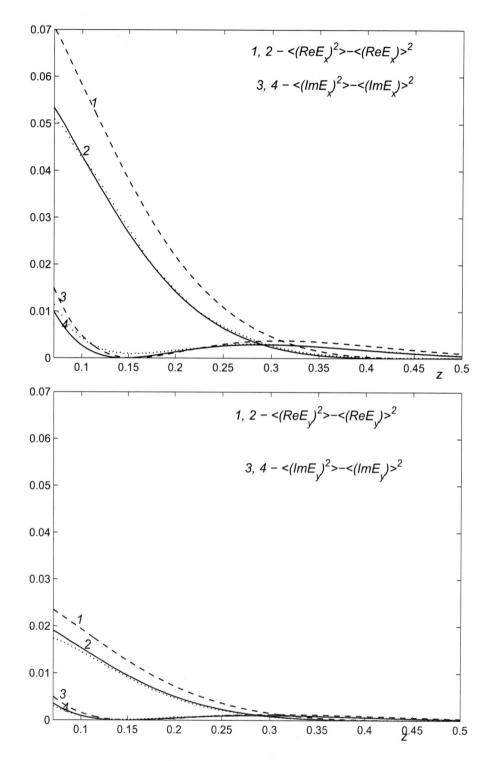

Figure 13. Variance of real and imaginary parts of components E obtained by: 1, 3 – Born expansion; 2, 4 – effective system; dots – numerical method.

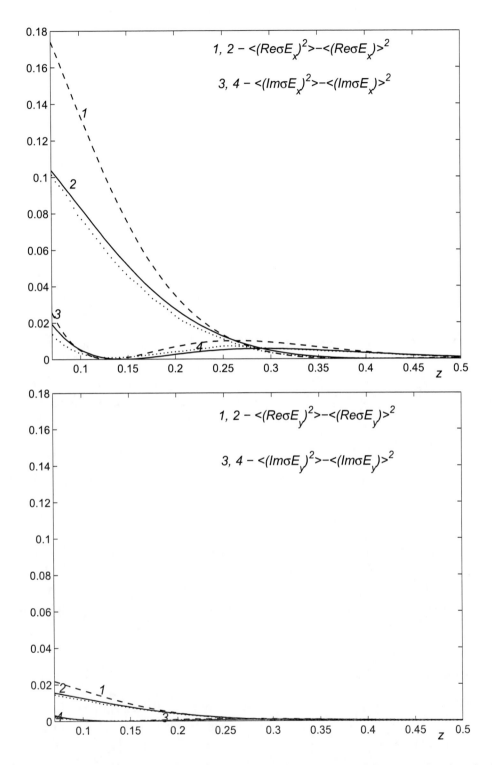

Figure 14. Variance of real and imaginary parts of components of current density obtained by: 1, 3 – Born expansion; 2, 4 – effective system; dots – numerical method.

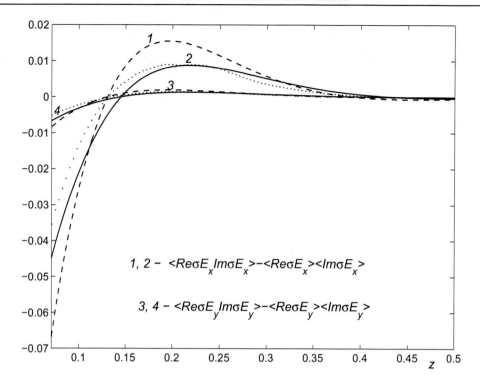

Figure 15. Covariance of real and imaginary parts of components of current density obtained by: 1, 3 – Born expansion; 2, 4 – effective system; dots – numerical method.

Conclusion

The results of numerical testing have shown that the approach proposed to estimate the impact of small-scale medium heterogeneities on the mean flow, the mean current density and the mean electric and magnetic field strengths is effective. We have obtained equations depending on the scale of smoothing for the effective coefficients. Within the approach multifractals are obtained if the minimum scale l_0 tends to zero. However only the theory of differential equations and the theory of random processes have been applied to the derivation of the effective formulas. In this approach we use physical parameters whose statistical properties can be measured directly (at least, in principle). To estimate the mean physical values only the mean conductivity and variance of conductivity are used.

References

[1] Sahimi, M. Flow phenomena in rocks: from continuum models, to fractals, percolation, cellular automata, and simulated annealing. *Reviews of Modern Physics*, **65** (1993),1393–1534.

[2] Mastryukov A. F., Mikhilenko B. G. Numerical solution of Maxwell's equations for anisotropic media using the Laguerre transform. *Russion Geology and Geophysics*, **49** (2008), 621–627.

[3] Epov M. I., Shurina E. P., Nechaev O. V. 3D forward modeling of vector field for induction logging. *Russian Geology and Geophysics*, **48** (2007), 770–774.

[4] Bogoliubov N. N., Shirkov D. V. *Introduction to the theory of quantized fields*. Moscow: Nauka, 1973.

[5] Dagan G. Higher-oder correction of effective permeability of heterogeneous isotropic formations of lognormal conductivity distribution. *Transport in Porous Media*, **12** (1993), 279–290.

[6] Shvidler M. I. *The statistical mechanics of porous media*. Moscow: Nedra, 1985 (in Russian).

[7] Rytov S. M., Kravtsov Yu. A., Tatarskii V. I. *Principles of statistical radiophysics*, III, IV. Berlin: Springer-Verlag, 1987.

[8] Landau L. D., Lifshitz E. M. *Electrodynamics of continuous media*. Oxford-Elmsford, New York: Pergamon Press, 1984.

[9] Matheron G. *Elements pourune theoriedes milieux poreux*. Paris: Masson, 1967.

[10] Gluzman S., Sornette D. Self-similar approximants of the permeability in heterogeneous porous media from moment equation expansions. *Transport in Porous Media*, **71** (2008), 75–97.

[11] Yukalov V. I. Self-semilar approximations for strongly interacting systems. *Physica A*, **167** (1990), 833–860.

[12] Yukalov V. I., Gluzman S. Self-semilar bootstrap of divergent series. *Phys. Rev. E*, **55** (1997), 6552–6570.

[13] Biswal B., Manwart C., Hilfer R. Three-dimensional local porosity analysis of porous media. *Physica A*, **255** (1998), 221–241.

[14] Krylov S. S., Lyubchich V. A. The apparent resistivity scaling and fractal structure of an iron formation. *Izvestiya. Physics of the Solid Earth*, **38** (2002), 1006–1012.

[15] Bekele A., Hudnall H. W., Daigle J. J., Prudente A., Wolcott M. Scale dependent variability of soil electrical conductivity by indirect measures of soil properties. *Journal of Terramechanics*, **42** (2005), 339–351.

[16] Mandelbrot B. B. *The fractal geometry of nature*. San Francisco: Freeman, 1982.

[17] Kolmogorov A. N. A refinement of previous hypotheses concerning the local structure of turbulence in a viscous incompressible fluid at high Reynolds number. *J. Fluid Mech.*, **13** (1962), 82–85.

[18] Monin A. Yaglom A. *Statistical fluid mechanics*, II. Cambridge: MIT Press, 1975.

[19] Meneveau, C. Sreenivasan K. R. The multifractal spectrum of the dissipation field in turbulent flows. *Nucl. Phys. B Proc. Suppl.*, **2** (1987a), 49–76.

[20] Mandelbrot B. B. Intermittent turbulence in self-similar cascades: divergence of high moments and dimension of the carrier. *J. Fluid Mech.*, **62** (1974), 331–358.

[21] Molchan G. M. Scaling exponents and multifractal dimensions for independent random cascades. *Commun. Math. Phys.*, **179** (1996), 681–702.

[22] Schertzer D., Lovejoy S. Physical modeling and analysis of rain and clouds by anisotropic scaling multiplicative processes. *J. Geophys. Res.*, **92** (1987), 9693–9714.

[23] Kida S. Log-stable distribution and intermittency of turbulence. *J. Phys. Soc. Jpn.*, **60** (1991), 5–8.

[24] Gnedenko B. V., Kolmogorov A. N. *Limit, distributions for sums of independent random variables*. Trans. and annotated by K. L. Chung. Cambridge: Addison-Wesley, 1954.

[25] Kuz'min G. A., Soboleva O. N. Subgrid modeling of filtration in porous self-similar media. *J. Appl. Mech. Tech. Phys.*, **43** (2002), 583–592.

[26] Kuz'min G. A., Soboleva O. N. Conformal symmetric model of the porous media. *Appl. Math. Lett.*, **14** (2001), 783–788.

[27] Soboleva 0. N. Effective permeability in a porous medium with log-stable statistics. *J. Appl. Mech. Technical Phys.*, **46** (2005), 891–900.

[28] Boufadel M. C., Lu S., Molz F. J., Lavallee D. Multifractal scaling of the intrinsic permeability. *Water Resours. Research*, **36** (2000), 3211–3222.

[29] Samorodnitsky G., Taqqu M. S. *Stable non-Gaussian random processes*. N. Y. London: Chapman – Hill., 1994.

[30] Uchaikin V. V., Zolotarev V. M. *Chance and Stability*. Utrecht, VSP: Stable Distributions and their Applications, 1999.

[31] Kurochkina E. P., Soboleva O. N., Epov M. I. Resistivity logging in a multi-scale medium with lognormal conductivity *Russian Geology and Geophysics*, **48** (2007), 851–856.

[32] Schwartz L., *Mathematics for the physical sciences*. Dover Publications, 2008.

[33] Ogorodnikov V. A., Prigarin S. M. Numerical *Modeling of random processes and fields: algorithms and applications*. The Netherlands: Utrecht, 1996.

[34] Marchuk G. I. *Methods of numerical mathematics*, II. Translated by J. Ruzicka, New York, Heidelberg, Berlin: Applications of Mathematics, Springer-Verlag, 1975.

[35] Chambers J. M., Mallows C., Stuck B. W. A method for simulating stable random variables. *J. Amer. Statist. Assoc.*, **71** (1976), 340–344.

[36] Kurochkina E. P., Soboleva O. N. Effective coefficients of quasi-steady Maxwell's equations with multiscale isotropic random conductivity. *Physica A*, **390** (2011), 231–244.

[37] Koshljakov N. S., Smirnov M. M., Gliner E. B. *Differential equations of mathematical physics*. Moscow: Fizmatgiz, 1962 (in Russian).

[38] Lebedev V. I. Difference analogies of orthogonal decjmpositions of basic differetial operators and some boundary value problems, I. *J. Comut. Maths. Math. Phys.*, **3** (1964), 449–465 (in Russian).

[39] Davydycheva S., Drushkin V., Habashy T. An efficient finite-difference scheme for electromalnetic logging in 3D anisotropic inhomogeneous media. *Geophysics*, **68** (2003), 1525–1536.

[40] Sleijpen G., Van der Vorst H., Modersitzki J. Differences in the effects of rounding errors in Krylov solvers for symmetric indefinite linear systems. *Matrix Anal. Appl.* 22 (2000), 726–751.

In: Classification and Application of Fractals
Editor: William L. Hagen

ISBN 978-1-61209-967-5
© 2012 Nova Science Publishers, Inc

Chapter 10

GENERATING EUCLIDEAN STRUCTURES USING IFS WITH MEMORY

Michael Frame, Brenda Johns on and Kathleen Meloney
Yale University, New Haven, CT, US

ABSTRACT

We find conditions under which the attractor of an n-IFS, an IFS determined by a prescribed set of allowed compositions of length $n + 1$, consists of a finite family of parallel lines, a modest step in the problem of characterizing the topological types of the attractors of n-IFS. A step in this analysis is showing that the attractor of the n-IFS with allowed compositions $R_1 \cup R_2$ is the union of the attractor of the n-IFS with allowed compositions R_1 and the attractor of the n-IFS with allowed compositions R_2, if the edge transition graphs of R_1 and R_2 are disconnected. Another is that the attractor is nonempty if and only if the edge transition graph contains a cycle. For attractors consisting of lines, the endpoints of the lines must constitute unions of cycles, and images of cycle points, on opposite edges of the unit square. These lines can be generated by n-IFS if and only if each endpoint lies in a distinct address length $(n + 1)$-square along that edge of the unit square.

1   Introduction

Recall the standard formulation for an iterated function system (IFS) [5], [12]. Denote by $C(\mathbb{R}^2)$ the compact subsets of $\mathbb{R}^2$, a complete metric space under the Hausdorff metric h. Given maps $T_1, \ldots, T_n : \mathbb{R}^2 \to \mathbb{R}^2$, contractions in the Euclidean metric, define a function $\mathcal{T} : C(\mathbb{R}^2) \to C(\mathbb{R}^2)$ by $\mathcal{T}(B) = \cup_{i=1}^{n}\{T_i(x) : x \in B\}$. The function $\mathcal{T}$ is a contraction map in h, so by the contraction mapping theorem, for any $B \in C(\mathbb{R}^2)$ the sequence $\mathcal{T}(B), \mathcal{T}^2(B), \mathcal{T}^3(B), \ldots$ converges to a unique $A \in C(\mathbb{R}^2)$, the *attractor* of the IFS $\mathcal{T}$. The set A is characterized by $\mathcal{T}(A) = A$. That is, A is the fixed point of $\mathcal{T}$.

In this formulation, the T_i are applied in all combinations, and so we could call this an *unrestricted* or *memoryless* IFS. The fixed point condition gives rise to a hierarchy of decompositions of A:

$$A = \bigcup_{i=1}^{n} T_i(A)$$
$$= \bigcup_{j=1}^{n} \bigcup_{i=1}^{n} T_j(T_i(A))$$
$$= \cdots.$$

The basic development is unaltered if some compositions of the T_i are forbidden, but the class of attractors is vastly expanded, as are the modalities for describing these attractors. Even with a drastically restricted collection of T_i, classifying the topological type of the attractor is daunting. We approach a restricted version of this problem: under what circumstances are the attractors families of lines?

2 IFS with Memory

An IFS has *memory* if only some compositions of the IFS transformations are allowed, all others being forbidden. This notion has some history and an extensive literature. Variations on this construction are called *graph-directed*, *recurrent*, *hierarchcal*, and *Markov* IFS. See [2], [3], [4], [5], [6], [10], [11], [14], [17], [18], [20], [21], [22], [23], for example. Distinctions between these types of IFS with memory can be found in [15].

We say an IFS has *n-step memory* if there is a collection $\mathcal{F}$ of *forbidden compositions*, all of length $\leq n + 1$, satisfying two properties.

- Every forbidden composition contains some element of $\mathcal{F}$.

- At least one forbidden composition of length $n + 1$ does not contain a forbidden composition of length j for all j, $1 \leq j \leq n$.

We call an IFS with n-step memory an n-IFS; a standard (memoryless) IFS is a 0-IFS.

Alternatively, an n-IFS can be characterized by a list $\mathcal{A}$ of *allowed compositions*. A bit of care is needed to alternate between characterizing an IFS by forbidden compositions and by allowed compositions, if compositions of different lengths are specified. To illustrate this point, consider an IFS with transformatiopns T_1 and T_2, and $\mathcal{F} = \{12, 211\}$. To find the allowed compositions, first rewrite $\mathcal{F}$ as consisting of length 3 compositions, $\mathcal{F} = \{112, 212, 121, 122, 211\}$. Then $\mathcal{A} = \{111, 221, 222\}$, the complementary set of length 3 compositions.

Another approach to studying IFS with memory is to recognize them as subshifts of finite type, an important topic, with a substantial literature, in dynamical systems. An excellent reference is [19]. Methods from the theory of subshifts can be applied to solve some problems about IFS with memory. For example, the memory reduction method of [7], described in Sect. 3, is an application of higher block shifts, in particular, Prop. 2.3.9 of [19]. However, some IFS with memory problems are better approached through understanding the geometry of the n-IFS attractors. We believe the issue of generating families of lines is one of these problems.

We build most IFS from four transformations

$$T_1(x, y) = \left(\frac{x}{2}, \frac{y}{2}\right) + (0, 0), \qquad T_2(x, y) = \left(\frac{x}{2}, \frac{y}{2}\right) + \left(\frac{1}{2}, 0\right), \qquad (1)$$

$$T_3(x, y) = \left(\frac{x}{2}, \frac{y}{2}\right) + \left(0, \frac{1}{2}\right), \qquad T_4(x, y) = \left(\frac{x}{2}, \frac{y}{2}\right) + \left(\frac{1}{2}, \frac{1}{2}\right).$$

Central to our analysis of n-IFS attractors is the *address* of a region in the attractor. The attractor of the memoryless IFS with transformations (1) is the filled-in unit square S. The *length n address regions* of S are

$$S_{i_n \ldots i_1} = T_{i_n} \circ \cdots \circ T_{i_1}(S).$$

We call these length n address regions n-*squares*. Note the order of the address digits agrees with the order of the composition of the T_i. The 1- and 2-squares are illustrated in Fig. 1.

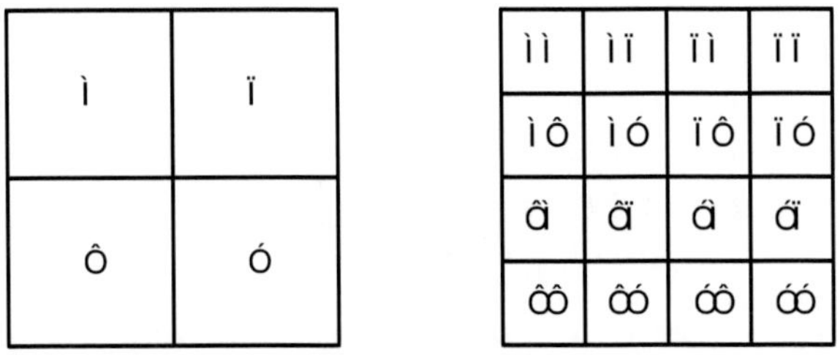

ì	ï
ô	ó

ìì	ïì	ïì	ïï
ìô	ìó	ïô	ïó
ỗ	ã̈	â̂	ã̈
ôô	ôó	óô	óó

Figure 1. The 1- and 2-squares.

A few examples of 1-, 2-, and 3-IFS attractors are given in Sect. 4.

For 1-IFS there are simple matrix and graphical representations. The matrix representation is

$$\begin{bmatrix} m_{11} & m_{12} & m_{13} & m_{14} \\ m_{21} & m_{22} & m_{23} & m_{24} \\ m_{31} & m_{32} & m_{33} & m_{34} \\ m_{41} & m_{42} & m_{43} & m_{44} \end{bmatrix}$$

where each $m_{ij} = 0$ or 1 according as the composition $T_i \circ T_j$ is forbidden or allowed. The graphical representation has four vertices, one for each T_i, with a directed edge to vertex j from vertex i if the composition $T_j \circ T_i$ is allowed. Because they can be viewed as describing which transitions from T_i to T_j are allowed, these representations are called the *transition matrix* and *transition graph* of the 1-IFS.

For 2-IFS the analog of the transition matrix is a $4 \times 4 \times 4$ array, which we also call a transition matrix. This can be expressed as a quadruple of 4×4 matrices presented in this order

$$[m_{ij1}][m_{ij2}][m_{ij3}][m_{ij4}]$$

where $m_{ijk} = 0$ or 1 according as the composition $T_i \circ T_j \circ T_k$ is forbidden or allowed.

Here and for longer memory IFS the sensible version of the transition graph is the *edge transition graph*. Given an $(n+1)$-tuple $i_{n+1} \ldots i_1$, the *initial n-tuple* is $i_n \ldots i_1$ and the *terminal n-tuple* is $i_{n+1} \ldots i_2$. For an n-IFS with allowed $(n+1)$-tuples R, the vertices of the edge transition graph are the n-tuples, $i_n \cdots i_1$ representing the composition $T_{i_n} \circ \cdots \circ T_{i_1}$, and there is an edge

$$i_n \cdots i_1 \to j_n \cdots j_1$$

if and only if $i_n \ldots i_1$ is the initial n-tuple and $j_n \ldots j_1$ is the terminal n-tuple of an element of R. That is, $j_{n-1} = i_n, j_{n-2} = i_{n-1}, \ldots, j_1 = i_2$ and the $(n+1)$-tuple $j_n i_n i_{n-1} \ldots i_1$ is allowed. For a 1-IFS this graph, the transition graph described above, has 4 vertices and at most 16 edges, not too difficult to interpret visually. For 2-IFS, the graph has 16 vertices and perhaps an impenetrable thicket of edges.

For later use we end with an observation of how the attractors A_1 and A_2 of n-IFS with allowed $(n+1)$-tuples R_1 and R_2 are related to the attractor A_3 of the n-IFS with allowed $(n+1)$-tuples $R_1 \cup R_2$. The first guess is that if $R_1 \cap R_2 = \emptyset$, then $A_3 = A_1 \cup A_2$. Often first guesses are wrong, and the next example shows this is the case here.

Example 2.1 *The attractor of the union need not be the union of the attractors.*

In the left and middle of Fig. 2 we see the attractors of the 1-IFS with transition matrices

$$\begin{bmatrix} 1 & 0 & 0 & 1 \\ 0 & 1 & 1 & 0 \\ 0 & 1 & 1 & 0 \\ 1 & 0 & 0 & 1 \end{bmatrix} \quad \text{and} \quad \begin{bmatrix} 0 & 1 & 1 & 0 \\ 1 & 0 & 0 & 1 \\ 1 & 0 & 0 & 1 \\ 0 & 1 & 1 & 0 \end{bmatrix}$$

Certainly the allowed pairs for these two 1-IFS have no common member, yet the right of Fig. 2 shows the attractor of the 1-IFS that allows the union of these sets of allowed pairs. Clearly, the attractor of the union does not equal the union of the attractors. So what is true? Under what conditions does the attractor of the union equal the union of the attractors?

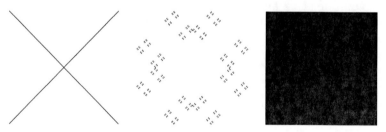

Figure 2. Attractors of Ex 2.1.

This raises the tangential issue: what is the relation between the attractors of n-IFS having complementary sets of allowed $(n+1)$-tuples?

We begin with some notation. For $j = 1, 2$ let

$$A_j^k = \left\{ S_{i_k \ldots i_1} : i_k \ldots i_{k-n}, i_{k-1} \ldots i_{k-n-1}, \ldots, i_{n+1} \ldots i_1 \in R_j \right\} \qquad (2)$$

So for example,

$$A_j^{n+1} = \left\{ S_{i_{n+1} \ldots i_1} : i_{n+1} \ldots i_1 \in R_j \right\},$$
$$A_j^{n+2} = \left\{ S_{i_{n+2} \ldots i_1} : i_{n+2} \ldots i_2 \in R_j \text{ and } i_{n+1} \ldots i_1 \in R_j \right\}$$

and so on.

Keeping in mind the order of addresses is identical with the order of the composition of transformations, for all $k \geq n+1$

$$A_j^{k+1} \subseteq A_j^k$$

because

$$S_{i_{k+1}\ldots i_1} \subset S_{i_{k+1}\ldots i_2}.$$

In fact,

$$S_{i_{k+1}\ldots i_2} = S_{i_{k+1}\ldots i_2 1} \cup S_{i_{k+1}\ldots i_2 2} \cup S_{i_{k+1}\ldots i_2 3} \cup S_{i_{k+1}\ldots i_2 4}.$$

If arbitrarily long compositions are allowed (Prop 6.1 gives necessary and sufficient conditions for the existence of arbitrarily long allowed compositions), each A_j^k is a nonempty compact set and

$$A_j = \bigcap_k A_j^k.$$

Recall A_j is the attractor of an n-IFS with allowed $(n+1)$-tuples R_j.

Lemma 2.1 *If $R_3 = R_1 \cup R_2$, then $A_1 \cup A_2 \subseteq A_3$.*

Proof. For each $k \geq n+1$ we show

$$A_1^k \cup A_2^k \subseteq A_3^k. \tag{3}$$

First recall Eq (2) for $j = 1, 2, 3$. Certainly, each of the $n+1$-tuples that belongs to R_1 or R_2 also belongs to R_3, establishing (3). The result follows by taking $k \to \infty$. $\qquad\square$

Now we find a simple condition under which $A_1 \cup A_2 = A_3$. We say the edge transition graphs of R_1 and R_2 are *disconnected* if no n-tuple is both terminal for an element of R_1 and initial for an element of R_2, or initial for an element of R_1 and terminal for an element of R_2.

Proposition 2.1 *Suppose the edge transition graphs of R_1 and R_2 are disconnected. Then $A_1 \cup A_2 = A_3$.*

Proof. For each $k \geq n+1$ we show

$$A_3^k \subseteq A_1^k \cup A_2^k. \tag{4}$$

Suppose $S_{i_k\ldots i_1}$ belongs to A_3^k and let $i_{n+1}\ldots i_1$ denote the initial $(n+1)$-tuple of $i_k\ldots i_1$. Then $i_{n+1}\ldots i_1 \in R_3 = R_1 \cup R_2$, so $i_{n+1}\ldots i_1$ belongs to one of R_1 and R_2, say it belongs to R_1. Now consider $i_{n+2}\ldots i_2$. The segment $i_{n+1}\ldots i_2$ is an initial n-tuple of $i_{n+2}\ldots i_2$ and a terminal n-tuple of $i_{n+1}\ldots i_1$. The hypothesis of disconnected edge transition graphs implies no initial (respectively, terminal) n-tuple of R_1 is also a terminal (respectively, initial) n-tuple of R_2. Consequently, $i_{n+2}\ldots i_2 \in R_1$. Continuing in this way, we deduce that each of

$$i_k\ldots i_{k-n}, \; i_{k-1}\ldots i_{k-n-1}, \; \ldots, \; i_{n+2}\ldots i_2, \text{ and } i_{n+1}\ldots i_1$$

belong to R_1. Then by (2), $S_{i_k\ldots i_1}$ belongs to A_1^k. Arguing similarly, we see that every square of A_3^k belongs to A_1^k or to A_2^k, establishing (4). The result follows by taking $k \to \infty$ and using Lemma 2.1. $\qquad\square$

3 Relations Between n-IFS for Different n

Conditions under which the attractor of a 1-IFS is also the attractor of a 0-IFS were given in [13]. A transformation T_i is a *rome* if for all j the composition $T_i \circ T_j$ is allowed. The attractor of a 1-IFS is also the attractor of a 0-IFS if

1. there is at least one rome, and

2. for each transformation T_{i_n}, there is an allowed composition $i_n \ldots i_1$, with T_{i_1} a rome.

Details can be found in [13].

Adapting the method of higher block codes (Prop. 2.3.9 of [19]), in [7] it is shown that for all $n > 1$ the attractor of any n-IFS is also the attractor of a 1-IFS, though of course more transformations may be needed.

For instance, suppose a 2-IFS forbids the compositions 14, 23, 32, 124, and 441. The attractor of this 2-IFS also is the attractor of the 1-IFS with 13 transformations $R_{ij} = T_i \circ T_j$ for $i, j = 1, 2, 3, 4$, and $ij \neq 14, 23, 32$. When we say the transformation R_{ij} can follow R_{kl}, we mean $j = k$ and the composition ijl is allowed. This corresponds to applying $T_i \circ T_j \circ T_l$, *not* to applying the composition $T_i \circ T_j \circ T_j \circ T_l$. In this example, the only exclusion, beyond those resulting from $j \neq k$, are R_{12} cannot follow R_{24} and R_{44} cannot follow R_{41}.

While it is true that the attractor of every n-IFS also is the attractor of a 1-IFS, some problems, including classifying Euclidean attractors, are better approached through memory relations than with one step of memory and a (perhaps much) larger collection of transformations.

In Sects. 3 and 5 of [8] we see that the attractor of an n-IFS is also the attractor of an s-IFS for all $s \geq n$. In addition, the transition rules of a 1-IFS can be embedded into those of a 2-IFS with attractor 4 copies of the 1-IFS attractor by taking transition matrices $b_{ijk} = a_{jk}$ for all i, j, k, and into those of a 3-IFS with attractor 16 copies of the 1-IFS attractor by taking transition matrices $c_{mijk} = a_{jk}$ for all m, i, j, k. Similarly, the transition rules of a 2-IFS can be embedded in those of a 3-IFS with attractor four copies of the 2-IFS attractor by taking transition matrices $c_{mijk} = b_{ijk}$ for all m, i, j, k. This is illustrated in Sect. 4.

4 A Gallery of n-IFS Attractors

Here we present a very small sample of attractors of 1, 2, and 3-IFS. As mentioned in Sect. 2, these n-IFS can be represented by transition matrices or transition graphs, but for $n > 1$ these can be tedious. An alternate approach is to note the empty $(n + 1)$-squares: for a 1-IFS the forbidden words are the addresses of the empty 2-squares, for a 2-IFS the forbidden words are the addresses of the empty 3-squares, and so on.

To illustrate the embedding results mentioned at the end of Sect. 3, note the right

Generating Euclidean Structures Using IFS with Memory   335

Figure 3. Attractors of 1-IFS.

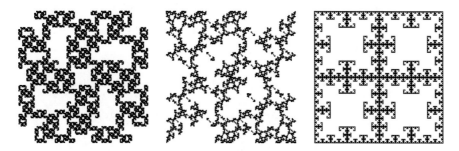

Figure 4. Attractors of 2-IFS.

attractor of Fig. 3 has transition matrix

$$\begin{bmatrix} 1 & 1 & 1 & 0 \\ 1 & 1 & 0 & 1 \\ 1 & 0 & 1 & 1 \\ 0 & 1 & 1 & 1 \end{bmatrix},$$

the right attractor of Fig. 4 has transition matrix

$$\begin{bmatrix} 1 & 1 & 1 & 0 \\ 1 & 1 & 1 & 0 \\ 1 & 1 & 1 & 0 \\ 1 & 1 & 1 & 0 \end{bmatrix} \begin{bmatrix} 1 & 1 & 0 & 1 \\ 1 & 1 & 0 & 1 \\ 1 & 1 & 0 & 1 \\ 1 & 1 & 0 & 1 \end{bmatrix} \begin{bmatrix} 1 & 0 & 1 & 1 \\ 1 & 0 & 1 & 1 \\ 1 & 0 & 1 & 1 \\ 1 & 0 & 1 & 1 \end{bmatrix} \begin{bmatrix} 0 & 1 & 1 & 1 \\ 0 & 1 & 1 & 1 \\ 0 & 1 & 1 & 1 \\ 0 & 1 & 1 & 1 \end{bmatrix}.$$

That is, denoting by a_{jk} the transition matrix of the right image of Fig. 3, the transition matrix of the right image of Fig. 4 is $b_{ijk} = a_{jk}$ for all i, j, k, and the transition matrix of the right image of Fig. 5 is $c_{mijk} = a_{jk}$ for all m, i, j, k. The middle images of Figs 4 and 5 have transition matrices $e_{mijk} = d_{ijk}$ for all m, i, j, k.

5 Topological Types of 0-IFS Attractors

An example of the variety of topological types of simple fractals is given on pgs 232 – 7 of [25], the attractors A of IFS consisting of three transformations of the form

$$T_i(x, y) = (r_i \cdot \cos(\theta_i) \cdot x - s_i \cdot \sin(\varphi_i) \cdot y, r_i \cdot \sin(\theta_i) \cdot x + s_i \cdot \cos(\varphi_i) \cdot y) + (e_i, f_i), \quad (5)$$

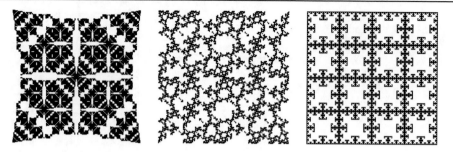

Figure 5. Attractors of 3-IFS.

where $r_i = \pm 1/2$, $s_i = \pm 1/2$, $\theta_i = \varphi_i = 0, \pi/2, \pi, 3\pi/2$, and e_i and f_i take the values 0, 1/2, and 1, chosen so that

$$T_1(A) \subset [0, 1/2] \times [0, 1/2], T_2(A) \subset [1/2, 1] \times [0, 1/2], T_3(A) \subset [0, 1/2] \times [1/2, 1].$$

Note that each of these transformations with $s_i = -1/2$ is equivalent to one with $r_i = -1/2$:

$$\begin{aligned}(+, -, 0) &\longleftrightarrow (-, +, \pi), & (+, -, \pi/2) &\longleftrightarrow (-, +, 3\pi/2), \\ (+, -, \pi) &\longleftrightarrow (-, +, 0), & (+, -, 3\pi/2) &\longleftrightarrow (-, +, \pi/2).\end{aligned} \quad (6)$$

Here $(-, +, \pi/2)$ represents $r_i = -1/2$, $s = 1/2$, and $\theta = \varphi = \pi/2$, for example. Thus there are 8 distinct transformations of this type, and so $8^3 = 512$ sets of IFS rules. On pgs 232 – 4 of [25] we find 224 different fractals generated by IFS of this type. Four examples are shown in Fig. 6, generated by these IFS tables.

r	s	θ	e	f
-.5	.5	0	.5	0
.5	-.5	0	.5	.5
.5	.5	$\pi/2$	.5	.5

r	s	θ	e	f
.5	.5	π	.5	.5
.5	.5	$\pi/2$	1	0
.5	.5	0	0	.5

r	s	θ	e	f
.5	.5	$\pi/2$	.5	0
-.5	.5	$\pi/2$	1	.5
-.5	.5	$-\pi/2$	0	.5

r	s	θ	e	f
.5	-.5	0	0	.5
.5	.5	$-\pi/2$	.5	.5
.5	.5	0	0	.5

(7)

Figure 6. Cantor set, dendrite, looped, hybrid; attractors of the IFS (7).

None of these is symmetric across the diagonal line $y = x$, so reflecting each of these across the diagonal gives an additional 224 fractals. The remaining $512 - 2 \cdot 224 = 64$ are symmetric across the diagonal. In fact, there are 8 distinct fractals (Fig. 7), each generated by 8 distinct IFS tables. All together we have $224 + 8 = 232$ topologically distinct fractals. Inspection of these fractals reveals that they fall into four topological types

- Cantor sets: totally disconnected, perfect sets.

- Dendrites: connected and simply connected sets.

- Looped: connected and containing non-contractible loops.

- Hybrid: disconnected but not totally disconnected, with simply-connected components.

By *loop* we mean a set homeomorphic to a circle. This list of types is not such a surprise: within these categories only one more possibility remains: disconnected and containing non-contractible loops.

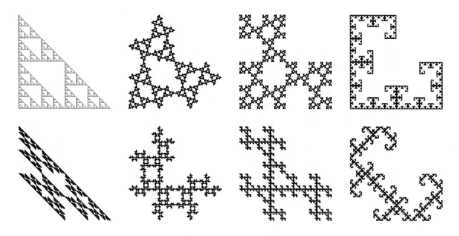

Figure 7. Fractals symmetric across the diagonal.

The fourth topological type is missed on pg 237 of [25]. To check that we have identified a new topological type, note that in the last IFS table of (7),

$$T_1(x, y) = (x/2, -y/2) + (0, 1/2), \quad T_3(x, y) = (x/2, y/2) + (0, 1/2).$$

Then
$$L = T_1(L) \cup T_3(L),$$

where L is the line segment with endpoints $(0, 0)$ and $(0, 1)$. Consequently, L is contained in the invariant set of this IFS, so the fractal is not totally disconnected. On the other hand, applied to the filled-in unit square, the portion of the fourth iterate contained in $[1/2, 1] \times [0, 1/2]$ is disconnected from the other portions. Consequently, this fractal is not connected.

Recalling (6), for this family of fractals we can always take $s = 1/2$, and so each of the three transformations can take one of eight forms, $(r, \theta) =$

$$(+, 0), (+, \pi/2), (+, \pi), (+, 3\pi/2), (-, 0), (-, \pi/2), (-, \pi), (-, 3\pi/2),$$

and so each of these IFS tables is determined by a point in an $8 \times 8 \times 8$ lattice. This collection is small enough that classification can be achieved by exhaustion (of cases, if not of the classifier). A map of those lattice points that give each topological type might reveal interesting patterns, but we were too exhausted to finish the map.

This approach is unlikely to produce manageable results for IFS with memory. Using the transformations (1), very roughly for 1-step memory there are $2^{16} = 65,536$ transition matrices. To be sure, some of these have empty attractors, but many, many do not. The situation is far more involved with 2-step memory: there are $2^{64} = 18,446,744,073,709,551,616$ transition matrices. A more thoughtful approach is needed.

6 Single Lines in 1-IFS and 2-IFS

Even restricted to the simple transformations (1), the general topological classification problem for IFS with memory seems out of reach at the moment. We restrict our attention to a simpler problem, identifying Euclidean structures, lines and combinations of lines, that can arise as the attractor of IFS with memory. We start with the simple cases of single lines, and continue with combinations and some more involved arrangements.

By a *line* we mean a line segment whose endpoints lie on opposite edges of the unit square, e.g., a line segment whose endpoints are $\{(0, a), (1, b)\}$ or $\{(a, 0), (b, 1)\}$, or a diagonal line with endpoints $(0, 0)$ and $(1, 1)$, or $(0, 1)$ and $(1, 0)$.

With 1-step memory we can produce 6 lines: given any pair of transformations T_i and T_j, use the transition graph with all combinations of T_i and T_j. Two examples are given in Fig. 8. To position each line, the unit square is shown in light grey.

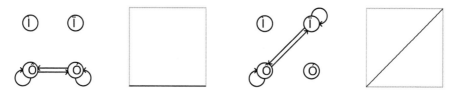

Figure 8. Two examples of lines generated by 1-IFS

In Fig. 9 we see two sets of two lines. In the language of Prop 2.1, the left image is the attractor of a 1-IFS with allowed 2-tuples $R_1 \cup R_2$, with $R_1 = \{11, 12, 21, 22\}$ and $R_2 = \{33, 34, 43, 44\}$. The transition graphs for R_1 and R_2 are disconnected, so as expected from Prop 2.1, the attractor of the 1-IFS with allowed 2-tuples $R_1 \cup R_2$ is the union of the attractors of the 1-IFS with allowed 2-tuples R_1 and with allowed 2-tuples R_2. For the right image, $R_1 = \{11, 12, 21, 22\}$ and $R_2 = \{11, 14, 41, 44\}$. The transition graphs of R_1 and R_2 are not disconnected, so the attractor of the 1-IFS with allowed 2-tuples $R_1 \cup R_2$ is a proper superset of the union of the attractors of the 1-IFSs with allowed 2-tuples R_1 and with allowed 2-tuples R_2

Note also that the endpoints of these lines are two of the four corners of the unit square. Each corner is the fixed point of one of the transformations (1). The fixed point of T_i has address i^∞.

What are the possibilities of generating single lines with 2-IFS? As pointed out in Sect. 3 of [8], the attractor of every 1-IFS also is the attractor of a 2-IFS. For example, the line

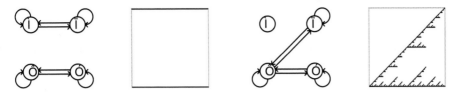

Figure 9. Left: two independent lines. Right: two lines not so independent.

on the right side of Fig. 8 is produced by the 1-IFS with transition matrix

$$\begin{bmatrix} 1 & 0 & 0 & 1 \\ 0 & 0 & 0 & 0 \\ 0 & 0 & 0 & 0 \\ 1 & 0 & 0 & 1 \end{bmatrix} \qquad (8)$$

and consequently by Prop 3.1 of [8], also by the 2-IFS with transition matrix

$$\begin{bmatrix} 1 & 0 & 0 & 1 \\ 0 & 0 & 0 & 0 \\ 0 & 0 & 0 & 0 \\ 1 & 0 & 0 & 1 \end{bmatrix} \begin{bmatrix} 1 & 0 & 0 & 1 \\ 0 & 0 & 0 & 0 \\ 0 & 0 & 0 & 0 \\ 1 & 0 & 0 & 1 \end{bmatrix} \begin{bmatrix} 1 & 0 & 0 & 1 \\ 0 & 0 & 0 & 0 \\ 0 & 0 & 0 & 0 \\ 1 & 0 & 0 & 1 \end{bmatrix} \begin{bmatrix} 1 & 0 & 0 & 1 \\ 0 & 0 & 0 & 0 \\ 0 & 0 & 0 & 0 \\ 1 & 0 & 0 & 1 \end{bmatrix}. \qquad (9)$$

Note that the 1-IFS transition matrix does not involve T_2 or T_3. Consequently, the middle two matrices of the 2-IFS (9) can be replaced by 0-matrices without altering the attractor.

Can 2-IFS generate single lines different from those produced by 1-IFS? For example, can a 2-IFS produce the line L with endpoints $(1/2, 0)$ and $(1/2, 1)$? First note an obvious condition:

If part of an attractor lies in the square with address ijk, then the
IFS must allow the composition $T_i \circ T_j \circ T_k$.

Referring to the left side of Fig. 10 we see the addresses of the eight 3-squares whose right boundaries lie on the line L. The 2-IFS allowing just these 8 triples has attractor the empty set, not the line L.

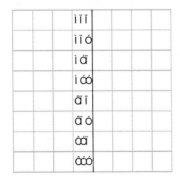 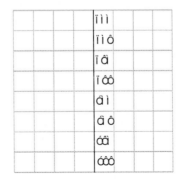

Figure 10. Length 3 addresses containing parts of the line between $(1/2, 0)$ and $(1/2, 1)$.

Perhaps this isn't such a surprise. After all, eight other 3-squares have left boundaries on the line L. The addresses of these squares are given on the right side of Fig. 10. The 2-IFS allowing these 8 triples and the previous 8 triples – the addresses of every 3-square intersecting L in any way – has as attractor ... the empty set, again. What's going on here?

In this case, the problem is clear. For example, consider the address 122, corresponding to the composition $T_1 \circ T_2 \circ T_2$. In this 2-IFS, which transformation can be applied after T_1? If the composition $T_i \circ T_1 \circ T_2 \circ T_2$ is allowed, then the triple $i12$ must be one of the 16 allowed triples of this 2-IFS. Because $i12$ is not an allowed triple for this 2-IFS, NO transformation can be applied after $T_1 \circ T_2 \circ T_2$. Checking the other 15 allowed compositions, we see that none can be followed by any T_i, so this 2-IFS has an empty attractor.

This leads to the natural question: can we find necessary and sufficient conditions on the allowed compositions to guarantee a nonempty attractor? Call a composition $T_{i_n} \circ \cdots T_{i_1}$ a *cycle* if

$$\cdots \circ T_{i_n} \circ \cdots T_{i_1} \circ T_{i_n} \circ \cdots T_{i_1} \circ T_{i_n} \circ \cdots T_{i_1} = (T_{i_n} \circ \cdots T_{i_1})^\infty$$

is allowed. The *nonempty attractor condition* is this.

Proposition 6.1 *A set $\mathcal{A}$ of allowed compositions gives rise to a nonempty attractor if and only if $\mathcal{A}$ contains a cycle.*

Proof. Suppose $\mathcal{A}$ contains the cycle $T_{i_n} \circ \cdots T_{i_1}$. Then the point with address $(i_n \cdots i_1)^\infty$ is the result of repeated applications of the composition $T_{i_n} \circ \cdots T_{i_1}$ to the unit square, so the attractor contains at least this point. That is, if $\mathcal{A}$ contains a cycle, the attractor is nonempty.

Now suppose the attractor is nonempty, so it must contain at least one point. The address of that point is an infinite string $s = \ldots i_p i_{p-1} \ldots i_1$. The number of transformations is finite (4 in our examples), so some n-tuple must occur twice:

$$i_{p_n} i_{p_{n-1}} \cdots i_{p_1} = i_{q_n} i_{q_{n-1}} \cdots i_{q_1}$$

with $p_n > q_n$. Then $i_{p_n} i_{p_{n-1}} \cdots i_{q_1}$ gives an allowed cycle, because it has the same initial and terminal n-segment. That is, if the attractor is nonempty, $\mathcal{A}$ contains a cycle. $\square$

In the next proposition, we see that to generate a single line, the endpoints of that line must be fixed points of two of the T_i, that is, corners of the unit square. Because all six of these lines can be generated by 1-IFS, 2-IFS cannot generate any additional lines in isolation.

Proposition 6.2 *Suppose the attractor $\mathcal{A}$ consists of a single line L. Then each of the endpoints of L must be a fixed point of one of the transformations T_i.*

Proof. Suppose L is a line with at least one endpoint, P, not a corner of the unit square. For definiteness, suppose P lies on the segment between $(0,0)$ and $(1,0)$, so P is the only point of L on that segment. Then the address of P is an infinite string $\pi = \ldots p_2 p_1$ of 1s and 2s, not all 1s or all 2s because P is neither $(0,0)$ nor $(1,0)$. Arguing as in the proof of Prop 6.1, for some $i > j, p_i = p_j$ and so π contains the fragment

$$p_i p_{i-1} \cdots p_{j+1} p_j = p_i p_{i-1} \cdots p_{j+1} p_i.$$

Generating Euclidean Structures Using IFS with Memory 341

Consequently the cycle $(p_i p_{i-1} \ldots p_{j+1})^\infty$ is allowed. Not all of $p_i, p_{i-1}, \ldots, p_{j+1}$ are equal, so the cycle consists of at least two points on the line between $(0,0)$ and $(1,0)$, impossible because the point P is the only point of L that lies on the segment between $(0,0)$ and $(1,0)$. □

Consequently, except for the 6 single lines generated by 1-IFS, no line in isolation can be generated by an n-IFS, for any $n > 1$.

7 Combinations of Lines in 1-IFS

We have seen that using the transformations (1), the n-IFS attractors that consist of a single line have as endpoints of the line the fixed points of two of the T_i. Thus there are 6 such lines, and all can be achieved as attractors of 1-IFS. As illustrated in Fig. 9, any two of these lines without common endpoint can be generated by taking the union of the sets of allowed pairs generating each of the lines. In this section we explore other combinations of lines that can be attractors for 1-IFS.

The endpoints of a family of lines comprise a finite set of points on two edges of the unit square. In Sect. 9 we shall see the collection of these endpoints on an edge of the unit square must consist of unions of cycles and their images, and that the maximum length of a cycle generated by an n-IFS is a simple function of n. In this section and the next, we try to discover patterns through examples.

71 Fixed points and their images

First, suppose the line endpoints are fixed points of some T_i, and images of these fixed points.

Example 7.1 *Two horizontal lines, or two vertical lines.*

We have seen that applying T_1 and T_2 in all combinations produces the line segment L_1 between $(0,0)$ and $(1,0)$. Given this line, $T_3(L_1)$ is the line segment between $(0,1/2)$ and $(1/2,1/2)$, and $T_4(L_1)$ is the line segment between $(1/2,1/2)$ and $(1,1/2)$. That is, $T_3(L_1) \cup T_4(L_1) = L_2$, the line segment between $(0,1/2)$ and $(1,1/2)$. See Fig. 11 for the transition graph that yields L_1 and L_2 as its attractor. The endpoints of L_1 are the fixed points with addresses 1^∞ and 2^∞; the endpoints of L_2 have addresses $3(1)^\infty$ and $4(2)^\infty$ Given the order of addresses agrees with the order of composition, by $3(1)^\infty$ we mean the address of the point obtained by applying T_3 to $(x,y) = \lim_{n \to \infty} T_1^{\circ n}(w,z)$, for any $(w,z) \in S$. Note that some points have two addresses. For example, $(0,1/2)$ has address $3(1)^\infty$ and $1(3)^\infty$.

Because T_3 and T_4 cannot be followed by any T_i, the line L_2 has no images in the attractor. If we allow the loops $T_3 \to T_3$ and $T_4 \to T_4$, the attractor contains a sequence of ever shorter line segments, at heights $3/4, 7/8, 15/16, \ldots, (2^n - 1)/2^n, \ldots$. See the left side of Fig. 12. Allowing all combinations of T_3 and T_4 gives an attractor consisting of the line segments between $(0, (2^n - 1)/2^n)$ and $(1, (2^n - 1)/2^n)$, for $n = 0, 1, 2, \ldots$, together with the line segment between $(0,1)$ and $(1,1)$. See the right side of Fig. 12.

Figure 11. A line and its image: transition graph, attractor, addresses.

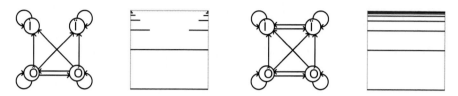

Figure 12. Some additions to the 1-IFS of Fig. 11.

Similar families of lines, one more consisting of two horizontal lines, and two consisting of two vertical lines each, are produced by three 1-IFS. In all these cases, the endpoints of the lines are a^∞ and b^∞, and $c(a)^\infty$ and $d(b)^\infty$.

Example 7.2 *Two parallel slanted lines.*

Suppose we take two endpoints, say 2^∞ and $4(2)^\infty$, from one pair of horizontal lines, and two endpoints 3^∞ and $1(3)^\infty$ from the other pair of horizontal lines. Draw the lines L_3 between 3^∞ and $4(2)^\infty$, and L_4 between $1(3)^\infty$ and 2^∞. Find the length 2 addresses through which these lines pass (right side of Fig. 13), and build the 1-IFS with these allowed pairs. Indeed, this 1-IFS has attractor the two slanted parallel lines L_3 and L_4. See Fig. 13.

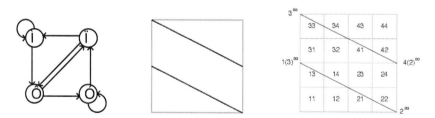

Figure 13. Slanted parallel lines: transition graph, attractor, addresses.

Other pairs of slanted parallel lines are the attractors of the three obvious variations on this 1-IFS. Can other pairs of lines be generated by 1-IFS?

72 2-cycles

In addition to a fixed point and its image, we can produce endpoints of two line segments on a side of the unit square by taking the points of a 2-cycle on that side of the square.

Example 7.3 *Lines bounded by 2-cycles.*

Consider the 2-cycles with addresses $(13)^\infty$ and $(31)^\infty$ on the left edge of the unit square, and $(24)^\infty$ and $(42)^\infty$ on the right edge of the unit square. Draw the horizontal lines, L_5 and L_6, connecting these endpoints, note the length 2 addresses they occupy (right side of Fig. 14), and construct the 1-IFS with these allowed pairs. This 1-IFS has attractor L_5 and L_6. See the left and center of Fig. 14.

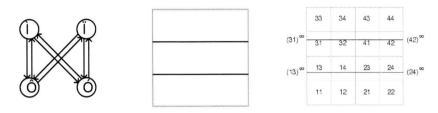

Figure 14. Lines bounded by 2-cycles: transition graph, attractor, addresses.

Ex 7.3 has an obvious variation, a 1-IFS with attractor consisting of two vertical line segments. Together with the pair of independent horizontal (left side of Fig. 9) and independent vertical lines of Sect. 6, Exs 7.1, 7.2, and 7.3 and their variations are all the pairs of lines that can be the attractors of 1-IFS. We shall prove this in Sect. 9. In Sect. 8 we shall see if we can produce more and different families of lines using 2-IFS.

8 Lines in 2-IFS

Recalling the relation between the 1-IFS transition matrix (8) and the 2-IFS transition matrix (9), any line or pair of lines produced by a 1-IFS also can be produced by a 2-IFS. To seek new families of lines for 2-IFS, we look for triples of lines. Most straightforward is to seek triples of points on opposite edges of the unit square.

Example 8.1 *Endpoints consisting of 3-cycles.*

The bottom edge of the unit square is generated by T_1 and T_2. With these transformations, two 3-cycles can be formed: $\{(112)^\infty, (121)^\infty, (211)^\infty\}$, and $\{(122)^\infty, (212)^\infty, (221)^\infty\}$. Consider the first of these. On the left side of Fig. 15 we see the three vertical lines between the points of this 3-cycle on the bottom of the unit square, and the corresponding points of the 3-cycle $\{(334)^\infty, (343)^\infty, (433)^\infty\}$ on the top edge. Superimposed on the unit square are the length 3 addresses. Noting the addresses through which these lines pass, construct the 2-IFS with these addresses as the allowed triples. The transition matrix is

$$\begin{bmatrix} 0 & 1 & 0 & 1 \\ 1 & 0 & 1 & 0 \\ 0 & 1 & 0 & 1 \\ 1 & 0 & 1 & 0 \end{bmatrix} \begin{bmatrix} 1 & 0 & 1 & 0 \\ 0 & 0 & 0 & 0 \\ 1 & 0 & 1 & 0 \\ 0 & 0 & 0 & 0 \end{bmatrix} \begin{bmatrix} 0 & 1 & 0 & 1 \\ 1 & 0 & 1 & 0 \\ 0 & 1 & 0 & 1 \\ 1 & 0 & 1 & 0 \end{bmatrix} \begin{bmatrix} 1 & 0 & 1 & 0 \\ 0 & 0 & 0 & 0 \\ 1 & 0 & 1 & 0 \\ 0 & 0 & 0 & 0 \end{bmatrix}.$$

Indeed, the 2-IFS with this transition matrix has attractor consisting of these three vertical lines.

Using the other pair of 3-cycles on the top and bottom edges of the unit square, or the pairs of corresponding 3-cycles on the left and right edges, we find four families of three vertical or horizontal lines whose endpoints constitute 3-cycles.

Because the transition matrix is easily read from the length 3 addresses through which the lines pass, for most of the remaining examples we shall not give the transition matrices.

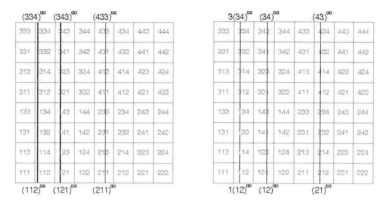

Figure 15. Left: 3-cycle. Right: 2-cycle and an image.

Example 8.2 *Endpoints consisting of a 2-cycle and an image of a 2-cycle point.*

Consider the 2-cycle $\{(12)^\infty, (21)^\infty\}$ along the bottom edge of the unit square, together with the corresponding 2-cycle $\{(34)^\infty, (43)^\infty\}$ along the top edge. These are shown on the right side of Fig. 15. Connecting these cycle points by vertical lines, the 2-IFS with allowed triples the length 3 addresses through which these lines pass does generate this pair of lines. Adding the points $1(12)^\infty$ and $3(34)^\infty$, and the line between them, we obtain a 2-IFS that generates this triple of lines.

Another triple of lines is obtained by replacing $1(12)^\infty$ and $3(34)^\infty$ with $2(21)^\infty$ and $4(43)^\infty$, but not with $2(12)^\infty$ and $4(34)^\infty$ or with $1(21)^\infty$ and $3(43)^\infty$, because, for example, $2(12)^\infty = (21)^\infty$. With these and the analogous constructions using horizontal lines, we find four families of three vertical or horizontal lines whose endpoints constitute 2-cycles and the image of one point of the 2-cycle.

Example 8.3 *Endpoints consisting of a 2-cycle and a fixed point.*

Consider the 2-cycles $\{(12)^\infty, (21)^\infty\}$ and $\{(34)^\infty, (43)^\infty\}$, and the fixed points 1^∞ and 3^∞. Connect these with vertical line segments and construct the 2-IFS with allowed triples given by the addresses of the 3-squares the lines cross. The attractor of this 2-IFS is these three vertical line segments. See the left side of Fig. 16. Replacing the fixed points 1^∞ and 3^∞ with 2^∞ and 4^∞ gives another family of three vertical lines; the analogous constructions with horizontal lines give two families of three horizontal lines.

The attractor of this 2-IFS is the union of the attractors of two 2-IFS, one attractor being the line between 1^∞ and 3^∞, the other two lines between the 2-cycles $\{(12)^\infty, (21)^\infty\}$ and $\{(34)^\infty, (43)^\infty\}$. The first of these has allowed 3-tuples

$$R_1 = \{111, 113, 131, 133, 311, 313, 331, 333\}$$

Generating Euclidean Structures Using IFS with Memory   345

and the second

$$R_2 = \{121, 123, 141, 143, 321, 323, 341, 343, 212, 214, 232, 234, 412, 414, 432, 434\}.$$

The initial and terminal 2-tuples of R_1 are $11, 13, 31, 33$, those of R_2 are $12, 21, 23, 32, 14, 41, 34, 43$. These have no common elements, so the edge transition graphs are disconnected, giving another example of Prop 2.1.

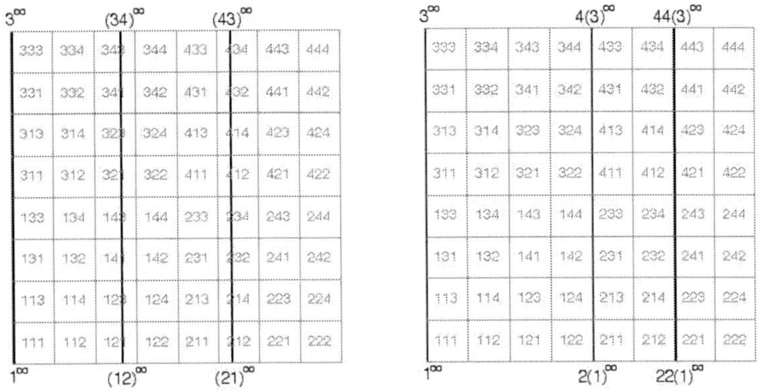

Figure 16. Left: a fixed point and a 2-cycle. Right: a fixed point and two images.

Example 8.4 *Endpoints consisting of a fixed point and two images.*

Consider the fixed points 1^∞ and 3^∞, and their images $2(1)^\infty$, $22(1)^\infty$, $4(3)^\infty$, and $44(3)^\infty$. Connect these with vertical line segments and construct the 2-IFS with allowed triples given by the addresses of the 3-squares the lines cross. The attractor of this 2-IFS is these three vertical line segments. See the right side of Fig. 16. The obvious left-right reflection, and analogous patterns of horizontal lines give a total of four families of three lines.

This example raises a natural question: can a 2-IFS generate four vertical lines, with fixed points and three of their images as endpoints? So far, all the families of lines generated by 2-IFS consist of three or fewer lines, so perhaps this construction is impossible, but why? The next example gives the answer.

Example 8.5 *A fixed point and three images does not work as endpoints for a 2-IFS.*

Begin with the fixed points 1^∞ and 3^∞ and their images $2(1)^\infty$, $22(1)^\infty$, $222(1)^\infty$, $4(3)^\infty$, $44(3)^\infty$, and $444(3)^\infty$. Derived from the addresses of the occupied 3-squares, the transition matrix is

$$\begin{bmatrix} 1 & 0 & 1 & 0 \\ 1 & 1 & 1 & 1 \\ 1 & 0 & 1 & 0 \\ 1 & 1 & 1 & 1 \end{bmatrix} \begin{bmatrix} 0 & 0 & 0 & 0 \\ 0 & 1 & 0 & 1 \\ 0 & 0 & 0 & 0 \\ 0 & 1 & 0 & 1 \end{bmatrix} \begin{bmatrix} 1 & 0 & 1 & 0 \\ 1 & 1 & 1 & 1 \\ 1 & 0 & 1 & 0 \\ 1 & 1 & 1 & 1 \end{bmatrix} \begin{bmatrix} 0 & 0 & 0 & 0 \\ 0 & 1 & 0 & 1 \\ 0 & 0 & 0 & 0 \\ 0 & 1 & 0 & 1 \end{bmatrix}.$$

As the left side of Fig. 17 shows, the attractor of this 2-IFS does not consist of only four vertical lines. The trouble here is that to allow the point $444(3)^\infty$, the triple 444 must be allowed. In a 2-IFS, allowing 444 necessarily allows 4^n for all n, and the point $444(3)^\infty$

gives rise to a sequence of points $(4)^n(3)^\infty$ converging to the fixed point 4^∞. Analogous arguments for other addresses on the line between $222(1)^\infty$ and $444(3)^\infty$ give an infinite sequence of vertical lines approaching the line between 2^∞ and 4^∞. Perhaps a fixed point and three images can be obtained with a 3-IFS.

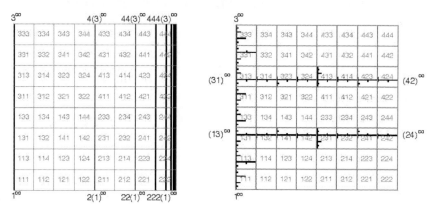

Figure 17. Left: a fixed point and three images. Right: a pair of fixed points and two perpendicular 2-cycle.

Example 8.6 *A vertical line with fixed point endpoints and two horizontal lines with 2-cycle endpoints does not work.*

In Ex 8.3 we produced three vertical lines, one between a pair of fixed points and two between a pair of 2-cycles. Can a 2-IFS produce a vertical line between a pair of fixed points, and two horizontal lines between a pair of 2-cycles? On the right side of Fig. 17 we see the answer is "no." More precisely, "yes," we can produce these three lines, but not without a cascade of infinitely many smaller segments. To understand this complication, note that the square with address 31 contains a line segment L between $(0, 2/3)$ and $(1/4, 2/3)$, the portion of the segment with endpoints $(31)^\infty$ and $(42)^\infty$ lying in that square. Then the square with address 331 contains the horizontal line segment $T_3(L)$. By this and similar arguments we obtain an infinite collection of decorations on the three length 1 segments with which we started.

Are these all the triples of lines generated by 2-IFS? Not quite: Ex 7.2 can be extended for 2-IFS.

Example 8.7 *Endpoints consisting of a 2-cycle and different fixed points.*

Take the pair of 2-cycles $\{(12)^\infty, (21)^\infty\}$ and $\{(34)^\infty, (43)^\infty\}$, and the fixed points 1^∞ and 4^∞. This differs from Ex 8.3 because there the fixed points could be joined by a vertical line, while here they cannot. Draw the line segments between 1^∞ and $(34)^\infty$, between $(12)^\infty$ and $(43)^\infty$, and between $(21)^\infty$ and 4^∞. The 2-IFS given by the length 3 addresses through which these lines pass has attractor consisting of these lines. See the left side of Fig. 18.

This suggests one more possibility, a modification of Ex 8.1.

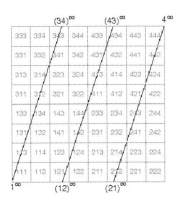 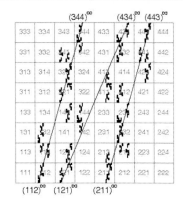

Figure 18. Left: a 2-cycle and two fixed points not on a vertical line. Right: an attempt using two 3-cycles.

Example 8.8 *Endpoints consisting of two 3-cycles that cannot be connected by vertical or horizontal segments.*

Take the 3-cycle $\{(112)^\infty, (121)^\infty, (211)^\infty\}$ on the bottom edge of the unit square, and the 3-cycle $\{(344)^\infty, (434)^\infty, (443)^\infty\}$ on the top edge of the unit square. Connect these with lines as indicated on the right side of Fig. 18 and as usual construct the 2-IFS using the occupied length 3 addresses. The attractor of this 2-IFS is superimposed on the lines, on the right side of Fig. 18. The attractor is not three lines. Why not?

A good first check that we have found the 2-IFS with the right allowed triples is to be sure each 3-square containing part of a line also contains points of the attractor. This is true, except for square 124. Although we have allowed that triple, no point of the attractor lies in that square because no point of the attractor lies in the square 24, and $S_{124} = T_1(S_{24})$.

That empty square is not the only problem with this example. In all the other cases, except Ex 8.6, which will turn out to be a special case of the issue we are going to raise now, we observe that all the line segments are parallel. This is essential because these families of lines are self-generating: each line in the family is composed of images of other lines of the family under the application of similarity transformations, which preserve slopes.

Consider the line segments of the left side of Fig. 18, for example. Say L_a is the line segment between 1^∞ and $(34)^\infty$, L_b between $(12)^\infty$ and $(43)^\infty$, and L_c between $(21)^\infty$ and 4^∞. Referring to the left side of Fig. 19, we see $T_3(L_b)$ is the part of L_a between $(1/6, 1/2)$ and $(1/3, 1)$, and also $T_2(L_b)$ is the part of L_c between $(2/3, 0)$ and $(5/6, 1/2)$. The remainder of L_a and L_c are generated by these parts of L_a and L_c. Referring to the midle of Fig. 19, $T_1(T_3(L_b))$ is the part of L_a between $(1/12, 1/4)$ and $(1/6, 1/2)$, and repeated applications of T_1 will fill in the rest of L_a. Similarly for the rest of L_c. Finally, referring to the right side of Fig. 19, we see $T_1(L_c)$ is the part of L_b between $(1/3, 0)$ and $(1/2, 1/2)$, and $T_4(L_a)$ is the part of L_b between $(1/2, 1/2)$ and $(2/3, 1)$. That is, each line of this family is obtained from the others by application of (slope-preserving) similarities, so all lines in a family must have the same slope.

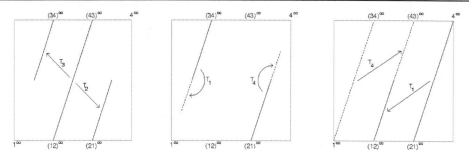

Figure 19. Self-generation of this family of three line segments.

9 General Conditions for Generating Lines

From the examples in Sections 6, 7, and 8, we are able to generalize and find conditions under which an n-IFS has attractor consisting of a finite collection of lines. We begin by describing n-IFS generating families of vertical lines. The situation for horizontal lines is similar. A few additional considerations are necessary to handle families of slanted lines.

The endpoints of these lines belong to cycles, or are images of cycle points, along an edge of the unit square. As a first step, we need to generate a given cycle, and nothing else, with an n-IFS using the T_i generating that edge of the unit square. An example will clarify the issue. Consider the 4-cycle

$$\{(1122)^\infty, (1221)^\infty, (2211)^\infty, (2112)^\infty\}.$$

For brevity we refer to this as the 4-cycle $(1122)^\infty$, because the addresses of al four points in the cycle can be read from this. To generate this by a 2-IFS, we must allow the triples

$$112, \quad 122, \quad 221, \quad 211 \tag{10}$$

obtained by taking the first through third entries on the left of $112211221122\ldots$, followed by the second through fourth, third through fifth, and fourth through sixth.

To see how this generates the 4-cycle, begin with 211, for example. Which transformation can be applied next? Keeping in mind the address order is the order of composition, the next transformation is applied on the left. We cannot apply T_1 because this would give 1211, and the triple 121 is not allowed. Applying T_2 gives 2211, acceptable because both triples 221 and 211 are allowed. Continuing, next must be T_1, giving 12211. Next is 112211, then 2112211, then 22112211, and so on. That is, allowing the triples (10), the 2-IFS generates the cycle $(1122)^\infty$.

Can we generate the 4-cycle $(1122)^\infty$ with a 1-IFS? Yes, but only with some complications: 1-IFS are determined by allowed pairs and generating this 4-cycle requires allowing all four pairs 11, 12, 21, and 22. Allowing the pair 11 in a 1-IFS generates the fixed point $(1)^\infty$, for example. In fact, by allowing all four pairs we obtain the 0-IFS $\{T_1, T_2\}$, which has attractor the unit interval on the x-axis.

Investigating the 1- and 2-squares reveals the reason. The points of the 4-cycle have coordinates

$$(1/5, 0), (2/5, 0), (3/5, 0), (4/5, 0),$$

Generating Euclidean Structures Using IFS with Memory   349

obtained by solving $T_1(T_1(T_2(T_2(x, y)))) = (x, y)$, $T_1(T_2(T_2(T_1(x, y)))) = (x, y)$, and so on. Each point of the 4-cycle lies in a distinct 2-square, but pairs of points of this 4-cycle lie in the same 1-square. To generalize this observation, we define a minimal q-cycle for an n-IFS.

Definition 9.1 *A q-cycle $(i_q \ldots i_1)^\infty$ is* minimal *for an n-IFS if that q-cycle is allowed for an n-IFS, but no subcycle, a string of the form $(i_a \ldots i_b)^\infty$, $q \geq a \geq b \geq 1$ with $a - b < q - 1$ is allowed for that n-IFS.*

Example 9.1 *A minimal 4-cycle for a 2-IFS.*

We have seen the 4-cycle $(1122)^\infty$ is allowed for the 2-IFS with allowed triples (10). We show this is a minimal 4-cycle for this 2-IFS. The length 1 subcycles are 1^∞ and 2^∞. For these to be allowed in this 2-IFS, the triples 111 and 222 would have to be allowed. The length 2 subcycles are $(11)^\infty$, $(12)^\infty$, $(21)^\infty$, and $(22)^\infty$. The first requires the triple 111, the second and third require 121 and 212, the fourth requires 222. The length 3 subcycles are $(112)^\infty$, $(122)^\infty$, $(221)^\infty$, and $(211)^\infty$. The first and fourth require the triple 121, the second and third require the triple 212. No subcycle of $(1122)^\infty$ is allowed in this 2-IFS, so this 4-cycle is minimal for this 2-IFS.

Because we are interested first in generating the endpoints of the line segments, we consider n-IFS with the transformations T_1 and T_2. Endpoints along other edges of the unit square are handled similarly, using different pairs of the T_i.

Proposition 9.1 *A q-cycle is minimal for an n-IFS if and only if no points of the cycle lie in the same n-square.*

Proof. Suppose $\{x_1, \ldots, x_q\}$ is a q-cycle, and denote the addresses of these cycle points by

$$x_1 = (a_{1,1}a_{1,2} \ldots a_{1,q})^\infty,$$
$$x_2 = (a_{2,1}a_{2,2} \ldots a_{2,q})^\infty,$$
$$\ldots \ldots \ldots \ldots \ldots \ldots \ldots \ldots \quad (11)$$
$$x_q = (a_{q,1}a_{q,2} \ldots a_{q,q})^\infty.$$

The cycle is the result of application of the appropriate transformations

$$x_2 = T_i(x_1), x_3 = T_j(x_2), \ldots$$

In terms of addresses, this is realized by inverse shift maps, with this effect of length q addresses

$$a_{2,1}a_{2,2} \ldots a_{2,q} = T_i(a_{1,1}a_{1,2} \ldots a_{1,q}) = ia_{1,1}a_{1,2} \ldots a_{1,q-1},$$
$$a_{3,1}a_{3,2} \ldots a_{3,q} = T_j(a_{2,1}a_{2,2} \ldots a_{2,q}) = ja_{2,1}a_{2,2} \ldots a_{2,q-1},$$
$$\ldots \ldots \ldots \ldots \ldots \ldots \ldots \ldots \ldots \ldots \ldots \ldots \ldots$$

That is, we have

$$a_{1,1} = a_{2,2}, \quad a_{1,2} = a_{2,3}, \quad a_{1,3} = a_{2,4}, \quad \ldots \quad (12)$$
$$a_{2,1} = a_{3,2}, \quad a_{2,2} = a_{3,3}, \quad a_{2,3} = a_{3,4}, \quad \ldots$$
$$\ldots \ldots \ldots \ldots \ldots \ldots \ldots \ldots \ldots \ldots \ldots \ldots$$

Note also the effect of the transformations $x_1 \to x_2 \to x_3 \to \cdots$ on the left-most address digits is

$$a_{1,1} \to a_{2,1} \to a_{3,1} \to \cdots \tag{13}$$

Now for the purpose of contradiction, suppose the q-cycle is minimal for an n-IFS, and that two points, x_i and x_j, of the cycle lie in the same n-square. The n-squares containing x_i and x_j have addresses

$$x_i : a_{i,1}a_{i,2}\ldots a_{i,n},$$
$$x_j : a_{j,1}a_{j,2}\ldots a_{j,n}$$

and $a_{i,1} = a_{j,1}$ because x_i and x_j belong to the same n-square. Say $j > i$. Then considering the first digits of the addresses and (13), we find

$$a_{i,1} \to a_{i+1,1} \to \cdots \to a_{j-1,1} \to a_{j,1} = a_{i,1},$$

where the equality follows from the assumption that x_i and x_j lie in the same n-square. That is, the points with addresses

$$\left(a_{i,1}a_{i+1,1}\cdots a_{j-2,1}a_{j-1,1}\right)^\infty,$$
$$\left(a_{i+1,1}a_{i+2,1}\cdots a_{j-1,1}a_{i,1}\right)^\infty,$$
$$\ldots\ldots\ldots\ldots\ldots\ldots\ldots\ldots\ldots$$
$$\left(a_{j-1,1}a_{i,1}\cdots a_{j-3,1}a_{j-2,1}\right)^\infty$$

constitute a $(j-i)$-cycle. Recalling that x_i and x_j belong to the q-cycle (11), because $j-i < q$, this contradicts the minimality of the q-cycle.

Now suppose each point x_i of a q-cycle lies in a distinct n-square. If this q-cycle is not minimal, then the allowed $(n+1)$-tuples give rise to a p-cycle for some $p < q$.

For the q-cycle, from (13) we have

$$a_{1,1} \to a_{2,1} \to \cdots \to a_{q,1} \to a_{1,1}.$$

We make the simplifying assumption that the p-cycle begins with $a_{1,1}$ and has the form

$$a_{1,1} \to a_{2,1} \to \cdots \to a_{p,1} \to a_{p+1,1} = a_{1,1}.$$

The more general situation simply involves more intricate bookkeeping. We know x_1 and x_{p+1} lie in the n-squares with addresses

$$a_{1,1}a_{1,2}\ldots a_{1,n} \quad \text{and} \quad a_{p+1,1}a_{p+1,2}\ldots a_{p+1,n}.$$

Also, we know $a_{p+1,1} = a_{1,1}$. We want to show

$$a_{p+1,2} = a_{1,2}, \quad a_{p+1,3} = a_{1,3}, \quad \ldots, \quad a_{p+1,n} = a_{1,n}$$

and so x_1 and x_{p+1} lie in the same n-square.

From (12) we know

$$a_{1,2} = a_{2,3} = a_{3,4} = \cdots = a_{p+1,p+2}. \tag{14}$$

Why should $a_{p+1,p+2} = a_{p+1,2}$? Recall the point with address $a_{p+1,1}a_{p+1,2}a_{p+1,3}\cdots$ belongs to a p-cycle, so

$$a_{p+1,1} = a_{p+1,p+1}, a_{p+1,2} = a_{p+1,p+2}, \dots. \tag{15}$$

Combining (14) and the second equality of (15), we obtain

$$a_{1,2} = a_{p+1,p+2} = a_{p+1,2}.$$

That is, x_1 and x_{p+1} share their first two address digits. Continuing, we find that x_1 and x_{p+1} occupy the same n-square, contradicting the hypothesis. Consequently, the q-cycle is minimal for this n-IFS. $\quad\square$

Not surprisingly, minimality is inherited for longer memory.

Corollary 9.1 *If a cycle is minimal with respect to n-step memory, then that cycle is minimal for all $m \geq n$.*

Proof. If no points of the cycle lie in the same n-square, then certainly no points of the cycle lie in the same m-square, for all $m \geq n$. $\quad\square$

Next, we find the set of lengths of n-IFS minimal cycles along each edge of the unit square.

Corollary 9.2 *For every k, $1 \leq k \leq 2^n$, there is a minimal n-IFS k-cycle along each edge of the unit square.*

Proof. Each edge of the unit square is determined by two of the T_i, so the number of n-squares along the edge is 2^n. By Prop 9.1 the maximum length of an n-IFS minimal cycle is 2^n.

We focus on the edge determined by T_1 and T_2. For every k, $1 \leq k \leq 2^n$, we want to construct a k-cycle with each point of the cycle belonging to a distinct n-square; that is, with the initial length n segment of the address of each cycle point being different from the initial length n segment of every other cycle point. We consider two cases, $1 \leq k \leq n$ and $n < k \leq 2^n$.

For $1 \leq k \leq n$, consider $x_1 = (21\ldots1)^\infty$, with $k-1$ consecutive 1s. Then certainly

$$x_1 = (21\ldots1)^\infty, x_2 = (121\ldots1)^\infty, \ldots, x_k = (1\ldots12)^\infty,$$

where each bracketed expression has length k, are points of a k-cycle, no two of which lie in the same n-square because no two cycle point addresses have the same initial length n segments.

For $n < k \leq 2^n$, we want a k-cycle for which the address of each point has a different inital length n segment. The case $k = 2^n$ is the classical binary order n de Bruijn sequence (see [24]), that is, a sequence of length 2^n that contains every length n string as a subsequence exactly once. This is done by exhibiting an Eulerian cycle in the n-IFS edge transition graph, also called a de Bruijn graph, a fragment of which is shown in Fig. 20. The full $n = 3$ de Bruijn graph is shown in Fig. 10.3 of [1].

For $n \leq k < 2^n$, a k-cycle each point of which has a unique initial length n segment can be obtained by *cutting down* a de Bruijn cycle. Diaconis and Graham [9] refer to this as Babai's Cutting Down Lemma. By this lemma, any length 2^n de Bruijn cycle can be cut down to a cycle of length k, for any k, $n \leq k \leq 2^n - 1$, every element of which has a unique initial length n segment. □

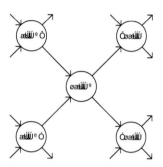

Figure 20. A portion of a de Bruijn graph.

Now we can begin our general results for families of vertical lines generated by n-IFS.

Theorem 9.1 *Given any cycle C along the bottom edge of the unit square, for large enough n there is an n-IFS with attractor the family of vertical lines that intersect the bottom of the unit square in the points of C.*

Proof. Write $C = \{(x_1, 0), (x_2, 0), \ldots, (x_k, 0)\}$. For large enough n, the n-squares along the bottom of the unit square are small enough that each $(x_i, 0)$ lies in a distinct n-square. Then by Prop 9.1 this cycle is minimal for an n-IFS. Let R_1 denote the allowed $(n+1)$-tuples for the n-IFS with attractor C. See the left side of Fig. 15 for an example. Let R_2 denote the set of $(n+1)$-tuples formed from the elements of R_1, replacing each T_1 with T_3, and each T_2 with T_4. Then the n-IFS with allowed $(n+1)$-tuples R_2 has attractor the points $\{(x_1, 1), (x_2, 1), \ldots, (x_k, 1)\}$. For each i, let L_i be the line segment with endpoints $(x_i, 0)$ and $(x_i, 1)$.

What are the addresses of the $(n+1)$-squares that intersect L_i? Suppose $(x_i, 0)$ lies in the $(n+1)$-square $a_1 \ldots a_n a_{n+1}$, where of course each a_j is 1 or 2. Let $\widehat{a}_j = 3$ if $a_j = 1$ and $\widehat{a}_j = 4$ if $a_j = 2$. Then L_i intersects the 2^{n+1} $(n+1)$-squares with addresses

$$a_1 \cdots a_n a_{n+1}, \; a_1 \cdots a_n \widehat{a}_{n+1}, \; a_1 \cdots \widehat{a}_n a_{n+1}, \; a_1 \cdots \widehat{a}_n \widehat{a}_{n+1}, \; \ldots, \; \widehat{a}_1 \cdots \widehat{a}_n \widehat{a}_{n+1}.$$

See the left side of Fig. 15 for an illustration.

Now consider the n-IFS with allowed $(n+1)$-tuples the addresses of the $(n+1)$-squares that intersect the L_i. We show this n-IFS has attractor the set of lines $\{L_1, \ldots, L_k\}$. Without loss of generality, suppose that $(x_i, 0)$ and $(x_j, 0)$ are consecutive points in the cycle C. Then for $k = 1$ or 2, $T_k(x_i, 0) = (x_j, 0)$. In terms of addresses,

$$a_1 \cdots a_{n+1} \to k a_1 \cdots a_n = b_1 b_2 \ldots b_{n+1}, \qquad (16)$$

where $(x_j, 0)$ lies in the $(n+1)$-square with address $b_1 b_2 \ldots b_{n+1}$. For each line L_i let

$$L_i^- = L_i \cap ([0,1] \times [0, 1/2]) \quad \text{and} \quad L_i^+ = L_i \cap ([0,1] \times [1/2, 1]).$$

We show that

$$T_k(L_i) = L_j^- \quad \text{and} \quad T_{\widehat{k}}(L_i) = L_j^+. \tag{17}$$

This follows from two observations. First, (16) implies that T_k takes each $(n+1)$-square intersecting L_i to an $(n+2)$-square intersecting L_j^-, and $T_{\widehat{k}}$ takes each $(n+1)$-square intersecting L_i to an $(n+2)$-square intersecting L_j^+. Second, all the transformations (1) take vertical lines to vertical lines, T_k takes the lower endpoint of L_i to the lower endpoint of L_j, and $T_{\widehat{k}}$ takes the upper endpoint of L_i to the upper endpoint of L_j. Then this set $\mathcal{L}$ of vertical lines is invariant under the n-IFS with these allowed $(n+1)$-tuples. This n-IFS generates the entire cycle C along the bottom edge of the square, so no proper subset of $\mathcal{L}$ is invariant under this n-IFS. Being the minimal invariant subset, $\mathcal{L}$ is the attractor of the n-IFS. □

Certainly, the corresponding theorem holds for families of horizontal lines.

Examples 8.2 and 8.4 show that vertical lines also can be generated over cycles and the images of some cycle points.

Theorem 9.2 *Given any finite set D consisting of a cycle C and images of some cycle points, along the bottom of the unit square, for large enough n there is an n-IFS with attractor the family of vertical lines that intersect the bottom of the unit square in the points of D.*

Proof. The proof is similar to that of Thm 9.1, but with these changes. First, take n large enough so that every point of D, not just every point of the cycle, lies in a distinct n-square.

The second point is best understood by an illustration. Suppose the cycle is $\{x_1, \ldots, x_k\}$ and the remaining points of D are $x_{k+1} = T_i(x_k)$ and $x_{k+2} = T_j(x_{k+1})$. Of course, there could be more points in this family of images of x_k, and there could be images of more points of the cycle. Under our assumption, and continuing with the notation of the proof of Thm 9.1,

$$L_{k+1} = T_i(L_k) \cup T_{\widehat{i}}(L_k) \quad \text{and} \quad L_{k+2} = T_j(L_{k+1}) \cup T_{\widehat{j}}(L_{k+1}).$$

Then the allowed $(n+1)$-tuples for this n-IFS are the addresses of the $(n+1)$-squares that intersect $L_1, \ldots, L_{k+2}$. Because neither image of x_{k+2} belongs to D, no transformations are applied to L_{k+2} to generate addtional allowed $(n+1)$-tuples. □

Next, applying Prop 2.1, unions of cycles and some of their images can be the lower endpoints of vertical lines comprising the attractor of an n-IFS, so long as the sets of allowed $(n+1)$-tuples determining these cycles and their images have disconnected edge transition graphs.

Now we see that all attractors consisting of families of vertical lines arise in this way. Before we begin this, suppose we try to build a family of vertical lines with endpoints not containing any cycle. If Thm 9.3 is correct, this is impossible. What goes wrong?

Example 9.2 *Trying to build a family of vertical lines without a cycle.*

On the left side of Fig. 21 we see the 3-squares in which we hope to construct vertical lines that are the attractor of a 2-IFS. The allowed triples of this 2-IFS must be the addresses of

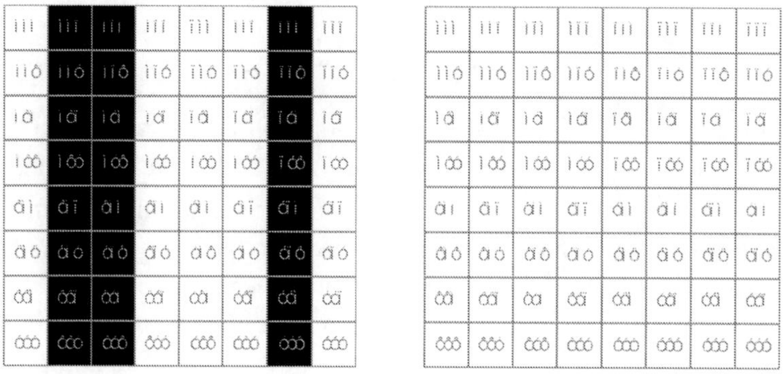

Figure 21. Left: Trying to build an attractor without a cycle. Right: The second iterate of this 2-IFS.

these 3-squares. To see that the lower endpoints of any lines contained in these 3-squares contain no cycle, first observe that neither fixed point 1^∞ or 2^∞ belong to the 3-squares 112, 121, or 221. The 2-cycle is $\{(12)^\infty, (21)^\infty\}$ lies in the 3-squares 121 and 212, so the 2-cycle is not a subset of the line endpoints. The 3-cycles are $\{(112)^\infty, (121)^\infty, (211)^\infty\}$ and $\{(122)^\infty, (212)^\infty, (221)^\infty\}$, neither of which is contained in the 3-squares 112, 121, and 221. Consequently, the endpoints of any triple of lines contained in the 3-squares shaded in the left side of Fig. 21 does not contain a cycle. In fact, the allowed triples contain no cycle, so by Prop 6.1 this 2-IFS has empty attractor. The second iterate of this 2-IFS transforms the set of filled squares on the left image of Fig. 21 into the right image of Fig. 21.

Theorem 9.3 *For any n-IFS attractor consisting of a family of vertical lines, those lines intersect the bottom of the unit square in a finite collection of points that consist of minimal cycles and images of cycle points. Moreover, every point of this finite collection must lie in a distinct n-square.*

Proof Denote by C the set of points at which the vertical lines intersect the bottom edge of the unit square.

First we observe that if C is a finite set, then every point of C belongs to a cycle or is the image of a cycle point. This follows from a slight elaboration of the proof of Prop 6.1. Given a point $a \in C$, the address of a is an infinite string of 1s and 2s, because a string of length k determines a square of side length 2^{-k}, not a single point. What form can an infinite address take? First, any finite initial string can be ignored: for example, both $(12)^\infty$ and $(12)^\infty 111212$ are addresses of the point $(1/3, 0)$. Second, recall that if the infinite string $\sigma = \ldots i_k i_{k-1} \ldots i_1$ is the address of the point (x, y), then $j_2 j_1 \sigma$ is the address of the point $T_{j_2}(T_{j_1}(x, y))$. Suppose an infinite address τ cannot be written as $\tau = \alpha \beta \gamma$ for any finite strings α and γ, and for the infinite cyclic string β; i.e., $\beta = (i_k \ldots i_1)^\infty$ for some k. Then τ must contain arbitrarily long strings of 1s or of 2s, say 1s. If τ is the address of the point $(x, 0)$, then the infinite collection of points $(x/2, 0), (x/4, 0), (x/8, 0), \ldots$ must also belong to C. Consequently, every point of C has an address that is cyclic, or the image of a cycle point.

To show the cycles are minimal, it suffices to show every point of C lies in a distinct n-square. Suppose x_1 and x_2 are points of C that lie in the same n-square. Suppose x_1

is a point of a k-cycle and has address $(i_k \ldots i_1)^\infty$. Then x_1 is the unique fixed point of the composition $T_{i_k} \circ \cdots \circ T_{i_1}$. What is the effect of $T_{i_k} \circ \cdots \circ T_{i_1}$ on x_2? First, note that $T_{i_k} \circ \cdots \circ T_{i_1}(x_2) \neq x_2$ because the fixed point x_1 is unique. Second, $T_{i_k} \circ \cdots \circ T_{i_1}(x_2) \neq x_1$ because this composition is a linear contraction, with contraction factor $1/2^n$, so the only point it takes to x_1 is x_1 itself. On the other hand,

$$|x_1 - T_{i_k} \circ \cdots \circ T_{i_1}(x_2)| = |T_{i_k} \circ \cdots \circ T_{i_1}(x_1) - T_{i_k} \circ \cdots \circ T_{i_1}(x_2)| = (1/2^n)|x_1 - x_2|$$

and so for $m = 1, 2, \ldots$, the points $(T_{i_k} \circ \cdots \circ T_{i_1})^m(x_2)$ constitute an infinite sequence of distinct points of C, ever more closely approaching x_1. This contradicts the finiteness of C. $\qquad\square$

(In the preceding proof, note that if x_1 and x_2 lie in different n-squares, then because both belong to the cycle C, some finite composition of the T_i takes x_2 to x_1 and so the argument generating the infinite sequence of points approaching x_1 cannot be applied.)

In Fig. 13 and the left side of 18 we see 1- and 2-IFS attractors consisting of slanted lines. The first are lines between fixed points and their images. This is different from Fig. 11 because there the fixed points, 1^∞ and 2^∞, are the endpoints of one of the horizontal lines in this attractor; in Fig. 13 the fixed points, 2^∞ and 3^∞, are not endpoints of a line in this attractor. Of course, the specification of allowed 2-tuples determines where the lines lie.

On the left side of Fig. 18, on top and bottom of the unit square, the endpoints comprise a fixed point and a 2-cycle, but the fixed points are not endpoints of a line in the attractor. So far, to construct n-IFS with attractors consisting of slanted lines, allow two of the fixed points, on a diagonal of the unit square, and an image of the fixed point on both the upper and lower edges of the unit square, or corresponding cycles on both the upper and lower edges.

Generalizing these observations, we see families of slanted lines are formed, for example, by cycles $\{(x_1, 0), \ldots, (x_q, 0)\}$ and $\{(x_1, 1), \ldots, (x_q, 1)\}$, along with the fixed points $(0, 0)$ and $(1, 1)$, with L_1 connecting $(0, 0)$ and $(x_1, 1)$, L_2 connecting $(x_1, 0)$ and $(x_2, 1), \ldots, L_{q+1}$ connecting $(x_q, 0)$ and $(1, 1)$. In addition to these cycles, images of some of these lines can belong to attractors of n-IFS. We need only take n large enough that no lines intersect the same $(n+1)$-square.

For instance, on the left side of Fig. 22 we have two images of fixed points on both the upper and lower edges. On the right side we have 2-cycles and the image of a cycle point, on both the upper and lower edges. But the 2-IFS determined by these allowed triples does not produce an attractor consisting of four lines. Why is this?

Recalling the proof of Prop 9.1, when we try to generate these four lines with a 2-cycle we encounter a problem because both 1^∞ and $1(12)^\infty$ lie in 2-square 11. A visual clue of this problem is that the part of the attractor in address 11 is repeated, scaled of course, in address 411. Using the 3-IFS with allowed 4-tuples the addesses of the 4-squares through which these lines pass, we obtain the attractor in Fig. 23.

Finally, following Ex 8.8, we see families of slanted lines of different slopes, or families of vertical or horizontal lines and slanted lines, cannot be the attractor of an n-IFS for any n. Are there other possibilities? For example, can we generate parallel lines using only some points of a cycle? For example, consider the 3-cycles on the top and bottom edges of the right side of Fig. 18. Trying to build two parallel lines, we select the lines

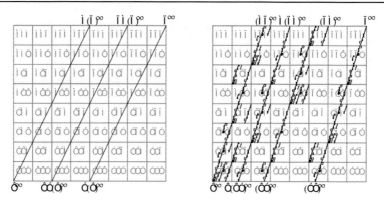

Figure 22. Left: Two slanted lines and the image of one, by a 2-IFS. Right: Three slanted lines and the image of one, attempted by a 2-IFS.

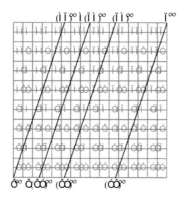

Figure 23. Three slanted lines and the image of one, by a 3-IFS.

between $(112)^\infty = (1/7, 0)$ and $(344)^\infty = (3/7, 1)$, and between $(211)^\infty = (4/7, 0)$ and $(443)^\infty = (6/7, 1)$. These lines pass through distinct 2-squares, so in principle we can generate these lines with a 2-IFS. On the left side of Fig. 24 we see the 3-squares through which these lines pass. Taking the addresses of these 3-squares for the allowed triples, on the right side of Fig. 24 we see the sixth iteration of this 2-IFS. The attractor is not a pair of parallel lines, but appears to be a Cantor set on each line. That this 2-IFS does not generate the full pair of lines is no surprise: two points from the 3-cycle cannot be generated alone, with a 2-IFS, or with a 3-IFS, or with an n-IFS for any $n \geq 2$. Thus any family of slanted lines must have endpoints consisting of complete cycles.

Now we have the ingredients to classify the families of slanted lines that can be attractors of n-IFS using the transformations (1).

Theorem 9.4 *The attractor of an n-IFS can consist of slanted lines, all of the same slope, with endpoints on opposite sides of the unit square. Two of these endpoints lie on a diagonal of the square, the others are images of fixed points, the points of a cycle, or some images of points of a cycle. The length of memory, n, must be taken large enough that no two lines pass through the same n-square.*

Proof. First, we have just seen that if the endpoints of a family of lines on an edge of the

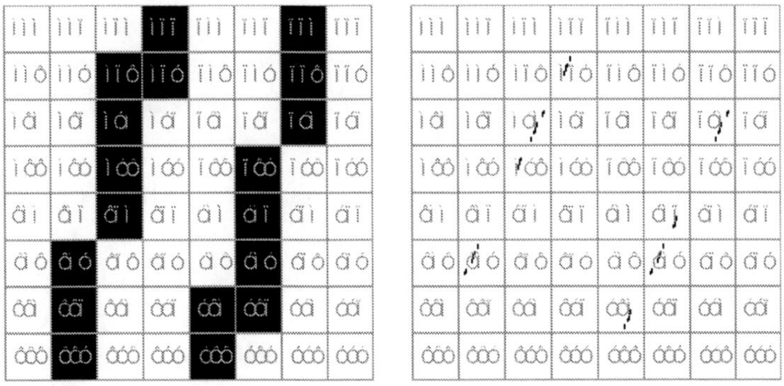

Generating Euclidean Structures Using IFS with Memory 357

Figure 24. Trying to build two parallel lines using two points of a 2-cycle. Left: first iteration. Right: sixth iteration.

unit square contain points of a cycle, then all points of that cycle must be endpoints. That is, there are no partial cycles.

Denote by $(x_1, 0), \ldots, (x_k, 0)$ and $(w_1, 1), \ldots, (w_k, 1)$ the line endpoints. First, suppose the x_i and the w_i constitute k-cycles, and no x_i or w_i is 0 or 1. Order these points so for $i = 1, \ldots, k$, x_i and w_i are endpoints of a line segment. The argument presented in Ex 8.8 shows why all lines have the same slopes, $1/|w_i - x_i|$ if the slopes are positive, $-1/|w_i - x_i|$ if the slopes are negative. Write the addresses of x_1 and w_1 as $(a_k \ldots a_1)^\infty$ and $(b_k \ldots b_1)^\infty$. Then the next points x_2 and w_2 have addresses $(a_1 a_k \ldots a_2)^\infty$ and $(b_1 b_k \ldots b_2)^\infty$. Suppose $a_1 = 1$ and $b_1 = 2$, where we are projecting the endpoints on the upper edge of the unit square to the lower edge. Then

$$|x_2 - w_2| = |T_1(x_1) - T_2(w_1)|$$
$$= |x_1/2 - (w_1/2 + 1/2)|. \tag{18}$$

Certainly, if the lines are parallel, $x_2 - w_2 = x_1 - x_1$. In particular, the signs of $x_2 - w_2$ and $x_1 - w_1$ must be the same. Say both are positive. Then Eq (18) becomes

$$x_1 - w_1 = 1$$

impossible if $0 < x_1 < 1$ and $0 < w_1 < 1$. Consequently, the assumption that $a_1 = 1$ and $b_1 = 2$ is wrong, and so $a_1 = b_1$. Continuing this way, we see that $a_2 = b_2, \ldots, a_k = b_k$. That is, the line segments with endpoints $(x_i, 0)$ and $(w_i, 1)$ are vertical, not slanted.

Consequently, if slanted lines are the attractor of an n-IFS with endpoints containing a cycle, then one set of endpoints must include a corner of the unit square. In order for the lines to have the same slopes, the other set of endpoints must include the diagonally opposite corner point.

As illustrated in Fig. 23, endpoints can consist of a cycle and images of cycle points, but they cannot consist of two cycles terminating families of line segments of different slopes, because each family of slanted lines must include a corner point, and if two lines of different slopes terminate at the same corner point, the problem illustrated in the right image of Fig. 9 shows the attractor will consist of fractals, not just line segments.

Finally, the condition that n be large enough that no two lines pass through the same n-square is familiar. $\qquad \square$

Summarizing, if the attractor of an n-IFS with transformations (1) consists of a finite collection of line segments with endpoints on opposite sides of the unit square, all the line segments are parallel, and the endpoints consist of cycle points and images of cycle points.

10 Classification of Topological Types of n-IFS Attractors

Knowing how to produce attractors consisting of finite collections of lines is very far from solving the topological type classification problem, but it is a first step. We do know at least one more thing, that the dimension of the attractor is useless as an assay of topological type. All the examples of Section 5 are made of 3 copies, each scaled by a factor of $r_i = 1/2$, and all satisfy Hutchinson's open set condition [16], so have Hausforff dimension d given by the Moran equation

$$r_1^d + r_2^d + r_3^d = 1 \tag{19}$$

that is, $d = \log(3)/\log(2)$. A parameter taking the same value for each member of a family is not a useful tool for distinguishing members of that family.

For 1-IFS using the transformations (1) and with transition matrix $M = [m_{ij}]$, the extension of the Moran equation is

$$\rho([r_i^d m_{ij}]) = 1, \tag{20}$$

where $\rho(N)$ is the spectral radius of the matrix N. Using the method developed in [7], the dimension of the attractor of any n-IFS can be computed using an analog of Eq (20). Nevertheless, from the observation about dimensions of 0-IFS, we expect dimension is too crude an invariant to capture the nuance of topological type.

However, we can make one simple observation: for IFS with memory using transformations (1), we can produce some topological types unknown in the examples of Sect. 5. The right sides of Figs 3, 4, and 5 show attractors having exactly 1, 4, and 16 loops; the left side of Fig. 25 has 3 loops. This last is obtained from the 2-IFS with the right attractor of Fig. 4 by removing the allowed triples 324, 342, 413, and 431. The non simply-connected examples of Sect. 5 all contain infinitely many loops. So far, we have been unable to find an example invalidating this conjecture.

Conjecture No n-IFS attractor consists of more than 1 connected component and infinitely many loops.

Figure 25. Left: a 2-IFS attractor with three loops. Center and right: Attractors of IFS using affinities instead of similarities.

In case the restriction to transformations (1) seems too rigid, we mention that once more general transformations are allowed, many more possibilities arise. This is the source of the

Generating Euclidean Structures Using IFS with Memory   359

title of [5]. Recall the Hausdorff distance $h(A, K)$ between compact subsets A and K of the plane is

$$h(A, K) = \inf\{\delta : A \subseteq K_\delta \text{ and } K \subseteq A_\delta\},$$

where

$$A_\delta = \{x : d(x, y) \leq \delta \text{ for some } y \in A\},$$

the δ-neighborhood of A. Then for any $\epsilon > 0$ and for any compact subset K of the plane, we can find an IFS with attractor A whose Hausdorff distance from K is less than ϵ. Roughly, when viewed at resolution ϵ, A and K are indistinguishible. For example, the center and right sets in Fig. 25 are generated by affinities of the form

$$T(x, y) = (x/2, 0) + (e, f) \quad \text{and} \quad T(x, y) = (0, y/2) + (e, f)$$

for appropriate translations e and f, 8 for the left side, 12 for the right. For this reason, we consider only IFS using the transformations (1).

Conclusion

We have classified the families of lines that are attractors of n-IFS by studying the cycles along the edges of the unit square, minimal for n-IFS. More complex structures may be understood by studying cycles internal to the square.

For example, on the left side of Fig. 26 we see the 4-cycle $(1243)^\infty$, consisting of points with addresses

$$(1243)^\infty, \ (2431)^\infty, \ (4312)^\infty, \text{ and } (3124)^\infty \tag{21}$$

These points are the attractor of the 2-IFS with allowed triples

$$124, \ 243, \ 431, \text{ and } 312 \tag{22}$$

the addresses of the 3-squares occupied by these cycle points.

On the right side of Fig. 26 we have drawn lines passing through these cycle points. Following the construction of Example 7.2, we draw the lines with endpoints $(1)^\infty$ and $2(4)^\infty$, $(2)^\infty$ and $4(3)^\infty$, $(4)^\infty$ and $3(1)^\infty$, and $(3)^\infty$ and $1(2)^\infty$.

Note, for example, the right endpoint of the first line is described as $2(4)^\infty$, not $4(2)^\infty$, although both are addresses of the point $(1, 1/2)$. The reason is that the square with address 244 is occupied, but 422 is not. The 2-IFS determined by the addresses of the 3-squares occupied by the lines in the right side of Fig. 26 allows $2(4)^\infty$ but not $4(2)^\infty$.

On the left side of Fig. 27 we see the attractor of the 2-IFS with allowed triples the addresses of the 3-squares occupied by the square with corners the 4-cycle points (21). That is, in addition to the allowed triples (22), we allow

$$231, \ 232, \ 412, \ 414, \ 324, \ 323, \ 143, \text{ and } 141$$

This 2-IFS does not fill in the entire square, but resembles a very sparse Cantor set on each side of the square. The reason for these gaps is clear: for example, the upper right corner of the 2-square 24 is empty, so the upper right corner of the 3-square 124 is empty.

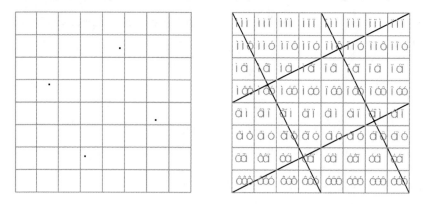

Figure 26. Left: Points of the 4-cycle $(1243)^\infty$. Right: These 4-cycle points, with some lines passing through them.

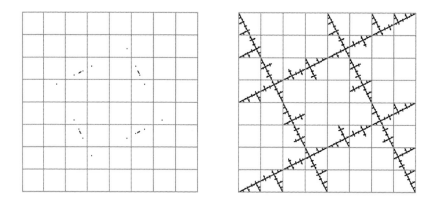

Figure 27. Left: The attractor of the 2-IFS of the square with corners the 4-cycle points. Right: The attractor of the 2-IFS determined by the lines on the right side of Fig. 26.

On the right side of Fig. 27 we see the attractor of the 2-IFS with allowed triples the addresses of all the 3-squares through which the lines of Fig. 26 pass. The reason for the decorations on these four lines comes from how the allowed compositions move some bits of lines to other lines. We can think of this attractor as built up from the 4-cycle (21), filling in the lines.

Of course, this attractor also is understood as that with set of allowed triples the union of that for the lines with endpoints $(1)^\infty$, $2(4)^\infty$, and $2(1)^\infty(4)^\infty$, and that for the lines with endpoints $(4)^\infty$, $1(2)^\infty$, and $4(3)^\infty$, $(2)^\infty$. Each pair of these lines contains the 4-cycle (21), so exactly how cycle information encodes n-IFS attractors is not clear, yet.

References

[1] J-P. Allouche, J. Shallit, *Automatic Sequences. Theory, Applications, Generalizations*, Cambridge Univ. Pr., Cambridge, 2003.

[2] D. Ashlock, J. Golden, *Chaos automata: iterated function systems with memory*, Physica D **181** (2003), 274–285.

[3] A. Barbé, *On a class of fractal matrices (III): limit structures and hierarchical iterated function systems* Int. J. Bifurcation and Chaos **5** (1995), 1119–1156.

[4] A. Barbé, F. von Haesler, G. Skordev, *Limit sets of restricted random substitutions*, Fractals **14** (2006), 37–47.

[5] M. Barnsley, *Fractals Everywhere*, 2nd ed. Academic Press, 1993.

[6] M. Barnsley, J. Elton, D. Hardin, *Recurrent iterated function systems*, Constructive Approximation **5** (1989), 3–31.

[7] R. Bedient, M. Frame, K. Gross, J. Lanski, B. Sullivan, *Higher block IFS 1: Memory reduction and dimension computation*, Fractals **18** (2010), 145–155.

[8] R. Bedient, M. Frame, K. Gross, J. Lanski, B. Sullivan, *Higher block IFS 2: Relations between IFS with different levels of memory*, to appear in Fractals.

[9] P. Diaconis, R. Graham, "Products of universal cycles," pgs 35 – 55 of *A Lifetime of Puzzles: A Collection of Puzzles in Honor of Martin Gardner's 90 th Birthday*, E. Demaine, M. Demaine, T. Rodgers, eds., A. K. Peters, 2008.

[10] G. Edgar, *Integral, Probabilty, and Fractal Measures*, Springer-Verlag, 1998.

[11] G. Edgar, R. Mauldin, *Multifractal decompositions of digraph recursive fractals*, Proc. Lond. Math. Soc. **65** (1992), 604–628.

[12] K. Falconer, *Fractal Geometry. Mathematical Foundations and Applications*, John Wiley, New York, 1990.

[13] M. Frame, J. Lanski, *When is a recurrent IFS attractor a standard IFS attractor?*, Fractals **7** (1999), 257–266.

[14] B. Hambly, S. Nyberg, *Finitely ramified graph-directed fractals, spectral asymptotics and the multidimensional renewal theorem*, Proc. Edinburgh Math. Soc. **46** (2003), 1–34.

[15] J. Hart, *Iterated function systems and recurrent iterated function systems*, Ph. D. thesis, Washington State University, Pullman, WA, 1996.

[16] J. Hutchinson, *Fractals and self-similarity*, Indiana Univ. Math. J. **30** (1981), 713–747.

[17] J. King, J. Geronimo, *Singularity spectrum for recurrent IFS attractors*, Nonlinearity **6** (1992), 337–348.

[18] J. Layman, T. Womack, *Linear Markov iterated function systems*, Computers & Graphics **14** (1990), 343–353.

[19] D. Lind, B. Marcus, *An Introduction to Symbolic Dynamics and Coding*, Cambridge Univ. Pr. 1995.

[20] P. Massopust, *Fractal functions, fractal surfaces, and wavelets*, Academic Press, 1994.

[21] L. Máté, *On full-like sofic shifts*, preprint.

[22] R. Mauldin, S. Graf, S. Williams, *Exact Hausdorff dimension in random recursive constructions*, Proc. Natl. Acad., Sci. USA **84** (1987), 3959–3961.

[23] R. Mauldin, S. Williams, *Hausdorff dimension in graph directed constructions*, Trans. Amer. Math. Soc. **309** (1988), 811–829.

[24] F. Mendivil, "Fractals, graphs, and fields," *Amer. Math. Monthly* **110** (2003), 503–515.

[25] H.-O. Peitgen, H. Jürgens, D. Saupe, *Chaos and Fractals. New Frontiers in Science*, 2nd ed, Springer-Verlag, 2004.

In: Classification and Application of Fractals
Editor: William L. Hagen

ISBN 978-1-61209-967-5
© 2012 Nova Science Publishers, Inc.

Chapter 11

TRACE THEOREMS ON SCALE IRREGULAR FRACTALS

Raffaela Capitanelli and Maria Agostina Vivaldi

Diparti mento di Scienze di Base e Applicate per l'Ingegneria,
Universita di Roma Sapienza, Roma, Italy

ABSTRACT

"Scale irregular fractals" are irregular objects that exhibit some fractal properties but do not satisfy any exact scaling relation. In this chapter, we state some trace results on "scale irregular fractals" and on prefractal structures approximating the limit objects. These results are crucial tools in the study of the asymptotic convergence of energy forms defined on prefractal structures approximating the "scale irregular fractals".

AMS Subject Classification: 46E35, 35A23, 28A80, 35J25.
Keywords: Trace, Lipschitz Boundary, Fractals, Boundary Value Problems.

1 Introduction

The mathematical theory of fractal bodies has been based mainly on self-similarity. Strict self-similarity, however, is a too stringent property to be realistic in physical applications so this has led to investigation of more general models which can be seen as deterministic or random mixtures of self-similar fractals. They are generated by an iteration procedure involving different families of Euclidean similarities. The procedure operates in a deterministic or random way: the jump from one family to another mimics the influence of the environment on the morphogenesis of the fractal structure. These irregular objects exhibit some fractal properties, are locally spatially homogeneous but do not satisfy any exact scaling relation. The limit object and some relevant analytic properties depend on the structural constants of the families and the asymptotic frequency of the occurrence of each family.

---

*E-mail address: raffaela.capitanelli@uniroma1.it
†E-mail address: vivaldi@dmmm.uniroma1.it

In this chapter, we state some trace results on "scale irregular fractals" and we focus our attention on two illustrative examples: "fluctuating Koch curves" and "mixtures of Sierpinski Gaskets". We also discuss trace results on prefractal structures approximating the limit mixture and, in particular, we investigate the sharp dependence of the constants appearing in the trace estimate on the step of the iteration procedure.

These results are crucial tools in the study of the asymptotic convergence of energy forms defined on prefractal structures approximating the limit mixture (see, for instance, [10], [11] and [12]). There is a huge literature about trace theorems in Lipschitz domains. We mention the pioneering contributions of E. Gagliardo [16], the books of J. Nečas [30] and P. Grisvard [17], and the more recent papers of F. Brezzi, G. Gilardi [8], D. Jerison, C. E. Kenig [18], Z. Ding [13]; for a more complete list, we refer to the bibliography quoted therein.

Their approach does not fit directly in our geometry where we are dealing with a increasing number of sides and graphs and we are interested in the sharp dependence of the constants appearing in the trace estimate in terms of the step of the iteration procedure.

In this work, either for the scale irregular fractals or for the prefractal structures approximating the limit fractal, we follow the approach that A. Jonsson and H. Wallin developed in the general framework of Besov spaces and d-sets ([21], [22], [33], see also [26]).

The main idea is that in order to obtain a restriction theorem on the boundary of a set it suffices to estimate from above the scaling of the measure with respect to the radius of the ball $B(x, r)$, that is, an estimate of the type $\mu(\partial\Omega \cap B(x, r)) \leq Cr^{\delta}$. We point out that this idea comes from D.R. Adams (see [2], [3], [4]) and we refer to the survey [24] for other comments and remarks.

The layout of the chapter is as follows. In the second section, we recall the definitions and the properties of the scale irregular fractals. In the third section, we prove trace theorems for the specific geometry of the prefractal and fractal sets. Theorem 3.2 deals with polygonal boundaries and it states – by an "elementary" proof– a restriction result with an estimate (weaker than estimate of Theorem 3.5) but sharper than the estimate that one can deduce in a straightforward way from the previous mentioned results (see, for instance, [18] and [13]). In Theorems 3.5 and 3.6, we prove sharp estimates by using delicate tools as the representation of the functions in terms of Bessel kernels and deep interpolation results due to Peetre (see [31]). Moreover, we also state a restriction result for the limite mixture (see Theorem 3.9). We note that the scale irregular fractal is not necessarily a d-set (actually, it verifies the weaker condition (7)) and it does not satisfy the assumptions of Theorem 1 in [19]. Finally, we think that the recent paper [20] of A. Jonsson in terms of atomic decompositions is not in the spirit of our approach. In the last section, we discuss some applications of the previous results: in particular, we recall some asymptotic results for energy forms defined on prefractal structures approximating the limit mixture.

2 Scale Irregular Fractals

In this section, we recall the definition of "scale irregular fractals" (see [7], [28], and [29]). Let $\mathcal{A}$ be a finite set. For $a \in \mathcal{A}$ let

$$\Psi^{(a)} = \{\psi_1^{(a)}, \ldots, \psi_{N_a}^{(a)}\} \tag{1}$$

be the family of contractive similitudes $\psi_i^{(a)} : \mathbb{R}^2 \to \mathbb{R}^2$, $i = 1, \ldots, N_a$ with contraction factor $\ell_a^{-1} \in (0, 1)$.

Let $\Xi = \mathcal{A}^\mathbb{N}$. We define for $\xi \in \Xi$ a left shift S on Ξ such that if $\xi = (\xi_1, \xi_2, \xi_3, \ldots)$, then $S\xi = (\xi_2, \xi_3, \ldots)$. For $\mathcal{O} \subset \mathbb{R}^2$, we set

$$\Phi^{(a)}(\mathcal{O}) = \bigcup_{i=1}^{N_a} \psi_i^{(a)}(\mathcal{O})$$

and

$$\Phi_n^{(\xi)}(\mathcal{O}) = \Phi^{(\xi_1)} \circ \cdots \circ \Phi^{(\xi_n)}(\mathcal{O}).$$

The fractal $F^{(\xi)}$ associated with the sequence ξ is defined by

$$F^{(\xi)} = \overline{\bigcup_{n=1}^{+\infty} \Phi_n^{(\xi)}(\Gamma)}$$

where Γ is a nonempty compact subset of $\mathbb{R}^2$. We remark that these fractals do not have any exact self-similarity, that is, there is no scaling factor which leaves the set invariant: however, the family $\{F^{(\xi)}, \xi \in \Xi\}$ satisfies the following relation

$$F^{(\xi)} = \Phi^{(\xi_1)}(F^{(S\xi)}). \tag{2}$$

Moreover, the spatial symmetry is preserved and the set $F^{(\xi)}$ is locally spatially homogeneous, that is, the volume measure $\mu^{(\xi)}$ on $F^{(\xi)}$ satisfies the following *locally spatially homogeneous condition* (3). Before describing this measure, we introduce some notations. For $\xi \in \Xi$, we set $i|n = (i_1, \ldots, i_n)$ and $\psi_{i|n} = \psi_{i_1}^{(\xi_1)} \circ \cdots \circ \psi_{i_n}^{(\xi_n)}$ and, for any set $\mathcal{O} \subset \mathbb{R}^2$, $\psi_{i|n}(\mathcal{O}) = \mathcal{O}^{i|n}$.

The volume measure $\mu^{(\xi)}$ is the unique Radon measure supported on $F^{(\xi)}$ such that

$$\mu^{(\xi)}(\psi_{i|n}(F^{(S^n\xi)})) = \frac{1}{N^{(\xi)}(n)} \tag{3}$$

(see Section 2 in [7]) where

$$N^{(\xi)}(n) = \prod_{i=1}^{n} N_{\xi_i}. \tag{4}$$

The fractal set $F^{(\xi)}$ and the volume measure $\mu^{(\xi)}$ depend on the structural constants of the families and the asymptotic frequency of the occurrence of each family. We denote by $h_a^{(\xi)}(n)$ the frequency of the occurrence of a in the finite sequence $\xi|n$, $n \geqslant 1$:

$$h_a^{(\xi)}(n) = \frac{1}{n} \sum_{i=1}^{n} \mathbf{1}_{\{\xi_i = a\}}, \quad a \in \mathcal{A}.$$

Let p_a be a probability distribution on $\mathcal{A}$, and suppose that ξ satisfies

$$h_a^{(\xi)}(n) \to p_a, \quad n \to +\infty,$$

where $0 \le p_a \le 1$, $\sum_{a \in \mathcal{A}} p_a = 1$; moreover,

$$|h_a^{(\xi)}(n) - p_a| \le \frac{g(n)}{n}, \quad a \in \mathcal{A} \ (n \ge 1),$$

where g is a regular increasing function on the real line, $g(0) = 1$, $g(n) \le g_o n^{\beta_0}$, $g_o > 1$, $0 \le \beta_0 < 1$.

If $\beta_0 = 0$, that is, we consider the case of the fastest convergence of the occurrence factors, the measure $\mu^{(\xi)}$ has the property that there exist two positive constants C_1, C_2, such that,

$$C_1 r^{d^{(\xi)}} \le \mu^{(\xi)}(B(P, r) \cap F^{(\xi)}) \le C_2 r^{d^{(\xi)}}, \quad \forall P \in F^{(\xi)}, \tag{5}$$

with

$$d^{(\xi)} = \frac{\sum\limits_{a \in \mathcal{A}} p_a \ln N_a}{\sum\limits_{a \in \mathcal{A}} p_a \ln \ell_a}, \tag{6}$$

where $B(P, r)$ denotes the Euclidean ball with center in P and radius $0 < r \le 1$ (see [7], [28] and [29]). According to Jonsson and Wallin (see [22]), we say that $F^{(\xi)}$ is a d-set with respect to measure $\mu^{(\xi)}$. The measure $\mu^{(\xi)}$ is equivalent to the restriction to $F^{(\xi)}$ of the Hausdorff measure $\mathcal{H}^d$, with $d = d^{(\xi)}$.

If instead $\beta_0 > 0$, then

$$C_1 r^{d^{(\xi)}+\delta} \le \mu^{(\xi)}(B(P, r) \cap F^{(\xi)}) \le C_2 r^{d^{(\xi)}-\delta}, \quad \forall P \in F^{(\xi)} \tag{7}$$

with $d^{(\xi)}$ as in (6) and for any $\delta > 0$.

We recall that these inequalities play an important role in trace results (see the following section).

We focus our attention on two illustrative examples: "fluctuating Koch curves" and "mixtures of Sierpinski Gaskets".

2.1 The "fluctuating Koch curves"

Let $\mathcal{A} = \{1, 2\}$: for $a \in \mathcal{A}$, let $2 < \ell_a < 4$, and, for each $a \in \mathcal{A}$, let

$$\Psi^{(a)} = \{\psi_1^{(a)}, \dots, \psi_4^{(a)}\} \tag{8}$$

be the family of contractive similitudes $\psi_i^{(a)} : \mathbb{R}^2 \to \mathbb{R}^2$, $i = 1, \dots, 4$, with contraction factor ℓ_a^{-1} :

$$\psi_1^{(a)}(z) = \frac{z}{\ell_a}, \quad \psi_2^{(a)}(z) = \frac{z}{\ell_a} e^{i\theta(\ell_a)} + \frac{1}{\ell_a},$$

$$\psi_3^{(a)}(z) = \frac{z}{\ell_a} e^{-i\theta(\ell_a)} + \frac{1}{2} + i\sqrt{\frac{1}{\ell_a} - \frac{1}{4}}, \quad \psi_4^{(a)}(z) = \frac{z-1}{\ell_a} + 1,$$

where

$$\theta(\ell_a) = \arcsin\left(\frac{\sqrt{\ell_a(4 - \ell_a)}}{2}\right), \tag{9}$$

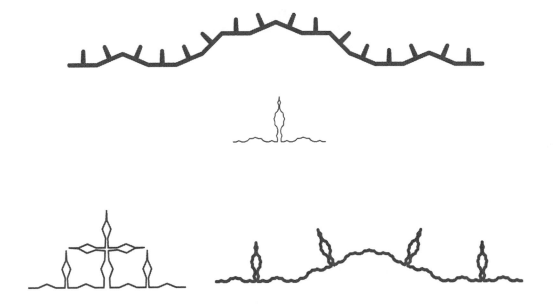

Figure 1. Forth Step Koch prefractals with $l_1 = 3.85$ and $l_2 = 2.85$.

and $\iota = (0,1)$, $z = (x_1, x_2)$.

The fluctuating Koch curve $K^{(\xi)}$ is the irregular fractal associated with the sequence ξ and the set $\Gamma = \{A, B\}$ where $A = (0,0)$ and $B = (1,0)$.

Let K be the line segment of unit length with A and B as endpoints. We set, for each n in $\mathbb{N}$, $K^{(\xi),n} = \Phi_n^{(\xi)}(K)$: $K^{(\xi),n}$ is the so-called n-th prefractal curve approximating the fluctuating Koch curve $K^{(\xi)}$ (see Fig. 1).

2.2 The "mixtures of Sierpinski Gaskets"

We consider the set of vertices of an equilateral triangle T of $\mathbb{R}^2$, that is, $\Gamma = \{A, B, C\}$ where $A = (0,0)$, $B = (1,0)$, and $C = (\frac{1}{2}, \frac{\sqrt{3}}{2})$.

Let $\mathcal{A} = \{2, 3\}$. For $a \in \mathcal{A}$, we put $\ell_a = a$ and take

$$\Psi^{(a)} = \{\psi_1^{(a)}, \ldots, \psi_{N_a}^{(a)}\} \qquad (10)$$

be the family of contractive similitudes $\psi_i^{(a)} : \mathbb{R}^2 \to \mathbb{R}^2$, $i = 1, \ldots, N_a$, with contraction factor ℓ_a^{-1}:

$$\psi_i^{(a)}(z) = b_i^{(a)} + \ell_a^{-1}(z - b_i^{(a)}),$$

$z \in \mathbb{R}^2$, $i = 1, \ldots, N_a$, which carry the triangle T into each one of the $N_a = \frac{\ell_a(\ell_a+1)}{2}$ upward facing smaller triangles obtained by decomposing T into ℓ_a^2 congruent equilateral triangles of side ℓ_a^{-1}.

The mixture of Sierpinski Gaskets $S^{(\xi)}$ is the irregular fractal associated with the sequence ξ and the set $\Gamma = \{A, B, C\}$.

Figure 2. Third step Sierpinski prefractal.

Let S be the unit simplex with vertices A, B and C. We set, for each n in $\mathbb{N}$, $S^{(\xi),n} = \Phi_n^{(\xi)}(S)$: $S^{(\xi),n}$ is the so-called n-th prefractal curve approximating the mixtures of Sierpinski Gaskets $S^{(\xi)}$ (see Fig. 2).

We denote by $F^{(\xi),n}$ the prefractal curve approximating the irregular scale fractal $F^{(\xi)}$, that is,

$$F^{(\xi),n} = \begin{cases} K^{(\xi),n} & \text{if } F^{(\xi)} = K^{(\xi)}, \\ S^{(\xi),n} & \text{if } F^{(\xi)} = S^{(\xi)}. \end{cases}$$

3 Trace Theorems

In this section, we state trace theorems for the specific geometry of the prefractal and fractal sets.

Let $L^2(\cdot, m)$ be the Lebesgue space with respect to a measure m on subsets of $\mathbb{R}^2$; we omit reference to m, unless better understanding suggests keeping full notation. Let D be an arbitrary open set of $\mathbb{R}^2$. We denote by $H^s(D)$, where $s \in \mathbb{R}_+$, the fractional Sobolev space (for the definitions and the main properties of Sobolev spaces, see [5] and [4]).

We deal with Sobolev spaces $H^s(F^{(\xi),n})$ on prefractal curves $F^{(\xi),n}$ defined in terms of local Lipschitz charts.

Remark 3.1. We observe that the "classical" definition of the Sobolev space $H^s(\partial D)$, on a Lipschitz boundary ∂D given in [5], [30] and [17] is slightly different from that given in [8]. When $0 \leq s < \frac{3}{2}$ the Sobolev space $W^{s,2}(\partial D)$ defined according to Brezzi and Gilardi coincides with the Sobolev space $H^s(\partial D)$, (see Theorem 2.24 in [8]). Indeed, when $s \geq \frac{3}{2}$, the space $H^s(\partial D)$ is smaller than the space $W^{s,2}(\partial D)$.

We now define Besov spaces $B^{2,2}_{s,d}(F^{(\xi)})$, $0 < s < 1$, on the mixture $F^{(\xi)}$. There are many equivalent definitions of these spaces (see for instance [32], [22] and [18]). We recall here the one which best fits our aims and we refer to [22] for a more general setting.

Let $F^{(\xi)}$ be the scale irregular fractal and $\mu^{(\xi)}$ its invariant measure (see (3)). By $B^{2,2}_{s,d}(F^{(\xi)})$, with $0 < s < 1$, we denote the space of all functions f such that

$$\|f\|^2_{B^{2,2}_{s,d}(F^{(\xi)})} = \|f\|^2_{L^2(F^{(\xi)},\mu^{(\xi)})} + \iint\limits_{|x-y|<1} \frac{|f(x) - f(y)|^2}{|x - y|^{2s+d}} \, d\mu^{(\xi)}(x) \, d\mu^{(\xi)}(y) < +\infty. \quad (11)$$

We recall that, for f in $L^1_{loc}(D)$, where D is an arbitrary open set of $\mathbb{R}^2$, the trace operator γ_0 is defined as

$$\gamma_0 f(x) := \lim_{r \to 0} \frac{1}{m_2(B(x,r) \cap D)} \int\limits_{B(x,r) \cap D} f(y) \, dy \quad (12)$$

(m_2 denotes the 2-dimensional Lebesgue measure) at every point $x \in \overline{D}$ where the limit exists (see, for example, page 15 in [22]).

The following theorems deal with the trace to the set $F^{(\xi),n}$ of Sobolev spaces $H^\sigma(\mathbb{R}^2)$. Set

$$\ell^{(\xi)}(n) = \prod_{i=1}^{n} \ell_{\xi_i}. \quad (13)$$

In the following Theorem 3.2 by an elementary proof we obtain an estimate (weaker than estimate in Theorem 3.5) but sharper than estimate that one can deduce in a straightforward way from the classical results (see, for instance, [18] and [13]). From now on, we denote the function u and its traces $\gamma_0 u$ by the same symbol.

We put

$$m_1(F^{(\xi),n}) = \begin{cases} \sum\limits_{i|n} l(\psi_{i|n}(K)) & \text{if } F^{(\xi),n} = K^{(\xi),n}, \\ \sum\limits_{i|n} l(\psi_{i|n}(S)) & \text{if } F^{(\xi),n} = S^{(\xi),n}, \end{cases} \quad (14)$$

where l denote the arc length and, for any Borel set $E \in \mathbb{R}^2$,

$$m_1(E) = m_1(E \cap F^{(\xi),n}).$$

Let Ω be the polygonal domain with vertices $E = (0,1), F = (1,1), G = (0,-1)$ and $H = (1,-1)$.

Theorem 3.2. *Let $u \in H^1(\Omega)$. Then,*

$$\|u\|^2_{L^2(F^{(\xi),n},m_1)} \leqslant c \, n \frac{N^{(\xi)}(n)}{\ell^{(\xi)}(n)} \|u\|^2_{H^1(\Omega)}, \quad (15)$$

where c does not depend on n.

We do the proof in the simpler geometry of a prefractal Koch curve $K^n = K^{(\xi),n}$ with $\ell_1 = \ell_2 = \ell \in (2,4)$; the proof for the prefractal mixtures being similar.

Proof. Let T^ℓ be the triangle with vertices $A = (0,0)$, $B = (1,0)$ and $C^\ell = \left(\frac{1}{2}, \frac{\sqrt{\ell(4-\ell)}}{2\ell}\right)$ and let K be the line segment of unit length with A and B as endpoints.

Then, by changing the variables, we obtain

$$\|u\|^2_{L^2(\psi_{i_1 i_2 \cdots i_n}(K))} = \ell^{-n}\|u \circ \psi_{i|n}\|^2_{L^2(K)}. \tag{16}$$

By using a "classical" trace theorem (see, for example, [16] or [30]) in the triangle T^ℓ, the following estimate holds $\forall\, g \in H^1(T^\ell)$

$$\|g\|^2_{L^2(K)} \leqslant c_1^*\{\|g\|^2_{L^2(T^\ell)} + \|\nabla g\|^2_{L^2(T^\ell)}\}. \tag{17}$$

By combining (16) and (17), we obtain

$$\|u\|^2_{L^2(\psi_{i|n}(K))} \leqslant c_1^*\ell^{-n}\left(\|u \circ \psi_{i|n}\|^2_{L^2(T^\ell)} + \|\nabla u \circ \psi_{i|n}\|^2_{L^2(T^\ell)}\right), \tag{18}$$

where $i|n = i_1 i_2 \ldots i_n$ and we have chosen in (17) $g = u \circ \psi_{i|n}$. We now change again the variables

$$\|u\|^2_{L^2(\psi_{i|n}(K))} \leqslant c_1^*\ell^{-n}\left(\ell^{2n}\|u\|^2_{L^2(\psi_{i|n}(T^\ell))} + \|\nabla u\|^2_{L^2(\psi_{i|n}(T^\ell))}\right). \tag{19}$$

As

$$\|u\|^2_{L^2(K^n)} = \sum_{i|n} \|u\|^2_{L^2(\psi_{i|n}(K))} \tag{20}$$

and the triangles $\psi_{i|n}(T^\ell)$ are not overlapping subsets of Ω, we obtain

$$\|u\|^2_{L^2(K^n)} \leqslant c_1^*\left\{\ell^n\|u\|^2_{L^2(\Omega_n)} + \ell^{-n}\|\nabla u\|^2_{L^2(\Omega_n)}\right\}, \tag{21}$$

where $\Omega_n = \cup_{i|n}\psi_{i|n}(T^\ell)$.

Now in order to estimate the first term appearing in the right end side of (21), we use that the Lebesgue measure of Ω_n vanishes as n goes to $+\infty$ and that the following Trudinger inequality holds (see Theorem 8.25 in [5])

$$\int_\Omega e^{\frac{u^2}{k_0^2}}\,dy \leqslant m_2(\Omega) + 1$$

where $k_0 = k_1\|u\|_{H^1(\Omega)}$ (from now on $m_2(E)$ denotes the 2-dimensional Lebesgue measure on $\mathbb{R}^2$ of the set E).

Then, $\forall\, k \in [1, +\infty)$, we obtain from Chebishev inequality,

$$\|u\|^2_{L^2(\Omega_n)} \leqslant k^2 m_2(\Omega_n) + \int_{\Omega_n \bigcap\{|u|\geqslant k\}} u^2\,dy \leqslant$$

$$\leqslant k^2 \cdot 4^n\ell^{-2n} + \left(\int_{\Omega_n \bigcap\{|u|\geqslant k\}} u^p\,dy\right)^{2/p} m_2(\Omega_n \bigcap\{e^{\frac{u^2}{k_0^2}} \geqslant e^{\frac{k^2}{k_0^2}}\})^{(1-2/p)} \leqslant$$

$$\leqslant k^2 \cdot 4^n\ell^{-2n} + 3^{\frac{p-2}{p}}\|u\|^2_{L^p(\Omega_n)}e^{-\frac{(p-2)k^2}{pk_0^2}}.$$

Then, with easy calculations, we obtain that

$$\|u\|^2_{L^2(\Omega_n)} \le k_1^2 4^n \ell^{-2n} \|u\|^2_{H^1(\Omega)} \frac{p}{p-2} \left(n \ln(\frac{\ell^2}{4}) + \ln(\frac{3^{\frac{p-2}{p}}(p-2)}{k_1^2 p}) + 1 \right)$$

and from (21) we achieve the result. $\qquad\square$

Remark 3.3. Concerning the sharpness of the constant in the trace inequality, we note that if we choose a cut smooth function f (for example, $f \in C^1(\mathbb{R}^2)$) that is equal to one in a ball of radius one centered in the point $A = (0,0)$ and supported in the concentric ball of radius two, we obtain that the quantity $\|f\|^2_{H^1(\mathbb{R}^2)}$ is bounded by a constant independent of n while $\|f\|^2_{L^2(F^{(\xi),n})}$ behaves like the length of the polygonal curve $\frac{N^{(\xi)}(n)}{\ell^{(\xi)}(n)}$: then, the estimate in the previous Theorem 3.2 could not be sharp. The sharp estimate will be provided by the following Theorem 3.5.

Before proving Theorem 3.5, we need to evaluate the arc length Lebesgue measure on $F^{(\xi),n}$ where $F^{(\xi),n}$ denotes the prefractal set approximating the fractal set $F^{(\xi)}$.

Theorem 3.4. *Let $F^{(\xi),n}$ be the n-th prefractal curve. Then,*

$$m_1(B(P,r) \cap F^{(\xi),n}) \leqslant C^\xi \frac{N^{(\xi)}(n)}{\ell^{(\xi)}(n)} r, \quad \forall P \in \mathbb{R}^2, \tag{22}$$

where C^ξ is independent of n, $B(P,r)$ denotes the Euclidean ball with center in P and radius $0 < r \le 1$ and m_1 is the arc length Lebesgue measure defined in (14).

Proof. Let h be the integer such that

$$\frac{1}{\ell^{(\xi)}(h)} < r \le \frac{\ell_{\xi_h}}{\ell^{(\xi)}(h)},$$

then $B(P,r) \subset B\left(P, \frac{\ell_{\xi_h}}{\ell^{(\xi)}(h)}\right)$.

When $h > n$,

$$m_1\left(B(P,r) \cap F^{(\xi),n}\right) \le m_1\left(B\left(P, \frac{\ell_{\xi_h}}{\ell^{(\xi)}(h)}\right) \cap F^{(\xi),n}\right) \le C_1^\xi r,$$

where C_1^ξ is independent of n.

We consider the case when $h \le n$. Let $\mathcal{T}$ be the open set satisfying the *open set condition* (see [14], [15]). We remark that in the case of fluctuating Koch curve $\mathcal{T}$ is the triangle with vertices A, B and $C^* = \left(\frac{1}{2}, \frac{\sqrt{\ell^*(4-\ell^*)}}{2\ell^*}\right)$ where $\ell^* = \min(\ell_1, \ell_2)$ and in the case of mixtures of Sierpinski Gaskets $\mathcal{T}$ is the triangle with vertices A, B and C. There are at most C_2^ξ triangles $\psi_{w|h-1}(\mathcal{T}) = \psi_{w_1}^{(\xi_1)} \circ \cdots \circ \psi_{w_{h-1}}^{(\xi_{h-1})}(\mathcal{T})$ (C_2^ξ independent of n) that have not-empty intersection with $B\left(P, \frac{\ell_{\xi_h}}{\ell^{(\xi)}(h)}\right)$. Then

$$m_1(B(P,r) \cap F^{(\xi),n}) \le m_1\left(B\left(P, \frac{\ell_{\xi_h}}{\ell^{(\xi)}(h)}\right) \cap F^{(\xi),n}\right) \le C_2^\xi \frac{N^{(\xi)}(n-h+1)}{\ell^{(\xi)}(n)} <$$

$$< C_2^\xi \frac{N^{(\xi)}(n-h+1)}{\ell^{(\xi)}(n)} \ell^{(\xi)}(h) r \le C_2^\xi \frac{\ell^{(\xi)}(h)N^{(\xi)}(n-h+1)}{N^{(\xi)}(n)} \frac{N^{(\xi)}(n)}{\ell^{(\xi)}(n)} r \le CC_2^\xi \frac{N^{(\xi)}(n)}{\ell^{(\xi)}(n)} r$$

and this concludes the proof. $\qquad\square$

Theorem 3.5. *Let $u \in H^\sigma(\mathbb{R}^2)$. Then, for $\frac{1}{2} < \sigma \leq 1$,*

$$\|\gamma_0\, u\|^2_{L^2(F^{(\xi),n},m_1)} \leqslant C_\sigma \frac{N^{(\xi)}(n)}{\ell^{(\xi)}(n)} \|u\|^2_{H^\sigma(\mathbb{R}^2)}, \tag{23}$$

where C_σ is independent of n.

Proof. We closely follow the proof of Lemma C on page 107 and Lemma 6 on page 149 in [22]. Any $u \in H^\sigma(\mathbb{R}^2)$ can be written in terms of Bessel kernels G_σ of order σ as $u = G_\sigma * g$, where $g \in L^2(\mathbb{R}^2)$, with $\|u\|^2_{H^\sigma(\mathbb{R}^2)} = \|g\|^2_{L^2(\mathbb{R}^2)}$ (see, for example, page 13 in [4], page 219 in [5] and also page 6 in [22]).

Then

$$\|\gamma_0\, u\|^2_{L^2(F^{(\xi),n})} = \int_{F^{(\xi),n}} \left| \int_{\mathbb{R}^2} G_\sigma(x - y)g(y)\, dy \right|^2 dm_1,$$

$$\|\gamma_0\, u\|^2_{L^2(F^{(\xi),n})} \leq \int_{F^{(\xi),n}} \left(\int_{\mathbb{R}^2} |G_\sigma(x - y)|^{2a} |g(y)|^2 dy \right) \left(\int_{\mathbb{R}^2} |G_\sigma(x - y)|^{2(1-a)}\, dy \right) dm_1, \tag{24}$$

where the number a, $0 < a < 1$, will be chosen later.

Now we use the estimates for the Bessel kernels (see, for example, page 11 and page 9 in [4] and also Lemma 1 on page 104 in [22]) and we obtain for a fixed (positive) number b:

$$\int_{\mathbb{R}^2} |G_\sigma(x - y)|^{2(1-a)}\, dy \leqslant$$

$$\leqslant \int_{|x-y| \leqslant b} |G_\sigma(x - y)|^{2(1-a)}\, dy + \int_{|x-y| \geqslant b} |G_\sigma(x - y)|^{2(1-a)}\, dy = I_1 + I_2,$$

where

$$I_1 \leqslant c_1 \int_{|x-y| \leqslant b} |x - y|^{2(1-a)(\sigma-2)}\, dy = c_1 \frac{2\pi b^{2(1-a)(\sigma-2)+2}}{2(1-a)(\sigma-2)+2}, \tag{25}$$

if

$$2 > 2(2 - \sigma)(1 - a), \text{ i.e. } a > \frac{1 - \sigma}{2 - \sigma} \tag{26}$$

and

$$I_2 \leqslant c_2 \int_{|x-y| \geqslant b} e^{-c_3 |x-y| 2(1-a)} dy = c_2 \frac{2\pi}{(2c_3(1-a))^2} e^{-2c_3(1-a)b} \{2c_3(1-a)b + 1\} \tag{27}$$

(note that c_1, c_2 and c_3 are positive constants that depend on b and on σ, but that do not depend on n). Hence, taking into account (25) and (27)

$$\int_{\mathbb{R}^2} |G_\sigma(x-y)|^{2(1-a)} \, dy \leqslant C_3, \tag{28}$$

where C_3 is a positive constant that depend on b and on σ, but that does not depend on n. On the other hand:

$$\int_{F^{(\xi),n}} \left(\int_{\mathbb{R}^2} |G_\sigma(x-y)|^{2a} g^2(y) \, dy \right) dm_1 = \int_{\mathbb{R}^2} g^2(y) \left(\int_{F^{(\xi),n}} |G_\sigma(x-y)|^{2a} \, dm_1 \right) dy.$$

We proceed as previously

$$\int_{F^{(\xi),n}} |G_\sigma(x-y)|^{2a} \, dm_1 \leqslant$$

$$\leqslant \int_{F^{(\xi),n} \bigcap \{|x-y| \leqslant b\}} |G_\sigma(x-y)|^{2a} \, dm_1 + \int_{F^{(\xi),n} \bigcap \{|x-y| \geqslant b\}} |G_\sigma(x-y)|^{2a} \, dm_1 = X_1 + X_2,$$

where

$$X_2 \leqslant c_2 \int_{F^{(\xi),n} \bigcap \{|x-y| \geqslant b\}} e^{-2c_3|x-y|a} \, dm_1 \leqslant c_2 e^{-2c_3 ba} \cdot \frac{N^{(\xi)}(n)}{\ell^{(\xi)}(n)}. \tag{29}$$

In the last inequality we have taken into account that the arc length measure of the curve $F^{(\xi),n}$ is $\frac{N^{(\xi)}(n)}{\ell^{(\xi)}(n)}$.

Now we have to estimate the term X_1 and we write (see e.g. [25])

$$X_1 \leqslant c_1 \int_{F^{(\xi),n} \bigcap \{|x-y| \leqslant b\}} |(x-y)|^{2a(\sigma-2)} \, dm_1 =$$

$$= \int_0^{+\infty} m_1 \{x \in F^{(\xi),n} \bigcap \{|x-y| \leqslant b : |x-y|^{2a(\sigma-2)} > r\} \, dr \tag{30}$$

and

$$m_1 \left\{ x \in F^{(\xi),n} \bigcap \{|x-y| \leqslant b : |x-y|^{2a(\sigma-2)} > r \right\} =$$

$$= m_1 \left\{ x \in F^{(\xi),n} | x-y| \leqslant \min\{r^{\frac{1}{2a(\sigma-2)}}, b\} \right\}.$$

By Theorem 3.4 (according to Lemma 1 on page 104 in [22] again) we derive from (30)

$$X_1 \leqslant c_1 \cdot c_0 \frac{N^{(\xi)}(n)}{\ell^{(\xi)}(n)} \left\{ \int_{b^{2a(\sigma-2)}}^{+\infty} r^{\frac{1}{2a(\sigma-2)}} \, dr + \int_0^{b^{2a(\sigma-2)}} b \, dr \right\}$$

hence

$$X_1 \leqslant c_1 \cdot c_0 \frac{N^{(\xi)}(n)}{\ell^{(\xi)}(n)} \left\{ \frac{2a(2-\sigma) \cdot b^{1+2a(\sigma-2)}}{1+2a(\sigma-2)} + b^{1+2a(\sigma-2)} \right\} \tag{31}$$

if

$$1 > 2a(2-\sigma), \text{ i.e. } a < \frac{1}{2(2-\sigma)}. \tag{32}$$

Hence, taking into account (29) and (31), we obtain

$$\int_{F^{(\xi),n}} |G_\sigma(x-y)|^{2a} \, dm_1 \leqslant C_4 \cdot \frac{N^{(\xi)}(n)}{\ell^{(\xi)}(n)}, \tag{33}$$

where C_4 is a positive constant that depend on b and on σ, but that does not depend on n.
Coming back to (24), taking into account (28) (26), (33), (32) and choosing

$$\frac{1-\sigma}{2-\sigma} < a < \frac{1}{2(2-\sigma)}$$

we obtain

$$\|\gamma_0 u\|^2_{L^2(F^{(\xi),n})} \leqslant C_\sigma \frac{N^{(\xi)}(n)}{\ell^{(\xi)}(n)} \int_{\mathbb{R}^2} g^2(y) \, dy = C_\sigma \frac{N^{(\xi)}(n)}{\ell^{(\xi)}(n)} \|u\|^2_{H^\sigma(\mathbb{R}^2)},$$

where $C_\sigma = C_3 \cdot C_4$ is independent of n. $\qquad\square$

Similar results to Theorem 3.5 have been stated by G. Berger [6] and Y. Achdou and N. Tchou [1] in a different geometry. Actually, we can say more. The following theorem states a trace result in the fractional Sobolev spaces $H^s(F^{(\xi),n})$ for functions belonging to the fractional Sobolev spaces $H^\sigma(\mathbb{R}^2)$.

Theorem 3.6. *Let $u \in H^\sigma(\mathbb{R}^2)$. Then, for $\frac{1}{2} < \sigma < \frac{3}{2}$,*

$$\|\gamma_0 u\|^2_{H^s(F^{(\xi),n})} \leqslant C_\sigma \left(\frac{N^{(\xi)}(n)}{\ell^{(\xi)}(n)} \right)^2 \|u\|^2_{H^\sigma(\mathbb{R}^2)}, \tag{34}$$

where C_σ is independent of n and $s = \sigma - \frac{1}{2}$.

Proof. The proof is similar to the proof of Theorem 3.5. The principal steps are the following. First we prove the following estimates for the Bessel kernels

$$\frac{1}{r} \iint_{|x-y|<r} |G_\alpha(x) - G_\alpha(y)|^\beta \, dm_1(x) \, dm_1(y) \leq c \left(\frac{N^{(\xi)}(n)}{\ell^{(\xi)}(n)} \right)^2 r^{\gamma\beta},$$

where $\beta > 0, 0 < \alpha < 2, 0 < \gamma = \frac{1}{\beta} - 2 + \alpha < 1$, and $r \leq 1$.

As any $u \in H^{\alpha}(\mathbb{R}^2)$ can be written in terms of Bessel kernels G_{α} of order α as $u = G_{\alpha} * g$, where $g \in L^2(\mathbb{R}^2)$, with $\|u\|^2_{H^{\alpha}(\mathbb{R}^2)} = \|g\|^2_{L^2(\mathbb{R}^2)}$ we obtain as second step

$$\frac{1}{r} \iint\limits_{|x-y|<r} |u(x) - u(y)|^2 \, dm_1(x) \, dm_1(y) \le c \left(\frac{N^{(\xi)}(n)}{\ell^{(\xi)}(n)} \right)^2 r^{2\alpha-1} \|g\|^2_{L^2(\mathbb{R}^2)},$$

where $0 < \alpha < 2$, and $r \le 1$.

Finally, we use the characterisation of Besov and Sobolev spaces in terms of real interpolation and the deep results of Peetre (see Theorem 1.3 in [31] and Theorem 7.1 in [21]). $\qquad\square$

Remark 3.7. We remark that the condition "$\sigma < \frac{3}{2}$" seems to be sharp. In fact according to the example of David in [18] we can construct a polygonal curve in $\mathbb{R}^2$ where the L^2 norm of the tangential derivative along the polygonal curve cannot be estimate only in terms of the ratio between the number of the sides and the contraction factor. We consider the following sawtooth line Γ_n with slopes alternating between 1 and -1, that is, $\Gamma_n = \{(x_1, x_2) \in \mathbb{R}^2 : x_2 = \frac{1}{2^n} h(2^n x_1)\}$, where $h(x_1) = x_1$ for $0 \le x_1 \le 1$, $h(x_1) = 2 - x_1$ for $1 < x_1 \le 2$ and h is 2-periodic. Actually, following David' example, we choice the function $g(x_1, x_2) = \tilde{g}(x_1, x_2) x_2 (\log(10/|x_2|))^{\beta}$ belonging to $H^{\frac{3}{2}}(\mathbb{R}^2)$ for $\beta \in (0, \frac{1}{2})$, where $\tilde{g}$ is a smooth function equal to one in a ball of radius one centered in the point $A = (0, 0)$ and supported in the concentric ball of radius three. It is easy to verify that the ratio between the number of the sides of Γ_n and the contraction factor is bounded by a constant independent of n instead the L^2 norm of the tangential derivative of the function $g(x, y)$ is at least $(\log(10 \cdot 2^n))^{\beta}$ and then it tends to $+\infty$ as $n \to +\infty$.

Remark 3.8. On the other hand, if we repeat twice Theorem 3.5, once for the function u and once for the gradient of the function u, we obtain

$$\|\gamma_0 u\|^2_{H^1(F^{(\xi),n})} \le C_{\sigma} \frac{N^{(\xi)}(n)}{\ell^{(\xi)}(n)} \|u\|^2_{H^{\sigma}(\mathbb{R}^2)}, \tag{35}$$

for $\frac{3}{2} < \sigma < 2$. Roughly speaking, if $\mathcal{S}$ is a Lipschitz curve, the restriction theorem to $H^1(\mathcal{S})$ for functions belonging to $H^{\frac{3}{2}}(\mathbb{R}^2)$ is the more difficult one and it requires some peculiar deep tools. Actually, in the framework of Besov spaces and d-sets, Theorem 1 in Chapter VI in [22] holds for any exponent but we point out that the identification of the Besov spaces $B_s^{2,2}$ and the Sobolev spaces H^s is delicate if $s \ge 1$ and the Markov property plays a crucial role. We remark that Lipschitz boundary does not satisfy this property.

We conclude this section by stating a restriction result on the set $F^{(\xi)}$ for Sobolev spaces $H^{\sigma}(\mathbb{R}^2)$. We remark that the limit mixture is not necessarily a d-set, but verifies (from (7))

$$\mu^{(\xi)}(B(P, r) \cap K^{\xi}) \le C_2 r^{d^{(\xi)}-\delta}, \quad \forall P \in \mathbb{R}^2 \tag{36}$$

with $\delta > 0$.

The following trace theorem can be proved as the previous Theorem 3.6 by replacing m_1 with the measure $\mu^{(\xi)}$ and the lenght dimension one with the upper dimension $d^{(\xi)} - \delta$.

Theorem 3.9. *Let* $u \in H^\sigma(\mathbb{R}^2)$. *Then, for* $1 - \frac{d^{(\xi)}}{2} < \sigma \leqslant 2 - \frac{d^{(\xi)}}{2}$,

$$\|\gamma_0 u\|^2_{B^{2,2}_{s,d}(F^{(\xi)})} \leqslant C_\sigma \|u\|^2_{H^\sigma(\mathbb{R}^2)}, \tag{37}$$

where $s = \sigma - 1 + \frac{d^{(\xi)} - \delta}{2}$ *and* $d = d^{(\xi)} - \delta$.

4 Asymptotics Results

In this section we discuss some applications of the previous trace theorems: in particular, we recall some asymptotic results for the energy forms defined on prefractal structures approximating the limit mixture.

Let $\Omega^{(\xi)}$ ($\Omega_n^{(\xi)}$) be the set bounded by Γ_0, Γ_1, Γ_2 and Γ_3 where $\Gamma_0 = \{x = (x_1, x_2) \in \mathbb{R}^2 : 0 < x_1 < 1, x_2 = -1\}$, $\Gamma_1 = \{x = (x_1, x_2) \in \mathbb{R}^2 : x_1 = 1, -1 < x_2 < 0\}$, $\Gamma_3 = \{x = (x_1, x_2) \in \mathbb{R}^2 : x_1 = 0, -1 < x_2 < 0\}$, and Γ_2 is $K^{(\xi)}$ ($K^{(\xi),n}$) (see Fig. 3). The curve $K^{(\xi)}$ and $K^{(\xi),n}$ have been defined in Section 2.1.

We consider the "prefractal" forms $a_n(\cdot, \cdot)$ on $L^2(\Omega^{(\xi)})$ by defining

$$a_n(u, u) = \begin{cases} \displaystyle\int_{\Omega_n^{(\xi)}} |\nabla u|^2 \, dx + c_n \int_{K^{(\xi),n}} |\gamma_0 u|^2 \, ds & \text{for } u|_{\Omega_n^{(\xi)}} \in V(\Omega_n^{(\xi)}), \\ +\infty & \text{otherwise,} \end{cases} \tag{38}$$

where $V(\Omega_n^{(\xi)}) = \{u_n \in H^1(\Omega_n^{(\xi)}) : \gamma_0 u_n = 0 \text{ on } \Gamma_0\}$. Moreover, we consider the "fractal" form $a(\cdot, \cdot)$ on $L^2(\Omega^{(\xi)})$ by defining

$$a(u, u) = \begin{cases} \displaystyle\int_{\Omega^{(\xi)}} |\nabla u|^2 \, dx + c_0 \int_{K^{(\xi)}} |\gamma_0 u|^2 \, d\mu^{(\xi)} & \text{for } u \in V(\Omega^{(\xi)}), \\ +\infty & \text{otherwise,} \end{cases} \tag{39}$$

where $V(\Omega^{(\xi)}) = \{u \in H^1(\Omega^{(\xi)}) : \gamma_0 u = 0 \text{ on } \Gamma_0\}$.

In [11], it is shown that the energies $\{a_n(\cdot, \cdot)\}$ defined in (38) converge – in the Mosco sense – to the energy $a(\cdot, \cdot)$ defined in (39) with a proper choice of the constants c_n (see [11], Theorem 5.2). It is also proved that the for every choice of $f \in L^2(\Omega^{(\xi)})$ the functions u_n minimizing the complete energy functional $J_n(v, v) = \frac{1}{2} a_n(v, v) - \int_{\Omega_n^{(\xi)}} fv \, dx$ solve

the mixed Dirichlet–Robin problem for the Laplace equation on $\Omega_n^{(\xi)}$. Moreover, the functions u_n converge to the function u that is the minimizer of the complete energy functional $J(v, v) = \frac{1}{2} a(v, v) - \int_{\Omega^{(\xi)}} fv \, dx$ and the solution of the mixed Dirichlet–Robin problem for

the Laplace equation on $\Omega^{(\xi)}$ (see also [9].)

We now recall the definition of convergence of forms introduced by Mosco in [27], denoted in the following M–convergence.

Definition 4.1. A sequence of forms $\{a_n(\cdot, \cdot)\}$ M-converges to a form $a(\cdot, \cdot)$ in $L^2(\Omega)$ if

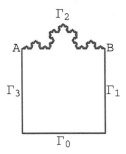

Figure 3. The domain $\Omega^{(\xi)}$.

(a) For every v_n converging weakly to u in $L^2(\Omega)$

$$\underline{\lim} \, a_n(v_n, v_n) \geq a(u, u) \quad \text{as} \quad n \to \infty. \tag{40}$$

(b) For every $u \in L^2(\Omega)$ there exists v_n converging strongly in $L^2(\Omega)$ such that

$$\overline{\lim} \, a_n(v_n, v_n) \leq a(u, u) \quad \text{as} \quad n \to \infty. \tag{41}$$

Theorem 4.2. *Let $c_n = c_0 \frac{\ell^{(\xi)}(n)}{4^n}$, then the sequence of forms $a_n(\cdot, \cdot)$ M–converges in the space $L^2(\Omega^{(\xi)})$ to the form $a(\cdot, \cdot)$.*

We stress the fact that key tools in the proof of Theorem 4.2 are the estimate (23) in Theorem 3.5 and the estimate (37) in Theorem 3.9.

We note that the energy forms defined in (38) and in (39) are the sum of two energies: a volume energy on the two dimensional domain and a layer energy on the prefractal or fractal curve. Both the energy forms are local but it could be interesting also consider non local layer energies. More precisely, for a set $F \subset \mathbb{R}^2$ that is a d-set with respect to a measure μ, we can consider non local Dirichlet forms of this type

$$E(u, u) = \iint\limits_{(F \times F) \setminus \diamond} \frac{(u(x) - u(y))^2}{|x - y|^{2s} \mu(B(x, |x - y|))} \, d\mu(x) \, d\mu(y),$$

where $s \in (0, 1)$, and $\diamond$ denotes the diagonal. These non local Dirichlet forms are strictly related to Besov spaces (see, for example, [23]), and the corresponding stochastic processes are stable-like jump processes.

In the spirit of the previous result, we can prove similar asymptotic results for non local forms using Theorems 3.6 and 3.9 as it will be shown in a forthcoming paper.

We conclude this section with another application. Recently in [12] we have proved a result in the framework of homogenization. This paper deals with a reinforcement problem for a plane domain $\Theta^{(\xi)}$ whose boundary is a deterministic or random "mixture" of self-similar Koch curves.

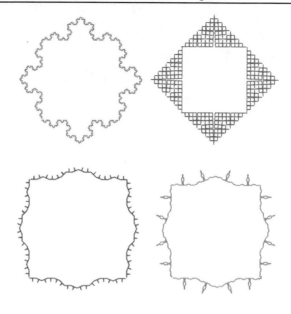

Figure 4. Some prefractal domains $\Theta^{(\xi),n}$.

More precisely, we consider the set $\Theta^{(\xi)}$ bounded by 4 scale irregular Koch curves $K_j^{(\xi)}$, $j = 1, 2, 3, 4$, with endpoints A and B, B and C, C and D, D and A respectively, where $A = (0,0)$, $B = (1,0)$, $C = (1,-1)$, and $D = (0,-1)$. Moreover, we consider the sets $\Theta^{(\xi),n}$ bounded by 4 approximating prefractal curves $K_j^{(\xi),n}$, $j = 1, 2, 3, 4$, starting from the segments K_j with endpoints A and B, B and C, C and D, D and A respectively (see Fig. 4).

We construct an ε–thin polygonal 2–dimensional fiber $\Sigma_\varepsilon^{(\xi),n}$, $n \in \mathbb{N}$, $0 < \varepsilon < 1$, around pre-fractal approximating domains $\Theta^{(\xi),n}$. The geometry of the fiber is regulated by the families of contractive similarities, whose n–iterations in the plane generate the Koch mixture as $n \to +\infty$. We denote by $\Theta_\varepsilon^{(\xi),n}$ the reinforced domain $\overline{\Theta^{(\xi),n}} \cup \Sigma_\varepsilon^{(\xi),n}$.

We introduce discontinuous coefficients a_ε^n in $\Theta_\varepsilon^{(\xi),n}$, which have jumps along the boundary of $\Theta^{(\xi),n}$

$$a_\varepsilon^n(x) = \begin{cases} c_n w_\varepsilon^n(x) & \text{if } x \in \Sigma_\varepsilon^{(\xi),n}, \\ 1 & \text{if } x \notin \Sigma_\varepsilon^{(\xi),n}. \end{cases} \qquad (42)$$

The singular weights w_ε^n and the constants c_n are chosen by taking into account the geometry of the fibers and the structural parameters of the fractal boundary. The related "weighted" energy functionals on $L^2(\Theta^*)$ are

$$F_\varepsilon^n[u] = \begin{cases} \displaystyle\int_{\Theta_\varepsilon^n} a_\varepsilon^n(x) |\nabla u|^2 \, dx & \text{if } u|_{\Theta_\varepsilon^{(\xi),n}} \in H_0^1(\Theta_\varepsilon^{(\xi),n}; w_\varepsilon^n), \\ +\infty & \text{otherwise,} \end{cases} \qquad (43)$$

where Θ^* is a regular domain containing the closure of the domains $\Theta_\varepsilon^{(\xi),n}$ and $H_0^1(\Theta_\varepsilon^{(\xi),n}; w_\varepsilon^n)$ is the weighted Sobolev space with respect to the measure $w_\varepsilon^n m_2$.

The asymptotic behavior of the energy functionals F_ε^n is studied while, simultaneously, the thickness of the fibers $\varepsilon = \varepsilon(n)$ and the conductivity a_ε^n of the fibers converges to 0 as $n \to +\infty$.

In Theorem 4.2 of [12], it is shown that, with an appropriate definition of the weights w_ε^n and with the choice of the parameters $c_n = c_0 \frac{\ell^{(\xi)}(n)}{4^n}$, the energy functionals in the Hilbert space $L^2(\Theta^*)$ M-converge to the energy functional

$$F[u] = \begin{cases} \displaystyle\int_{\Theta^{(\xi)}} |\nabla u|^2 \, dx + c_0 \int_{\partial\Theta^{(\xi)}} |\gamma_0 u|^2 \, d\mu^{(\xi)} & \text{if } u|_{\Theta^{(\xi)}} \in H^1(\Theta^{(\xi)}), \\ +\infty & \text{if } u \in L^2(\Theta^*) \setminus H^1(\Theta^{(\xi)}), \end{cases} \tag{44}$$

where $\gamma_0 u$ denotes the trace of u on the boundary of $\Theta^{(\xi)}$ and $\mu^{(\xi)}$ is the "intrinsic" measure on $\partial\Theta^{(\xi)}$. Once again key tools in proving Theorem 4.2 of [12] are Theorems 3.5 and 3.9.

References

[1] Y. Achdou, N. Tchou, Neumann conditions on fractal boundaries. *Asymptot. Anal.*, **53** (2007), No. 1-2, 61–82.

[2] D. R. Adams, Traces of potentials arising from translation invariant operators. *Ann. Scuola Norm. Sup. Pisa (3)*, **25** (1971), 203–217.

[3] D. R. Adams, A trace inequality for generalized potentials. *Studia Math.* **48** (1973), 99–105.

[4] D. R. Adams, L. I. Hedberg, *Function Spaces and Potential Theory*. Berlin: Springer-Verlag, 1966.

[5] R. A. Adams, *Sobolev Spaces*, New York: Academic Press, 1975.

[6] G. Berger, Eigenvalue distribution of elliptic operators of second order with Neumann boundary conditions in a snowflake domain. *Math. Nachr.* **220** (2000), 11–32.

[7] M. T. Barlow, B. M. Hambly, Transition density estimates for Brownian motion on scale irregular Sierpinski gasket. *Ann. Inst. H. Poincaré*, **33** (1997), 531–556.

[8] F. Brezzi, G. Gilardi , *Finite Elements Mathematics*. In Finite Element Handbook, Eds. Kardestuncer H., D. H. Norrie, New York: McGraw–Hill Book Co., 1987.

[9] R. Capitanelli, Transfer across scale irregular domains. *Series on Advances in Mathematics for Applied Sciences* **82** (2009), 165–174.

[10] R. Capitanelli, Asymptotics for mixed Dirichlet–Robin problems in irregular domains. *J. Math. Anal. Appl.*, **362** (2010), No. 2, 450–459.

[11] R. Capitanelli, Robin boundary condition on scale irregular fractals. *Commun. Pure Appl. Anal.* **9** (2010), No. 5, 1221–1234.

[12] R. Capitanelli, M. A. Vivaldi, Insulating Layers and Robin Problems on Koch Mixtures, to appear on Journal of Differential Equations. doi:10.1016/j.jde.2011.02.003

[13] Z. Ding, A proof of the trace theorem of Sobolev spaces on Lipschitz domains. *Proc. Amer. Math. Soc.*, **124** (1996), No. 2, 591–600.

[14] K. J. Falconer, *The geometry of fractal sets*. Cambridge tracts in mathematics, 1985.

[15] J. E. Hutchinson, Fractals and selfsimilarity. *Indiana Univ. Math. J.* **30** (1981), 713–747.

[16] E. Gagliardo, Caratterizzazioni delle tracce sulla frontiera relative ad alcune classi di funzioni in n variabili, *Rend. Sem. Mat. Univ. Padova*, **27** (1957), 284–305.

[17] P. Grisvard, *Elliptic problems in nonsmooth domains* . Boston: Pitman, 1985.

[18] D. Jerison , C.E. Kenig, The inhomogeneous Dirichlet problem in Lipschitz domains. *J. Funct. Anal.* **130** (1995), 161–219.

[19] A. Jonsson, Besov spaces on closed subsets of $\mathbb{R}^n$. *Trans. Amer. Math. Soc.* **341** (1994), No. 1, 355–370.

[20] A. Jonsson, Besov spaces on closed sets by means of atomic decomposition. *Complex Var. Elliptic Equ.* **54** (2009), No. 6, 585–611.

[21] A. Jonsson, H. Wallin, A Whitney extension theorem in L_p and Besov spaces. *Ann. Inst. Fourier (Grenoble), Université de Grenoble. Annales de l'Institut Fourier* , **28** (1978), No. 1, vi, 139–192.

[22] A. Jonsson, H. Wallin, Function spaces on subsets of $\mathbb{R}^n$. *Math. Rep.* **2** (1984), No. 1, xiv+221.

[23] T. Kumagai, Function spaces and stochastic processes on fractals. *Fractal geometry and stochastics III*, 221–234, *Progr. Probab.*, 57, *Birkhauser, Basel*, 2004.

[24] M. R. Lancia, M. A. Vivaldi, Teoremi di traccia per domini irregolari. *Seminari Scientifici del Dipartimento di Metodi e Modelli Matematici per Le Scienze Applicate* . Quaderno n. 4. 1999.

[25] J. Maly, W.P. Ziemer, *Fine regularity of solutions of elliptic partial differential equations*. Mathematical Surveys and Monographs, 51. RI: American Mathematical Society, Providence, 1997.

[26] V. G. Maz'ja, *Sobolev spaces*. Springer Series in Soviet Mathematics, Translated from the Russian by T. O. Shaposhnikova, Berlin: Springer-Verlag, 1985, xix+486.

[27] U. Mosco, Convergence of convex sets and of solutions of variational inequalities. *Adv. Math.*, **3** (1969), 510–585.

[28] U. Mosco, Harnack inequalities on scale irregular Sierpinski gaskets, Nonlinear problems in mathematical physics and related topics, II, *Int. Math. Ser.*, **2** (2002), 305–328.

[29] U. Mosco, Gauged Sobolev inequalities, *Appl. Anal.*, **86** (2007), No. 3, 367–402.

[30] J. Necăs, *Les méthodes directes en théorie des équationes elliptiques*. Masson: Paris, 1967.

[31] J. Peetre, On the trace of potentials. *Ann. Scuola Norm. Sup. Pisa Cl. Sci. (4)* **2** (1975), No. 1, 33–43.

[32] H. Triebel, *Fractals and spectra. Related to Fourier analysis and function spaces*. (Monographs in Mathematics: Vol. 91), Basel: Birkhäuser Verlag 1997.

[33] H. Wallin, The trace to the boundary of Sobolev spaces on a snowflake. *Mskr. Math.* **73** (1991), 117–125.

In: Classification and Application of Fractals
Editor: William L. Hagen

ISBN978-1-61209-967-5
© 2012 Nova Science Publishers, Inc.

Chapter 12

PHYSICS ON THE NET FRACTALS

Zygmunt Bak

Institute of Physics, Jan Dlugosz University, Csestochowa, Poland

PACS 61.43.Hv, 63.22.+m, 02.20.Hj.
Keywords:generalized self-similarity, modulated fractals , net fractals.
AMS Subject Classification: 28A80, 06D99, 81Q35.

1. Introduction

The idea of a fractal has become an effective tool in the analysis of common features of many complex processes observed in physics, biology, chemistry or earth sciences. Any fractality of a physical system can be generated two-ways, it can arise due to the fractality of underlying medium (material) or due to the fractality of the process. This means that physical quantities can have fractal (power-law) characteristics even if the material itself does not need to have fractal microstructure. The hallmark of fractality of a geometrical set is a hierarchical organization of its elements, described by discrete scaling laws, which makes the fractal self-similar or self-affine. This effect can be observed at both classical and quantum levels. For example in quantum systems under some conditions, e.g. at the critical energy separating localized and extended states, the wave functions are shown to have fractal structure (see [1] and references therein). Evidence for that comes from the measurements of the participation numbers N_q (see [2]) in some systems.

As the essence of fractality is associated with the hierarchical organization of its elements, thus, by changing the hierarchy we can generate different families of fractals. For a proper description of real systems it is important to account for such hierarchies which reflect the symmetry of examined quantity. That's why, motivated by the ideas coming from experimental data we focus our study on construction of a new family of fractal structures which we believe will be useful in interpretation of some physical processes. In the following I will present the concept of *net fractals* which in the logarithmic scale become

---

*E-mail address: z.bak@ajd.czest.pl

384 Z. Bak

isomorphic with some crystal lattices. The idea of net fractals is a generalization of conventional fractals which allows description of some fractals in the spirit of solid state theory. The inspiration for the study came from high energy physics [3], [4] where anomalous scaling of factorial moments in particle production was observed. Contrary to conventional self-similar fractals, for which there is a unique scaling factor (scaling is isotropic), in the high energy physics data, the shrinking ratios of multiparticle final states can be different for each phase space direction [3]. Below we will present results of the studies that exploit such a generalized form of similarity.

2. Net Fractals

Although physical systems modeled by fractals are non-translation-invariant it is a well-known fact that the self-similar fractals as well as the physical quantities on fractal substrates show log-periodicities [5], [6]. This opens a possibility to describe the symmetries of some self-affine fractals in the way that is reminiscent of conventional formalism developed for crystalline systems. Being inspired by such an idea we will review the studies of fractal scaling symmetry, which are similar in spirit to the solid state theory. We believe that such a description allows a better understanding of important symmetry features by analogy with the well-known classical crystals.

Motivated by this idea we refer to the concept of net fractals immersed in a Euclidean space which exhibits different scaling ratios in each dimension. We say that $K \subset R^n$ satisfies the scaling law S, or is a (infinite) self-similar fractal if $K = S : K$. In the following we assume that self-affinity (generalized self-similarity) is realized via linear mappings which transform a point $\vec{r} = (x_1, x_2, x_3) \in K \subset R^3$ into the point $\vec{r}' = (x_1', x_2', x_3')$ according to the formula

$$x_i' = S_{i1}x_1 + S_{i2}x_2 + S_{i3}x_3 \quad 1, 2, 3. \tag{1}$$

In the vector formalism the linear self-similar transformation (1) can be written as $\vec{r}' = \hat{S}\vec{r}$, where $\hat{S}$ is the matrix of transformation (1). If we orient coordinate axes along the eigenvectors of matrix $\hat{S}$ (i.e., $\vec{x} = (x_1, x_2, x_3) \xrightarrow{S} (\xi, \eta, \rho)$, then the mapping (1) reduces to the transformation $S : (\xi, \eta, \rho) \to (\lambda_1\xi, \lambda_2\eta, \lambda_3\rho)$. Moreover, in these coordinates we can treat any linear mapping S as the superposition of three independent mappings $(S = S_1 \circ S_2 \circ S_3)$, where S_i fulfill relations $S_1 : (\xi, \eta, \rho) = (\lambda_1\xi, \eta, \rho)$, $S_2 : (\xi, \eta, \rho) = (\xi, \lambda_2\eta, \rho)$ etc. for any $\vec{x} = (\xi, \eta, \rho) \in K$.

Evidently one can create a richer family of linear transformations which reflects the generalized form of multiscale similarity [3],

$$S^{(m,n,l)} = (S_1)^m \circ (S_2)^m \circ (S_3)^l, \tag{2}$$

where $(S_i)^n$ denotes n-tuple superposition of transformation S_i. In conventional fractals the self-similar relation is fulfilled only if $n = m = l$. However, as we have mentioned above in some physical systems the scaling factors in each direction can be different. This means that component S_i of the self-affine transformation can be applied independently. As the consequence of that a more general scaling transformation $S^{(m,n,l)} : K = K$ with $n \neq m \neq l$ can be formulated. A geometrical set K which fulfills the relation $S^{(m,n,l)} : K = K$

for all all n, m, l being positive or negative integers) is called the *net fractal* . The action of $S^{(m,n,l)}$ transforms any point $\vec{x} \in R^3$ according to the formula

$$S^{(m,n,l)} : \vec{x} = (\lambda_1^m \xi, \lambda_2^n \eta, \lambda_3^l \rho) = (e^{ma}\xi, e^{nb}\eta, e^{lc}\rho),\qquad(3)$$

where $a, b, c = ln\lambda_i$

The family of net fractals can be created two-ways. Firstly, we take some subset $K \subset R^3$ and create the set $F_{\vec{\lambda}}$

$$F_{\vec{\lambda}} = \bigcup_{m,m,l} S^{(m,n,l)} : K.\qquad(4)$$

where $\vec{\lambda}$ represents the set of scaling factors in different directions $\vec{\lambda} = (\lambda_1, \lambda_2, \lambda_3)$. It is evident, for such an infinite-size fractal, that any mapping $S^{(m,n,l)}$ fulfills the generalized self-similarity condition

$$S^{(m,n,l)} : F_{\vec{\lambda}} = F_{\vec{\lambda}},\qquad(5)$$

There is an alternative procedure which allows us to construct fractals fulfilling relations (5). To show that suppose we have some 1D, infinite-size fractals $F_1, F_2, F_3 \subset R^1$ each with the self-affine mapping S_i such that $S_i : F_i = F_i$ and for any $x^{(i)} \in F_i \subset R^1$ we have $S_i : x^{(i)} = \lambda_i x^{(i)}$. Let us construct the Cartesian product

$$F_{123} = F_1 \times F_2 \times F_3.\qquad(6)$$

In view of the relation (6 one to one correspondence between a set of linear scaling transformations inherent to 3D net fractals and some 3D crystals can be shown. It is evident that in the logarithmic scale any infinite 1D fractal $F_1 \subset R^1$ can be mapped onto a 1D crystal. Indeed, for any linear S_1 by definition we have $S_1 : F_1 = F_1$ and for any $x_o \in F_1$ we have $S_1 : x_o = \lambda_1 x_o = x_1$. Consequently $(S_1)^{\pm m} : x_o = (\lambda_1)^{\pm m} x_o = x_m$. Using the logarithmic scale we obtain $log(\frac{x_m}{x_o}) = m \cdot ln\lambda_1$, which means that in the log-scale the set of self-similar transformations on F_1 forms a 1D crystal lattice with the lattice spacing given by $a_1 = ln\lambda_1$. A detailed study of the relations between 1D discrete scale invariance and log-periodicity can be found in [7]. As we consider infinite and deterministic fractals with multiscale generalized self-similarity it is easy to show an isomorphism between net fractals and some crystal lattices. It simply suffices to picture a net fractal in the logarithmic coordinates. This fact allows a simple interpretation of net fractal symmetry (both translational and rotational) in terms of conventional crystallography [8]. It reflects the fact that self-affine fractals can be created by simple transformations. Indeed, any affine transformation is a combination of just shifts, rotations, scalings and shears.

2.1. Example: Cantor Set as a Crystal Chain in the Log-Scale Picture

As an example of the log scale crystallography of the fractals let us consider a triadic Cantor set /CS/. The Cantor set is created by repeatedly deleting the open middle thirds of the interval [0,1]. Iteration procedure that leads to formation of the Cantor set is pictured in Fig. 1.

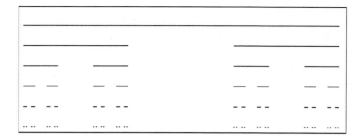

Figure 1. The first few steps in constructing the Cantor set.

Let us now picture this procedure in the logarithmic, log_3-scale coordinates. We take the interval $[0, 1]$ as an infinite sum of subsets T_n

$$[0,1] = \bigcup_{n=0}^{\infty} T_n = \bigcup_{n=0}^{\infty} [3^{-n-1}, 3^{-n}]. \tag{7}$$

In the log_3 scale each $T_n = [3^{-n-1}, 3^{-n}]$ is transformed into the interval $t_n = [-n-1, -n]$ being the unit cell of the half-infinite, log_3-scale crystal.

The set obtained after the first step of the Cantor procedure (pictured in the log_3-scale) is the union of two intervals $[-\infty, -1] \cup [log_3 2 - 1, 0]$. After the second step the log picture of the Cantor procedure is given by the union of intervals $[-\infty, -2] \cup [log_3 2 - 2, -1] \cup [log_3 2 - 1, log_3 7 - 2] \cup [log_3 8 - 2, 0]$ etc. As we can see, at every step k the number of segments is doubled, and the picture of points that belong to the interval $[-n - 1, -n]$ is identical with the picture of those points of the CS $_{log}$ that belong to the preceding unit interval $[-n, -n + 1]$ at the preceding stage of Cantor construction. The only exception is the appearance of subset $CS_{log}^{[-1,0]}$, i.e., of the points of the CS set that, in the log_3-scale, fall into the $[-1, 0]$ interval. At every step of the Cantor procedure, there arise essential changes in the appearance of the $CS_{log}^{[-1,0]}$ subset, which always gains a novel and more complicated structure. The log_3 picture of the first few steps of the CS_{log} construction is presented in the Fig. 2 [9].

We can summarize the log-scale construction as follows. After every step of the Cantor construction there arises a new structure of the $CS_{log}^{[-1,0]}$ subset, while the remaining part of the picture is identical with the one obtained after the preceding step of the Cantor procedure, but moved to the left by one unit segment. This means that each unit interval $[-n - 1, -n]$ undergoes the same reduction scheme (but with some delay with respect to the number of steps). Consequently, when the Cantor procedure is continued *ad infinitum* each unit segment $[-n-1, -n]$ becomes identical. In effect the Cantor set (in the log scale) is mapped onto a semi-infinite 1D crystal lattice. Evidently, if we consider the infinite-size CS_∞ the log_3 scale picture of it covers the unlimited 1D crystal lattice. We should point here, that the "unit cell" of this lattice has a complex, Cantor like, structure (see Fig 2).

It is evident that when we assume a net fractal in the log scale (e.g. a net fractal generated according to the rule (6) we obtain multiscale periodicity and full 3D crystal symmetry. Indeed, suppose that all members of the fractal $\hat{G}$ fulfill the self-affine mapping $S^{(m,n,l)} : \hat{G} \to \hat{G}$ in which any point $x = (\xi, \eta, \nu) \in \hat{G}$ is transformed according to the

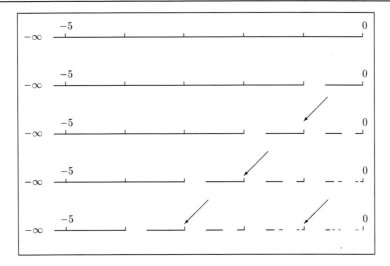

Figure 2. The first few steps of Cantor set construction pictured in the logarithmic $\log_3$ scale.

formula (3). In view of the arguments presented above, as expressed with the use of the multi-logarithmic scale, the family of mappings $S^{(m,n,l)} \in G_s^*$ (G_s^* denotes the dual space) is isomorphic with a 3D crystal lattice. In other words, the set of parameters $\vec{\lambda}$, which describes the self-affine mappings of a given fractal structure, forms a conventional 3D crystal lattice with lattice constants $a_i = \ln\lambda_i$. In this way we have proved isomorphic relation $S^{(m,n,l)} \to (ma_1, na_2, la_3)$ between G_s^* and some crystal lattice; the very same refers to the placement of the characteristic building blocks of the fractal G. As the eigendirections of the linear transformations are always orthogonal to each other, the isomorphism is limited to the crystal lattices of cubic, rhombic or tetragonal symmetry. This is why we call it "net fractals". As to the practical applications of the "net fractals" let us point out that physical systems with fractal properties which show different shrinking (scaling) ratios in different directions have been reported in the high-energy physics [4]. This characteristic feature of a physical system can be modelled only with the use of "net fractal" concept.

3. Modulated Net Fractals

The concept of net fractals gives inspiration for development of more complex fractal structures, namely the modulated fractals. To outline the concept of modulated fractals we should recall the fact that in nature there exist some solids for which the spatial ordering has a more complex symmetry. These are supercrystals with composition, or shift (displacement) modulations. In the simplest case of the 1D (position) modulation the supercrystal lattice sites are distributed according to the formula [10]

$$\bar{\kappa}_n \approx \kappa_1^o + (n-1)\cdot a - \gamma\cdot \sin[2\pi/b(\kappa_1^o + (n-1)\cdot a)], \qquad (8)$$

where κ_1^o is the zeroth-order position of the chain.

In this type of modulated crystals the ions occupy positions, which show modulatory displacements from the lattice sites predicted by the conventional crystal structure. The modulation period of the distance between neighboring atoms is either commensurate or incommensurate with respect to the basic structure. In any supercrystal the X-ray diffraction pattern spots form a $(n+d)$-dimensional reciprocal lattice [11]. In the theory of supercrystals the modulated commensurate or incommensurate phases are described by $(n+d)D$ Euclidean superspace containing nD sub-space called position space, while d denotes number of independent oscillations superimposed onto conventional lattice site position formula (compare (8)). The source of shift modulations is the presence of more than one ordering mechanism, each favoring different periodicity.

In the following we focus our attention on ((n+k)D supercrystal i.e nD crystals with k displacive oscillations of lattice sites. It can be proven that supercrystals can be regarded as some nonorthogonal projections of conventional (n+k)-dimensional crystals) onto a nD ($n \leq 3$) position space [11]. This means that modulated fractals can be assumed as the projection of conventional fractals within (n+k)D superspace, onto nD position space. This points the way for a simpler description of physical systems.

As the net fractals are isomorphic with some conventional crystals (in the log scale) there should be some fractal structures that are isomorphic (in the log scale they are identical) with the supercrystals. We call them *modulated fractals* /MF/. For physicists it is important to point out that some physical quantities exhibit log periodic oscillations of the form (8). This means that the new structures can be useful in a theoretical description of these systems.

We define a 1D modulated fractal as a set represented in the log-scale by a 1D modulated crystal (supercrystal). It can be proved that 1D modulation of the crystal lattice along the line L can be obtained as an intersection of the 2D periodic structure ((1+1)D supercrystal) with the line L, provided that the line L is not parallel to a unit cell edge of the 2D superstructure. Thus, any 1D supercrystal represents a non-orthogonal projection of some conventional 2D crystal onto line L. The additional dimension is called the internal dimension of the modulated crystal [12]. This means that any modulated fractal can be regarded as a projection of some other fractal having higher dimensionality.

As an illustration let us consider a fractal F_α^β represented (in the log scale picture) by a (1+1)D supercrystal, which in real scale is the cartesian product $F_\alpha^\beta = F_\alpha \times F_\beta$ of fractals F_α and F_β. Suppose further that in real scale F_α and F_β have dimensionalities α and β with $0 \leq \alpha, \beta \leq 1$. By the analogy with the supercrystals we denote F_α^β as the $(\alpha+\beta)D$ (super)-fractal. Its (log scale) projection onto the line L which is not parallel to an edge of the (1+1)D unit cell gives us the modulated fractal. Recently such projections of fractals have been applied in description of superconductivity that arises due to subquantum medium coherence [52]. Another example of a displacive modulation arises for these heteroepitaxial systems, for which the elemental constituents exhibit significant ionic radii mismatch. In such quasi-2D systems a (2+1)D or (2+2)D [10] overlayer superstructure is to be expected.

The fully deterministic superstructure of a 1D MF is characterized by two characteristic periods (as seen in the log-scale), one associated with the basic fractal scaling and the other given by the (in general incommensurate) oscillation period. Thus, the MF shows simultaneously the feature of a simple deterministic fractal and multiperiodicity characteristic of multifractals. In supercrystals as a rule we observe systems in which lattice modulations are composed only of a few sinusoidal oscillations. However, we can consider a more general case when the modulation is given by an arbitrary function $f(x)$. Its Fourier decomposition is presented by infinite series of sinusoidal modulations. Suppose we follow the concept of modeling supercrystals [11], [12] and ascribe an extra dimension to each component of the Fourier's expansion, then the arbitrary deviation $f(x)$ from the perfect crystal is represented by a fractal F^∞ immersed in the R^∞ space.

Concluding, the principal difference between MF and an ordinary fractal can be summarized as follows: in any MF $\hat{S}G$ the action of self-affine transformation $S_{SF}^{(m,n,l)} \in \hat{S}G_s^*$ transforms any point $\vec{x} = (\xi, \eta, \rho) \in \hat{S}G$ according to the formula:

$$S_{SF}^{(m,nl)} : \vec{x} \to (e^{\bar{\kappa}_n}\xi, e^{\bar{\mu}_m}\eta, e^{\bar{\sigma}_l}\rho), \tag{9}$$

where $\bar{\kappa}_n, \bar{\mu}_m, \bar{\sigma}_l$ are given by expressions of the type (8), while for the ordinary fractals this transformation reads

$$S^{(n,m,l)} : \vec{x} \to (e^{na}\xi, e^{mb}\eta, e^{lc}\rho). \tag{10}$$

From Eq. (9) results that the members of MF $\vec{x} \in \hat{S}G$ don't occupy sites predicted for the ordinary fractal but oscillate in a regular manner around sites of ideal structure. Moreover, one would expect that both Hausdorff dimension and lacunarity of the MF, at least in the case of commensurate modulation, are exactly the same as they are in the basic fractal structure.

As the net fractals and MF are isomorphic with the crystal or supercrystal lattices respectively the symmetry can be deduced from the crystallographic considerations. Evidently, the crystallographic symmetry is reflected by the isomorphism in the symmetry of net fractals and MF. Indeed, let us consider action of a symmetry element $\hat{O}_\alpha^i$ being a member of crystallographic group $\hat{O}_\alpha$, which characterizes the symmetry of the self-dual space G_s^*. By definition for any $S^{(m,n,l)} \in G_s^*$ we have $O_\alpha^i : S^{(m,n,l)} = S^{(m',n',l')} \in G_s^*$. Thus, we have $O_\alpha^i : S^{(m,n,l)} : G = G$ and $S^{(m,n,l)} : G = G$, which means that the crystal symmetry reflects the internal symmetry of the fractal G or a MF (if the supercrystal symmetry is taken into account). Consequently, invariance of the fractal structure under symmetry transformations $\vec{r'} \to \hat{R}_\alpha\vec{r}$ leads to the expectation that various physical properties share this symmetry. Evidently the symmetry of a geometrical fractal can be different from the symmetry of a physical system that has fractal structure. However, in the case of a fractal that is decorated by single species (atoms) with isotropic interactions between them these symmetries should be exactly the same.

From the general point of view the most evident area of perspective applications of the "crystalline" symmetry is the problem of real-space renormalization. The other is any Laplacian dynamical problem (diffusion, vibration or electron propagation) on a fractal (e.g., Sierpinski or Vicsek [13]) lattice. As it was shown in [13] with any irreducible representation of the point group that describes symmetry of a fractal there is associated an eigenenergy of the system. This means that the rotational symmetry of the problem allows prediction of new nesting properties [14]. Our result says that the mathematical apparatus elaborated by crystallographers (e.g., tables of irreducible representations) can serve as a convenient tool for classifying spectra of dynamical systems on fractals. The same refers to the studies of thermodynamical behaviour of the fractal (or MF) system G . Suppose we have a thermodynamical potential, let's say the free energy, assumed to be a function of some order parameters. If we expand it in powers of logarithms of the order parameters (i.e.

we construct the fractal counterpart of the Landau-Ginzburg functional it will contain only the symmetry invariants of the self-dual space G_s^*. The latter again is classified in terms of crystal symmetry elements. We believe that such kind of expansion applies in the case of the systems with a superstatistical behaviour [16]. Let us present now the theory on the of phase transitions in the net fractal or MF systems.

4. Phase Transitions on the Net and Modulated Fractals

Phenomenological Landau-Ginzburg theory has proven to be a powerful tool in description of second-order phase transitions in the bulk systems. There are many physical systems, e.g. ferroelectrics, for which the Landau-Ginzburg theory /LGT/ gives reasonable results. Growing interest in systems of reduced dimensionality has led to the question whether an extension of LGT onto fractal, superlattices, quantum dot or wire systems is possible. The main difficulty arises from the essential inhomogeneity of the low dimensional systems and different scaling relations of fractal systems. Nevertheless, a few approaches to the problem have been proposed [17], [18] [19]. In our paper an alternative approach, which for discussed above net fractals or MF resembles conventional formulation of LGT of phase transition derived for 3D systems.

Phase transitions and scale invariance are closely related, let us remember that fluctuation theory of phase transitions is based on the scaling ideas. In the following we assume that an ordered system, with complex microscopical organization, shows fractal scaling symmetry. The fractality of a physical system can be generated two-ways, it can arise due to the fractality of underlying medium or due to the fractality of the process. The latter mechanism can arise in the vicinity of phase transition when accumulated fluctuations of the order parameter form fractal patterns. In quantum systems under some conditions, e.g. at the critical energy separating localized and extended states, the wave functions are shown to have fractal structure (see [1] and references therein).

The fractality of the condensed matter systems manifests itself many ways, it can arise from the geometrical structure of nanocomposites [20], ferroelectric domain patterns [21] or fractality of relaxation processes [22]. We limit our considerations to the systems with the hereditary fractality due to the geometrical structure. As the example of such a system can serve perovskite ferroelectric relaxor systems of the formula $AB'_x B''_{1-x} O_3$ (e.g. $PbMg_{1/3}Nb_{2/3}O_3$). In the temperatures below onset of polarization, experiments reveal presence of nanometer scale regions, in which the ratio of B' and B" cations is 1:1 irrespective of stoichiometric composition. Specific heat measurements as well as neutron experiments confirm fractality of this microstructure [20]. There are many other ferroelectric systems that show fractality, fractal structure often arises in thin ferroelectric films, used in fabrication of capacitors, due to the diffusion limited aggregation or fatigue processes [23]. Another example form the high-temperature superconductors (of perovskite structure) [24].

In real systems we usually observe a collection of fractal clusters of non-uniform structure and the aggregate density decreases with aggregate size. For considerable range of length scales the correlation between mass and radius of aggregate is given by [14]

$$m(r) = m_o(r/a)^{d_m}, \tag{11}$$

where d_m is the mass dimension (in general different from the topological dimension D) that characterizes the space embedding the fractal cluster and a is a constant.

Cooperative processes in a fractal system give rise to an inhomogenous order parameter (e.g. polarization). The fractal geometry of the medium appears in the power-law behaviour of the order parameter which we assume as the vector field $\vec{P}$. The order parameter can have fractal characteristics even if the material itself does not need to have fractal microstructure. This effect can be observed at both classical and quantum levels. In some physical systems the wave-functions show fractal structure as results form the measurements of the participation numbers N_q (see [2], [15])

$$N_q = \left(\int |\psi(r)|^2 dr \right)^{-1} \propto L^{(q-1)D_q}. \tag{12}$$

In any case it is natural to assume that scaling of the effective polarization P(r,T) follows a self-similar scaling relation.

$$P(r,T) = P_o(T)(r/a)^{d_p}, \tag{13}$$

here T is the temperature.

The essential property of fractal systems (and consequently of physical processes on them) is the spatial inhomogeneity. Nevertheless, the properties of fractal systems show universal features despite of the distinction of the structure. It suggests that the classical idea to describe phase transitions in an unified way is applicable in these systems. The simplest approach to the modeling of the second order phase transitions is the famous Landau-Ginzburg theory. It is well known that LGT is valid systems with higher dimensions, while fractals are systems of reduced dimensionality. Thus, there may arise a question whether LGT is applicable in this case as well. To any physical systems various definitions of dimension can be proposed. In description of collective behavior of many particle systems we shall be interested in a geometrical dimension, i.e., the dimension of the Euclidean space embedding a particle and/or a spectral (dynamical) dimension, which is related to the collective excitations of the system. The spectral dimension d_s is defined via the density of states [24] $n(\epsilon)d\epsilon \approx (\epsilon-\epsilon_o)^{d_s/2-1}d\epsilon$, where the spectral dimension d_s can take any real (i.e. also fractional) value. It can be proven that in the case of non translation-invariant structures the spectral dimension is the best generalization of the Euclidean dimension of the system when dealing with dynamical or thermodynamical properties. Indeed, let us consider how the dimensionality enters the thermodynamical quantities. For an ideal Fermi/Bose gas the grand potential reads

$$\ln \Xi = \int_0^\infty n(\varepsilon) \ln \left(1 \pm e^{-\beta \varepsilon} \right) d\varepsilon, \tag{14}$$

From Eq. (14) one can easily see that all the information about the dimensionality of the actual system enters thermodynamical formulas *via* the density of states $n(\varepsilon)$. Thus, as we can see from the definition of $n(\epsilon)$ the thermodynamical evolution of any system depends on its spectral dimension. It is well known that low-dimensional systems (e.g. so called quasi-2 dimensional ones or fractals) can exhibit the spectral dimensions $d_s > 3$ [24]. Since we can have $d_s > 3$ one would expect that for some fractals LGT should give even a better description of phase transitions in some fractals then in the case of bulk (3D) systems.

392   Z. Bak

The LGT bases on the thermodynamical expansion of the free energy density in the powers of order parameter. In conventional systems the free energy functional F_p usually takes the form of [17], [19]

$$F_p = F_o + \frac{1}{2} \int_W \left[g|\nabla P|^2 + aP^2 + \frac{b}{2}P^4 \right] d^D r. \tag{15}$$

In the presence of fractality, the classical density of free energy needs to be integrated by an appropriate modelization of the microstructure, which alters the basic formulation of the LGT [19]. The other problem arises from the nonlocality [17] and nonhomogeneity of the order parameter in the fractal systems. There were a few attempts to handle the problem which account for these effects [17], [19], however, these formulations lead to calculational complications. Since we limit our considerations to the *net fractals* and MF we can give much simpler description which gives deeper insight in the physics of the phase transitions in the fractal systems.

As we have shown above the *net fractals* are isomorphic (in the logarithmic scale) with some crystal lattices, which means that when expressed in the logarithmic coordinates the mass density of a *net fractal* becomes uniform. In the case of ferroelectric crystals usually the polarization $\vec{P}$ plays the role of the order parameter. However, within the LGT the choice of the order parameter is somewhat arbitrary and the other physical quantity can be used instead, provided that it gives a better and simpler description of the phase transitions. In the fractal system, due to its nonuniform distribution the polarization is not a good candidate for the order parameter. Instead of $\vec{P}$, let us consider the quantity $\eta^{-1}(T) = -ln\left[P\left(a/r\right)^{d_m}\right] = -ln(P_o) = const$ (compare Eq. (13)). Evidently at T_c the order parameter η is undefined so we complete the definition of η by setting $\eta = 0$ for $T \geq T_c$ as results from the limit $\eta(T_c) = \lim_{P \to 0^+} \eta = 0$. With this choice, in view of the arguments above we can claim that in the log scale, both mass density and η on the *net fractals* show uniform distribution. This opens the possibility to describe the second order phase transitions in a spirit that is similar to the LGT for bulk systems.

The free energy functional of a bulk system (15) depends directly on the order parameter P. In homogenous bulk systems the polarization is also homogenous and its local value can be calculated within mean-field approximation. In the fractal systems, due to the inhomogeneity, any "mean field approximations" can be applied to calculate physical characteristics. However, as we have shown above, the physical quantities of the *net fractals* become uniform when presented in the logarithmic scale. Basing on this fact we will expand the free energy functional in power series of the logarithm of the polarization $\eta^{-1}(T) = -ln(P_o)$.

Since in the log scale the fractal systems become uniform the gradient term $|\nabla P|^2$ in Eq. (15) can be neglected. However in the case of nonuniform systems the counterpart of the gradient term in Eq. (15) requires special attention. The reason is that although any *net fractal* is represented by some crystal lattice the correlations between neighbouring sites depend on the situation they exert in real scale fractal. In the log-scale scenario some nearest neighbour lattice sites may represent fractal points that are not the nearest neighbours in the fractal geometry. This means that correlation between some lattice points is broken while among other it remains strong. As a matter of fact the average correlations represent rather some kind of a "percolation network" then some real crystal. Thus it means that one would expect that the order parameter variations are governed by the fractional dynamics

Physics on the Net Fractals 393

[17]. As it was shown in [17] within this approximation the gradient term $|\nabla P|^2$ in Eq. (15) should be replaced by the fractional gradient term $\Sigma_i |\nabla_{\xi_i}^\mu \eta_i \nabla_{\xi_i}^\mu \eta_i|$, where $\nabla_{\xi_i}^\mu$ represents the Riemann-Louville fractional derivative of order μ with respect to the log coordinate ξ_i [47]. Concluding, the free energy of the ordered *net fractal* system, written in the logarithmic coordinates, takes the form of [15]

$$F_\eta = F_o^\lambda + \frac{1}{2} \int_{W_\lambda} \left[g_\lambda \sum_i |\nabla_{\xi_i}^\mu \eta_i \nabla_{\xi_i}^\mu \eta_i| + a_\lambda \eta^2 + \frac{b_\lambda}{2} \eta^4 \right] d^D \vec{\xi}, \qquad (16)$$

where the superscript λ reminds us that respective constants are taken in the log scale and satisfy the conventional assumption of the LGT. The isomorphism between *net fractals* and some crystal lattices makes the integration go over the conventional Euclidean space of integral dimension D. Since we have proved that in the log scale the "order parameter" η is uniform then, provided that there is no ferroelectric domain structure, the gradient term in Eq. (16) can be neglected. In view of the above the equilibrium values of η can be obtained from the free energy condition $\delta F_\eta = 0$, which is equivalent to $(\partial F_\eta / \partial \eta) = 0$. Thus, in view of Eq. (16) this leads to

$$a_\lambda \eta + b_\lambda \eta^3 = 0, \qquad (17)$$

with conventional solutions $\eta = 0$ and $\eta = \sqrt{-a_\lambda/b_\lambda}$. Since $\eta \propto P(T)$, in view of Eq. (13) the equilibrium value of polarizations is given by

$$P(r, T) = exp \left[-\sqrt{-b_\lambda/a_\lambda} \right] \cdot (r/a)^{d_p}. \qquad (18)$$

Since our approach resembles the bulk formulation of the LGT it can be immediately extended onto antiferroelectric or multisublattice ordering.

4.1. Phase Transitions on the Modulated Fractals

Till now we have considered the simplest case when in the log scale the mass and order parameters are uniform. In real fractals, e.g. these formed by stochastic growth (i.e. diffusion limited aggregation), fractal structure shows logarithmic periodicities (see [25] and ref. therein). Independently, on the structure of the underlying medium the log periodic oscillations can arise due to the multifractality of the ordering process. The other source of the logarithmic periodicity of $\vec{\eta}$ is the multisublatttice (e.g. antiferroelectric or helical) structure of the fractal order parameter. In many systems like the DLA structures, rupture, earthquake, and financial crushes single mode, log-periodic oscillations with amplitude of the order of 10% have been reported [26]. The presence of log periodic oscilllations can be seen in the studies of statistical properties of DNA sequences in the bacterial chromosomes [27]. The sinusoidaly modulated log-periodicity was observed also in the specific heat of many boson systems [28], stock [29], [30] or crude oil [31] market data. The appearance of sinusoidal modulation superimposed onto log-periodicity was found in many economic data and interpreted as the precursor effect of financial crisis or speculation bubble. All these examples indicate modulated fractality of the underlying geometry and/or modulated fractality of the system properties.

In any case one would expect that in the log scale the order parameters can exhibit periodic (in the $\xi = log(R)$ space) oscillations. Below we will show, that contrary to the previous approaches to the LGT on fractals our approach is valid in description of these systems. Let us discuss the simplest case of the log periodicity, where only one (single mode) log periodic oscillatory contribution η_Q to the constant value η_o is present. In other words we assume that the wave-vector dependent susceptibility $\chi(q)$ (in the log scale) shows two maxima at $q_1 = 0$ and $q_2 = Q$. Evidently, some uncorrelated local fluctuations $\mu(\xi)$ are always present. Concluding we assume that the logarithm of the polarization density (i.e., our order parameter η) is given by.

$$\vec{\eta} = \vec{\eta}_o + \vec{\eta}_Q + \vec{\mu}(\xi). \tag{19}$$

As usual we expand the functional of the free energy density, in power series of the order parameters. Since $\vec{\eta}$ is dominated by the correlated fluctuations determined by the $q_1 = 0$ and $q_2 = Q$ wave-vectors, we will use the Fourier transform of the free-energy density [15]. Assuming coupling between correlated components to arise due to the local fluctuations the effective functional of the free energy density is given by [15]

$$F_p = \tfrac{1}{2} \int d^3\xi \left[\tfrac{1}{\chi_K}\eta_o^2 + \tfrac{1}{\chi_Q}\eta_Q^2 + \tfrac{1}{2}g_o\,\eta_o^4 + \tfrac{1}{2}g_Q\,\eta_Q^4 + \; + g_Q^o\,\eta_o^2\,\eta_Q^2 + \Gamma_Q^o\,(\vec{\eta}_o \cdot \vec{\eta}_Q)^2 \right]. \tag{20}$$

The expression in the brackets is the free energy per unit volume of the log space. We should mention here that the fractional gradient term of Eq. (15) in the case of order parameter (19) has three contributions which can be absorbed into the first three terms of Eq. (20). Although we consider here only the terms up to the fourth order in the polarization amplitude it is not hard to include higher order terms.

To gain information about behavior of the system described by free energy (20) we must minimize it with respect to the order parameters. From the free energy minimum condition $\delta F_m = 0$:

$$\frac{\partial F_p}{\partial \eta_o} = \frac{\partial F_p}{\partial \eta_Q} = 0 \tag{21}$$

we obtain a set of equations that determines equilibrium values of the coupled parameters of order. The detailed discussion of the solution can be found in [15]. Generally, depending on the relation between parameters of the free energy expansion both components of the oscillation are damped , one of them survives or both components of the oscillation coexist. Although we have considered here only the terms of expansion up to fourth powers of order parameters , the qualitative picture of the system will not be changed significantly when six order terms are included.

5.   Fluctuations

The phase transitions arise due to the accumulation of fluctuations in the ordered system. Basing on the specific symmetry of the net fractals and MF some general conclusion about nature of fluctuations in such systems can be drawn. In real systems the positions of species which form the fractal structure often deviate from an ideal mathematical fractal. In conventional physical systems, in which the allowed positions of its elements are uniformly distributed within R^3-space the position fluctuations are given by the Gaussian (normal) distribution. In an ideal self-similar fractal its elements are non-uniformly distributed, however, in the logarithmic scale the uniform distribution of allowed positions is restored. In real systems, modeled by fractals, positions of their elements can fluctuate, thus the log-uniform distribution is perturbed. If all other assumptions of the Laplace theorem are fulfilled then the probability distribution of fluctuations of real fractal structures is given by the well-known log-normal distribution

$$P(x) = \frac{1}{\sqrt{2\pi}\sigma x} exp\left(-\frac{(lnx - \mu)^2}{2\sigma^2}\right).$$ (22)

Consequently, one would expect that the probability distribution of many physical quantities on self-similar fractals follows the log-normal distribution. In support of this let us recall the real systems that exhibit fractal behaviour like isothermal aggregation [32] or stock fluctuations [33] with the log-normal statistics. In the ideal MF, even in the log-periodic scale, the distribution of points is non-uniform. However, if we introduce the variable $\mu = \mu_o + lnx + \alpha cos(\beta lnx)$ (as results from Eq. (8) the uniform density of points is restored again. Thus under some additional assumptions, necessary for the Laplace theorem to be valid, one would expect that fluctuations in real MF systems are given by the sine-log-normal distribution of the type.

$$P(x)dx = \frac{1}{\sqrt{2\pi}\sigma x} exp\left(-\frac{(lnx + \alpha cos(\beta lnx) - \mu_o)^2}{2\sigma^2}\right).$$
$$d(lnx + \alpha cos(\beta lnx)).$$

The above considerations prove the log-normal distribution of the fluctuation probability is the hallmark of the log-periodicity. Moreover, the appearance of sinusoidal modification in the probability distribution of the form of (23) indicates the presence of the modulated fractality in the examined system.

6. Magnetic Interactions on the Net Fractal System

The main focus of theoretical studies of fractal systems is on understanding the combined effect of the underlying topology on magnetic interactions. As a rule, because of complexity, such relation is studied numerically, however, numerical calculations do not provide a simple understanding of the parameters that control the process. That is why even simplified analytical models are still attractive. Below, an analytical approach to magnetic coupling on net fractals will be given.

Direct magnetic interactions on fractals are rather weak due to the small wavefunction overlapping the ionic wavefunction. The long-range RKKY indirect coupling, is expected to dominate. The effective RKKY interaction arises due to scattering of mobile electrons on the magnetic moments of impurity ions. From the other side the restrictive assumption of an isotropic surrounding, that validates the RKKY model doesn't hold [34], [35]. Therefore different concepts accounting for the effect of reduced geometry on magnetic interactions are still under debate [34] - [37]. However, as we have shown in preceding chapters for the net fractal system in log coordinates the homogeneity of the system is restored. This opens a possibility to describe the symmetries of some magnetic systems in the way that is reminiscent of the conventional RKKY formalism developed for crystalline systems. Moreover, as it was pointed in [8] real fractals allow manifestation of the RKKY-like indirect magnetic coupling. Motivated by this fact we present a study of indirect interaction, which is similar in spirit to the RKKY approach in the solid state theory.

Consider a "net fractal" cluster, consisting of N localized magnetic moments. Let us discuss now the non-homogeneities of the spin density and exchange integrals on a "net fractal". We assume that the density of magnetic moments follows the averaged mass density distribution $m(\vec{r})$ which for real fractals (in any direction D_i) scales on the average as $m((0, 0, ...x_i, 0...)) = m_i \cdot |x_i|^{d_i}$.

The symmetry of the fractal is reflected also in the symmetry of exchange interactions. As we have restricted our considerations to the net fractals in the log scale only of it is pictured onto some crystalline lattice. This does not mean that system becomes uniform, indeed, in the fractal system the electron mobility is restricted to the directions allowed by the internal geometry. This means that we have mapped the magnetic fractal onto a crystal lattice, in which the spins form a percolation clusters separated from the surrounding by the broken (dead) magnetic bonds.

To account for the existence of "dead bonds" we should model the (log scale) fractal system by some percolation lattice. It is a well-known fact that the structure of the equation of particle motion on the percolating network has the form of diffusion equation (or the linearized equation of motion for ferromagnetic spins) [38]. This allows the mobility of charge carriers to be modeled by the diffusion process. The diffusion on the fractal system as a rule involves the possibility of fractional dynamics [39], [40], [41]. Following this idea we will apply the model of fractional dynamics to the modeling of itinerant electron states in the fractal systems.

Introduction of fractional time derivatives into the modeling of such a dynamical system generates intrinsic damping of vibrations [40]. This effect mimics the damping due to the random interaction with environment. The fractional time derivative accounts for damping of the particle motion, while fractional space derivatives describe the reduced dimensionality of the system. Within this approach the generalized fractional Hamiltonian of the free electron system can be formally written as [40]. [41]:

$$_\xi D^\beta u = \lambda \cdot (_t D^\alpha) u \tag{23}$$

in a 1D system, and

$$(\Delta)^\beta = \lambda \cdot (_t D^\alpha) u = 0 \tag{24}$$

in the higher dimensional systems, with Δ being the Laplace operator. One should note here that parameters α and β model different features of the dynamical system. The order of fractional time derivative α describes the damping, while the value of β reflects the reduced dimensionality of the fractal set. In conclusion, the values of α and β are determined by different mechanisms and their values are, in general, different. We assume that the spatial derivatives are modeled by the right-side Rieman-Louville fractional derivative $_\xi D_0^\alpha$ while the time derivatives $_t D^\alpha$ we take as the Caputo ones [47]. Please note here that the Rieman-Louville derivative $_\xi D_0^\alpha$ (in view of the relation $\xi = lnx$) is nothing but the Hadamard derivative $_\xi D_0^\alpha$ when the real space coordinates x are used (for details see Kilbas et al. [47] p. 110)

The direct effect of the fractional dynamics is the unconventional spectrum of itinerant systems. In many low dimensional or fractal systems it acquires a more general form

$$n(\epsilon)d\epsilon \approx (\epsilon - \epsilon_o)^{d_s/2-1}d\epsilon \tag{25}$$

where the scaling factor d_s is a fraction and is called the spectral (or dynamical) dimension [24].

As it is evident that the indirect magnetic interactions are governed by the values of the effective spectral dimension [37]. With the use of formula (25) the analytical expression for the RKKY-like exchange integral in the case of arbitrary spectral dimension αD can be found [37])

$$J(\xi) = J_o\,\xi^{\alpha-2}\cdot$$
$$\left[J_{\alpha/2-1}(\kappa\xi)\,Y_{\alpha/2-1}(\kappa\xi) + J_{\alpha/2}(\kappa\xi)\,Y_{\alpha/2}(\kappa\xi)\right]. \tag{26}$$

where $\xi = ln\ x$, while $J_\nu(x)$ and $Y_\nu(x)$ are the Bessel and the Neumann functions respectively [24].

Therefore we can conclude that when the "net fractal" system is pictured in the logarithmic coordinates calculated the indirect exchange interactions it exhibits RKKY-reminiscent features. The exchange integrals show conventional sign reversing oscillatory behaviour. The leading term in the exchange integrals $J(\xi)$ decays with the interspin separation ξ (measured in the log scale) as $J(\xi) \propto \xi^{-\alpha}$. This means that the envelope of the $J(\xi)$ is governed by the spectral dimension α. The obtained formula explains the trends in the variation of exchange integrals as the function of interspin (measured in the log-scale) separation.

7. Fracton Excitations in the Net Fractal Systems

Fractal concepts describe not only the static geometrical features of these systems but also their dynamical properties and interactions. There are many oscillating processes and phenomena based on non-homogenous interactions. Our focus is on the dynamics of mechanical oscillations on a "net fractal" system.

As before (see chapter 7) we assume that effective forces follow the common power law scaling with the separation [42]

$$\sigma(\lambda\varepsilon) = \lambda^{-\alpha}\sigma(\varepsilon). \tag{27}$$

Suppose we perturb the fractal cluster locally and consider the local vibrations still pictured in log coordinates. As it was shown in [43], [44] such an approach gives proper description of the net fractal excitations. Although the mass/spin distribution in the log picture of the net fractal is uniform the dynamics of excitations is affected by existence of "dead bonds". This causes that in the system there arises the fractional dynamics and the excited states of the system are given by solutions of the Eqs (23-24) [43], [44]. The fractional time derivative accounts for the random interactions with the surrounding, while fractional space derivatives describe the reduced dimensionality of the system. One should note here that parameters α and β model different features of the dynamical system. The order of fractional time derivative α describes the damping, while the value of β reflects the reduced dimensionality of the fractal set. Following the approaches of [45], [46] we assume that time derivatives $_tD_*^\alpha$ are that of Caputo [47], while the space derivatives are the right-side Riemann-Louville ones.

Finding the solution of Eq. (23) for arbitrary values $\alpha \neq \beta$ in its more general form is impossible, however, under some additional assumptions we can find some specific solutions. Indeed, suppose we can separate the variables ξ and t. This means we assume that $u(\xi, t) = u_1(\xi) \cdot u_2(t)$. Provided that our solutions meet such an assumption we can rewrite Eq. (23) in the form of

$$\frac{1}{u_1(\xi)} \, _\xi D_0^\beta u_1(\xi) - \frac{1}{c^\alpha u_2(t)} \, _t D_*^\alpha u_2(t) = 0. \tag{28}$$

The equation (28) is equivalent to the two independent differential equations of the single variable ξ or t.

$$_\xi D_0^\beta u_1(\xi) = -\kappa^2 \, u_1(\xi) = (i\kappa)^2 \, u_1(\xi), \tag{29}$$

and

$$_t D_*^\alpha u_2(t) = \kappa^2 \, c^\alpha u_2(t), \tag{30}$$

where $\kappa \in \Re$. Solutions of both equations (29) and (30) can be easily written (for details see [47]) with the use of the generalized Mittag-Leffler functions $E_{\alpha,\beta}(\xi, t)$

$$E_{\alpha,\beta}(z) = \sum_{n=0}^\infty \frac{z^n}{\Gamma(\alpha n + \beta)}. \tag{31}$$

The solution of Eq. (30) has the unique solution (for details see [47] p. 230)

$$u_2(t) = b_o \, E_{\alpha,1}(-\kappa^2 c^\alpha t^\alpha) + b_1 \, t \, E_{\alpha,2}(-\kappa^2 c^\alpha t^\alpha), \tag{32}$$

provided that $b_o = u_2(0^+)$ and $b_1 = (_\xi D_0^1)u_2(0^+)$ (we assume that $1 \leq \alpha, \beta \leq 2$.

To find the solution of Eq. (29) let us consider the following equation

$$_\xi D_0^{\beta/2} u_1(\xi) \pm i\kappa \, u_1(\xi) = 0 \tag{33}$$

with the unique solution.

$$u_1(\xi) = u_o \, \xi^{\beta/2-1} E_{\beta/2,\beta/2}(\pm i\kappa \xi^{\beta/2}), \tag{34}$$

provided that $D^{\beta/2-1}u(0^+)$ is known. Thus, in view of the above the product

$$u(\xi, t) = u_1(\xi) \cdot u_2(t) \tag{35}$$

is the solution of (28) [47].

We can easily see that the action of the operator $_\xi D_0^{\beta/2}$ onto the both sides of Eq. (33) converts it into (29). This means that the solutions (34) are the solutions of Eq. (29). However, a solution (34) is a function of purely imaginary arguments, thus it is a complex function. To find the real and imaginary part of it let us recall some properties of the Mittag-Leffler functions. As we know the generalized Mittag-Leffler function fulfills the relation [47]

$$e_\alpha^{\lambda\xi} = \xi^{\alpha-1} E_{\alpha,\alpha}(\lambda\xi^\alpha) \xrightarrow{\alpha \to 1} e^{\lambda\xi} \tag{36}$$

and can be regarded as the counterparts of exponential functions defined in a space of fractional dimension. That is why the functions $e_\alpha^{\lambda\xi}$ are called α-exponentials [47]. The latter solution is consistent with the ideal fracton solution pictured as the logarithmic phonons. Plots of some functions $E_{\alpha,\alpha}$ that describe fracton vibrations are presented in figs 3.

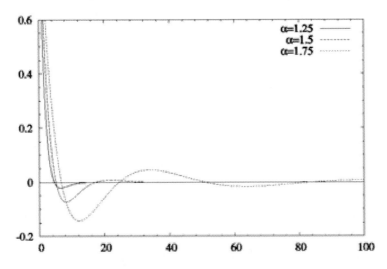

Figure 3. The spatial variation of the $E_{\alpha,\alpha}(x)$ for different values of α.

When differential equations are used in modeling of physical systems often instead of initial conditions that preserve uniqueness of the solution some boundary conditions are imposed on them. Boundary conditions may not preserve uniqueness of the solutions but always represent some quantity which has clear physical interpretation. Generally it is better to impose some boundary condition than discuss meaning of the quantity $D^{\beta-1}u(0^+)$. We assume that a fractal is immersed in a matrix which has larger force constants than those of a fractal. This means that ends of the fractal are tethered by the surrounding material of the matrix. We should point out here that for the fractal systems two kinds of tethering can be defined. One can impose the constraints on both internal and external (hull) boundaries of the fractal or on the hull of the fractal cluster only. However, within continuous approximation we have used above only the hull constraints can be defined. Fortunately, this is the situation observed in the real systems [48].

Suppose that our finite system, extended over continuous manifold M, is limited by the boundary ∂M. Let us set the typical constraints of the form $u(\xi,t)|_{\partial M} = 0$. This is just the case of a fractal tethered at the boundary [49]. Such a situation can be observed in real systems with fractal morphology in which the amplitudes of vibrations fall sharply at the fractal edges [48]. This means that the vibration modes are confined entirely to the interior of the fractal cluster. We should remind the reader here that the boundary constraints of the type $u(\xi,t)|_{\partial M} = 0$ are not equivalent to the Cauchy boundary conditions necessary to provide the uniqueness of solution. The latter are defined in the discussion of Eqs (34) and

(32). This means that there are many vibration modes which fulfill the tethering $u(\xi, t)|_{\partial M}$ =0. Let us discuss the allowed modes of fracton vibration with boundary constraints of the type $u(\xi, t)|_{\partial M}$ =0.

It can be easily seen that in case of symmetrical M (e.g. $\xi \in [-L, L]$) the solution (35) can satisfy our $u(\xi, t)|_{\partial M}$ =0 boundary constraints. Indeed, our boundary constraints are equivalent to $E_{\alpha,\alpha}(i\kappa|L|) = 0$. This means that the number of allowed vibration eigenmodes is equal to the number of zeroes ξ_n of the solution (35) (i.e. $E_{\alpha,\alpha}(\xi_n) = 0$)), of the generalized Mittag-Leffler functions. Thus, the allowed values of κ (κ is the counterpart of wave-vector in conventional systems) become quantized, $\kappa_n \propto \xi_n/L$. As we know the Mittag-Leffler functions have a finite and odd number of zeroes; thus the condition $u(\xi, t)|_{\partial M} = 0$ implies that only a finite number of vibration eigenmodes within a finite fractal system is possible.

The study of the effect of boundaries on the preselection of allowed fracton eigenmodes is of great importance from both fundamental and practical point of view. The fractal diffraction gratings [50] or fractal antennas [51] show attractive performances as the elements of different electronic devices. The main focus of this research is on resonant transmission/absorption slits/apertures. The results presented above can give practical indications for the engineering of devices with required localized fracton resonances in systems built of collections of fractal clusters or some fractal slit patterns, which exhibit fractal Fabry-Perot resonant absorption. We should point out here that all the results obtained above hold also for modulated fractals provided we use the uniform coordinate $\mu = lnx + \alpha cos(\beta lnx)$ (compare discussion after Eq. (22)) and Hadamard derivative $_\mu\hat{D}_0^\alpha$. It is worth pointing out here that formalism presented above is nothing but a form of the 1D fractional quantum mechanics similar to that of Laskin [41].

Conclusions

The analysis outlined above places the symmetry of linear deterministic fractals within a common framework of the solid state symmetry. This agrees with our intuition that, as a direct consequence of the ubiquitous self-similarity, fractals can be created by simple transformations. Generally, any affine transformations are combinations of just shifts, rotations, scalings and shears. Of course when we have limited ourselves to the linear scaling transformations only.

From practical point of view symmetry results in the reduction of complexity and may even allow discovering novel structural properties of hierarchical systems. Recently, logarithmic periodic oscillatory deviations in the behaviour of physical observables have been reported for several systems ([52], and references therein). Our MF structures can either explain or give a simpler picture of these effects. Along with the fractals for which symmetry of their self-affine transformations is isomorphic with the symmetry of conventional crystals there arises the concept of the MFs. These hierarchical systems show position modulations of their components with respect to that one expected for conventional fractals. The symmetry of self-affine transformations of the MFs can be directly related to the symmetry of supercrystals. We expect that the idea of the MFs can stimulate methods of characterization of fractal structures.

References

[1] Mildenberger, A.; F. Evers, F.; Mirlin, A.D. Phys.Rev. 2002 **B66**, 033109/1-4.

[2] Mendez-Bermudez, J.A.; Kottos, T. Phys. Rev. 2005 **B72**, 064108/1-4.

[3] Bialas, A.; Peschanski, R. Nucl. Phys. 1986 **B273**, 703-710, ibid 1988 **B308**, 851-863.

[4] Yuanfang, W.; Yang, Z.; Lianshou, L. Phys. Rev. 1995 **D51**, 6576-6579 .

[5] Zhou, W.X.; Sornette, D. Phys. Rev. 2002, E66, 046111/1-8.

[6] Sahimi, M.; Arbabi S. Phys. Rev. Lett. 1996 **77**, 3689-3682.

[7] Sornette, D. Phys. Rep. 1998 **297**, 239-270.

[8] Bak Z, Jaroszewicz R. Eur. J. Phys. 2009 **B64**, 231-235.

[9] Bak, Z. Materials Science-Poland 2008 **26**, 913-919.

[10] Bak, Z.; Gruhn W. J. Alloys Compounds 1995 **219**, 296-298.

[11] Wolff, P.M. Acta Crystallgr. 1974 **A30**, 777-785.

[12] J. Kocinski, Commensurate and Incommensurate Phase Transitions, PWN Warsaw/Elsevier, Amsterdam, 1990, pp. 63-71.

[13] Schwalm, W.A.; Schwalm, M.K.; Giona, M. 1997 Phys. Rev. **E 55**, 6741-6743.

[14] Alexander S. Phys. Rev. 1984 **B 29**, 5504-5508.

[15] Bak, Z. Phase Transitions 2007 **80**, 79-87.

[16] Souza, A.M.C.; Tsallis, C. Physica A 2004 **342**, 132-138.

[17] Milovanov, A. V.; Rasmussen, J.J. Phys. Lett. 2005 **A 337**, 75-80.

[18] Kim, C.K.; Rakhimov, A.; Yee, J. H. Phys. Rev. 2005 **B71**, 024518/1-6.

[19] Tarasov, V. E.; Zaslavsky G. M. Physica 2005 **354A**, 249-261.

[20] Gvasaliya, S. N.; Lushnikov, S. G.; Moriya, Y.; Kawaji, H.; Atake, T. Physica 2001 **B305**, 90-95

[21] Tadic, B. Eur. J. Phys. 2002 **B28**, 81-89.

[22] Galyarova, N. M.; Gorin, S. V.; Dontsova, L. I. Mat. Res. Innovat. 1999 3, 30-41.

[23] Li, X.; Liu, J.; Zhao, J.; Lu, D.; Xuan, J.; Gu, H. Phys. Lett. 1995 **A200**, 445-449.

[24] Bak, Z. Phys. Rev. 2003 **B68**, 064511/1-9.

[25] Blumenfeld, R.; Ball R. C. Phys. Rev. 1993 **E47**, 2298-2302 .

[26] Gluzman, S.; Sornette, D. Phys. Rev. 2002 **E 65** 036142/1-19.

[27] Jose, M. V.; Govezensky, T.; Bombadilla J. R. Physica 2005 **A 351** 477-498.

[28] de Oliveira, I. N.; Lyra, M. L.; Albuquerque, E. L.; da Silva, L. R. J. Phys. (Cond. Matter) 2005 **17** 3499-3508.

[29] Gnacinski, P.; Makowiec, D. Physica 2004 **A 344**, 322-325.

[30] Ide, K.; Sornette, D. Physica 2002 **A 307**, 63-106.

[31] Alvarez-Ramirez, J.; Ibarra-Valdez, C.; Araceli, B.; Rodriguez, E. Physica 2005 **A 349**, 625-640.

[32] Kellermann, K.; Craievich, A. F. Phys. Rev. 2003 **B 67**, 085405/1-7.

[33] Antoniou, I.; Ivanov, Vi. V. ; Ivanov, V. V.; Zrelov, P. V. Physica 2004 **A 331**, 617-638.

[34] Bouzerar, R.; Bouzerar, G.; Ziman, T. Phys. Rev. 2006 **B73**, 024411/1-8.

[35] Mahadevan. P.; Zunger, A.; Sarma, D.D. Phys. Rev. Lett. 2004 **93**, 177201/1-4.

[36] Balcerzak, T. J. Mag. Mag. Mater. 2007 **310**, 1651-1653.

[37] Bak, Z.; Jaroszewicz, R.; Gruhn, W. J. Mag. Mag. Mater. 2000 **213**, 340-348.

[38] Nakayama, T.; Yakubo, K.; Orbach, R.L. Rev. Mod. Phys. 1994 **66**, 381-443.

[39] Hanyga, A. Proc. Roy. Soc. (London) 2001 **A 457**, 2993-3005 .

[40] Ryabov, Ya. E.; Puzenko, A. Phys. Rev. 2002 **B 66**, 184201/1-8.

[41] Laskin, N. Phys. Rev. 2002 **E 66**, 056108/1-7.

[42] Balakin, A. S.; Susarrey, O.; Bravo, A. Phys. Rev. 2006 **E 64**, 066131/1-4.

[43] Bak, Z. Materials Science -Poland 2007 **25**, 491-496.

[44] Bak, Z. Acta Phys. Pol. 2008 **A113**, 541-544.

[45] Povstenko, Y. Z. J. Therm. Stresses 2005 **28**, 83-102.

[46] Gorenflo, R.; Iskanderov, A.; Luchko, Y. Fractional Calculus & Applied Analysis 2000 **3**, 76-89.

[47] A. A. Kilbas, H. M. Srivastava, and J. J. Trujillo: "Theory and Applications of fractional Differential Equations", Elsevier, Amsterdam 2006.

[48] Yakubo, K.; Nakayama, T. Phys. Rev. 1989 **B 40**, 517-523.

[49] Mukherjee, S.; Nakanishi, H. Physica 2001 **A 294**, 123-138.

[50] Wen, W,; Zhou, L.; Hou, B.; Chan, C. T.; Sheng, P.; Phys. Rev. 2005 **B 72**, 153406/1-4.

[51] Jaggard, D. L.; Jaggard, A. D. Wave Motion 2001 **34**, 281-289.

[52] Agop, M.; Ioannou, P. D.; Nica P. J Math. Phys. 2005 **46**, 062110/1-24.

In: Classification and Application of Fractals
Editor: William L. Hagen

ISBN978-1-61209-967-5
© 2012 Nova Science Publishers, Inc.

Chapter 13

FRACTAL PROPERTIES OF SOLUTIONS OF DIFFERENTIAL EQUATIONS

Mervan Pavsic, Darkov Zubrinic and Vesnav Zupanovic
Department of Mathematics FER, University of Zagreb, Zagreb, Croatia

Keywords: box dimension, Minkowski content, spiral, chirp, spiral chirp, multiple spiral, bifurcation, Schrödinger equation, Euler equation, Riemann-Weber equation, Hartman-Wintner equation, Bessel equation, half lienar equation, Liénard equation, Poincaré map, Hopf bifurcation, weak focus, logistic map, oscillation, fractal oscillatority, clothoid, Fresnel integral, elliptic BVPs, weak solution, nonregularity, singular dimension, Besov space, Sobolev space

AMS Subject Classification: 37C45, 37G10, 37G15, 34C15, 28A12, 28A75, 35Q55, 37G35, 28A12, 34C15, 26A30, 26A42, 26A18, 28A80

1. Introduction

1.1. A Short Overview

We give a survey of recent results by the authors and their collaborators concerning fractal analysis of trajectories of dynamical systems, oscillatory solutions of ODE's and singular sets of elliptic PDE's.

The idea of fractal dimension, i.e. noninteger dimension, has a long history, going back to the very beginning of the 20th century (Hermann Minkowski 1903, Felix Hausdorff 1919, Georges Bouligand 1928). The notion of box dimension is related to Minkowski and Bouligand. There are many other names for box dimension appearing in the literature, usually meaning the upper box dimension. One can encounter other equivalent names such as box counting dimension, Minkowski-Bouligand dimension, the Cantor-Minkowski order, Minkowski dimension, Bouligand dimension, Borel logarithmic rarefaction, Besicovitch-Taylor index, entropy dimension, Kolmogorov dimension, fractal dimension, capacity dimension, and limit capacity.

The notion of Minkowski content is not so widely known as the box dimension. We mention here the work by Lapidus and He [29] dealing with this notion, in the context of the generalized Weyl-Berry conjecture. Also Benoît Mandelbrot uses the notion of lacunarity, which is defined as the reciprocal value of the Minkowski content.

An important motivation for our work was the monograph by Claude Tricot [66]. From this monograph we learned that due to the fact that a realistic smooth curve has a non-vanishing width, it looks as an object of almost a planar nature. We are interested in curves which are nonrectifiable (i.e. of infinite length) near the point of accumulation. Some classes of spirals and chirps are examples of such objects. Rectifiable curves could be classified using their length. On the other hand nonrectifiable curves could be distinguished using their box dimension.

As a simple example, we may consider two planar systems with spiral trajectories Γ as in Figure 1, having different "concentrations" at the origin. This difference of concentrations can be explained by the notion of box dimension. So, the corresponding box dimensions of spirals on Figure 1 are $4/3$ and $12/7$. As we see, the second spiral on Figure 1 is almost two dimensional in the sense of box dimension.

Since 1970s dimension theory for dynamics has evolved into an independent field of mathematics. Its main goal is to measure complexity of invariant sets and measures using fractal dimensions. Here we deal with box dimensions only, since the Hausdorff dimension of trajectories that we consider is always trivial.

There are many important contributions by distinguished specialists dealing with the study of Hausdorff dimension of chaotic phenomena. Since 1970s thermodynamic formalism, developed by Sinai, Ruelle, and Bowen, resulted in Hausdorff dimension of the Smale horseshoe and in a lots of results about Hausdorff dimension of Julia and Mandelbrot sets. Since 1980 physicists started to estimate and compute fractal dimensions of strange attractors (Lorenz, Henon, Chua, etc.), in order to measure their complexity. Fractal dimensions are estimated also for attractors of infinite-dimensional dynamical systems. For applications of fractal dimensions to dynamics see a survey article Županović and Žubrinić [84] and the references therein.

In Section 2. we considered a different problem, namely, we tried to compute the box dimension of a trajectory itself and to study its dependence on the bifurcation parameter. Well known bifurcations can be described in the new way using fractal analysis. It has been done for the Hopf-Takens bifurcation, see Žubrinić and Županović [75, 79], and also for the saddle-node and period doubling bifurcations of discrete one-dimensional systems, see [18], and some generalizations, see Horvat Dmitrović [31]. Bifurcations of continuous dynamical systems, which do not involve spiral trajectories, can be studied using the time 1 map, see [32], and the results analogous to the discrete systems are obtained. An interesting phenomenon has been found in the continuous and discrete case, namely that the box dimension of trajectory, viewed as a function of bifurcation parameter, has a nontrivial jump at the point of bifurcation. Box dimension depends on the quantity and the quality of the objects born in the bifurcation.

Graphs of solutions of a class of second order differential equations are also considered in the case where the corresponding system has weak focus. Relation between box dimension of the graph and of the trajectory has been found, see [55, 34].

We also consider spiral trajectories in $\mathbb{R}^3$ contained in Lipschitz or Hölder surfaces, see [76, 78]. Box dimension of spiral trajectories and their projections into coordinate planes, which are spirals or chirps, has been computed. A class of systems in $\mathbb{R}^3$ has been found

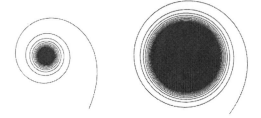

Figure 1. Spirals $r = \varphi^{-1/2}$ and $r = \varphi^{-1/6}$ have box dimensions $4/3$ and $12/7$ respectively.

for which the box dimension of spiral trajectories depends essentially on the coefficients of the system. Such dependence cannot occur for planar systems.

Section 3. deals with a class of nonlocal Schrödinger Cauchy problems, which represent a natural extension of planar dynamical systems of weak focus type. We deal with a type of problems closely related to those considered in Section 2.. We payed particular attention to the study of fractal properties of multiple spirals in $\mathbb{R}^4$ and their projections to $\mathbb{R}^3$, that we call spiral chirps. We show among others that any dynamical system in $\mathbb{R}^n$ can be naturally interpreted in terms of the corresponding Schrödinger Cauchy problem. In other words, the study of dynamical systems in $\mathbb{R}^n$ is a special case of the study of Schrödinger Cauchy problems. In particular, the 16th Hilbert problem can be interpreted in terms of a class of nonlocal Schrödinger Cauchy problems.

Secion 4. deals with the notion of fractal oscillatory near a point for real functions of a real variable, which has been introduced in Pašić [49], and studied for the second-order linear differential equations of Euler type as the basic model, see also Pašić [48] and Wong [68]. This includes the nonrectifiability of graph of oscillatory solutions and the computation of their box dimension. Recently it has been extended to the Riemann-Weber equation (see Pašić [50]), and in a more general setting to second-order linear differential equations of Hartman-Wintner type (see Kwong, Pašić, Wong [39]) as well as to second-order half-linear and nonlinear differential equations (see Pašić, Wong [54], and Wong [69]). In all these articles chirp-like solutions have been studied.

Section 5. is devoted to the study of nonregularity of weak solutions of a class of elliptic boundary value problems involving p-Laplace operator. The nonregularity of a family of elliptic boundary value problems is studied via the notion of singular dimension of a Banach space (or just a set) of measurable real functions, which is defined as the supremum of Hausdorff dimension of singular sets of functions. The phenomenon of the loss of regularity of weak solutions of p-Laplace equations for $p = 2$ is treated as well.

Section 6. is devoted to the study of fractal properties of Euler spiral, also known as the clothoid or the Cornu spiral. It appears among others in optimal control theory. We have computed box dimension of generalized clothoids and obtained a new proof of the asymptotic expansion of generalized Fresnel integrals.

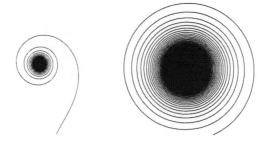

Figure 2. Spirals $r = \varphi^{-1/2}$ and $r = 6\varphi^{-1/2}$ both have box dimensions $4/3$, but different Minkowski contents.

1.2. Definitions

Let us introduce some basic definitions that we shall need in the sequel. Assume that $A \subset \mathbb{R}^N$ is a bounded set. By the *Minkowski sausage* of radius ε around A we mean the ε-neighbourhood of A, and denote it by $A_\varepsilon := \{y \in \mathbb{R}^N : d(y, A) < \varepsilon\}$, where $d(y, A)$ is Euclidean distance from y to A. It represents a generalization of the notion of ball, in which case the center A is just a single point.

The *lower s-dimensional Minkowski content* of A, $s \geq 0$, is defined by:

$$\mathcal{M}_*^s(A) := \liminf_{\varepsilon \to 0} \frac{|A_\varepsilon|}{\varepsilon^{N-s}}, \tag{1.1}$$

and analogously the *upper s-dimensional Minkowski content* $\mathcal{M}^{*s}(A)$. Note that in general these values are in $[0, \infty]$. The corresponding lower box dimension of A is defined by

$$\underline{\dim}_B A := \inf\{s \geq 0 : \mathcal{M}_*^s(A) = 0\},$$

and analogously the upper box dimension,

$$\overline{\dim}_B A := \inf\{s \geq 0 : \mathcal{M}^{*s}(A) = 0\}.$$

If both of these values coincide, the common value is denoted by $\dim_B A$, and is called box dimension of A. For various properties of box dimensions see for example Falconer [22]. Clearly, $\underline{\dim}_B A \leq \overline{\dim}_B A$, and the inequality may be strict, see [22]. Moreover, it is possible to construct a class of fractal sets $A \subset \mathbb{R}^N$ for which $\underline{\dim}_B A = 0$ and $\overline{\dim}_B A = N$, see [74].

If A is such that $d = \dim_B A$ exists and $0 < \mathcal{M}_*^d(A) \leq \mathcal{M}^{*d}(A) < \infty$, we say that A is *nondegenerate*. Otherwise A is said to be *degenerate*.

If $\mathcal{M}_*^s(A) = \mathcal{M}^{*s}(A)$ for some $s \geq 0$, this common value is called *s-dimensional Minkowski content* of A, and is denoted by $\mathcal{M}^s(A)$. If for some $d \geq 0$ we have that $\mathcal{M}^d(A) \in (0, \infty)$, then we say that the set A is *Minkowski measurable*. In this case clearly $d = \dim_B A$.

The name of box dimension stems from the following: if we have an ε-grid in $\mathbb{R}^N$ composed of closed N-dimensional boxes with side ε, and if $N(A, \varepsilon)$ is the number of boxes of the grid intersecting A, then $\overline{\dim}_B A = \limsup_{\varepsilon \to 0} \frac{\log N(A, \varepsilon)}{\log(1/\varepsilon)}$, and analogously for $\underline{\dim}_B A$.

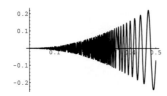

Figure 3. $(2, 4)$-chirp has box dimension $7/5$.

1.3. Basic Examples

A basic example of fractal sets with nontrivial box dimension is a-string defined by $A = \{k^{-a} : k \in \mathbb{N}\}$, where $a > 0$, introduced by Lapidus, see e.g. [40]. Here $\dim_B A = 1/(1 + a)$.

An important rôle in this article is played by the following Tricot's formulas, see [66, p. 121].

(i) Box dimension of the spiral in the plane defined in polar coordinates by $r = m \varphi^{-\alpha}$, $\varphi \geq \varphi_1 > 0$, where $\varphi_1, m > 0$ and $\alpha \in (0, 1]$ are fixed, is equal to $2/(1 + \alpha)$. See Figures 1 and 2.

(ii) Assuming that $0 < \alpha < \beta$, box dimension of the graph of the function $f_{\alpha,\beta}(x) = x^\alpha \sin(x^{-\beta})$, $x \in (0, 1]$, called (α, β)-chirp, is equal to $2 - (\alpha + 1)/(\beta + 1)$. See Figure 3.

On Figure 2 we have two spirals with box dimensions both equal to $4/3$, but still we clearly see that their "concentrations" at the origin are different. Therefore we use a subtler tool, the Minkowski content $\mathcal{M}^d(\Gamma)$ of spirals Γ. For the spiral Γ defined by $r = m\varphi^{-\alpha}$, $\alpha \in (0, 1)$, its value is equal to expression (2.4) in Theorem 2.1. The respective values of Minkowski content for spirals on Figure 2 are approximately 7 and 77, that is, the right-hand spiral has almost eleven times larger Minkowski content than the left-hand spiral.

1.4. Notation

Given any two real functions $a(t)$ and $b(t)$ of real variable, such that $a(t) \geq 0$ and $b(t) \geq 0$ for all t, we write $a(t) \simeq b(t)$ as $t \to 0$ (as $t \to \infty$), and say that $a(t)$ and $b(t)$ are *comparable*, if there exist positive constants C and D such that $C\, a(t) \leq b(t) \leq D\, a(t)$ for all t sufficiently close to $t = 0$ (sufficiently large). We write $a(t) \sim b(t)$ if $a(t)/b(t) \to 1$ as $t \to 0$ (as $t \to \infty$).

Two sequences $(a_n)_{n \geq 1}$ and $(b_n)_{n \geq 1}$ of positive real numbers are said to be *comparable*, and we write $a_n \simeq b_n$ as $n \to \infty$, if

$$A \leq a_n/b_n \leq B,$$

for some positive constants A and B and n sufficiently large.

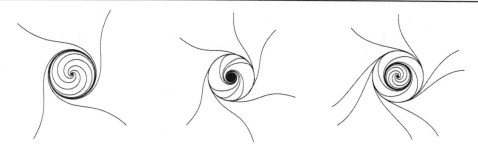

Figure 4. Portraits of the solutions of (2.2) for parameters $a_0 < 0$, $a_0 = 0$, $a_0 \in (0,1)$.

Figure 5. Portraits of the solutions of (2.2) for parameters $a_0 = 1$, $a_0 > 1$.

2. Fractal Properties of Trajectories of Dynamical Systems

Let us first illustrate the connection between box dimension of spiral trajectories of focus type and limit cycle type and bifurcations which appear from the focus and from the limit cycle. As an example we consider a planar vector field described with the following system in polar coordinates:

$$\begin{cases} \dot{r} &= r(r^4 - 2r^2 + a_0), \\ \dot{\varphi} &= 1. \end{cases} \tag{2.2}$$

By varying parameter a_0 we obtain the phase portraits exhibited on Figures 4 and 5. Here we consider spirals Γ near the corresponding limit cycles, and we call them spirals of the limit cycle type. We notice that the spiral of the limit cycle type on Figure 5 tends slower to its limit cycle than the corresponding spirals of the limit cycle type on Figure 4. We point out that for $a_0 = 1$, Figure 5, we have the limit cycle of algebraic multiplicity two, while for $a_0 < 1$, Figure 4, the corresponding limit cycle has multiplicity one. Algebraic multiplicity of the limit cycle is defined as the multiplicity of the root $r = a$ of the right-hand side of (2.2). In the case of $a_0 = 1$ the box dimension of spirals of limit cycle type is $3/2$, while in all the other cases it is equal to 1, see Theorem 2.5.

2.1. Hopf Bifurcation of Planar Systems

In this section we present two different approaches to study fractal properties of the spiral trajectories of the weak focus related to Hopf bifurcation.

Fractal Properties of Solutions of Differential Equations 411

By a planar spiral of focus type we mean the graph of a function $r = f(\varphi)$, $\varphi \geq \varphi_1$, such that $f(\varphi) \to 0$ as $\varphi \to \infty$ and $k \mapsto f(\varphi + 2k\pi)$ is decreasing for each φ.

Following [75] we classify spirals in three different ways.

(a) Spirals of *focus type* are defined as above. Spirals of *limit cycle type* are defined by $r = 1 - f(\varphi)$, or $r = 1 + f(\varphi)$, with f as above. Note that $r = 1 - f(\varphi)$ tends to the limit cycle $r = 1$ from inside, and $r = 1 + f(\varphi)$ from outside.

(b) We introduce *power and exponential* spirals of focus type by $r = \varphi^{-\alpha}$ and $r = e^{-\varphi}$, respectively. Analogous spirals of limit cycle type are $r = 1 - \varphi^{-\alpha}$ and $r = 1 - e^{-\varphi}$. Similarly for the spirals tending to the limit cycle from outside. It is clear that the definition can be given in the more general form.

(c) We say that a spiral is *nondegenerate* if its Minkowski content is nondegenerate. Analogously for degenerate spirals.

For example, $r = \varphi^{-1}$, $\varphi \geq \varphi_1 > 0$, defines the degenerate power spiral of focus type. It is nonrectifiable, and of box dimension 1. Nonrectifiability in this case is reflected in the fact that 1-dimensional Minkowski content of the spiral is infinite. By $r = 1 - e^{-\varphi}$ we obtain degenerate exponential spiral of limit cycle type.

By defining $r = \varphi^{-\alpha}$ we obtain nondegenerate power spirals of focus type for $\alpha \in (0, 1)$, and $\dim_B \Gamma = 2/(1 + \alpha)$. If $r = 1 - \varphi^{-\alpha}$, we obtain nondegenerate power spirals of limit cycle type, for any $\alpha > 0$, and $\dim_B \Gamma = \frac{2+\alpha}{1+\alpha}$. It can be shown that these spirals are not only nondegenerate, but Minkowski measurable, see [75].

Now we cite a result from [75] which is a generalization of the Tricot's formula, see [66, p. 121] for spiral $r = \varphi^{-\alpha}$, $\varphi \geq \varphi_1 > 0$, where $\alpha \in (0, 1]$. Here is a simplified, but equivalent form of [75, Theorem 6]. It can be found in [37].

Theorem 2.1. (Minkowski measurable spirals, [37, Theorem 2]) *Assume that* f : $[\varphi_1, \infty) \to (0, \infty)$ *is a decreasing,* C^2 *function converging to zero, and* $\varphi_1 > 0$. *Assume that there exists the limit*

$$m := \lim_{\varphi \to \infty} \frac{f'(\varphi)}{(\varphi^{-\alpha})'}. \tag{2.3}$$

Let there be a positive constant C *such that* $|f''(\varphi)| \leq C\varphi^{-\alpha}$ *for all* $\varphi \geq \varphi_1$. *Let* Γ *be the graph of the spiral* $r = f(\varphi)$ *with* $\alpha \in (0, 1)$, *and define* $d := 2/(1+\alpha)$. *Then* $\dim_B \Gamma = d$, *the spiral is Minkowski measurable, and moreover,*

$$\mathcal{M}^d(\Gamma) = m^d \pi (\pi\alpha)^{-2\alpha/(1+\alpha)} \frac{1+\alpha}{1-\alpha}. \tag{2.4}$$

2.1.1. Fractal Analysis of Normal Forms of Systems with Pure Imaginary Eigenvalues

We consider planar polynomial vector fields of the form

$$\begin{aligned}
\dot{x} &= P(x, y) \\
\dot{y} &= Q(x, y).
\end{aligned}$$

Figure 6. The cases of $a_0 < 0$, $a_0 = 0$, and $a_0 > 0$ for system (2.6).

Let us introduce the notion of *strong focus* as a singular point (equilibrium point) where the matrix of the linear part of the vector field has both real and imaginary parts of eigenvalues nonzero. A *weak focus* is a singular point of focus type where both eigenvalues are pure imaginary and nonzero.

A standard model where the Hopf bifurcation occurs, written in the form of a vector field, is

$$X = \left(-y + a_0 x + x(x^2 + y^2)\right) \frac{\partial}{\partial x} + \left(x + a_0 y + y(x^2 + y^2)\right) \frac{\partial}{\partial y}, \quad (2.5)$$

or equivalently,

$$\begin{cases} \dot{x} &= -y + a_0 x + x(x^2 + y^2), \\ \dot{y} &= x + a_0 y + y(x^2 + y^2). \end{cases} \quad (2.6)$$

In polar coordinates it has the form

$$\begin{cases} \dot{r} &= r(r^2 + a_0), \\ \dot{\varphi} &= 1. \end{cases}$$

Changing the parameter a_0 from positive values across zero to negative ones, the phase portrait changes from strong focus through weak focus to strong focus surrounded with a limit cycle. See Figure 6.

In the classical work by Takens [64] a more general situation is considered permitting multiple parameters and multiple limit cycles. Therefore it is called the *Hopf-Takens bifurcation*. Takens considers a p-parameter family of planar vector fields X such that $X(0) = 0$. It is required that when all parameters are equal to zero then the field X has eigenvalues on the imaginary axes and nonzero. He proved that if X is of codimension l (see Guckenheimer and Holmes [27]), then all possible nearby phase portraits and related bifurcations of X can be separated into two models $X_{\pm}^{(l)}$, where

$$X_{\pm}^{(l)} = \left(-y \frac{\partial}{\partial x} + x \frac{\partial}{\partial y}\right) \pm \left((x^2 + y^2)^l + a_{l-1}(x^2 + y^2)^{l-1} + \cdots + a_0\right) \left(x \frac{\partial}{\partial x} + y \frac{\partial}{\partial y}\right),$$

Fractal Properties of Solutions of Differential Equations 413

$(a_0, \ldots, a_{l-1}) \in \mathbb{R}^l$, is a normal form of X. This system is called the *standard model for Hopf Takens bifurcation*. In polar coordinates it reads as:

$$\begin{cases} \dot{r} &= r(r^{2l} + \sum_{i=0}^{l-1} a_i r^{2i}), \\ \dot{\varphi} &= 1. \end{cases} \tag{2.7}$$

The case of $l = 1$ corresponds to the classical situation of Hopf bifurcation.

Here we consider a standard model of Hopf-Takens bifurcation from the point of view of fractal geometry. We are interested in fractal properties of trajectories near singular points and limit cycles.

If a spiral trajectory is tending to a strong focus, it can be shown that it is of exponential type. We describe our results for standard Hopf-Takens model:

(1) any spiral trajectory tending to a weak focus is of power type,

(2) any spiral trajectory tending to a limit cycle of multiplicity one is of exponential type, while a spiral trajectory tending to a limit cycle of multiplicity $m > 1$ is of power type.

In order to state our results it is necessary to introduce a few additional notions. We say that a spiral $r = f(\varphi)$ of focus type is *comparable with the spiral $r = \varphi^{-\alpha}$ of power type, or power spiral* if

$$\underline{C}\varphi^{-\alpha} \leq f(\varphi) \leq \overline{C}\varphi^{-\alpha} \tag{2.8}$$

for some $\underline{C}, \overline{C} > 0$, and for all $\varphi \in [\varphi_1, \infty)$. Analogously for spirals with negative orientation, that is, $\underline{C}|\varphi|^{-\alpha} \leq f(\varphi) \leq \overline{C}|\varphi|^{-\alpha}$ for $\varphi \in (-\infty, \varphi_1]$.

A spiral $r = f(\varphi)$ of focus type is *comparable with the exponential spiral $r = e^{-\beta\varphi}$* if

$$\underline{C}e^{-\beta\varphi} \leq f(\varphi) \leq \overline{C}e^{-\beta\varphi}$$

for some $\underline{C}, \overline{C} > 0$ and $\beta > 0$, and for all $\varphi \in [\varphi_1, \infty)$. Analogously for spirals with negative orientation, that is, for $\varphi \in (-\infty, \varphi_1]$ and $\beta < 0$.

Remark 2.2. Theorems 2.3 and 2.5 below can be extended to more general vector fields, that is, fields with normal form (2.7). For this it suffices to use Takens [64, Theorem 1.5 and Remark 1.6] with $p = l$. Indeed, in this case the whole configuration of closed integral curves near the singularity of p-parametric family X, $p = l$, is differentiably equivalent to the corresponding configuration in $X_{\pm}^{(l)}$, so that box dimensions of the corresponding trajectories are preserved.

Theorem 2.3. (The case of focus, [75, Theorem 9]) *Let Γ be a part of a trajectory of (2.7) near the origin.*

(a) If $a_0 \neq 0$, then the spiral Γ is of exponential type, that is, comparable with $r = e^{a_0\varphi}$, and hence

$$\dim_B \Gamma = 1.$$

(b) Let k be a fixed integer, $1 \leq k \leq l$, $a_l = 1$ and $a_0 = \cdots = a_{k-1} = 0$, $a_k \neq 0$. Then Γ is comparable with the spiral $r = \varphi^{-1/2k}$, and

$$d := \dim_B \Gamma = \frac{4k}{2k + 1}.$$

The spiral Γ is Minkowski measurable, and its d-dimensional Minkowski content is equal to an explict constant.

Theorem 2.3 shows the connection between the multiplicity k of focus of (2.7) and the box dimension of a spiral trajectory tending to the focus.

Remark 2.4. Notice that from Theorem 2.3 and the fact that the initial vector field X and its normal form $X_{\pm}^{(l)}$ are locally diffeomorphic, it follows that the Hopf-Takens bifurcation of codimension l, assuming additionally that $k = l$ (i.e. when we have birth of l limit cycles), occurs with box dimension equal to $4l/(2l+1)$. In particular, for $l = 1$ we have classical Hopf bifurcation of two dimensional flows, for which at the moment of bifurcation the spiral trajectory has dimension equal to $4/3$. Remark also that the larger box dimension of a spiral trajectory at the moment of bifurcation, the more limit cycles are born. The same result can be obtained using the Poincaré map, see Section 2.1.2..

Theorem 2.5. (The case of limit cycle, [75, Theorem 10]) *Let the system (2.7) have a limit cycle $r = a$ of multiplicity m, $1 \le m \le l$. By Γ_1 and Γ_2 we denote the parts of two trajectories of (2.7) near the limit cycle from outside and inside respectively.*
(a) Then Γ_1 and Γ_2 are comparable with exponential spirals $r = a \pm e^{-\beta\varphi}$ when $m = 1$, $\beta \ne 0$ (depending only on the coefficients a_i, $0 \le i \le l-1$);
(b) Γ_1 and Γ_2 are comparable with power spirals $r = a \pm \varphi^{-1/(m-1)}$ when $m > 1$.
In both cases we have

$$d := \dim_B \Gamma_i = 2 - \frac{1}{m}, \quad i = 1, 2.$$

For $m = 1$ we have degenerate spirals, while for $m > 1$ the spirals are Minkowski measurable.

Theorem 2.5 shows the connection between multiplicity of limit cycles of (2.7) and the box dimension of the corresponding spiral trajectories.

Example 2.6. Let us consider system (2.7) for $l = 1$, see Figure 6.

(1) For $a_0 > 0$ we have exponential spirals of focus type with box dimension 1. In this case they are rectifiable.

(2) For $a_0 = 0$ we have power spirals of focus type with box dimension $4/3$.

(3) For $a_0 < 0$ a limit cycle appears, and box dimension of all trajectories is equal to 1.

It is to be noted that the box dimension of any trajectory near the origin is nontrivial (that is, larger than one) only for $a_0 = 0$, since then a periodic orbit is born.

Example 2.7. Let us consider system (2.7) for $l = 2$ and $a_1 = -2$. Figures 4 and 5 show the following:

(1) If $a_0 < 0$ all box dimensions are equal to 1, and the spirals are of exponential type.

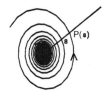

Figure 7. Poincaré map.

(2) If $a_0 = 0$ then $\dim_B \Gamma_1 = 4/3$, and the spirals are of power type; here Γ_1 is a part of any trajectory near the origin. The part near the limit cycle $r = \sqrt{2}$ has box dimension equal to 1, and it is of exponential type.

(3) If $a_0 \in (0, 1)$ we have two limit cycles of multiplicity one, and all box dimensions are equal to 1. Spirals are of exponential type.

(4) If $a_0 = 1$ then we have limit cycle $r = 1$ of multiplicity two, and all trajectories near the limit cycle (either inside or outside) have box dimensions equal to $3/2$. These are spirals of power type. Trajectories inside the limit cycle, but near the origin, have box dimension equal to 1 (exponential case).

(5) If $a_0 > 1$ then box dimensions of all trajectories are equal to 1 (exponential case).

As above, here we also have box dimension of trajectory near the origin of nontrivial value only for $a_0 = 0$, since then a periodic orbit is born. We shall encounter the same phenomenon in a class of one-dimensional discrete systems.

2.1.2. Poincaré Map and Fractal Dimension of Trajectories

In this section we use Poincaré map of a weak focus and a limit cycle to analyse fractal properties of the corresponding spiral trajectories. Also we cite a reverse result where we obtain some information about Poincaré map using box dimension of the trajectory.

If Γ is a spiral of limit cycle type (tending to a limit cycle Γ_0, say from inside), to each point x of Γ_0 we attach an axis $\sigma = \sigma(x)$ through this point, perpendicular to the limit cycle, oriented inwards, and with origin at x. The set of all such axes σ will be denoted by Σ_c. Let $P_\sigma : (0, \varepsilon_\sigma) \cap \Gamma \to (0, \varepsilon_\sigma) \cap \Gamma$ be the Poincaré map corresponding to any axis $\sigma \in \Sigma_c$ and defined by Γ. We have $P_\sigma(x_k) = x_{k+1}$, where the sequence $(x_k) = (0, \varepsilon_\sigma) \cap \Gamma$ in σ is arranged in decreasing order. By P_σ^k we denote k-fold composition of P_σ. We take $\varepsilon_\sigma > 0$ small enough, so that $P_\sigma^k(s) \to 0$ as $k \to \infty$ for all $s \in (0, \varepsilon_\sigma) \cap \Gamma$. If the family of Poincaré maps $\{P_\sigma : \sigma \in \Sigma_c\}$ is such that there exists $\beta > 0$ satisfying

$$d_k(s_\sigma) := P_\sigma^k(s_\sigma) - P_\sigma^{k+1}(s_\sigma) \simeq k^{-1-\beta}, \quad k \to \infty, \tag{2.9}$$

where $s_\sigma = \max(\Gamma \cap (0, \varepsilon_\sigma))$ (here the maximum is taken on the σ-axis), we say that Γ is the *limit cycle spiral of the Poincaré power* β. In other words, there exist two positive constants $A_\sigma < B_\sigma$ such that for any axis $\sigma \in \Sigma_c$ there holds $A_\sigma k^{-1-\beta} \leq d_k(s_\sigma) \leq B_\sigma k^{-1-\beta}$ for all k.

Remark 2.8. It is easy to see that condition (2.9) implies that the family of Poincaré maps has the following property for all $\sigma \in \Sigma_c$:

$$P_\sigma^k(s_\sigma) \simeq k^{-\beta}, \quad k \to \infty. \tag{2.10}$$

Indeed, $P_\sigma^k(s_\sigma) = \sum_{i=k}^\infty d_i(s_\sigma) \simeq \sum_{i=k}^\infty i^{-1-\beta} \simeq \int_k^\infty x^{-1-\beta} dx \simeq k^{-\beta}$, as $k \to \infty$. This is the reason why the limit cycle spirals satisfying (2.9) are said to be of power β.

Similarly, if Γ is a spiral of focus type, we consider the set Σ_0 of all axes σ through the focus. Let $P_\sigma : (0, \varepsilon_\sigma) \cap \Gamma \to (0, \varepsilon_\sigma) \cap \Gamma$ be the Poincaré map corresponding to any fixed axis σ (we assume that $s = 0$ corresponds to the focus for any axis σ). If the family of Poincaré maps $\{P_\sigma : \sigma \in \Sigma_0\}$ is such that there exists $\beta > 0$ such that (2.9) holds for any fixed $\sigma \in \Sigma_0$, we say that Γ is the *focus spiral of Poincaré power* β.

Remark 2.9. The definition of the power spiral $r = \varphi^{-\beta}$ in (2.8) is analogous to these definitions of spirals of the Poincaré power β. We call them power spirals of power β.

Let us recall that the function $d(s) = P(s) - s$, where $P(\cdot)$ is the Poincaré map and s small enough, is the displacement function. Note that if the Poincaré map corresponds to a limit cycle spiral with respect to an axis $\sigma \in \Sigma_c$ oriented inwards (if the spiral is inside the limit cycle), or outwards (if the spiral is outside the limit cycle) then in both cases we have $d(s) < 0$. The same holds for spirals of focus type.

Here we cite a result from [79] which is a main theorem in fractal analysis of the Poincaré map. From the asymptotics expansion od the Poincaré map of the weak focus and the limit cycle we obtain box dimension of the trajectory.

Theorem 2.10. (Poincaré map, [79, Theorem 1]) *Let Γ be a spiral trajectory of a planar vector field of class C^1. Let $P_\sigma(s)$ be the Poincaré map with respect to an axis σ, and assume that it has the form $P_\sigma(s) = s + d_\sigma(s)$ for each σ, where the displacement function $d_\sigma(\cdot) : (0, r_\sigma) \to (-\infty, 0)$ is monotonically nonincreasing, such that $-d_\sigma(s) \simeq s^\alpha$ as $s \to 0$, for a constant $\alpha > 1$ independent of σ. Then Γ is the spiral of power $1/(\alpha - 1)$. Furthermore,*

(a) if Γ is a focus spiral associated with a system (2.12) such that $p(x, y) = O(r^2)$ and $q(x, y) = O(r^2)$ as $r = \sqrt{x^2 + y^2} \to 0$, then

$$\dim_B \Gamma = \begin{cases} 2 - \frac{2}{\alpha} & \text{for } \alpha > 2, \\ 1 & \text{for } 1 < \alpha \le 2, \end{cases} \tag{2.11}$$

and Γ is Minkowski nondegenerate for $\alpha \neq 2$, and Minkowski degenerate for $\alpha = 2$;

(b) if Γ is a limit cycle spiral, then

$$\dim_B \Gamma = 2 - \frac{1}{\alpha},$$

and it is Minkowski nondegenerate.

In the proof of Theorem 2.10(b) we exploited the well known *flow-box* theorem, dealing with diffeomorphic equivalence of phase portraits, i.e. mapping trajectories onto trajectories. In particular, the function realizing the equivalence is lipeomorphism. Recall that

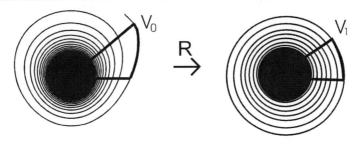

Figure 8. Flow-sector theorem: weak focus flow in sectors near the singular point is lipeomorphically equivalent to the annulus flow.

for two closed sets U and V we say to be diffeomorphic (lipeomorphic) if there exists a diffeomorphism (bilipschitz map) between their open neighbourhoods.

In the proof of Theorem 2.10(a) we exploited the following analog of flow-box theorem that we call *flow-sector theorem*. It shows that in any sufficiently small sector with vertex at the weak focus the dynamics is lipeomorphically equivalent to that of the annulus flow in a sector, by $\dot{r} = 0$, $\dot{\varphi} = 1$ in $\mathbb{R}^2 \setminus \{0\}$ in polar coordinates (r, φ). The result seems to be new even for analytic systems (2.12) such that near the singularity the flow is of spiral type. Let

$$\begin{cases} \dot{x} &= -y + p(x, y) \\ \dot{y} &= x + q(x, y), \end{cases} \quad (2.12)$$

where $p(x, y)$ and $q(x, y)$ are given C^1 functions such that $|p(x, y)| \leq C(x^2 + y^2)$ and $|q(x, y)| \leq C(x^2 + y^2)$ for some positive constant C and for (x, y) near the origin.

Theorem 2.11. (Flow-sector theorem, [79, Theorem 3]) *Let $U_0 \subset \mathbb{R}^2$ be an open sector with the vertex at the origin, such that its opening angle is in $(0, 2\pi)$, and the boundary of U_0 consists of a part of a trajectory and of intervals on two rays emanating from the origin Assume that*

$$p(x, y) = O(r^2), \quad q(x, y) = O(r^2) \quad as \; r = \sqrt{x^2 + y^2} \to 0. \quad (2.13)$$

If the diameter of U_0 is sufficiently small, then system (2.12) restricted to U_0 is lipeomorphically equivalent to the system

$$\begin{cases} \dot{r} &= 0 \\ \dot{\varphi} &= 1, \end{cases} \quad (2.14)$$

defined on the sector $V_0 = \{(r, \varphi) : 0 < r < 1, \; 0 < \varphi < \pi/2\}$ in polar coordinates (r, φ).

The main result dealing with box dimension of spiral trajectories near the weak focus of (2.12) is contained in Theorem 2.10. It enables us to apply it to the Hopf bifurcation, see Theorem 2.13.

In the following theorem we establish the connection between the asymptotic behaviour of iterates of the Poincaré map associated with a spiral, and the box dimension of the spiral. It complements our results from Section 2.1.. The main Theorem 2.10 will be a consequence of Theorem 2.12 below dealing with continuous systems, and of a result from [18] dealing with one-dimensional discrete systems, see Section 2.4..

418 Mervan Pavsic, Darkov Zubrinic and Vesnav Zupanovic

Theorem 2.12. ([79, Theorem 4]) *Let Γ be a spiral trajectory of a planar vector field of class C^1.*

(a) If Γ is a focus spiral trajectory of power $\beta > 0$, associated with the system described by (2.12), such that $p(x, y) = O(r^2)$ and $q(x, y) = O(r^2)$ as $r = \sqrt{x^2 + y^2} \to 0$, then

$$\dim_B \Gamma = \begin{cases} \frac{2}{1+\beta} & \text{for } \beta \in (0, 1), \\ 1 & \text{for } \beta \geq 1. \end{cases} \tag{2.15}$$

Furthermore, Γ is Minkowski nondegenerate when $\beta \neq 1$, and it is Minkowski degenerate for $\beta = 1$.

(b) If Γ is a limit cycle spiral trajectory of power $\beta > 0$, then

$$\dim_B \Gamma = \frac{2+\beta}{1+\beta}, \tag{2.16}$$

and the spiral is Minkowski nondegenerate.

Now we can prove the Poincaré map Theorem 2.10.

Proof of Theorem 2.10. The fact that the spirals are of power $\beta = 1/(\alpha - 1)$ follows immediately from [18, Theorem 1] or Theorem 2.31, with $f(s) = -d(s)$ there. The claims in (a) and (b) then follow from Theorem 2.12. $\qquad\square$

The next theorem deals with the Hopf bifurcation of the analytic systems of the form

$$\begin{cases} \dot{x} &= ax - y + p(x, y) \\ \dot{y} &= x + ay + q(x, y). \end{cases} \tag{2.17}$$

Let p and q be analytic functions such that

$$p(x, y) = \sum_{k=2}^{\infty} p_k(x, y), \quad q(x, y) = \sum_{k=2}^{\infty} q_k(x, y), \tag{2.18}$$

where p_k and q_k are homogeneous polynomials of degree k. In polar coordinates (r, φ) it has the form

$$\frac{dr}{d\varphi} = \sum_{k=2}^{\infty} s_k(\varphi) r^k. \tag{2.19}$$

Let $r(\varphi, r_0)$ be the solution of (2.19) such that $r = r_0$ for $\varphi = 0$. For r small enough we can write

$$r(\varphi, r_0) = r_0 + \sum_{k=2}^{\infty} u_k(\varphi) r_0^k, \tag{2.20}$$

with $u_k(0) = 0$ for $k \geq 2$. The Poincaré map for (2.17) near the focus is defined by

$$P(r_0) = r(2\pi, r_0) = r_0 + \sum_{k=2}^{\infty} u_k(2\pi) r_0^k.$$

The coefficient $u_k(2\pi)$ in the above expansion is called *k-th Lyapunov coefficient* of the weak focus, $k \geq 2$. We denote the first nonzero Lyapunov coefficient by V_k. It can be shown that in such V_k the index k is always odd, see e.g. Dumortier, Llibre, Artés [16, p. 124], or Roussarie [60, Lemma 8]. Lyapunov coefficients were first introduced in [43]. The integer $m = \frac{k-1}{2}$ is called the multiplicity of a focus. If $m > 0$ than this focus is a multiple or weak focus.

There are some other names for the Lyapunov coefficient, like the Lyapunov number, Lyapunov constant or Lyapunov quantity. Also, there are different definitions, but all of them are mutually equivalent and the coefficients are equal up to some positive constant, see [16, p. 126]. Lyapunov coefficients are important in solving the stability problem of a plane system which is a perturbation of a linear focus at the origin, but computation of the Lyapunov coefficients is quite a difficult task. There exist different ways of their computation. A new algorithm for the computation of the Lyapunov coefficients, based on Gasull, Torregrosa [25] and relying on ideas of Françoise [24], is exposed in [16]. There one can find some numerical examples, see [16, pp. 144–145], to which our dimension result in Theorem 2.10 applies. In [16, Section 4.7] there is a list of articles dealing with the computation of Lyapunov coefficients.

From Theorem 2.10(a) follows the fact that the spirals near the weak focus always have box dimension equal to $4/3$ at parameter of Hopf bifurcation. It means that the classical Hopf theorem could be extended by dimensional result. Furthermore, we have proved analogous facts in [18] for saddle-node and period-doubling bifurcations of one-dimensional discrete systems, see Section 2.4..

Theorem 2.13. (Hopf bifurcation, [79, Theorem 5]) *Assume that $p(x, y)$ and $q(x, y)$ are analytic functions as in (2.18). Let Γ be a spiral trajectory near the origin of system (2.17), where $a = 0$. If the first nonzero Lyapunov coefficient is V_3, then the Hopf bifurcation occurs at the origin of the system (2.17) at $a = 0$, the spiral is of power $1/2$, and*

$$\dim_B \Gamma = \frac{4}{3}.$$

Furthermore, Γ is Minkowski nondegenerate.

Proof. The claim follows from Theorem 2.10(a) with $\alpha = 3$.

It is possible to extend this theorem for $l > 1$, which corresponds to Hopf-Takens bifurcation. Box dimension of a spiral trajectory depends on the nonzero Lyapunov coefficient. The following theorem extends Theorem 2.13.

Theorem 2.14. (Analytic systems, [79, Theorem 7]) *Let Γ be a spiral trajectory near the origin of system (2.12), where $p(x, y)$ and $q(x, y)$ are analytic functions as in (2.18). If the first nonzero Lyapunov coefficient is V_{2k+1}, then*

$$\dim_B \Gamma = 2 \left(1 - \frac{1}{2k + 1} \right).$$

Furthermore, Γ is Minkowski nondegenerate.

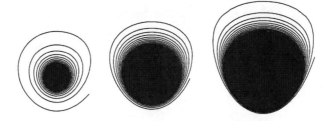

Figure 9. Three spirals of Liénard system (2.23) for $k = 1, 2, 3$, with box dimensions $4/3$, $8/5$, $12/7$, see (2.22).

In the simple model (2.5) the box dimension depends only on the exponents of the system, see [75], but in general it is not true because the Lyapunov coefficients depend on the coefficients of the system. In Caubergh, Dumortier [5, Theorem 5] a relation between Lyapunov coefficients and Hopf-Takens bifurcation has been established. Application of the results proved in [5] for Liénard systems

$$\begin{cases} \dot{x} &= -y + \sum_{i=1}^{N} a_{2i} x^{2i} + \sum_{i=k}^{N} a_{2i+1} x^{2i+1} \\ \dot{y} &= x, \end{cases} \quad (2.21)$$

and generalized Liénard systems can be found in Caubergh, Françoise [6]. For such systems it is very simple to compute Lyapunov coefficients in terms of the coefficients of the system. As an example we can extend [6, Proposition 8] by a dimensional result. We recall that [6, Proposition 8] is a result cited from Christopher, Lloyd [11]. For the sake of simplicity we state the following result under less general conditions than in [6, Proposition 8].

Theorem 2.15. (Liénard system, [79, Theorem 6]) *Let $a_{2k+1} \ne 0$ in (2.21), that is, a_{2k+1} is the first nonzero coefficient corresponding to an odd exponent of x. Then any spiral trajectory Γ, viewed near the origin, is of power $\frac{1}{2k}$ and has box dimension equal to*

$$\dim_B \Gamma = 2\left(1 - \frac{1}{2k+1}\right). \quad (2.22)$$

Furthermore, Γ is Minkowski nondegenerate.

An illustration of Theorem 2.15 is provided by Figure 9, where three spiral trajectories are shown, corresponding to the Liénard system

$$\begin{cases} \dot{x} &= -y + \left(\sum_{i=1}^{k} x^{2i}\right) + x^{2k+1} \\ \dot{y} &= x, \end{cases} \quad (2.23)$$

with $k = 1, 2, 3$ respectively. The terms with even exponents do not influence the box dimension of spirals.

Fractal Properties of Solutions of Differential Equations

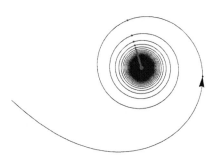

Figure 10. Poincaré sequence of weak focus.

The Liénard system in Theorem 2.15 concerns a class of planar systems in which $p(x, y)$ is a polynomial in x and $q(x, y) = 0$. One can ask if it is possible to construct polynomials $p(x, y)$ and $q(x, y)$ as in (2.18) such that a spiral trajectory of (2.12) near the origin does not have its box dimension of the form $2\left(1 - \frac{1}{2k+1}\right)$ or 1, that is, from the set

$$D_0 = \{\frac{4k}{2k+1} : k \in \mathbb{N}\} = \left\{\frac{4}{3}, \frac{8}{5}, \frac{12}{7}, \frac{16}{9}, \frac{20}{11}, \dots\right\}.$$

The answer is no. Moreover, even for analytic functions $p(x, y)$ and $q(x, y)$ as in (2.18), only dimensions from the set D_0 can be obtained. This is a consequence of Theorem 2.14, which follows immediately from Theorem 2.10(a).

Remark 2.16. From Theorem 2.10 and its consequences we know that for analytic fields with normal form (2.7) each spiral trajectory Γ of limit cycle type has box dimension of the form $\dim_B \Gamma = 2 - \frac{1}{m}$, where m is algebraic multiplicity of the spiral, that is, from the set

$$D_1 = \{2 - \frac{1}{m} : m \in \mathbb{N}\} = \left\{1, \frac{3}{2}, \frac{5}{3}, \frac{7}{4}, \frac{9}{5}, \dots\right\}.$$

Notice that in Theorem 2.14 the number k is a multiplicity of a weak focus of (2.12). The next theorem from Horvat Dmitrović [31] is the reverse of Theorem 2.14 since the multiplicity of a weak focus is a consequence of the box dimension. See Figure 10.

Theorem 2.17. (Multiplicity of a weak focus, [31, Theorem 10]) *Let Γ be a spiral trajectory near the origin of analytic system (2.12). Let $P : [0, r) \to \mathbb{R}$ be the Poincaré map associated to Γ near the origin. Let the sequence $S(x_1) = (x_n)_{n\geq 1}$ defined by $x_{n+1} = P(x_n)$ (stable focus) or $x_{n+1} = P^{-1}(x_n)$ (unstable focus), $x_1 \in (0, r)$ has box dimension of the form $\dim_B S(x_1) = 1 - \frac{1}{\alpha}$, with $\alpha \geq 3$ odd. Then*

$$P''(0) = \dots = P^{(\alpha-1)}(0) = 0, P^{(\alpha)}(0) \neq 0,$$

that is, the origin is a weak focus with multiplicity $m = \frac{\alpha-1}{2}$.

In Horvat Dmitrović [31] it is proved also the reverse connection between the box dimension of a spiral trajectory near a limit cycle and the multiplicity of a limit cycle with respect to the one stated in Theorem 2.10.

We consider a planar system $\dot{x} = f(x)$, with f analytic. Let γ be a limit cycle of the analytic system. Let P be the Poincaré map near the limit cycle γ, and d is a displacement map $d(s) = P(s) - s$. If $P'(0) \neq 1$ ($d'(0) \neq 0$), then γ is said to be a *hyperbolic limit cycle*. The stability of a hyperbolic limit cycle is determined by the sign of $d'(0)$, in the following way: if $d'(0) < 0$, then the cycle is stable, and if $d'(0) > 0$, it is unstable. If $d'(0) = 0$, that means that $P'(0) = 1$. Then γ is said to be a *nonhyperbolic limit cycle*, and $x = 0$ is a nonhyperbolic fixed point of a map P.

Let $P(s)$ be the Poincaré map for a cycle γ of the planar analytic system, and let $d(s) = P(s) - s$ be the displacement function. Then if

$$d(0) = d'(0) = \ldots = d^{(k-1)}(0) = 0, \ \ d^{(k)}(0) \neq 0, \tag{2.24}$$

then γ is called a multiple limit cycle of multiplicity k. If $k = 1$, then γ is called a simple limit cycle or hyperbolic. The nonhyperbolic limit cycle has multiplicity $k \geq 2$.

Notice that (2.24) is equivalent to $P'(0) = 1$, $P''(0) = \ldots = P^{(k-1)}(0) = 0$, $P^{(k)}(0) \neq 0$.

Let Γ be a spiral trajectory near a limit cycle of a planar vector field of class C^1. Let P be the associated Poincaré map. We denote by $Po(x_1) = \{x_n\}_{n \geq 1}$ the Poincaré sequence defined by

$$x_{n+1} = \begin{cases} P(x_n), & \gamma \text{ stable} \\ P^{-1}(x_n), & \gamma \text{ unstable} \end{cases} \tag{2.25}$$

with x_1 near γ. If Γ is a limit cycle spiral trajectory of power $\beta > 0$, then in [79, Theorem 4] it was proved that

$$\dim_B \Gamma = 1 + \dim_B Po(x_1). \tag{2.26}$$

Now we state a result which enables to derive the multiplicity of a limit cycle from its box dimension.

Theorem 2.18. (Multiplicity of limit cycles, [31, Theorem 11]) *Let Γ be a spiral trajectory near a limit cycle γ of an analytic planar vector field, and P the Poincaré map near the limit cycle γ. Let us assume that there exists the decreasing sequence $S(x_1) = (x_n)_{n \geq 1}$ defined by (2.25) with $x_1 \in (0, r)$ which has box dimension of the form $\dim_B S(x_1) = 1 - \frac{1}{\alpha}$, $\alpha > 1$.*
Then

$$P''(0) = \ldots = P^{(\alpha-1)}(0) = 0, \ \ P^{(\alpha)}(0) \neq 0$$

that is, the multiplicity of the limit cycle is equal to α.

Remark 2.19. The limit cycle under the assumptions of the Poincaré map Theorem 2.10 is of multiplicity α.

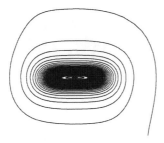

Figure 11. Nilpotent focus of $\dot{x} = y + xy^2$, $\dot{y} = -x^3$.

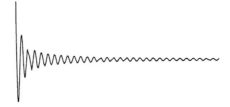

Figure 12. A function which is oscillatory at infinity.

Remark 2.20. The future work could be in the direction of degenerate focuses with nilpotent linear part or without linear part. Some of them could be obtained by transformation of the weak focuses. Also, the interesting objects that we intend to study are spiral trajectories near polycycles. The connection between box dimension and the emergence of the limit cycles might be of interest in the context of the 16th Hilbert problem, which asks for a uniform upper bound on the number of limit cycles which can appear from a polynomial planar system.

2.2. Oscillatory and Phase Dimension

This section reviews recent results conecting the fractal oscillatority of solutions of second-order nonlinear equations with spiral oscillatory of trajectories of the corresponding planar vector fields. The aim is to introduce tools for measuring fractal oscillatority of solutions of autonomous differential equations near $t = \infty$ using box dimension. Our idea is to analyse fractal properties of solutions of second-order nonlinear autonomous differential equations $\ddot{x} = g(x, \dot{x})$ using phase plane analysis. We assume that $g(0, 0) = 0$, and consider non-constant solutions $x(t)$ such that $x(t) \to 0$ and $\dot{x}(t) \to 0$ as $t \to \infty$. Here we consider differential equations whose related planar systems have linear part near the origin with pure imaginary eigenvalues, and with spiral trajectories. We propose the concept of oscillatory and phase dimensions of such solutions of the second-order nonlinear differential equation.

Let $x : [t_0, \infty) \to \mathbb{R}$, $t_0 > 0$, be a continuous function. We say that x is an *oscillatory function* (near $t = \infty$) if there exists a sequence $t_k \to \infty$ such that $x(t_k) = 0$, and functions $x|_{(t_k, t_{k+1})}$ intermittently change sign for $k = 1, 2, \ldots$. Let us define $X : (0, 1/t_0] \to \mathbb{R}$ by

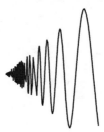

Figure 13. $(1,1)$-chirp $X(\tau)$.

Figure 14. A spiral $t \mapsto (x(t), \dot{x}(t))$.

$X(\tau) = x(1/\tau)$. We say that the function $X(\tau)$ is oscillatory near the origin if $x = x(t)$ is oscillatory near $t = \infty$. We measure the rate of oscillatority of $x(t)$ near $t = \infty$ by the rate of oscillatority of $X(\tau)$ near $\tau = 0$. More precisely, the *oscillatory dimension* $\dim_{osc}(x)$ (near $t = \infty$) is defined as the box dimension of the graph of $X(\tau)$ near $\tau = 0$.

Assume now that x is of class C^1. We say that x is a *phase oscillatory* function if the following stronger condition holds: the set $\Gamma = \{(x(t), \dot{x}(t)) : t \in [t_0, \infty)\}$ in the plane is a spiral converging to the origin. By a spiral here we mean the graph of a C^1-function $r = f(\varphi)$ in polar coordinates for which $f(\varphi) > 0$ and $f(\varphi) \to 0$ as $\varphi \to \infty$ (or $\varphi \to -\infty$). Note that in this definition the spiral may have self-intersections, but this will not be the case in this paper. The *phase dimension* $\dim_{ph}(x)$ of the function $x(t)$ is defined as the box dimension of the corresponding spiral $\Gamma = \{(x(t), \dot{x}(t)) : t \in [t_0, \infty)\}$.

We consider these two notions in the context of solutions x of autonomous second-order differential equations, for which the corresponding function X is chirp-like, that is, behaving "similar" to the function $X_\alpha(\tau) = \tau^\alpha \cos 1/\tau$. The graph of X_α has box dimension $\frac{3-\alpha}{2}$, according to Tricot's formula, see Section 1.3.. In [55] it was shown that then also the graph of X has the same box dimension. On the other hand if the trajectory $t \mapsto (x(t), \dot{x}(t))$ is "similar" to a spiral of power α, their box dimensions are $2/(1+\alpha)$, see Theorem 2.1.

In the sequel, (α, β)-*chirp-like functions* are defined by $y = p(t)\sin(q(t))$ or $y = p(t)\cos(q(t))$, such that $p(t) \simeq t^\alpha$, $q(t) \simeq t^{-\beta}$, $q'(t) \simeq t^{-\beta-1}$ as $t \to 0$.

The following theorem shows that from asymptotic behaviour of a spiral trajectory in the phase plane it is possible to derive information about concentration of the graph of the corresponding oscillatory function at infinity.

Theorem 2.21. (Chirp-spiral comparison, [55, 36]) *Let* $\alpha \in (0,1)$, *and assume that* $x : [t_0, \infty) \to \mathbb{R}$, $t_0 > 0$, *is a* C^1-*function such that the planar curve* $\Gamma = \{(x(t), \dot{x}(t)) :$

$t \in [t_0, \infty)\}$ *is a spiral* $r = f(\varphi)$, $\varphi \in (\varphi_0, \infty)$, $\varphi_0 > 0$ *in polar coordinates, near the origin, such that* $f(\varphi)$ *is monotonically decreasing,* $f(\varphi) \sim \varphi^{-\alpha}$, $f^{(j)}(\varphi) \sim (-1)^j \alpha(\alpha + 1)(\alpha + 2)...(\alpha + j - 1)\varphi^{-\alpha - j}$, $j = 1, ..., 5$ *as* $\varphi \to \infty$, *and* $\dot{\varphi}(t) \sim 1$, $\varphi^{(j)}(t) = o(t^{-j+1})$, $j = 2, ..., 5$ *as* $\varphi \to \infty$, *where* $\varphi(t)$ *is a continuous function defined by* $\tan \varphi(t) = \frac{\dot{x}(t)}{x(t)}$.

Define $X(\tau) = x(1/\tau)$. *Then* $X = X(\tau)$ *is* $(\alpha, 1)$-*chirp-like, and*

$$\dim_B G(X) = \frac{3 - \alpha}{2}, \tag{2.27}$$

where $G(X)$ *is the graph of the function* X. *Furthermore,* $G(X)$ *is Minkowski nondegenerate.*

Let us illustrate an application of Theorem 2.21 to Liénard equation:

$$\ddot{x} + g'(x)\dot{x} + x = 0 \tag{2.28}$$

is equivalent to the Liénard system

$$\begin{cases} \dot{x} &= -y - g(x) \\ \dot{y} &= x, \end{cases} \tag{2.29}$$

mentioned in the Section 2.1.2.. We consider analytic functions of the form

$$g(x) = \sum_{i=1}^{\infty} a_{2i}x^{2i} + \sum_{i=k}^{\infty} a_{2i+1}x^{2i+1}, \tag{2.30}$$

where $k \geq 1$.

Theorem 2.22. (Liénard equation, [55, Theorem 6]) *Let* $k \geq 1$ *and* $a_{2k+1} \neq 0$ *in (2.30). Then the oscillatory and phase dimensions of any solution* $x(t)$ *of (2.28) near the origin are equal to*

$$\dim_{osc}(x) = \frac{3}{2} - \frac{1}{4k}, \quad \dim_{ph}(x) = 2\left(1 - \frac{1}{2k + 1}\right).$$

The proof of this theorem exploits box dimension results for spiral trajectories in Section 2.1., together with Theorem 2.21.

Remark 2.23. Theorem 2.21 says that power spirals of power α generate $(\alpha, 1)$-chirp-like functions. The reverse theorem can also be proved, that is, $(\alpha, 1)$-chirp-like functions generate power spirals of power α, see [36]. As an interesting consequence of the reverse theorem we have that the phase dimension of the Bessel equation is equal to $4/3$. For more information concerning the fractal analysis of solutions of the Bessel equation see Section 4.4..

2.3. Lipschitz and Hölder Spirals in $\mathbb{R}^3$

Here we consider spiral trajectories in $\mathbb{R}^3$, which we assume to be contained in a two-dimensional surface. Their box dimension depends on properties of the surface as well.

We have noticed some new phenomena regarding box dimensions of spacial spiral trajectories. For example, there are systems for which the box dimension essentially depends on their coefficients, and not only on the exponents.

According to properties of the surface we distinguish the following two types of spirals. If the surface is defined by $z = r^\beta$, then for $\beta \geq 1$ we obtain Lipschitzian surface, while for $\beta \in (0,1)$ we obtain Hölderian surface. The corresponding spiral in $\mathbb{R}^3$, defined in cylindrical coordinates (r, φ, z) by

$$r = \varphi^{-\alpha}, \quad z = r^\beta, \quad \varphi \geq \varphi_1 > 0,$$

is said to be *Lipschitz-focus* spiral if $\beta \geq 1$, and *Hölder-focus* if $\beta \in (0,1)$.

For a spiral defined by

$$r = 1 - \varphi^{-\alpha}, \quad z = |1 - r|^\beta$$

we say to be a *Lipschitz-cycle spiral* if $\beta \geq 1$, and a *Hölder-cycle spiral* if $\beta \in (0,1)$, see Figure 16. For more general spirals in $\mathbb{R}^3$ see [78].

We deal with a class of systems such that the linear part in Cartesian coordinates has a pure imaginary pair and a simple zero eigenvalues. Its normal form in cylindrical coordinates is as follows (for normal forms see Guckenheimer and Holmes [27]):

$$\begin{aligned} \dot{r} &= c_1 r^3 + \cdots + c_m r^{2m+1} \\ \dot{\varphi} &= 1 \\ \dot{z} &= d_2 z^2 + \cdots + d_n z^n. \end{aligned} \quad (2.31)$$

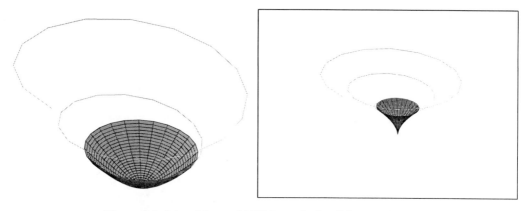

Figure 15. Lipschitz and Hölder spirals of focus type.

Let us formulate two results from [78] in a simplified form, presented in [76].

Theorem 2.24. (The case of focus in $\mathbb{R}^3$, [76, Theorem 2.1]) *Let Γ be a part of a trajectory of (2.31) near the origin in $\mathbb{R}^3$. Assume that k and p are minimal positive integers such that $c_k \neq 0$ and $d_p \neq 0$. Assume also $c_k d_p > 0$.*

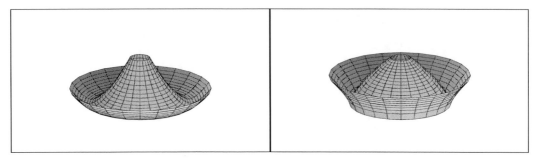

Figure 16. Surfaces containing cycle spirals of Lipschitz and Hölder types.

If $2k+1 \geq p$ then Γ is a Lipschitz-focus and nondegenerate spiral with

$$\dim_B \Gamma = \frac{4k}{2k+1}.$$

If $2k+1 < p$ then Γ is a Hölder-focus and nondegenerate spiral with

$$\dim_B \Gamma = 2 - \frac{2k+p-1}{2kp}.$$

Theorem 2.25. (The case of a limit cycle in $\mathbb{R}^3$, [76, Theorem 2.2]) *Let the system (2.31) have a limit cycle $r = a$ of multiplicity j, $1 \leq j \leq m$. By Γ_1 and Γ_2 we denote the parts of two trajectories of (2.31) near the limit cycle from outside and inside respectively.*
For $j = 1$ we have

$$\dim_B \Gamma_i = 1, \quad i = 1, 2.$$

For $j > 1$ we have the following alternative:
(a) if $j \geq p$ then Γ_i are Lipschitz-cycle nondegenerate spirals and

$$\dim_B \Gamma_i = 2 - \frac{1}{j}, \quad i = 1, 2;$$

(b) if $j < p$ then Γ_i are Hölder-cycle nondegenerate spirals and

$$\dim_B \Gamma_i = 2 - \frac{1}{p}, \quad i = 1, 2.$$

In the proof we use among others the fact that the box dimension and the nondegeneracy of a set are not affected by bilipschitz mappings.

The proof of Theorems 2.24 and 2.25 rests on the application of extensions of Theorems 2.26 and 2.27 below, see [78, Theorems 2 and 3] for the general form. The idea is to represent the spiral Γ of Theorem 2.24 in the form $r = f(\varphi)$, $\varphi \geq \varphi_1$, $z = g(r)$, where $f(\varphi)$ behaves like $\varphi^{-\alpha}$ with $\alpha := 1/(2k)$ for large φ, and $g(r)$ behaves like r^β with $\beta := 2k/(p-1)$ for small r. In proving Theorem 2.25 we have to change $g(r)$ to $g(|a-r|)$.

Theorem 2.26. (Lipschitzian spirals, [76, Theorem 3.1]) *(a) Let Γ be a Lipschitz-focus spiral in $\mathbb{R}^3$ defined in cylindrical coordinates by*

$$r = \varphi^{-\alpha}, \quad \varphi \geq \varphi_1, \quad z = g(r\cos\varphi, r\sin\varphi), \tag{2.32}$$

where $g : \mathbb{R}^2 \to \mathbb{R}$ is any given Lipschitz function defined in Cartesian coordinates. If $\alpha \in (0,1)$ then Γ is nondegenerate and $\dim_B \Gamma = \frac{2}{1+\alpha}$.

(b) Let Γ be a Lipschitz-cycle spiral defined by $r = 1$, $\varphi \geq \varphi_1$, $z = \varphi^{-\alpha}$. Then for any $\alpha > 0$ we have that Γ is nondegenerate and $\dim_B \Gamma = \frac{2+\alpha}{1+\alpha}$.

Theorem 2.27. (Hölderian spirals, [76, Theorem 3.2]) *Assume that $\beta \in (0,1)$. (a) Let Γ be a Hölder-focus spiral defined by $r = \varphi^{-\alpha}$, $\varphi \in [\varphi_1, \infty)$, $z = r^\beta$, and $\alpha \in (0,1)$. Then Γ is nondegenerate and $\dim_B \Gamma = \frac{2-\alpha(1-\beta)}{1+\alpha\beta}$.*

(b) Let Γ be a Hölder-cycle spiral defined by $r = 1 - \varphi^{-\alpha}$, $\varphi \in [\varphi_1, \infty)$, $z = |1 - r|^\beta$, $\alpha > 0$. Then Γ is nondegenerate and $\dim_B \Gamma = \frac{2+\alpha\beta}{1+\alpha\beta}$.

Example 2.28. We found a class of systems such that the box dimension of trajectories depends in nontrivial way on the coefficients of the system. Let us provide the following example:

$$\dot{r} = a_1 rz, \quad \dot{\varphi} = 1, \quad \dot{z} = b_2 z^2$$

Denoting the vertical projection of the spacial spiral into the horizontal plane by Γ we obtain the following surprising dimension result, depending just on the coefficients:

$$\dim_B \Gamma = \frac{2}{1 + a_1/b_2}, \quad \frac{a_1}{b_2} \in (0,1)$$

It is obtained by explicit computation using Tricot's formula, since the planar spiral has the form $r = C_1 \cdot (-b_2\varphi + C_2)^{-a_1/b_2}$ near the origin. Note that that the spacial spiral is contained in the Lipschitz surface $z = C_3 r^{b_2/a_1}$.

Remark 2.29. Let us consider the spiral Γ defined by $r = \varphi^{-\alpha}$, $z = r^\beta$. Only one of two projections of the spiral Γ onto horizontal and vertical planes, has box dimension equal to $\dim_B \Gamma$. More precisely, in the Lipschitz case, that is when $\beta \geq 1$, the horizontal projection spiral has the same box dimension as the initial spiral. In the Hölder case, that is when $\beta \in (0,1)$, only the vertical projection has the same box dimension, see [76, 78].

The projection of the spiral

$$r = \varphi^{-\alpha}, \quad z = r^\beta, \quad \varphi \in [\varphi_1, \infty),$$

for fixed $\alpha \in (0,1), \beta \in (0,1)$, onto the (y, z)-plane, is easily seen to be the curve

$$y = z^{1/\beta}\sin(z^{-1/\alpha\beta}), \quad z \in (0, z_1]$$

which is a chirp. Box dimension of the graph of this function is equal to $2 - \frac{\alpha(1+\beta)}{1+\alpha\beta}$ according to standard Tricot's formula.

Remark 2.30. The results about space spirals can be applied to the fractal analysis of a class of nonautonomous second order differential equations with corresponding 3-dimensional system whose trajectories are contained in the Lipschitz or Hölder surface. Furthermore, fold-Hopf bifurcation of the system (2.31) can be considered.

Fractal Properties of Solutions of Differential Equations 429

2.4. Bifurcations of Discrete Dynamical Systems

Our motivation for studying discrete systems was to be able to compute box dimension of spiral trajectories of more general continuous dynamical systems via their Poincaré mapping. We completed that study for the weak focus and presented the results in Section 2.1.2.. Now we describe the results from [18] which have been exploited for the fractal analysis of the Poincaré map in Section 2.1.2., and also we present some generalizations from [31].

The following theorem deals with solution sequences $(x_n)_{n\geq 1}$ converging monotonically to zero for which it is possible to compute box dimension directly from the equation.

Theorem 2.31. ([18, Theorem 1]) *Let $\alpha > 1$ and let $f : (0, r) \to (0, \infty)$ be a monotonically nondecreasing function such that $f(x) \simeq x^\alpha$ as $x \to 0$, and $f(x) < x$ for all $x \in (0, r)$. The sequence $S(x_1) := (x_n)_{n\geq 1}$ is defined by*

$$x_{n+1} = x_n - f(x_n), \quad x_1 \in (0, r).$$

Then

$$x_n \simeq n^{-1/(\alpha-1)} \quad \text{as } n \to \infty.$$

Furthermore,

$$\dim_B S(x_1) = 1 - \frac{1}{\alpha},$$

and the set $S(x_1)$ is Minkowski nondegenerate.

Now we consider what happens with box dimension when the saddle-node or the period doubling bifurcations occur. The following two bifurcation results are well known, see Devaney [15]. The novelty is additional information about the box dimension of trajectories.

The following theorem is a generalization of [18, Theorem 3] where the result is shown for $\alpha = 2$.

Theorem 2.32. (Saddle-node bifurcation, [18, Theorem 7]) *Suppose that a function $F : J \times (x_0 - r, x_0 + r) \to \mathbb{R}$, where J is an open interval in $\mathbb{R}$, is such that $F(\lambda_0, \cdot)$ is of class C^3 for some $\lambda_0 \in \mathbb{R}$, and $F(\cdot, x)$ of class C^1 for all x. Assume that*

$$
\begin{aligned}
F(\lambda_0, x_0) &= x_0, \\
\frac{\partial F}{\partial x}(\lambda_0, x_0) &= 1, \\
\frac{\partial^2 F}{\partial x^2}(\lambda_0, x_0) &< 0, \\
\frac{\partial F}{\partial \lambda}(\lambda_0, x_0) &\neq 0.
\end{aligned}
\tag{2.33}
$$

Then λ_0 is the point where saddle-node bifurcation occurs. There exists $r_1 \in (0, r)$ such that for any sequence $S(\lambda_0, x_1) = (x_n)_{n\geq 1}$ defined by $x_{n+1} = F(\lambda_0, x_n)$, $x_1 \in (x_0, x_0 + r_1)$, we have $|x_n - x_0| \simeq n^{-1}$ as $n \to \infty$,

$$\dim_B S(\lambda_0, x_1) = \frac{1}{2},$$

and $S(\lambda_0, x_1)$ is Minkowski nondegenerate. Analogous result holds if $x_1 \in (x_0 - r_1, x_0)$, where in (2.33) we have the opposite sign.

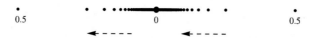

Figure 17. 1-string defined by $x_{n+1} = x_n - x_n^2$, has box dimension $1/2$.

Figure 18. $1/4$-string defined by $x_{n+1} = x_n - x_n^5$, has box dimension $4/5$.

Our Teorem 2.31 is crucial for the proof of Theorem 2.32, which means that from the following theorem we can obtain also the generalization of Theorem 2.32, see [31].

Theorem 2.33. (Box dimension in the case $F'(x_0) = 1$, [31, Theorem 2]) *Let $F : (x_0 - r, x_0 + r) \to \mathbb{R}$ be a map of class $C^{\alpha+1}$, where $\alpha \in \mathbb{N}$, $\alpha \geq 3$ such that $F(x_0) = x_0$, $F'(x_0) = 1$, and*

$$F^{(k)}(x_0) = 0, k = 2, \ldots, \alpha - 1, \ F^{(\alpha)}(x_0) < 0.$$

Then there is $r_1 > 0$ such that for the sequence $S(x_1) = (x_n)_{n \geq 1}$ defined by $x_{n+1} = F(x_n)$, $x_1 \in (x_0, x_0 + r_1)$ we have $|x_n - x_0| \simeq n^{-\frac{1}{\alpha - 1}}$ as $n \to \infty$ and

$$\dim_B S(x_1) = 1 - \frac{1}{\alpha}.$$

Moreover, the set $S(x_1)$ is Minkowski nondegenerate.

It is worth noticing that in the following theorem we have the same phenomenon as in the Hopf bifurcation in planar continuous case: the box dimension jumps just before the periodic orbit is born.

Theorem 2.34. (Period-doubling bifurcation, [18, Theorem 8]) *Let $F : J \times (x_0 - r, x_0 + r) \to \mathbb{R}$ be a function of class C^2, where J is an open interval in $\mathbb{R}$, and $F(\lambda_0, \cdot)$ is of class C^4 for some $\lambda_0 \in J$. Assume that*

$$F(\lambda_0, x_0) = x_0,$$
$$\frac{\partial F}{\partial x}(\lambda_0, x_0) = -1,$$
$$\frac{\partial^2 F}{\partial x^2}(\lambda_0, x_0) \neq 0,$$
$$\frac{\partial^2 (F^2)}{\partial \lambda \partial x}(\lambda_0, x_0) \neq 0, \quad \frac{\partial^3 (F^2)}{\partial x^3}(\lambda_0, x_0) \neq 0.$$

Then λ_0 is the point where period-doubling bifurcation occurs. There exists $r_1 \in (0, r)$ such that for any sequence $S(\lambda_0, x_1) = (x_n)_{n \geq 1}$ defined by

$$x_{n+1} = F(\lambda_0, x_n), \quad x_1 \in (x_0 - r_1, x_0 + r_1), \quad x_1 \neq x_0,$$

Fractal Properties of Solutions of Differential Equations 431

we have $|x_n - x_0| \simeq n^{-1/2}$ as $n \to \infty$,

$$\dim_B S(\lambda_0, x_1) = \frac{2}{3},$$

and $S(\lambda_0, x_1)$ is Minkowski nondegenerate.

A generalization of Theorem 2.34 can be obtained analogously as for Theorem 2.32, see [31]. Theorems 2.32 and 2.34 can be applied to the study of the logistic map.

Corollary 2.35. (Logistic map, [18, Corollary 1]) *Let $F(\lambda, x) = \lambda x(1-x)$, $x \in (0, 1)$, and let $S(\lambda, x_1) = (x_n)_{n \geq 1}$ be a sequence defined by initial value x_1 and $x_{n+1} = F(\lambda, x_n)$.*

(a) For $\lambda_0 = 1$, taking $x_1 > 0$ sufficiently close to $x_0 = 0$, we have that $x_n \simeq n^{-1}$ as $n \to \infty$, and

$$\dim_B S(1, x_1) = \frac{1}{2}.$$

(b) (Onset of period-2 cycle) For $\lambda_0 = 3$ the corresponding fixed point is $x_0 = 2/3$. For any x_1 sufficiently close to x_0 we have that $|x_n - x_0| \simeq n^{-1/2}$, and

$$\dim_B S(3, x_1) = \frac{2}{3}.$$

(c) For any $\lambda \notin \{1, 3\}$ and x_1 such that the sequence $S(\lambda, x_1)$ is convergent, we have that $\dim_B S(\lambda, x_1) = 0$.

(d) (Onset of period-4 cycle) If $\lambda_0 = 1 + \sqrt{6}$ then for any x_1 sufficiently close to period-2 trajectory $A = \{a_1, a_2\}$ we have that $d(x_n, A) \simeq n^{-1/2}$ as $n \to \infty$, and

$$\dim_B S(1 + \sqrt{6}, x_1) = \frac{2}{3}.$$

(e) (Period-3 cycle) Let $\lambda_0 = 1 + \sqrt{8}$ and let a_1, a_2, a_3 be fixed points of F^3 such that $0 < a_1 < a_2 < a_3 < 1$, $F(a_1) = a_2$, $F(a_2) = a_3$, and $F(a_3) = a_1$. Then there exists $\delta > 0$ such that for any initial value

$$x_1 \in (a_1 - \delta, a_1) \cup (a_2 - \delta, a_2) \cup (a_3, a_3 + \delta)$$

we have $d(x_n, \{a_1, a_2, a_3\}) \simeq n^{-1}$ as $n \to \infty$, and

$$\dim_B S(1 + \sqrt{8}, x_1) = \frac{1}{2}.$$

All trajectories appearing in this corollary are Minkowski nondegenerate.

We conjecture that the values of box dimensions of trajectories corresponding to all period-doubling bifurcation parameters λ_k where 2^k-periodic points occur, are equal to $2/3$ (see [18]).

Remark 2.36. It can be shown that the time 1 map of the standard planar normal form showing saddle-node bifurcation generates a discrete dynamical system with the same fractal properties as the system in Theorem 2.32, see [32]. In other words, saddle-node bifurcation of the planar continuous system also "appears" with box dimension equal to $1/2$.

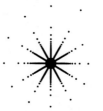

Figure 19. Neimark-Sacker bifurcation, rational rotation number case.

Figure 20. Neimark-Sacker bifurcation, irrational rotation number case.

Remark 2.37. The results could be extended to 2-dimensional discrete dynamical systems undergoing Neimark-Sacker bifurcation, which is a Hopf bifurcation for maps, see [32]. Roughly speaking, it seems that we have two different behaviours for the normal form, the rational and irrational rotation number cases. The rational 2-dimensional discrete case would have fractal behaviour like 1-dimensional discrete case, but the irrational one would have fractal behaviour like 2-dimensional continuous case. See Figures 19 and 20.

3. Fractal Analysis of Trajectories of Some Schrödinger Cauchy Problems

3.1. Introduction

This section is based on Milišić, Žubrinić and Županović [46], where one can find detailed proofs. We are interested in fractal properties of trajectories corresponding to the Cauchy problem for the following nonlinear Schrödinger equation:

$$\begin{cases} u_t(t,x) = i\Delta u(t,x) - \gamma u(t,x) \int_\Omega |u(t,x)|^2 dx \\ u(t,x) = 0 \quad \text{for } x \in \partial\Omega, \, t \in (t_{min}, t_{max}) \\ u(0,x) = u_0(x) \quad \text{for } x \in \Omega. \end{cases} \quad (3.34)$$

This initial-boundary value problem will be called the nonlinear Schrödinger problem, and denoted by (NLS) for short. Here i is the imaginary unit, γ is a given real number, Ω a bounded domain in $\mathbb{R}^N$, $N \geq 1$, and $u_0 : \Omega \to \mathbb{C}$ is a given initial function which we assume to be in $L^2(\Omega, \mathbb{C})$. By Δ we denote the standard Laplace operator, $\Delta u = \sum_{j=1}^N \frac{\partial^2 u}{\partial x_j^2}$. The solutions $u : (t_{min}, t_{max}) \to L^2(\Omega, \mathbb{C})$ of (3.34) can be considered as trajectories in the Hilbert space $L^2(\Omega, \mathbb{C})$ where $L^2(\Omega, \mathbb{C}) = L^2(\Omega) + iL^2(\Omega)$ is the complexification of the real space $L^2(\Omega)$. The values of t_{min} and t_{max} for the considered model depend on γ and $\|u_0\|_{L^2}$, and $0 \in (t_{min}, t_{max})$. We are dealing with unbounded time intervals (t_{min}, t_{max}) either of the form (t_{min}, ∞), or $(-\infty, \infty)$, or $(-\infty, t_{max})$.

The presence of the integral term in (3.34) indicates that the NLS Cauchy problem is of nonlocal type. Analogous problems have been considered for the modified cubic wave equation,

$$u_{tt} - \Delta u + c \left(\int_\Omega u(t, x)^2 dx \right) u = 0,$$

with homogeneous Dirichlet boundary condition, in Cazenave, Haraux and Weissler [8, 9, 10], in their study of completely integrable abstract wave equations. We consider the Schrödinger equation (3.34) as an approximation of the equation $u_t = i\Delta u - \gamma u |u|^2$ following the approach of [10, p. 130].

3.1.1. Interpretation of the Schrödinger Cauchy problem in ℓ_2

Our aim is to study (3.34) using the corresponding infinite system of nonlinear ordinary differential equations. To this end we exploit the Fourier series expansion, using the decomposition with respect to a Hilbert basis in $L^2(\Omega)$. Let (φ_j) be the orthonormal basis of eigenfunctions of the Laplace operator $-\Delta$, endowed with zero boundary data, and let (λ_j) be the corresponding sequence of eigenvalues. It is well known that $0 < \lambda_1 < \lambda_2 \le \lambda_3 \le \ldots$ (see e.g. Brezis [4, ersatz (12) on p. 209]). By writing a solution $u(t, x)$ in the form:

$$u(t, x) = \sum_{j=1}^\infty z_j(t) \varphi_j(x), \tag{3.35}$$

the NLS Cauchy problem (3.34) formally reduces to a lattice Schrödinger equation on the Hilbert space of quadratically summable sequences of complex numbers, which we denote by $\ell_2(\mathbb{C})$. In this way we obtain an infinite system of nonlinear ODE's:

$$\dot{z}_j = -i\lambda_j z_j - \gamma z_j \|z\|^2, \ j = 1, 2, \ldots, \tag{3.36}$$

with initial condition $z_j(0) = z_{j0}$, $j \ge 1$. Here $\| \cdot \|$ is the standard Euclidean norm, $(z_{j0}) \in \ell_2(\mathbb{C})$, and $z(t) = (z_j(t))_j \in \ell_2(\mathbb{C})$ for each $t \in (t_{min}, t_{max})$. Clearly, $z_{j0} = \langle u_0, \varphi_j \rangle = \int_\Omega u_0 \varphi_j dx$. For a given $u_0 \in L^2(\Omega, \mathbb{C})$, we interpret the NLS problem (3.34) as the lattice Schrödinger equation (3.36). To any given $v \in L^2(\Omega, \mathbb{C})$ we assign $z = (z_j)_j \in \ell_2(\mathbb{C})$, where $v = \sum_j z_j \varphi_j$ is the Fourier expansion of v. This assignment is an isometric isomorphism.

Taking $z = (z_1, 0, 0, \ldots)$, and assuming for simplicity that $\lambda_1 = 1$, $\gamma = 1$, the system (3.36) reduces to the following system in the plane:

$$\begin{cases} \dot{x} &= y - x(x^2 + y^2) \\ \dot{y} &= -x - y(x^2 + y^2). \end{cases} \tag{3.37}$$

Here we denote $x(t) = \operatorname{Re} z(t)$, $y(t) = \operatorname{Im} z(t)$, real and imaginary part of the complex number $z(t)$, respectively. System (3.37) can be simplified using the polar coordinates

(r, θ):

$$\dot{r} = r^3, \quad \dot{\theta} = -1, \qquad (3.38)$$

and it can be easily solved. Since $\dot{\theta} \neq 0$, the origin $r = 0$ is the only critical point. Since $\dot{\theta} < 0$ the flow is oriented clockwise around the origin. The phase portrait of (3.37) is shown by Figure 21.

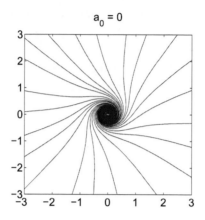

Figure 21. Phase portrait of the system (3.37).

We shall need the Sobolev spaces $H_0^1(\Omega, \mathbb{C})$ and $H_0^1(\Omega, \mathbb{C}) \cap H^2(\Omega, \mathbb{C})$ (in the sequel we omit $\mathbb{C}$) equipped with the corresponding norms defined by

$$\|u_0\|_{H_0^1}^2 = \sum_j \lambda_j |\langle u_0, \varphi_j \rangle|^2, \quad \|u_0\|_{H_0^1 \cap H^2}^2 = \sum_j \lambda_j^2 |\langle u_0, \varphi_j \rangle|^2. \qquad (3.39)$$

See Henry [30].

All the results in this section hold if in (3.34) we replace the Laplace operator Δ by $-\Delta$. The j-th component of the trajectory, viewed as a spiral in $\mathbb{C}$, only changes the orientation for each j from negative to positive, see (3.40) below.

3.1.2. Explicit Solutions of the NLS Problem

We consider the NLS initial-boundary value problem (3.34). It is possible to calculate the explicit solutions of problem (3.34), see [46, Lemma 1.1]. First we have

$$\begin{cases} r_j(t) = \frac{|\langle u_0, \varphi_j \rangle|}{\rho_0} (2\gamma t + \rho_0^{-2})^{-1/2}, \\ \theta_j(t) = -\lambda_j t + \arg \langle u_0, \varphi_j \rangle, \end{cases} \qquad (3.40)$$

where $z_j(t) = r_j(t) \exp(i \theta_j(t))$, and hence, assuming $\gamma \neq 0$,

$$u(t, x) = (2\gamma t + \|u_0\|_{L^2}^{-2})^{-1/2} \sum_{j=1}^{\infty} \frac{\langle u_0, \varphi_j \rangle}{\|u_0\|_{L^2}} e^{-i \lambda_j t} \varphi_j(x), \qquad (3.41)$$

Fractal Properties of Solutions of Differential Equations 435

where u_0 is a prescribed initial function. Here we notice that the sign of the parameter γ affects the maximal interval of the solution. More precisely, following the terminology introduced in Cazenave [7, Remark 3.1.6(ii)], we say that the solutions given by formula (3.41) are *positively (negatively) global* if $\gamma > 0$ ($\gamma < 0$). The solutions are global for $\gamma = 0$ for any initial value u_0, while for $\gamma \neq 0$ the global solution exists only when $u_0 = 0$.

Note that from (3.41) it directly follows the invariance property of the solutions of the NLS problem. More precisely, if U_0 is the span of a given subset of $\{\varphi_j : j \geq 1\} \subset L^2(\Omega)$, then the assumption $u_0 \in \overline{U}_0$ implies that $u(t) \in \overline{U}_0$ for all $t > 0$, where the closure is taken in $L^2(\Omega)$. Moreover, the invariance property can be reformulated as follows: each closed subspace $\overline{U}_0$, where U_0 is spanned by a subset of eigenfunctions φ_j of $-\Delta|_{H_0^1 \cap H^2}$, is invariant for the nonlinear evolution operator $T(t)$, $T(t)u_0 = u(t)$, see (3.41), associated with the problem (3.34): $T(t)\overline{U}_0 \subseteq \overline{U}_0$. In other words, a trajectory that starts in $\overline{U}_0$ remains in this space forever. In particular, if $u_0 \in \text{span}\{\varphi_1, \ldots, \varphi_k\}$, then $\Gamma(u_0) \subset \text{span}\{\varphi_1, \ldots, \varphi_k\}$. This means that if u_0 is such that $\langle u_0, \varphi_j \rangle = 0$ for all but finitely many j's, then (3.34) is essentially a finite-dimensional problem.

3.2. Minkowski Sequence Associated to a Trajectory

Let $\Gamma(u_0)$ be a trajectory generated by $u_0 \neq 0$, corresponding to $t \geq 0$ if the solution $u(t)$ is positively global, and to $t \leq 0$ if it is negatively global. Denote by $\Gamma_j(u_0)$ the orthogonal projection of $\Gamma(u_0) \subset L^2(\Omega)$ onto the φ_j-component, where φ_j are the eigenfunctions of $-\Delta|_{H_0^1 \cap H^2}$, see Section 3.1.1.. Here, we recall that the solution of (3.34) can be written in form $u(t, x) = \sum_{j=1}^{\infty} z_j(t)\varphi_j(x)$. Now, $\Gamma_j(u_0)$ can be viewed as a curve in the complex plane defined by $z_j(t) = \langle u(t), \varphi_j \rangle = r_j(t) \exp(i\theta_j(t))$.

Figure 22 shows an example of $\Gamma_j(u_0)$ for $j = 1, 2, 3, 4$, where $z_j(t) = (t + 1)^{-1/2} c_j e^{-i\lambda_j t}$ with $\lambda_j = j^2$ and $c_j = \langle u_0, \varphi_j \rangle$. For some chosen eigenvalues λ_j and Fourier coefficients c_j we plotted $z_j(t)$ on the time interval $[0, \infty)$. The trajectory $t \mapsto z(t)$ of the considered system in $\ell_2(\mathbb{C})$ is equal to a sequence $t \mapsto (z_j(t))_{j \in \mathbb{N}}$. In this way Figure 22 should indicate the projection of the considered trajectory into $\mathbb{C}^4$. Note that the corresponding (nontrivial) spirals tend to the origin with polynomial speed, which is due to the fact that the origin is the weak focus.

The following result describes some properties of the sequence of d-dimensional Minkowski contents of curves $\Gamma_j(u_0)$ for $d = 4/3$, $j \in \mathbb{N}$. Recall that if a set $A \subset \mathbb{R}^k$ is such that $d = \dim_B A > 0$, then $\mathcal{M}^s(A) = \infty$ for $0 \leq s < d$ and $\mathcal{M}^s(A) = 0$ for $s > d$. Hence, since $\dim_B \Gamma_j(u_0) = 4/3$ for all j, see Theorem 3.1(a) below, it has sense to consider $\mathcal{M}^d(\Gamma_j(u_0))$ for $d = 4/3$ only.

The sequence of Minkowski contents $(\mathcal{M}^d(\Gamma_j(u_0)))_j$ will be called the *Minkowski sequence* associated to the trajectory $\Gamma(u_0)$.

More generally, for any trajectory $\Gamma(u_0)$ in $L^2(\Omega)$ the corresponding *Minkowski sequence* is defined by $(\mathcal{M}^{d_j}(\Gamma_j(u_0)))_j$, where $d_j = \dim_B \Gamma_j(u_0)$, provided the box dimension exists for each j.

The following result relates the Minkowski sequence of the trajectory $\Gamma(u_0)$ with Lebesgue and Sobolev norms of initial function u_0.

Theorem 3.1. ([46, Theorem 11]) *Assume that in (3.34) we have* $\gamma \neq 0$.

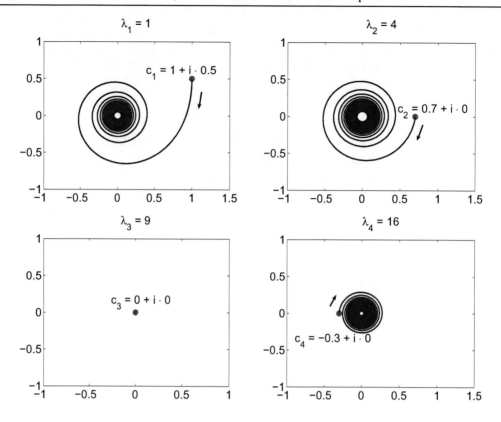

Figure 22. Projections $\Gamma_j(u_0)$ of $\Gamma(u_0)$, $j = 1, 2, 3, 4$.

(a) For any $u_0 \in L^2(\Omega)$ and j such that $\langle u_0, \varphi_j \rangle \neq 0$ we have $\dim_B \Gamma_j(u_0) = 4/3$. Moreover,

$$\mathcal{M}^{4/3}(\Gamma_j(u_0)) = 3\pi^{1/3} \left(\frac{\lambda_j |\langle u_0, \varphi_j \rangle|^2}{|\gamma| \|u_0\|_{L^2}^2} \right)^{2/3}.$$

In particular,

$$\mathcal{M}^{4/3}(\Gamma_j(u_0)) \leq 3\pi^{1/3} \left(\frac{\lambda_j}{|\gamma|} \right)^{2/3},$$

and equality is achieved if and only if $u_0 \in \operatorname{span}\{\varphi_j\}$, $u_0 \neq 0$. Furthermore, we have the following asymptotic behaviour

$$\max_{u_0 \in L_2(\Omega)} \mathcal{M}^{4/3}(\Gamma_j(u_0)) \simeq j^{4/(3N)} \quad \text{as } j \to \infty.$$

(b) For any $u_0 \in H_0^1(\Omega)$, $u_0 \neq 0$, we have the following identity relating the Minkowski sequence of the trajectory with Lebesgue and Sobolev norms of the initial function:

$$\sum_{j=1}^{\infty} [\mathcal{M}^{4/3}(\Gamma_j(u_0))]^{3/2} = \frac{\sqrt{27\pi}}{|\gamma|} \left(\frac{\|u_0\|_{H_0^1}}{\|u_0\|_{L^2}} \right)^2.$$

In particular,

$$\sum_{j=1}^{\infty} [\mathcal{M}^{4/3}(\Gamma_j(u_0))]^{3/2} \geq \frac{\sqrt{27\pi}}{|\gamma|} \lambda_1,$$

and equality is achieved if and only if $u_0 \in \text{span}\{\varphi_1\}$, $u_0 \neq 0$.

(c) For $u_0 \in H_0^1(\Omega) \cap H^2(\Omega)$, $u_0 \neq 0$, besides (3.42) we have the following identity:

$$\sum_{j=1}^{\infty} \lambda_j [\mathcal{M}^{4/3}(\Gamma_j(u_0))]^{3/2} = \frac{\sqrt{27\pi}}{|\gamma|} \left(\frac{\|u_0\|_{H_0^1 \cap H^2}}{\|u_0\|_{L^2}} \right)^2,$$

and

$$\mathcal{M}^{4/3}(\Gamma_j(u_0)) = o(j^{-4/(3N)}) \quad \text{as } j \to \infty.$$

In particular,

$$\sum_{j=1}^{\infty} \lambda_j [\mathcal{M}^{4/3}(\Gamma_j(u_0))]^{3/2} \geq \frac{\sqrt{27\pi}}{|\gamma|} \lambda_1^2,$$

and equality is achieved if and only if $u_0 \in \text{span}\{\varphi_1\}$, $u_0 \neq 0$.

3.3. Box Dimension of the Trajectory

For calculation of the box dimension of trajectories of NLS initial-boundary value problem (3.34) we use the well known fact that the box dimension is invariant with the respect to the bilipshitz mappings see Falconer [22, Corollary 2.4(b)] or Tricot [66, p. 121].

For a given trajectory $\Gamma(u_0)$ in $L^2(\Omega)$ we define its box dimension via finite-dimensional approximations:

$$\dim_B \Gamma(u_0) = \lim_{k \to \infty} \dim_B \Pi_k(\Gamma(u_0)),$$

where Π_k is the orthogonal projection of the Lebesgue space onto $\text{span}\{\varphi_1, \dots, \varphi_k\}$, or equivalently, from $\ell_2(\mathbb{C})$ onto $\mathbb{C}^k$, corresponding to the first k components of $\ell_2(\mathbb{C})$. The above limit exists due to the monotonicity property of box dimension, see Falconer [22, p. 37].

Definition 3.2. By the *multiple spiral* (or n-spiral) Γ_{2N} we mean a curve in $\mathbb{C}^N = \mathbb{R}^{2N}$ defined by

$$\Gamma_{2N} = \{(t^{-\alpha_1} e^{i\lambda_1 t}, \dots, t^{-\alpha_N} e^{i\lambda_N t}) \in \mathbb{C}^N : t \geq t_0\}, \tag{3.42}$$

where $t_0 > 0$.

The following result will be fundamental for the computation of box dimension of trajectories of problem (3.34). Its consequence is that the box dimension of multiply oscillating trajectories in $\mathbb{C}^N$ is always less then 2. The claim extends the formula for the box dimension of planar spirals due to Tricot [66, p. 121] to oscillating curves in $\mathbb{C}^N$. Since the result seems to be interesting for itself, we state it in a slightly more general form.

Theorem 3.3. ([46, Theorem 15]) *Let $\alpha_k > 0$ and $\lambda_k \neq 0$ be given numbers, $k = 1, \ldots, N$. Then the corresponding multiple spiral Γ_{2N} defined by (3.42) has box dimension*

$$\dim_B \Gamma_{2N} = \max \left\{ 1, \frac{2}{1 + \min_k \alpha_k} \right\}. \tag{3.43}$$

The curve Γ_{2N} is Minkowski nondegenerate if and only if $\min_k \alpha_k \neq 1$. It is rectifiable if and only if $\min_k \alpha_k > 1$.

Now, using the result given by Theorem 3.3 it is possible to derive the following result.

Theorem 3.4. ([46, Theorem 17]) *Assume that $u_0 \neq 0$ and $\gamma \in \mathbb{R} \setminus \{0\}$. Let $\Gamma(u_0)$ be the trajectory of (3.34) viewed in a bounded neighbourhood of the origin. Then $\dim_B \Gamma(u_0) = 4/3$.*

As we saw before, assuming that $\gamma \neq 0$, the natural projection of any trajectory $\Gamma(u_0)$ into its j-th component, that is, into the eigenspace spanned by φ_j, can be viewed as a spiral curve in the Gauss plane either of box dimension $4/3$ if $\langle u_0, \varphi_j \rangle \neq 0$, or zero if $\langle u_0, \varphi_j \rangle = 0$ (in this case the projection is just a point).

3.4. Applications

In this section we introduce the notion of a spiral chirp. The aim is to study the box dimension of spiral chirps and associated trajectories of several classes of NLS problems with different nonlinearities of nonlocal type. To achieve this goal, we shall use the results on box dimensions of trajectories obtained in previous sections. Finally, in Subsection 3.4.3. we show that any finite-dimensional polynomial dynamical system can be interpreted in terms of a NLS problem of nonlocal type, see Proposition 3.8.

3.4.1. Box Dimension of Spiral Chirps

Results on box dimension of trajectories of NLS initial-boundary value problem (3.34) in the previous section can be used in order to calculate the box dimension of some other curves, for example spiral chirps. More precisely, in the preceding subsection we have studied the box dimension of projections of Γ_0 into subspaces of the form $\text{span}_{\mathbb{C}} \{\varphi_1, \ldots, \varphi_N\}$, that is, on $\mathbb{R}^{2N}$. Now we would like to consider projections onto $\mathbb{R}^{2N-1}$.

The basic model in $\mathbb{R}^3$ is the curve that we call a *spiral chirp*:

$$\Gamma = \left\{ (t^{-\alpha} \cos t, t^{-\alpha} \sin t, t^{-\alpha} \cos t) \in \mathbb{R}^3 : t \geq t_0 \right\}, \tag{3.44}$$

with positive α and t_0. It is obtained by projecting the double spiral $\Gamma_4 = \{(t^{-\alpha} e^{it}, t^{-\alpha} e^{it}) : t \geq t_0\}$ in $\mathbb{C}^2 = \mathbb{R}^4$ into $\mathbb{R}^3$. A more general model could be the following.

Definition 3.5. By an (α, β)-*spiral chirp* we mean a curve in $\mathbb{R}^3$ defined by:

$$\Gamma_3 = \left\{ (t^{-\alpha} \cos(\lambda_1 t), t^{-\alpha} \sin(\lambda_1 t), t^{-\beta} \cos(\lambda_2 t)) \in \mathbb{R}^3 : t \geq t_0 \right\}, \tag{3.45}$$

for fixed positive α, β, t_0, and $\lambda_j \neq 0$, $j = 1, 2$.

The (α, β)-chirp is obtained by projecting the double spiral $\Gamma_4 = \{(t^{-\alpha}e^{i\lambda_1 t}, t^{-\beta}e^{i\lambda_2 t}) : t \geq t_0\}$ from $\mathbb{C}^2 = \mathbb{R}^4$ into $\mathbb{R}^3$. Of course, the box dimension of Γ_3 will remain the same if we replace the last component in (3.45) with $t^{-\beta} \sin \lambda_2 t$.

Theorem 3.6. ([46, Theorem 18]) *Let α and β be fixed positive numbers such that either*

$$\alpha \leq \beta, \text{ or } \beta \leq \alpha < 1, \text{ or } 1 < \beta \leq \alpha.$$

Furthermore, assume that λ_1, λ_2 are nonzero real numbers, and $t_0 > 0$. Then for the spiral chirp Γ_3 defined by expression (3.45) we have:

$$\dim_B \Gamma_3 = \max\left\{1, \frac{2}{1 + \min\{\alpha, \beta\}}\right\}. \tag{3.46}$$

The curve Γ_3 is rectifiable if and only if $\min\{\alpha, \beta\} > 1$.

3.4.2. Hopf Bifurcation for Other Types of Nonlinearities

Up to now we have studied the initial-boundary value problem for Schrödinger equation

$$u_t = i\Delta u - \gamma u V(\|u\|_{L^2(\Omega)}), \tag{3.47}$$

where $\Omega \subset \mathbb{R}^N$ and $\gamma \in \mathbb{R}$. More precisely, we considered the nonlinearity $V(\rho) = \rho^2$. In this section we consider some polynomial types of nonlinearity given by the expression

$$V(\rho) = \rho^{2l} + a_{l-1}\rho^{2(l-1)} + \cdots + a_1\rho^2 + a_0, \tag{3.48}$$

where a_i, $i = 0, \ldots, l-1$ are prescribed real parameters. The NLS equation (3.47) corresponds to the *standard generic generalized Hopf bifurcation*, see Takens [64], also called the *standard Hopf-Takens bifurcation*. The following result extends Theorem 3.4.

Theorem 3.7. ([46, Theorem 20]) *Let $\Gamma(u_0)$ be a part of a trajectory of (3.47) near the origin with $V(\rho)$ defined by (3.48), such that $u_0 \neq 0$ and $\gamma \neq 0$. Assume that $k := \min\{j : a_j \neq 0\} \geq 1$. Then*

$$\dim_B \Gamma(u_0) = \frac{4k}{2k+1}. \tag{3.49}$$

We have studied trajectories with spiral components $z_j(t)$ converging to zero with equal rate $\alpha > 0$, that is, $|z_j(t)| \simeq t^{-\alpha}$ as $t \to \infty$, for all j, see for example expressions (3.44) or (3.45). It is possible to construct dynamical systems yielding solutions in which the spiral components converge to zero with different rates, see [46].

3.4.3. Box Dimension of Trajectories of Some Nonintegrable Schrödinger Problems

Many dynamical systems important for applications, like for example Liénard systems, weakly damped systems etc., are not explicitly solvable. However, in spite of their nonintegrability it is possible to get the information about the box dimension of the corresponding trajectories, see [85] for some examples. Here, using the connection between the nonlinear system of ODE's and the Schrödinger equation pointed out already in Section 3.1.1., we

440   Mervan Pavsic, Darkov Zubrinic and Vesnav Zupanovic

would like to study the box dimension of trajectories of some nonintegrable Schrödinger evolution problems. For this purpose let us consider the following polynomial planar system:

$$\dot{x} = f(x,y) = \lambda_1 y + p(x,y)$$
$$\dot{y} = g(x,y) = -\lambda_1 x + q(x,y), \tag{3.50}$$

where we define $p(x,y) = f(x,y) - \lambda_1 y$ and $q(x,y) = g(x,y) + \lambda_1 x$. Here λ_1 is the first eigenvalue of $-\Delta|_{H_0^1 \cap H^2}$ corresponding to an open and bounded domain $\Omega \subset \mathbb{R}^N$. An initial point in the phase space is prescribed by $x(0) = x_{10}$ and $y(0) = y_{10}$. Naturally, we say that the system (3.50) is *equivalent* to (3.52) if there is a bijection between the corresponding solution sets. In this sense it is easy to see that the problem (3.50) is equivalent to Schrödinger equation (3.52) with $F: \mathbb{C}\varphi_1 \to \mathbb{C}\varphi_1$ defined by

$$[F(u)](\xi) = \Big(p(\langle \operatorname{Re} u, \varphi_1 \rangle, \langle \operatorname{Im} u, \varphi_1 \rangle) + iq(\langle \operatorname{Re} u, \varphi_1 \rangle, \langle \operatorname{Im} u, \varphi_1 \rangle) \Big) \varphi_1(\xi), \quad \xi \in \Omega, \tag{3.51}$$

and with initial function $u(0) = z_{01}\varphi_1$, where $z_{01} = x_{10} + iy_{10} \in \mathbb{C}$. Note that the nonlinear term $F(u)$ is of nonlocal type, i.e. F it is not defined in pointwise manner, but via scalar products, that is, by means of two integrals. The solution of (3.52) in this case has the form $u(t,x) = z(t)\varphi_1(x)$, where $z(t) = x(t) + iy(t)$, and the components satisfy (3.50). Note that (3.52) is of nonlocal type, and in general not explicitly solvable. Moreover, it is interesting that the study of *any* system of ODE's in $\mathbb{R}^n$ can be considered as the study of a Schrödinger problem.

We can assume without loss of generality that n is even, $n = 2k$ (if $n = 2k - 1$, we can add a new trivial equation $\dot{x}_{2k} = 0$, $x_{2k}(0) = 0$).

Proposition 3.8. ([46, Proposition 23]) *Any Cauchy problem for the system* $\dot{x} = f(x)$, *where* $x \in \mathbb{R}^n$ *with* n *even and* $f: \mathbb{R}^n \to \mathbb{R}^n$ *is a Lipschitz function, can be naturally interpreted as a Schrödinger problem of the form*

$$u_t = i\Delta u - F(u), \tag{3.52}$$

where $F: X_k \to X_k$, $k = n/2$, $X_k = \operatorname{span}_{\mathbb{C}}\{\varphi_1, \dots, \varphi_k\} \subseteq L^2(\Omega)$, *and* $F(u)$ *is defined explicitly by*

$$[F(u)](\xi) = \left[F\left(\sum_{j=1}^{k} z_j \varphi_j \right) \right](\xi) = \sum_{j=1}^{k} \left(p_j(x,y) + iq_j(x,y) \right) \varphi_j(\xi), \quad \xi \in \Omega, \tag{3.53}$$

with $x_j = \operatorname{Re} z_j = \langle \operatorname{Re} u, \varphi_j \rangle$, $y_j = \operatorname{Im} z_j = \langle \operatorname{Im} u, \varphi_j \rangle$, $p_j(x,y) = f_j(x,y) - \lambda_j y_j$, $q_j(x,y) = g_j(x,y) + \lambda_j x_j$, $j = 1, \dots, k$.

Remark 3.9. Note that the operator $F: X_k \to X_k$ in Proposition 3.8 is *nonlocal*, since (3.53) contains integral terms $x_j = \langle \operatorname{Re} u, \varphi_j \rangle = \int_{\Omega} (\operatorname{Re} u)(\xi)\varphi_j(\xi)\, d\xi$ on the right-hand side, $j = 1, \dots, k$, and similarly for y_j. The operator F can be naturally extended to $\tilde{F}: L^2(\Omega) \to L^2(\Omega)$ by defining $\tilde{F}(u) = \tilde{F}\left(\sum_{j=1}^{\infty} z_j \varphi_j \right) = \sum_{j=1}^{k} \big(p_j(x,y) + iq_j(x,y) \big)\varphi_j$, where $u = \sum_{j=1}^{\infty} z_j \varphi_j$ is the Fourier expansion of $u \in L^2(\Omega)$ with respect to the orthonormal basis (φ_j) of eigenfunctons of $-\Delta|_{H_0^1 \cap H^2}$. The claim in the proposition holds (with the same proof) for nonautonomous systems of ODE's as well. In this case the nonlinearity in (3.52) takes the form $F(t, u)$.

Remark 3.10. Proposition 3.8 enables us to reformulate the second part of the 16th Hilbert problem in terms of the Schrödinger equation (3.52) as follows. Let $F(u)$ be defined by (3.51), where $p(x, y)$ and $q(x, y)$ are any given polynomials in 2 real variables, such that the maximum of their degrees d is at least 2. Assuming that the value of $d \geq 2$ is fixed, find an upper bound for the number of limit cycles in the class of Schrödinger equations (3.52) in terms of d. For $d = 2$ (that is, for the class of quadratical planar systems of ODE's) it is even not known if the bound can be finite.

As an example, consider the following nonlocal Schrödinger evolution problem on a given bounded domain Ω in $\mathbb{R}^N$:

$$u_t = i\Delta u + \left[\sum_{j=k}^m a_{2j} \left(\int_\Omega (\mathrm{Re}\, u)\varphi_1 \, dx \right)^{2j} + \sum_{j=k}^m a_{2j+1} \left(\int_\Omega (\mathrm{Re}\, u)\varphi_1 \, dx \right)^{2j+1} \right] \varphi_1,$$

(3.54)

where $1 \leq k \leq m$.

Theorem 3.11. ([46, Theorem 26]) *Let $\Gamma(u_0)$ be the trajectory in $L^2(\Omega)$ of the Schrödinger evolution problem (3.54) viewed near the zero function in $L^2(\Omega)$, and generated by initial condition $u(0) = z_0\varphi_1$, with $z_0 = x_0 + iy_0 \neq 0$, and k is a positive integer. Assume that the coefficients $a_2, a_4 \ldots, a_{2m}$ and $a_{2k+1}, a_{2k+3}, \ldots, a_{2m+1}$ are real. Assume that $a_{2k+1} \neq 0$, i.e. a_{2k+1} is the first nonzero coefficient corresponding to an odd exponent of the integral term in (3.54). Then*

$$\dim_B \Gamma(u_0) = 2 - \frac{2}{2k+1}.$$

(3.55)

Problem (3.54) with the indicated initial condition is equivalent to the following Liénard system in the plane, see Proposition 3.8:

$$\dot{x} = f(x, y) = \quad \lambda_1 y + \sum_{j=1}^m a_{2j} x^{2k+1} + \sum_{j=k}^m a_{2j+1} x^{2j+1}$$

(3.56)

$$\dot{y} = g(x, y) = -\lambda_1 x.$$

The claim of Theorem 3.11 follows immediately from [79, Theorem 6].

Remark 3.12. Note that under the condition of $a_{2k+1} \neq 0$ the Liénard problem (3.56) is not explicitly solvable for $(x_0, y_0) \neq (0, 0)$. Therefore, the corresponding Schrödinger evolution problem (3.54) is also not explicitly solvable for $u(0) \neq 0$. In fact, they are equivalent to each other, due to Proposition 3.8.

4. Fractal Oscillations of Second-Order Differential Equations

4.1. Notation, Definition and Motivation

Everywhere in this part, we use the following notation. Let $T > 0$ be an arbitrarily given real number, $y = y(x)$ be a continuous function on $[0, T]$ and let $\Gamma(y)$ be the graph of $y(x)$

defined as usual by $\Gamma(y) = \{(x_1, x_2) \in \mathbb{R}^2 : x_1 \in (0, T], x_2 = y(x_1)\}$. When $y(x)$ is a solution of a differential equation, we assume that $y \in C([0, T]) \cap C^2((0, T])$.

According to some classic books from fractal geometry, see [22], [44] and [66], we know that the graph $\Gamma(y)$ can be considered as a fractal curve in $\mathbb{R}^2$. Its fractality is described by its box dimension $\dim_B \Gamma(y) =: s$ and the corresponding s-dimensional upper Minkowski content $\mathcal{M}^{*s}(\Gamma(y))$, see Section 1.2.:

$$\dim_B \Gamma(y) = \lim_{\varepsilon \to 0} \left(2 - \frac{\log |\Gamma_\varepsilon(y)|}{\log \varepsilon}\right),$$

$$\mathcal{M}^{*s}(\Gamma(y)) = \limsup_{\varepsilon \to 0} (2\varepsilon)^{s-2} |\Gamma_\varepsilon(y)|,$$

and its lower s-dimensional content $\mathcal{M}^s_*(\Gamma(y))$. Indeed, in all results of this section the upper and lower box dimensions of $\Gamma(y)$ coincide, and the common value is denoted by $\dim_B \Gamma(y)$. Moreover, the graphs $\Gamma(y)$ are Minkowski nondegenerate, i.e. $0 < \mathcal{M}^{*s}(\Gamma(y)) \leq \mathcal{M}^{*s}(\Gamma(y)) < \infty$. Please, note that in papers cited in this section the notation $\dim_M$ (Minkowski dimension) is used instead of $\dim_B$ (box dimension).

Here $\Gamma_\varepsilon(y)$ denotes the ε-neighbourhood of the graph $\Gamma(y)$ in $\mathbb{R}^2$ defined by

$$\Gamma_\varepsilon(y) = \{(t_1, t_2) \in \mathbb{R}^2 : d((t_1, t_2), \Gamma(y)) \leq \varepsilon\}, \; \varepsilon > 0,$$

and $d((t_1, t_2), \Gamma(y))$ denotes the Euclidean distance from (t_1, t_2) to $\Gamma(y)$, and $|\Gamma_\varepsilon(y)|$ denotes the Lebesgue measure of $\Gamma_\varepsilon(y)$.

We say that a function $y(x)$ *oscillates* near $x = 0$ if there is a decreasing sequence $a_n \in (0, T]$ such that $y(a_n) = 0$ and $a_n \to 0$ as $n \to \infty$, see Figure 23.

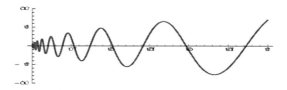

Figure 23. A function $y(x)$ which oscillates near $x = 0$.

However, if $y(x)$ oscillates near $x = 0$ and for some $s \in [1, 2)$ and $T_0 \in (0, T]$ we have:

$$\dim_B \Gamma(y \big|_{(0, T_0]}) = s \text{ and } \Gamma(y \big|_{(0, T_0]}) \text{ is Minkowski nondegenerate},$$

then we say that $y(x)$ is *fractal oscillatory* near $x = 0$, see [49, Definition 1.4]. Here $y \big|_{(0, T_0]}$ denotes the function restriction of $y(x)$ on the interval $(0, T_0]$.

The main motivation to study the fractal oscillations of solutions of second-order diferential equations we found in a study on the rectifiable oscillations of solutions of second-order diferential equations published in Pašić [48] and Wong [68]. More precisely, the *length* of a continuous function $y(x)$ is defined by:

$$\text{length}(\Gamma(y|_{[0, T_0]})) = \sup \sum_{i=1}^m \|(t_i, y(t_i)) - (t_{i-1}, y(t_{i-1}))\|_2,$$

Fractal Properties of Solutions of Differential Equations

where the supremum is taken over all partitions $0 = t_0 < t_1 < \ldots < t_m = T_0$ of the interval $[0, T_0]$ and $\| \cdot \|_2$ denotes the Euclidean norm in $\mathbb{R}^2$, see for instance [21] and [22]. Of course, if $y \in C^1((0, T])$, then

$$\text{length}\left(\Gamma(y|_{(0,T_0]})\right) = \lim_{\varepsilon \to 0} \int_\varepsilon^{T_0} \sqrt{1 + y'^2(x)}dx.$$

It was firstly shown that there are two classes of functions $y(x)$ and $z(x)$, which oscillate near $x = 0$ and at the same time the graphs $\Gamma(y)$ and $\Gamma(z)$ have respectively finite and infinite length. That is, $\text{length}\left(\Gamma(y)\right) < \infty$, $\text{length}\left(\Gamma(z)\right) = \infty$ and $y(x)$, $z(x)$ are explicitly given respectively by:

$$y(x) = c_1\sqrt{x}\,\cos(\rho \ln x) + c_2\sqrt{x}\,\sin(\rho \ln x), \tag{4.57}$$

$$z(x) = c_1 x\,\cos(\frac{1}{x}) + c_2 x\,\sin(\frac{1}{x}), \tag{4.58}$$

where $c_1, c_2 \in \mathbb{R}$ and $\rho > 0$. Also, $y(x)$ and $z(x)$ are the fundamental systems of all solutions of the following two differential equations respectively:

$$y'' + \frac{\lambda}{x^2}y = 0, \ x > 0, \tag{4.59}$$

$$z'' + \frac{1}{x^4}y = 0, \ x > 0, \tag{4.60}$$

where $\lambda = \rho^2 + 1/4$. It is natural to pose the following question:

(i) since $\text{length}\left(\Gamma(y)\right) < \infty$ it is clear that $\dim_B \Gamma(y) = 1$; what can we say about the finiteness of $\mathcal{M}^{*1}(\Gamma(y))$?

(ii) since $\text{length}\left(\Gamma(z)\right) = \infty$ we have that $\dim_B \Gamma(y) \geq 1$; what can we say about $s = \dim_B \Gamma(y)$ and $\mathcal{M}^{*s}(\Gamma(y))$?

(iii) the main coefficient in equations (4.59) and (4.60) is given respectively by $f(x) = \lambda x^{-2}$ and $f(x) = \lambda x^{-4}$; what we can say about $s = \dim_B \Gamma(y)$ and $\mathcal{M}_*^s(\Gamma(y))$ $\mathcal{M}^{*s}(\Gamma(y))$ for all solutions $y(x)$ of the linear differential equation $y'' + \lambda x^{-\sigma}y = 0$, which generalizes equations (4.59) and (4.60) because $f(x) = \lambda x^{-\sigma}$, where $\sigma \geq 2$? How does $\dim_B \Gamma(y)$ depend on the parameter σ?

The answers to previous questions have been published in Pašić [49] and it is recalled in the next section.

In order to show that a real function $y(x)$ is fractal oscillatory near $x = 0$ with the box dimension equal to a given real number $s \in (1, 2)$, it is enough to verify that $y(x)$ satisfies the following two inequalities:

$$c_1\varepsilon^{2-s} \leq |\Gamma_\varepsilon(y\,|_{(0,T_0]})| \quad \text{and} \quad |\Gamma_\varepsilon(y\,|_{(0,T_0]})| \leq c_2\varepsilon^{2-s}, \ \varepsilon \in (0, \varepsilon_0), \tag{4.61}$$

for some $T_0 \in (0, T]$ and $\varepsilon_0 > 0$. Therefore, we have explored a criterion on the function $y(x)$ by giving some sufficient conditions on $y(x)$ such that previous two inequalities are fulfilled in respect to arbitrarily given real number $s \in (1, 2)$.

By sumarizing [47, Lemma 2.1] and [47, Lemma 2.2], the folowing criterion have been used in order to establish (4.61). More precisely, we say that a function $m = m(\varepsilon)$ is an index function on $(0, \varepsilon_0]$ for some $\varepsilon_0 > 0$, if $m : (0, \varepsilon_0] \to \mathbb{N}$, $m(\varepsilon)$ is decreasing and $\lim_{\varepsilon \to 0} m(\varepsilon) = \infty$. Let $y \in C([0, T]) \cap C^1((0, T])$ and $a_n \in (0, T]$ be a decreasing sequence of consecutive zeros of $y(x)$ such that $a_n \to 0$ when $n \to \infty$. Let $k(\varepsilon)$ and $m(\varepsilon)$ be two index function satisfying:

$$|a_n - a_{n+1}| \leq \varepsilon \quad \text{for all } n \geq k(\varepsilon) \text{ and } \varepsilon \in (0, \varepsilon_0); \tag{4.62}$$

$$|a_n - a_{n+1}| > 4\varepsilon \quad \text{for all } n \leq m(\varepsilon) \text{ and } \varepsilon \in (0, \varepsilon_0); \tag{4.63}$$

$$y(x) \text{ is convex-concave function on } (a_{n+1}, a_n) \text{ for all } n \leq m(\varepsilon). \tag{4.64}$$

If for a given $s \in (1, 2)$, the function $y(x)$ satisfies:

$$c_1 \varepsilon^{2-s} \leq \sum_{n \geq k(\varepsilon)} \max_{x \in [a_{n+1}, a_n]} |y(x)|(a_n - a_{n+1}), \quad \varepsilon \in (0, \varepsilon_0), \tag{4.65}$$

$$c_2 \varepsilon^{2-s} \geq a_{m(\varepsilon)} \max_{x \in [0, a_{m(\varepsilon)}]} |y(x)| + \varepsilon \sum_{n=2}^{m(\varepsilon)} \int_{a_n}^{a_{n-1}} |y'(x)| \, dx, \quad \varepsilon \in (0, \varepsilon_0), \tag{4.66}$$

where $c_1, c_2, \varepsilon_0 > 0$, then $y(x)$ is fractal oscillatory near $x = 0$ with the box dimension of $\Gamma(y)$ equal to s.

The main difficulty in the application of previous criterion has been in the statement (4.64) since it causes that the zeros a_n of $y(x)$ coincide with its inflexion points. In such a case, it is not easy to provide two index functions $k(\varepsilon)$ and $m(\varepsilon)$ satisfying (4.62) and (4.63) if a_n is the sequence of inflexion points of $y(x)$.

Very recently in Pašić and Tanaka [52], the inequalities (4.65) and (4.66) are recovered but without any assumption of the type like (4.63) and (4.64), which allows the application of (4.65) and (4.66) in the fractal oscillations of more complicated differential equations.

4.2. Fractal Oscillations of Euler Type Equations

In this section we consider the Euler type linear differential equations of second-order:

$$y'' + \frac{\lambda}{x^\sigma} y = 0, \ x > 0, \tag{4.67}$$

where $\sigma \geq 2$ and $\lambda > 0$. It is clear that this equation in particular for $\sigma = 2, \lambda > 1/4$ and $\sigma = 4, \lambda = 1$ has the fundamental system of all solutions in explicit forms given respectively by (4.57) and (4.58). However, for other values of $\sigma > 2$, equation (4.67) does not allow the explicit form of its solution.

The first result on the fractal oscillations of solutions of linear differential equations was published in Pašić [49] and it is as follows.

Theorem 4.1. ([49, Theorem 1.1]) *Let* $y \in C([0, 1]) \cap C^2((0, 1])$ *be a nontrivial solution of the Euler type equation* (4.67), *where* $\sigma \geq 2$, $\lambda > 0$, *and* $\lambda > 1/4$ *as* $\sigma = 2$. *Then we have:*

Fractal Properties of Solutions of Differential Equations   445

(i) *if* $2 \leq \sigma < 4$, *then* $y(x)$ *is fractal oscillatory near* $x = 0$ *with the box dimension of* $\Gamma(y)$ *equal to* 1;

(ii) *if* $\sigma = 4$, *then* $y(x)$ *is not fractal oscillatory near* $x = 0$ *because* $\dim_B \Gamma(y) = 1$ *and* $\mathcal{M}^{*1}(\Gamma(y)) = \infty$;

(iii) *if* $\sigma > 4$, *then* $y(x)$ *is fractal oscillatory near* $x = 0$ *with the box dimension of* $\Gamma(y)$ *equal to* $s = 3/2 - 2/\sigma$.

With the help of statement (i), the rectifiable oscillations of solutions $y(x)$ of equation (4.67) for $\sigma \in [2, 4)$ (length $(\Gamma(y)) < \infty$) is verified once again but using the one dimesional Minkowski content $\mathcal{M}^{*1}(\Gamma(y))$.

The statement (ii) shows that graph $\Gamma(y)$ of every nontrivial solution $y(x)$ of (4.67) for $\sigma = 4$ is degenerate in the sense of the Minkowski content, more precisely, $\mathcal{M}^{*1}(\Gamma(y)) = \infty$.

The statement (iii) gives a refinement of the unrectifiable oscillations near $x = 0$ of solution $y(x)$ of equation (4.67) for $\sigma > 4$ (length $(\Gamma(y)) = \infty$). Also, it is clear that the main coefficient $f(x) = x^{-\sigma}$ of equation (4.67) is singlar near $x = 0$ and the order of growth for $f(x)$ near $x = 0$ explicitly influences the order of growth for the box dimension: $\dim_B \Gamma(y) = 3/2 - 2/\sigma$.

Regarding the statements (i) and (iii) of the previous theorem, we observe that the case $\sigma = 4$ is a critical case between fractal oscillations of eqation (4.67) with integer and fractional values of its fractal dimension.

The explicit nature of the coefficient $f(x) = x^{-\sigma}$ of equation (4.67) has been replaced with more general structure of Hartman-Wintner asymptotic condition near $x = 0$ considered in Kwong, Pašić and Wong [39], which is recalled in the next section.

4.3. Fractal Oscillations of Hartman-Wintner Type Equations

In this section, we consider the linear differential equations in a more general form than in the previous section:

$$y'' + f(x)y = 0, \ x \in (0, T], \tag{4.68}$$

where the coefficient $f(x)$ for some $T_0 \in (0, T]$ satisfies:

$$f \in C^2((0, T_0]), \ f(x) > 0 \text{ in } (0, T_0] \text{ and } f(0+) = \infty, \tag{4.69}$$

and the following Hartman-Wintner asymptotic condition:

$$\frac{1}{\sqrt[4]{f(x)}} \left(\frac{1}{\sqrt[4]{f(x)}} \right)'' \in L^1(0, T_0). \tag{4.70}$$

It is important, see [39, Lemmas 2.1 and 2.2]), that from (4.69) and (4.70) follows $\sqrt{f} \notin L^1(0, T_0)$ as well as the following asymptotic formulas of all solutions $y(x)$ of equation (4.68): there are $A \neq 0$, $B \in \mathbb{R}$ and $g_0, g_1 \in C(0, T]$ such that

$$y(x) = \frac{1}{\sqrt[4]{p(x)q(x)}} \Big[A \sin \left(\varphi(x) \right) + h_0(x) \Big], \tag{4.71}$$

$$y'(x) = -\sqrt[4]{\frac{q(x)}{p(x)^3}}\Big[A\cos\left(\varphi(x)\right) + h_1(x)\Big], \tag{4.72}$$

where

$$\varphi(x) = \int_x^{T_0} \sqrt{\frac{q(\tau)}{p(\tau)}}d\tau + B \quad \text{and} \quad \lim_{x\to 0} h_0(x) = \lim_{x\to 0} h_1(x) = 0.$$

In order to prove (4.71) and (4.72), we use some asymptotic methods presented in [12] and [28].

The asymptotic formulas (4.71) and (4.72) allow us to apply the inequalities (4.65) and (4.66) to all solutions $y(x)$ of equation (4.68) and to prove the following result.

Theorem 4.2. ([39, Theorem 1.4]) *Let* $y \in C([0, T]) \cap C^2((0, T])$ *be a nontrivial solution of equation* (4.68), *where* $f(x)$ *satisfies the conditions* (4.69) *and* (4.70). *If*

$$f(x) \simeq x^{-\sigma} \text{ near } x = 0, \sigma > 4, \tag{4.73}$$

then $y(x)$ *is fractal oscillatory near* $x = 0$ *with the box dimension of* $\Gamma(y)$ *equal to* $s = 3/2 - 2/\sigma$.

There are many classes of functions $f(x)$ satisfying the conditions (4.69) and (4.70). First of all, it is the coefficient of Euler type $f(x) = x^{-\sigma}$ provided $\sigma > 2$. Hence, previous theorem generalizes the observation given in Section 2. For other types of $f(x)$ satisfying (4.69) and (4.70) we refer the reader to [39].

In (4.73), the coefficient $f(x)$ satisfies a kind of pointwise asymptotics which depends on the parameter σ. Instead of that, in [51, Theorem 1.1] it is considered a more general asymptotic behaviour of $f(x)$ which is of integral type and also depending on parameter σ.

4.4. Fractal Oscillations of Bessel Type Equations

It is known that the classic Bessel differential equation,

$$t^2 z'' + t z' + (t^2 - \nu^2)z = 0, \ t > 1/T,$$

is oscillatory near $t = \infty$. On the oscillations near $t = \infty$ of several kinds of differential equations we refer reader to [2], [28], [41] and [63]. By the transformation $z(t) = y(1/t)$, this equation is transformed into the Bessel type equation oscillating near $x = 0$:

$$y'' + \frac{1}{x}y' + (\frac{1}{x^4} - \frac{\nu^2}{x^2})y = 0, \ x > 0. \tag{4.74}$$

The equation (4.74) is a particular case of the following one:

$$y'' + \frac{\mu}{x}y' + (\frac{\lambda}{x^\rho} - \frac{\nu^2}{x^2})y = 0, \ x \in (0, T], \tag{4.75}$$

where $\mu \in \mathbb{R}$, $\lambda > 0$, $\rho > 2$ and $\nu \in \mathbb{R}$. Very recently in Pašić and Tanaka [51], the authors have considered the fractal oscillations near $x = 0$ of the Bessel type linear differential equation (4.75) in the dependence on the parameters μ and ρ. It was verified that every

Fractal Properties of Solutions of Differential Equations

447

nontrivial solution of equation (4.75) is fractal oscillatory near $x = 0$ with the dimensional number

$$s = \frac{3}{2} + \frac{\mu - 2}{\rho},$$

where μ, λ and ρ in addition satisfy: $2\mu \leq \rho$ and $2\mu + \rho > 4$.

As the first consequence, it follows that every nontrivial solution of equation (4.74) is also fractal oscillatory near $x = 0$ with the dimensional number $s = \frac{5}{4}$.

As the second consequence, the chirp-like behaviour equation:

$$y'' + \frac{\beta - 2\alpha + 1}{x}y' + \left(\frac{\beta^2}{x^{2\beta+2}} - \frac{\alpha(\beta - \alpha)}{x^2}\right)y = 0, \ x > 0, \tag{4.76}$$

where $\alpha, \beta > 0$, it is also a particular case of equation (4.75) for $\mu = \beta - 2\alpha + 1$, $\lambda = \beta^2$, $\rho = 2\beta + 2$ and $v^2 = \alpha(\beta - \alpha)$. Hence, it follows that every nontrivial solution of equation (4.76) is also fractal oscillatory near $x = 0$ with the dimensional number

$$s = 2 - \frac{\alpha + 1}{\beta + 1}.$$

In a more general setting, a function $y(x)$ is said to have chirp-like behaviour if it has the following form: $y(x) = r(x)\sin\varphi(x)$ or $y(x) = r(x)\cos\varphi(x)$, where $r(x) > 0$ and $\varphi(x) > 0$ in $(0, T]$, $r(0) = 0$ and $\varphi(0+) = \infty$. Also, a linear differential equation is said to be of chirp-like behaviour if its fundamental system of all solutions is $y(x) = c_1y_1(x) + c_2y_2(x)$, where $y_1(x)$ and $y_2(x)$ are two functions of the previous type. For more details about the chirp-like behaviour of functions and the corresponding differential equations, we refer reader to Pašić and Tanaka [52].

4.5. Fractal Oscillations of Half-Linear Differential Equations

In this section, we recall some results on fractal oscillations of solutions of half-linear differential equation recently published in Pašić and Wong [54]:

$$(|y'|^{p-2}y')' + f(x)|y|^{p-2}y = 0, \ x \in (0, T], \tag{4.77}$$

where the coefficient $f(x)$ satisfies (4.69) and the following Hartman-Wintner type condition:

$$f^{\frac{1}{p}} \notin L^1(0, 1) \text{ and } f^{-\frac{1}{pq}}\left[f^{-\frac{1}{p^2}}\right]'' \in L^1(0, 1). \tag{4.78}$$

Obviously the condition (4.78) generalizes the related ones in (4.70) from $p = 2$ to $p > 1$. Also, it is clear that in particular for $p = 2$, equation (4.77) equals the linear equation (4.68) considered in previous sections.

Theorem 4.3. ([54, Theorem 6]) *Let $f(x)$ satisfy (4.69) and (4.78), and let $f(x) \simeq \lambda x^{-\alpha}$ near $x = 0$, where $\alpha > p$. Let $y(x)$ be a solution of equation (4.77).*

(i) *if $p < \alpha < p^2$, then $y(x)$ is fractal oscillatory near $x = 0$ with the box dimension of $\Gamma(y)$ equal to $s = 1$;*

(ii) *if $\alpha > p^2$, then $y(x)$ is fractal oscillatory near $x = 0$ with the box dimension of $\Gamma(y)$ equal to $s = 2 - \frac{1}{q} - \frac{p}{\alpha}$.*

Obviously this theorem generalizes [39, Theorem 1.4] from the case when $p = 2$ to more general case $p > 1$. The proof of the previous theorem published in [54], is based on the following two steps. Firstly, every solution $y(x)$ of equation (4.77) could be rewritten in the form:

$$y(x) = (p-1)^{\frac{1}{pq}} f^{-\frac{1}{pq}}(x) V^{\frac{1}{p}}(x) w(\varphi(x)),$$

$$|y'|^{p-2} y' = -(p-1)^{-\frac{1}{pq}} f^{\frac{1}{pq}}(x) V^{\frac{1}{q}}(x) |w'(\varphi(x))|^{p-2} w'(\varphi(x)),$$

where the function $w = w(t)$ $(t > 0)$ is the so-called generalized sine function which is determined by:

$$(|w'(x)|^{p-2} w'(x))' + (p-1)|w(x)|^{p-2} w(x) = 0, \ w(0) = 0, \ w'(0) = 1,$$

$$|w'(t)|^p + |w(t)|^p \equiv 1 \text{ for all } t > 0.$$

Secondly, it is shown that the functions $V(x)$ and $\varphi(x)$ satisfy the equations:

$$\varphi'(x) = \frac{-1}{(p-1)^{\frac{1}{p}}} f^{\frac{1}{p}}(x) + \frac{1}{p} \frac{f'(x)}{f(x)} |w'(\varphi(x))|^{p-2} w'(\varphi(x)) w(\varphi(x)),$$

$$V'(x) = \left[(p-1)^{\frac{1}{p}} f^{-\frac{1}{p}}(x)\right]' |y'|^p + \left[(p-1)^{-\frac{1}{q}} f^{\frac{1}{q}}(x)\right]' |y|^p.$$

Moreover, it is crucial to show that $V(x)$ and $\varphi(x)$ satisfy the following asymptotic conditions:

$$\varphi'(x) < 0 \text{ for all } x \in (0, 1] \quad \text{and} \quad \lim_{x \to 0+} \varphi(x) = \infty,$$

$$0 < \lim_{x \to 0+} V(x) < +\infty.$$

Methodology for the derivation of previous asymptotic formulas is fundamentally different than shown for the linear equation. Previous asymptotic formulas allow us to use the inequalities (4.65) and (4.66) to every solution $y(x)$ of equation (4.77).

4.6. Two-Point Fractal Oscillations of Linear Differential Equations

Recently in Pašić and Wong [53], authors studied a class of linear differential equations whose solutions oscillate at the same time near both boundary points of the interval $[0, 1]$. More precisely, a continuous function $y(x)$ is said to be 2-point oscillatory in $[0, 1]$ if there are two sequences $a_n, b_n \in (0, 1)$ such that $y(a_n) = y(b_n) = 0$, a_n is decreasing, b_n is increasing, $a_n \to 0$ and $b_n \to 1$ as $n \to \infty$, see Figure 24.

The simplest equation that supports this kind of oscillation is the so-called Riemann-Weber version of the Euler equation:

$$y'' + \frac{1}{x^2}\left(\frac{1}{4} + \frac{\lambda}{|\ln x|^2}\right) y = 0, \ x \in (0, 1), \tag{4.79}$$

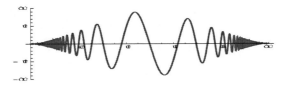

Figure 24. A function $y(x)$ which is 2-point oscillatory in $(0, 1)$.

where $\lambda > 1/4$. By using some elementary methods, it is easy to check that the fundamental system of all solutions of equation (4.79) is given by (see Pašić [50])

$$y_1(x) = [x \ln(1/x)]^{1/2} \cos(\rho \ln \ln(1/x)), \ x \in (0, 1),$$
$$y_2(x) = [x \ln(1/x)]^{1/2} \sin(\rho \ln \ln(1/x)), \ x \in (0, 1).$$

Obviously, the functions $y_1(x)$ and $y_2(x)$ are 2-point oscillatory in $[0, 1]$ and therefore, it is also true for all solutions of equation (4.79).

Next, the 2-point oscillations in $[0, 1]$ are also verified for the equation:

$$y'' + \frac{c(x)}{(x - x^2)^\sigma} y = 0, \ x \in (0, 1), \tag{4.80}$$

where $\sigma > 2$ and $c(x)$ is a smooth and positive function in $[0, 1]$. Moreover, the following theorem has been proved.

Theorem 4.4. ([53, Theorem 6.3]) *Let* $f \in C^2((0,1))$, $f(x) > 0$ *on* $(0, 1)$, $f(0+) = f(1-) = \infty$, *and let* $f(x)$ *satisfy the Hartman-Wintner condition (4.70), where* $T_0 = 1$. *Let* $\sigma > 4$ *and*

$$f(x) \simeq (x - x^2)^{-\sigma} \text{ near } x = 0 \text{ and } x = 1,$$

that is, there exist constants $\lambda_0 > 0$, $\lambda_1 > 0$, *and* $\delta \in (0, 1)$ *such that*

$$\tfrac{\lambda_0}{(x-x^2)^\sigma} \leq f(x) \leq \tfrac{\lambda_1}{(x-x^2)^\sigma} \text{ for all } x \in (0, \delta) \cup (1 - \delta, 1).$$

Then equation $y'' + f(x)y = 0$ *is 2-point oscillatory in* $[0, 1]$ *such that*

$$\dim_B \Gamma(y) = 3/2 - 2/\sigma \text{ and } \Gamma(y) \text{ is Minkowski nondegenerate}.$$

Thus, one can say that under the conditions on $f(x)$ as in previous theorem every solution $y(x)$ of equation $y'' + f(x)y = 0$ is 2-point fractal oscillatory in $[0, 1]$.

5. Nonregularity of Weak Solutions of Elliptic Equations

The aim of this section is to review some recent results devoted to the study of nonregularity of weak solutions of elliptic equations. More precisely, we are interested in singularities of weak solutions of elliptic equations. It is based on [70, 77, 80, 81, 82], where one can find more information and detailed proofs.

5.1. Singular Dimension of Solution Set of Some Classes of Elliptic Equations

It is well known that the following boundary value problem (BVP)

$$\begin{cases} -\Delta_p u &= F(x) \quad \text{in } \mathcal{D}'(\Omega), \\ u &\in W_0^{1,p}(\Omega), \end{cases} \tag{5.81}$$

possesses the unique weak solution, provided $F \in L^{p'}(\Omega)$, $1 < p < \infty$, $p' = p/(p-1)$. This BVP is nonlinear since we deal with the nonlinear differential operator $\Delta_p u = \text{div}\left(|\nabla u|^{p-2}\nabla u\right)$, i.e. with the usual p-Laplacian. We view (5.81) not as an individual elliptic problem, but as a family of BVP's indexed by F's in the indicated Lebesgue space. In [70] the problem of generating singularities of weak solutions of (5.81) in only one point has been studied, while [81] (see also Theorem 5.2 below) was devoted to the problem of generating singularities on general fractal subsets of Ω.

We are interested in finding the supremum of Hausdorff dimension of singular sets (to be defined in a moment) of the family of weak solutions of (5.81). The complete solution of this problem in the case of $p > 2$ can be seen in Theorem 5.1(c). The case of $p = 2$ has been solved in [33, Theorem 2], or see Theorem 5.1(b) below. The case of $1 < p < 2$ is an open problem, for which we have only a partial answer, see Theorem 5.1(a).

Let $f : \Omega \to \mathbb{R}$ be a given Lebesgue measurable function, where $\Omega \subseteq \mathbb{R}^N$ is a fixed open set. We say that $a \in \Omega$ is a singular point of f if there exist positive constants C and γ such that $f(x) \geq C|x - a|^{-\gamma}$ a.e. in a neighbourhood of a. Let Sing f be the set of all singular points of f, which we call the singular set of f. Note that the definition is meaningful since $f = g$ a.e. in Ω implies Sing $f = $ Sing g.

Assume that a vector space X (or just a nonempty set) of measurable functions $f : \Omega \to \mathbb{R}$ is given. We are interested in those functions in X which have the Hausdorff dimension of their singular sets as large as possible. We define the *singular dimension of* X by

$$\text{s-dim } X = \sup\{\dim_H(\text{Sing } f) : f \in X\}. \tag{5.82}$$

A function $f \in X$ (if it exists) such that Sing $f \neq \emptyset$ is said to be *maximally singular* in X if the above supremum is achieved with f. These notions have been introduced in [71] and [72]. The theory of singularities of functions in various Banach spaces goes back to 1950s, see for example a short surveys in [71] and [83].

The following result deals with the singular dimension of the solution set of BVP (5.81).

Theorem 5.1. ([82, Theorem 4]) *Assume that Ω is a bounded open set in $\mathbb{R}^N$ and $1 < p < \infty$, $pp' \leq N$. Let $X(\Omega, p)$ be the set of weak solutions $u \in W_0^{1,p}(\Omega)$ of (5.81) corresponding to all $F \in L^{p'}(\Omega)$.*
(a) If $1 < p < 2$ then

$$N - pp' \leq \text{s-dim } X(\Omega, p) \leq N - 2p. \tag{5.83}$$

(b) If $p = 2$ then s-dim $X(\Omega, 2) = N - 4$.
(c) If $p > 2$ then

$$\text{s-dim } X(\Omega, p) = N - pp'.$$

The proof of Theorem 5.1 exploits a regularity result for solutions of (5.81) involving Besov spaces due to Simon [61]. Let us recall some results about singular dimension of function spaces. Here the positive part of a real number r is denoted by $r_+ = \max\{0, r\}$.

In [71, Theorem 2] it has been shown that the Bessel potential space $L^{\alpha,p}(\mathbb{R}^N)$ has the singular dimension equal to $(N - \alpha p)_+$, provided $\alpha > 0, 1 < p < \infty$, that is,

$$\text{s-dim } L^{\alpha,p}(\mathbb{R}^N) = (N - \alpha p)_+.$$

If in addition to this we assume $\alpha p < N$, then these spaces contain maximally singular functions v, that is, such that $\dim_H(\text{Sing } v) = N - \alpha p$, see [33]. From the Calderón theorem, see Adams and Hedberg [1], we derive that analogous result holds also for Sobolev spaces $W^{k,p}(\mathbb{R}^N)$, provided $1 < p < \infty$:

$$\text{s-dim } W^{k,p}(\mathbb{R}^N) = (N - kp)_+,$$

and for $kp < N$ there exist maximally singular Sobolev functions. The case of $p = 1$ is an open problem. The result is interesting even in the case of $k = 0$, that is, for classical Lebesgue spaces $L^p(\mathbb{R}^N)$, $1 \leq p < \infty$:

$$\text{s-dim } L^p(\mathbb{R}^N) = N,$$

and there exist maximally singular Lebesgue functions, see [72]. Similar results have been obtained for Besov spaces $B_\alpha^{p,q}(\mathbb{R}^N)$, Lizorkin-Triebel spaces $F_\alpha^{p,q}(\mathbb{R}^N)$, provided $p, q \in (1, \infty)$:

$$\text{s-dim } B_\alpha^{p,q}(\mathbb{R}^N) = (N - \alpha p)_+, \quad \text{s-dim } F_\alpha^{p,q}(\mathbb{R}^N) = (N - \alpha p)_+, \tag{5.84}$$

and for $0 < \alpha p < N$ there exist maximally singular functions in these spaces, see [77]. The result (5.84) for Besov spaces has been partly extended to the case of $q = \infty$, see [82, Theorem 1]:

$$\text{s-dim } B_\alpha^{p,\infty}(\mathbb{R}^N) \leq (N - \alpha p)_+, \tag{5.85}$$

which was important in the proof of Theorem 5.1, see (5.88). Also for the Hardy space $H^1(\mathbb{R}^N)$ we have that its singular dimension is equal to N. Furthermore, maximally singular functions in all these spaces have been constructed explicitly.

5.2. Sketch of the Proof of Theorem 5.1

Let $X(\Omega, p)$ be the set of weak solutions of BVP (5.81) generated by all right-hand sides $F \in L^{p'}(\Omega)$. Note that the solution set $X(\Omega, p)$ is not a vector space for $p \neq 2$, since the operator Δ_p is nonlinear in this case. We do not know if $X(\Omega, p)$ possesses maximally singular functions in this case.

The crucial role in the proof of lower bounds in Theorem 5.1 is played by the following result, which enables us to generate singularities of weak solutions of (5.81) on fractal sets, see [81, Theorem 2]. By $d(x, A)$ we denote the Euclidean distance from x to A, that is, $d(x, A) = \inf\{|x - a| : a \in A\}$, while $\overline{\dim}_B A$ is the upper box dimension of A, see Falconer [22].

Theorem 5.2. ([81, Theorem 1]) *Assume that $F(x) = d(x, A)^{-\gamma}$, where $A \subset \Omega$ is a given nonempty set such that $\overline{\dim}_B A < N - pp'$ (in particular, $pp' < N$), and choose γ so that*

$$p < \gamma < \frac{1}{p'}(N - \overline{\dim}_B A). \tag{5.86}$$

Then $F \in L^{p'}(\Omega)$ and there exist positive constants C and D such that for the corresponding weak solution of (5.81) we have the following a priori estimate:

$$u(x) \geq C \cdot d(x, A)^{-(\gamma-p)/(p-1)} - D \quad a.e. \ in \ \Omega \tag{5.87}$$

In particular, u has singularity at least of order $\frac{\gamma-p}{p-1}$ on A, and $A \subseteq \text{Sing } u$.

The fact that $F \in L^{p'}(\Omega)$ in Theorem 5.2 follows from the right-hand side inequality in (5.86). The left-hand side inequality in (5.86) is responsible for generating singularities, see (5.87). In this way we see that it is possible to choose the set A such that $\overline{\dim}_B A < N - pp'$, with $\overline{\dim}_B A$ as close as we wish to $N - pp'$. This proves the lower bounds in Theorem 5.2 for any $p > 1$, since we can choose A to be generalized Cantor sets, so that $\dim_B A = \dim_H A$.

To obtain the upper bound in Theorem 5.1, a regularity result due to de Thélin [65] was exploited: if $F \in L^{p'}(\Omega)$, $1 < p \leq 2$, and Ω is bounded, then for the weak solution of (5.81) we have $u \in W^{2,p}_{loc}(\Omega)$. Hence, $X(\Omega, p) \subset W^{2,p}_{loc}(\Omega)$. From this we conclude s-dim $X(\Omega, p) \leq$ s-dim $W^{2,p}_{loc}(\Omega) = N - 2p$. This proves the upper bound in (5.83).

The case of $p = 2$ in Theorem 5.1 has been treated in [81, Theorem 2].

To prove the case of $p > 2$ in Theorem 5.1, we exploit a regularity result due to Simon, see [61, Chapter V, Theorem 1] or [65, Remark 4]: if $F \in L^{p'}(\Omega)$, and Ω is bounded, then for the weak solution of (5.81) we have $u \in B^{p,\infty}_{p',loc}(\Omega)$. In other words, $X(\Omega, p) \subseteq B^{p,\infty}_{p',loc}(\Omega)$. We conclude that

$$\text{s-dim } X(\Omega, p) \leq \text{s-dim } B^{p,\infty}_{p',loc}(\Omega) \leq N - p'p, \tag{5.88}$$

where we have exploited (5.85).

Remark 5.3. It would be interesting to know the precise value of s-dim $X(\Omega, p)$ for $1 < p < 2$ in Theorem 5.1(a).

Remark 5.4. The following result provides a good repertoire of Lebesgue integrable functions that are singular on a prescribed fractal set. Assume that $A \subset \mathbb{R}^N$ is a set for which there exists $d = \dim_B A$, and $\mathcal{M}^d_*(A) > 0$. Then for any $\varepsilon > 0$,

$$\int_{A_\varepsilon} d(x, A)^{-\gamma} dx < \infty \quad \Longleftrightarrow \quad \gamma < N - d,$$

where A_ε is the ε-neighbourhood of A; see [80, Theorem 4.1(c)]. The above equivalence is not true in general if A is such that $\mathcal{M}^d_*(A) = 0$. Namely, it is possible to construct a class of fractal sets A such that $d = \dim_B A$ exists, and $\mathcal{M}^d_*(A) = 0$, for which we have the following surprising equivalence: $\int_{A_\varepsilon} d(x, A)^{-\gamma} dx < \infty \Leftrightarrow \gamma \leq N - d$, see [74, Theorem 4.2].

5.3. Construction of Maximally Singular Weak Solutions for Laplace Equations

Here we consider the BVP (5.81) for $p = 2$, that is,

$$\begin{cases} -\Delta u &= F(x) \quad \text{in } \mathcal{D}'(\Omega), \\ u &\in H_0^1(\Omega), \end{cases} \tag{5.89}$$

with $F \in L^2(\Omega)$. Assume that $N \geq 5$ and let $(A_k)_{k \geq 1}$ be a sequence of compact subsets of Ω such that

$$\overline{\dim}_B A_k < N - 4, \quad \lim_{k \to \infty} (\dim_H A_k) = N - 4.$$

Such a sequence can be easily constructed using generalized Cantor sets, see [81, p. 364]. Let $\varepsilon_k > 0$ be small enough, so that $\overline{(A_k)_{\varepsilon_k}} \subset \Omega$ for any $k \geq 1$. Let us choose a sequence of numbers γ_k such that $2 < \gamma_k < \frac{1}{2}(N - 2)$, and define the corresponding sequence of functions $F_k \in L^2(\Omega)$ by $F_k(x) = d(x, A_k)^{-\gamma_k}$ for $x \in (A_k)_{\varepsilon_k}$ and $F_k(x) = 0$ for $x \in \Omega \setminus (A_k)_{\varepsilon_k}$. Then the function

$$F(x) = \sum_{k=1}^{\infty} 2^{-k} \frac{F(x)}{\|F_k\|_{L^2}}$$

generates a maximally singular weak solution u of (5.89) in the class of right-hand sides F in $L^2(\Omega)$. In other words, $\dim_H(\text{Sing } u) = N - 4$. See [81, pp. 364 and 365] for details.

5.4. Extended Singular Dimension

Given a measurable function $f : \Omega \to \mathbb{R}$ it is natural to define the *extended singular set* of f by

$$\text{e-Sing } f = \{a \in \Omega : \limsup_{r \to 0} \frac{1}{r^N} \int_{B_r(a)} f(x)\, dx = +\infty\}$$

The name is justified due to the obvious inclusion $\text{Sing } f \subseteq \text{e-Sing } f$. Here $B_r(a)$ is the open ball in $\mathbb{R}^N$ of radius r centered at a. It is easy to see that the extended singular set contains for example logarithmic and iterated logarithmic singularities of f. Similarly as in (5.82) we can then define the *upper singular dimension* of a space (or just a nonempty set) of functions X by

$$\text{s-}\overline{\dim}\, X = \sup\{\dim_H(\text{e-Sing } f) : f \in X\}.$$

This definition has been introduced in [71], see also [73]. We can also deal with *extended maximally singular functions* $f \in X$, defined by $\dim_H(\text{e-Sing } f) = \text{s-}\overline{\dim}\, X$. It can be easily seen that the analogues of Theorems 1 and 2 hold with s-dim X replaced by s-$\overline{\dim}\, X$. It suffices to change $\text{Sing } f$ with $\text{e-Sing } f$ in the proofs.

For the solution set $X = X(\Omega, p)$ appearing in Theorem 5.1 we have s-dim $X =$ s-$\overline{\dim}\, X$. For example, if in Theorem 5.1 we have $p \geq 2$ and $pp' = N$, then $\dim_H(\text{e-Sing } u) = 0$ for each $u \in X(\Omega, p)$. In particular, in the limiting case of $pp' = N$ the set of logarithmic singularities of any weak solution of BVP (5.81) generated by $F \in L^{p'}(\Omega)$ is small in the sense of Hausdorff dimension, and moreover, it is equal to zero.

5.5. Loss of Regularity of Weak Solutions of p-Laplace Equations for $p \neq 2$

In [70, Theorem 4] we have proved the following result dealing with generating singularities of weak solutions of (5.81). Note that the condition $p < N$ is natural in Theorem 5.5, since for $p \geq N$ the Sobolev functions do not possess singularities in the sense introduced above (an immediate consequence of the Sobolev imbedding theorem).

Theorem 5.5. ([70, Theorem 4]) *Assume that $p < N$, $p < \gamma < 1 + N/p'$, and $a \in \Omega$ is fixed. If $F \in L^{p'}(\Omega)$ has the order of singularity γ at $a \in \Omega$, then the corresponding weak solution u of (5.81) has the order of singularity equal to $\frac{\gamma-p}{p-1}$ at a.*

We are especially interested in the case when $\frac{\gamma-p}{p-1} > \gamma$, that is, when the solution u is "more singular" at $a \in \Omega$ than the input function F in (5.81). In this sense we speak about the loss of regularity of weak solution at $a \in \Omega$ with respect to the regularity of input function at a given point.

We say that the family of elliptic BVPs (5.81) indexed by $F \in L^{p'}(\Omega)$ has a *loss of regularity at a given point* $a \in \Omega$ if there exists $F \in L^{p'}(\Omega)$ which is singular at a, such that the corresponding weak solution has larger order of singularity at this point than the right-hand side F.

Assume that F has a singularity of order $\gamma > 0$ at a point $a \in \Omega$. It is easy to see that the loss of regularity of (5.81) at a point $a \in \Omega$ cannot occur for $p \geq 2$. Indeed, defining the difference of the orders of singularities of output and input functions (that is, of u and F) by

$$\delta = \delta(F) := \frac{\gamma - p}{p - 1} - \gamma = \frac{\gamma(2 - p) - p}{p - 1}, \tag{5.90}$$

we see that $\delta(F) < 0$ if $p \geq 2$. Hence, Theorem 5.5 implies that in this case the weak solution u of (5.81) has a singularity at a with corresponding order which is smaller than γ, i.e. the solution gets "more regular" at a.

Assuming that $\Omega = B_R(0)$ is a ball of radius R centered at the origin, let $F(x) = C|x|^{-\gamma}$, $C > 0$, such that γ satisfies the assumptions of Theorem 5.5. Then the corresponding weak solution of (5.81) can be written explicitly as

$$u(x) = \left(\frac{C}{m + N}\right)^{p'-1} \frac{|x|^{-\mu} - R^{-\mu}}{\mu}, \tag{5.91}$$

where $\mu = (\gamma-p)/(p-1) > 0$, see [70, Lemma 1]. As we see, it has the form $u(x) \simeq |x|^{-\mu}$ near $x = 0$.

Remark 5.6. A regularity result stated in Pucci and Servadei [59, Theorem 2.4] shows that the condition $p < \gamma$ in Theorem 5.5 cannot be relaxed. Compare with Lemma 5.10 for $m = -p$. Also conversely, Theorem 5.5 shows that the condition $a(x) \in L^{N/p(1-\varepsilon)}(\Omega)$ in [59, Theorem 2.4] cannot be relaxed.

5.5.1. Generating Singularities of Weak Solutions of (5.81) in the Case of $1 < p < 2$

Let us assume that $1 < p < 2$. We have $\delta = 0$ for

$$\gamma_c = \frac{p}{2 - p} \in (p, \infty), \tag{5.92}$$

Fractal Properties of Solutions of Differential Equations   455

which we call the *critical exponent* for the loss of regularity. Note that $\gamma_c > p$ and $\gamma_c \to \infty$ as $p \to 2-0$. The value of δ, where δ is defined by (5.90), will be called the *loss of regularity* at $a \in \Omega$ associated with F, if $\delta > 0$. The value of $|\delta|$ will be called the *gain of regularity* at a associated with F, if $\delta < 0$. If $\gamma = \gamma_c$ (i.e. $\delta = 0$) then there is no change of regularity. If for example $p = 2$, then we have $\delta = -2$, so the gain of regularity is equal to 2. The case of $p = 2$ is the only one in which the gain of regularity $|\delta|$ does not depend on γ.

Theorem 5.5 implies the following result concerning the property of loss of regularity of (5.81) at a given point $a \in \Omega$.

Theorem 5.7. ([83, Theorem 2.1]) *Assume that* $1 < p < 2$, $N > 2\gamma_c$, *where* γ_c *is the critical exponent defined by (5.92). Let* $a \in \Omega$ *be fixed, and denote by* $\mathcal{F}(a)$ *the family of all functions* $F \in L^{p'}(\Omega)$ *such that there exists* γ, $\gamma \in (\gamma_c, 1 + N/p')$, *for which* $F(x) \simeq |x - a|^{-\gamma}$ *in a neighbourhood of* a.

(a) For each $F \in \mathcal{F}(a)$ *the corresponding weak solution* u *of (5.81) has a loss of regularity at the point* a. *More precisely, the order of singularity of* u *at* a *is* $(\gamma - p)/(p - 1)$, *which is larger than* γ.

(b) The supremum of losses of regularity at $a \in \Omega$, *corresponding to all* $F \in \mathcal{F}(a)$, *is equal to*

$$\sup_{F \in \mathcal{F}(a)} \delta(F) = \frac{N}{\gamma_c} - 2. \tag{5.93}$$

Now we would like to study the loss of regularity of (5.81) on subsets of Ω.

Let A be a given nonempty subset of Ω. We say that (5.81) has a *loss of regularity on* A if there exists $F \in L^{p'}(\Omega)$ with a singularity at least of order $\gamma = \gamma(a) > 0$ at each $a \in A$, such that the corresponding weak solution u has a loss of regularity for all points $a \in A$. In other words, the order of singularity of u at any $a \in A$ is larger than $\gamma = \gamma(a)$.

We would like to see how large a subset $A \subset \Omega$ can be in the sense of Hausdorff dimension, on which (5.81) has a loss of regularity. The following theorem shows that there exist subsets A of Ω on which (5.81) has a loss of regularity, and such that their Hausdorff dimension is arbitrarily close to $N - p'\gamma_c$. Its proof exploits Theorem 5.2.

Theorem 5.8. ([83, Theorem 2.6]) *Assume that* $1 < p < 2$ *and* $N > p'\gamma_c$. *Let* $\mathcal{L}$ *be the family of all subsets of* Ω *on which the p-Laplace equation (5.81) has a loss of regularity. Then*

$$\sup\{\dim_H A : A \in \mathcal{L}\} \geq N - p'\gamma_c. \tag{5.94}$$

Remark 5.9. It would be interesting to know the precise value of the supremum on the left-hand side of (5.94). It could be named the *loss of regularity dimension* for the family of BVPs (5.81).

5.5.2. Loss of Hölder Regularity of Weak Solutions of (5.81) in the Case of $p > 2$

Our discussion of the loss of Hölder regularity for $p > 2$ is based on the following lemma.

Lemma 5.10. ([70, Lemma 1]) *Let* Ω *be a ball of radius* $R > 0$ *in* $\mathbb{R}^N$ *centered at the origin. Assume that* $1 < p < \infty$, $m > \max\{-p, -N\}$, *and let* $F(x) = C|x|^m$, $C > 0$.

Then the p-Laplace equation (5.81) possesses a unique weak solution in $W_0^{1,p}(\Omega)$, and it is given by

$$u(x) = \left(\frac{C}{m+N}\right)^{p'-1} \frac{R^\mu - |x|^\mu}{\mu}, \tag{5.95}$$

where $\mu = (m+p)/(p-1)$.

As we see, the weak solution u in Lemma 5.10 is of the form $u(x) = u(0) + D|x|^\mu$, where $D < 0$. If $m < 0$ then we have a gain of regularity at $x = 0$: in this case the function F is singular at $x = 0$, while the solution is uniformly bounded due to $\mu > 0$.

Assume therefore that $m \geq 0$. We say that the p-Laplace equation (5.81) has *a loss of Hölder regularity at $x = a$* if there exists a continuous function F of the form $F(x) \simeq |x - a|^m$ as $x \to a$, such that the corresponding weak solution u of (5.81) satisfies the condition $|u(x) - u(a)| \simeq |x - a|^\mu$ as $x \to a$, with $\mu < m$.

It is natural to measure the loss of Hölder regularity at a given point $a \in \Omega$ by the following quantity, which we call the loss of regularity of (5.81) associated with F having the form $F(x) \simeq |x - a|^m$:

$$\delta(F) := m - \mu. \tag{5.96}$$

According to Lemma 5.10 for $F(x) = C|x|^m$ we have $\mu = (m+p)/(p-1)$, therefore

$$\delta(F) = m - \frac{m+p}{p-1} = \frac{m(p-2) - p}{p-1}. \tag{5.97}$$

Note that for $1 < p \leq 2$ we have $\delta(F) < 0$, that is, we have a gain of Hölder regularity since $m < \mu$. The case of $p = 2$ is the only one in which $\delta(F)$ does not depend on m: here $\delta(F) = -2$, that is, we have a gain of regularity equal to 2. We are interested in the case of $\delta(F) > 0$, which is equivalent to (5.98) below.

Theorem 5.11. ([83, Theorem 3.3]) *Let $\Omega = B_R(0)$ and assume that $p > 2$. Denote by $\mathcal{F}(0)$ the set of all functions $F : \Omega \to \mathbb{R}$ of the form $F(x) = C|x|^m$ with m satisfying*

$$m > m_c := \frac{p}{p-2}, \tag{5.98}$$

(a) For any $F \in \mathcal{F}(0)$ the corresponding weak solution of (5.81) has a loss of Hölder regularity at $x = 0$.

(b) The supremum of losses of Hölder regularity of (5.81) in the class of right-hand sides of (5.81) from $\mathcal{F}(0)$ is equal to infinity:

$$\sup_{F \in \mathcal{F}(0)} \delta(F) = \infty.$$

5.6. Absence of Hypoellipticity of the p-Laplacian for $p \neq 2$

It is well known that the classical Laplace operator is hypoelliptic, that is, if $-\Delta u = F(x)$ in the weak sense, and $F \in C^\infty(\Omega)$, then for the weak solution we also have $u \in C^\infty(\Omega)$. However, this is not the case for general p-Laplace operators. This also represents a phenomenon of the loss of regularity of (5.81), but different from the ones discussed in previous sections, though related.

Fractal Properties of Solutions of Differential Equations   457

Example 5.12. ([83, Example 4.1]) Let $p \in (1, \infty) \setminus \{1 + 1/n : n \text{ is odd}\}$. Then the corresponding p-Laplace operator is not hypoelliptic.

This can be easily proved using Lemma 5.10, see [82] for details.

Remark 5.13. We do not know if the p-Laplace operator is hypoelliptic for p's of the form $p = 1 + \frac{1}{n}$, where $n \geq 3$ is an odd integer. The case of $n = 1$ corresponds to the classical Laplace operator, which *is* hypoelliptic.

6.   Fractal Properties of Generalized Fresnel Integrals

Clothoids appear in various branches of mathematics, physics and engineering. This planar curve is related to finding the shortest path for a car-like robot to go from one point to another with given initial and final tangent angles and curvatures. It is also used in civil engineering for modelling the road shape, highways, and railways. It finds applications in the problem of manoeuvering aircraft between straight and curved segments of the trajectory. Also, the clothoids arise in the description of diffraction phenomena in optics. For applications in optimal control theory see Degtiarova-Kostova and Kostov [14], [38] and Boissonnat, Cérézo, Degtiarova-Kostova, Kostov, Leblond [3], and the references therein. Clothoid splines are used in computer aided geometric design applications, see Walton and Meek [45].

The p-clothoid is a planar curve defined parametrically by Fresnel integrals:

$$x(t) = \int_0^t \cos(s^p)\, ds, \quad y(t) = \int_0^t \sin(s^p)\, ds, \tag{6.99}$$

where $t \in \mathbb{R}$ and $p > 1$. Its graph is symmetric with respect to the origin, and consists of two spirals converging to their focusses in the first and third quadrants, see Figure 25. The standard clothoid obtained for $p = 2$ is a curve with the curvature at any point proportional to the arc length, measured from the origin. More precisely, the curvature at the point $(x(t), y(t))$ is equal to $2t$, while the arc length is equal to t. This implies that the clothoid consists of two nonrectifiable spirals as $t \to \pm\infty$.

Our aim is to compute the box dimension of the clothoid. It is interesting to say that Hausdorff dimension of the clothoid and all spirals studied in previous sections is equal to 1, because of countable stability of Hausdorff dimension. This means that Hausdorff dimension, contrary to box dimension, does not distinguish nonrectifiable C^1-spirals.

The focus of the generalized Euler spiral when $t \to \infty$ has the following coordinates:

$$\begin{cases} a &= \int_0^\infty \cos(s^p)\, ds = \frac{1}{p}\Gamma(1/p)\cos(\pi/2p) \\ b &= \int_0^\infty \sin(s^p)\, ds = \frac{1}{p}\Gamma(1/p)\sin(\pi/2p). \end{cases} \tag{6.100}$$

Here $\Gamma(z)$ is the gamma function, see [19, p. 13, Vol. I]. It is easy to see that these improper integrals converge due to $p > 1$, see [35, Lemma 2 in Section 3]. For $p = 2$ the coordinates (6.100) of the focus of the clothoid have been computed by Euler already in 1781, see [20].

Theorem 6.1. (Box dimension of the p-clothoid, [35, Theorem 2]) *Let Γ_p be the p-clothoid defined by (6.99), $p > 1$. Then $d = \dim_B \Gamma_p = 2p/(2p-1)$. Furthermore, Γ_p is Minkowski measurable and*

$$\mathcal{M}^d(\Gamma_p) = (2p - 1)\left(p(p-1)^{p-1}\right)^{-2/(2p-1)} \pi^{1/(2p-1)}. \tag{6.101}$$

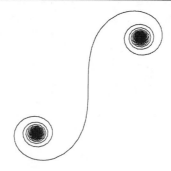

Figure 25. Standard clothoid Γ_2.

In the proof exploited the following asymptotics for the p-clothoid (6.99):

$$\Gamma_p \cdots \begin{cases} x(t) = a + \frac{1}{p\,t^{p-1}} \sin(t^p) + O(t^{-2p+1}) \\ y(t) = b - \frac{1}{p\,t^{p-1}} \cos(t^p) + O(t^{-2p+1}), \end{cases} \quad (6.102)$$

as $t \to \infty$, where a and b are defined in (6.100). It follows from (6.103) and (6.104) below by setting $N = 0$.

The basis for our analysis provides the following asymptotic expansion of Fresnel integrals associated to generalized Euler spirals. The result can be obtained using known expansions of Fresnel integrals based on complex variables and the gamma function, see Erdélyi [19, pp. 149-150, Vol. II]. In [35] we proposed a new, very short and elementary proof. Box dimension of the p-clothoid in Theorem 6.1 is a consequence of Theorem 2.1 together with the asymptotic expansion in Theorem 6.2.

Theorem 6.2. (Asymptotic expansion of the generalized Fresnel integral, [35], [19]) *Let $x(t)$ and $y(t)$ be generalized Fresnel integrals defined by (6.99), $p > 1$, and $a = \lim_{t \to \infty} x(t)$, $b = \lim_{t \to \infty} y(t)$. Then for any nonnegative integer N we have*

$$\begin{cases} x(t) = a + A_N(t) \sin(t^p) - B_N(t) \cos(t^p) + O(t^{-(2N+3)p+1}) \\ y(t) = b - B_N(t) \sin(t^p) - A_N(t) \cos(t^p) + O(t^{-(2N+3)p+1}), \end{cases} \quad (6.103)$$

when $t \to \infty$, where

$$\begin{cases} A_N(t) = \sum_{k=0}^{N} (-1)^k a_{2k} t^{-(2k+1)p+1} \\ B_N(t) = \sum_{k=0}^{N} (-1)^k a_{2k+1} t^{-(2k+2)p+1} \\ a_n = p^{-n-1}(p-1)(2p-1)\ldots(np-1), \ n \geq 1 \text{ and } a_0 = p^{-1}. \end{cases} \quad (6.104)$$

The box dimension of the classical clothoid is a special case of Theorem 6.1 for $p = 2$.

Theorem 6.3. (Box dimension of the clothoid, [37, Theorem 1]) *Let Γ be the clothoid defined by (6.99) for $p = 2$. Then $\dim_B \Gamma = 4/3$. Furthermore, Γ is Minkowski measurable and*

$$\mathcal{M}^{4/3}(\Gamma) = 3 \cdot 2^{-2/3} \pi^{1/3}.$$

It is possible to extend Theorem 6.1 to even more general clothoids. Let $q : (0, \infty) \to \mathbb{R}$ be a given function such that $q(t) \sim t^p$, $p > 1$, when $t \to \infty$. By the clothoid generated by the control function q, or the q-clothoid Γ_q, we mean a planar curve defined parametrically by

$$\Gamma_q \cdots \begin{cases} x(t) &= \int_0^t \cos(q(s)) \, ds \\ y(t) &= \int_0^t \sin(q(s)) \, ds. \end{cases} \tag{6.105}$$

See [35] for an extension of Theorem 6.2 to this case, and the corresponding dimension result.

The following result concerns oscillatory dimensions of component functions of the p-clothoid. From the analysis of the reflected function $X(\tau)$ below we see that it is a chirp-like function. For definitions of the reflected function and the oscillatory dimension see Section 2.2..

Theorem 6.4. (Oscillatory dimension of the component function, [35, 34]) *Assume that $p > 1$ and let $x(t)$ and $y(t)$ be the component functions of the p-clothoid defined by (6.99). Then the oscillatory dimension of both of them is equal to $(2 + p)/(1 + p)$. Furthermore, the graphs of the corresponding reflected functions $X(\tau)$ and $Y(\tau)$ are Minkowski nondegenerate.*

The proof relies on the following result.

Lemma 6.5. ([35, Lemma 3]) *Let $g : (0, \tau_0) \to \mathbb{R}$ be a smooth function, and $G(g)$ its graph in $\mathbb{R}^2$. If $h : (0, \tau_0) \to \mathbb{R}$ is a Lipschitz function then*

$$\underline{\dim}_B G(g + h) = \underline{\dim}_B G(g), \quad \overline{\dim}_B G(g + h) = \overline{\dim}_B G(g).$$

In particular, if there exists $\dim_B G(g)$, then $\dim_B G(g + h) = \dim_B G(g)$. Furthermore, if the graph $G(g)$ is Minkowski nondegenerate, then the same holds for $G(g + h)$.

Remark 6.6. Note that box dimensions $\dim_B \Gamma_p$ and $\dim_B G(X)$ coincide only for the classical clothoid where $p = 2$, and in this case the common value is equal to $4/3$, see [37]. Indeed,

$$\dim_B \Gamma_p - \dim_B G(X) = \frac{2 - p}{p^2 - 1},$$

and note that $\dim_B \Gamma_p < \dim_B G(X)$ for $p > 2$.

Acknowledgements. We express our gratitude to the members of the Seminar for differential equations and nonlinear analysis at the University of Zagreb for their help, in particular to Lana Horvat Dmitrović, Luka Korkut, Josipa-Pina Milišić, Goran Radunović, Maja Resman, and Domagoj Vlah.

References

[1] D.R. Adams, L.I. Hedberg, *Function Spaces and Potential Theory*, Springer, 1999.

[2] R.P. Agarwal, S.R. Grace, D. O'Regan, *Oscillation Theory for Second Order Linear, Half-Linear, Superlinear and Sublinear Dynamic Equations* . Kluwer Academic Publishers, London, 2002.

[3] J.-D. Boissonnat, A. Cérézo, E.V. Degtiarova-Kostova, V.P. Kostov, Leblond J., Shortest plane paths with bounded derivative of the curvature, C. R. Acad. Sci. Paris, 329 (1999), 613–618.

[4] H. Brezis, *Analyse fonctionnelle*, Masson, 1983.

[5] M. Caubergh, F. Dumortier, Hopf-Takens bifurcations and centers, J. Differential Equations, 202 (2004), no. 1, 1–31.

[6] M. Caubergh, J.P. Françoise, Generalized Liénard equations, cyclicity and Hopf-Takens bifurcations, Qualitative Theory of Dynamical Systems 6 (2005), 195–222.

[7] T. Cazenave, *Semilinear Schrödinger Equations* , Courant Lecture Notes in Mathematics 10, 2004.

[8] T. Cazenave, A. Haraux, F.B. Weissler, Une équation des ondes complètement intégrable avec non-linéarité homogène de degré trois, C. Rend. Acad. Sci. Paris, 313 (1991), 237–241.

[9] T. Cazenave, A. Haraux, F.B. Weissler, Detailed asymptotics for a convex Hamiltonian system with two degrees of freedom, J. Dynam. Diff. Eq., 5 (1993), 155-187.

[10] T. Cazenave, A. Haraux, F.B. Weissler, A class of nonlinear completely integrable abstract wave equations, J. Dynam. Diff. Eq., 5 (1993), 129–154.

[11] C.J. Christopher, N. G. Lloyd, Small-amplitude limit cycles in Liénard systems, Nonlinear Differential Equations Appl. 3, No 2, (1996), 183-190.

[12] W.A. Coppel, *Stability and asymptotic behavior of differential equations* , D. C. Heath and Co., Boston, 1965.

[13] E.D. Davies, *Spectral Theory and Differential Operators* , Cambridge University Press, 1995.

[14] E.V. Degtiarova-Kostova, V.P. Kostov, Suboptimal paths in a planar motion with bounded derivative of the curvature, C. R. Acad. Sci. Paris 321, (1995), 1441–1447.

[15] R.L. Devaney, *An Introduction to Chaotic Dynamical Systems* , The Benjamin/Cummings, New York, 1986.

[16] F. Dumortier, J. Llibre, J.C. Artés, *Qualitative Theory of Planar Differential Systems* , Springer-Verlag Berlin Heidelberg, (2006).

[17] Y. Dupain, M. Mendès France, C. Tricot, Dimension de spirales, Bull. Soc. Math. France 111 (1983), 193–201.

Fractal Properties of Solutions of Differential Equations 461

[18] N. Elezović, V. Županović, D. Žubrinić. Box dimension of trajectories of some discrete dynamical systems, Chaos, Solitons & Fractals 34 (2007) 244–252.

[19] A. Erdélyi, W. Magnus, F. Oberhettinger, F.G. Tricomi, *Higher Transcendental Functions*, Volumes I and II, McGraw-Hill, New York, Toronto and London, 1953.

[20] Euler Integrals and Euler's Spiral – Sometimes called Fresnel Integrals and the Clothoide or Cornu's Spiral (unsigned), Amer. Math. Monthly 25 (1918), 276–282.

[21] L.C. Evans, R.F. Gariepy, *Measure Theory and Fine Properties of Functions*, CRC Press, New York, 1999.

[22] K. Falconer, *Fractal Geometry*, Chichester: Wiley, 1990.

[23] H. Federer, *Geometric Measure Theory*, Springer-Verlag, 1969.

[24] J.-P. Françoise, Successive derivatives of a first return map, application to the study of quadratic vector fields, Ergodic Theory Dyn. Syst. 16 (1996), 87–96.

[25] A. Gasull, J. Torregrosa, A new approach to the computation of the Lyapunov Constants, Computational and Applied Mathematics, Vol. 20, N. 1-2, (2001), 1-29.

[26] D. Gilbarg, N.S. Trudinger, *Elliptic Partial Differential Equations of Second Order*, Springer Verlag, 1983.

[27] J. Guckenheimer, P. Holmes, *Nonlinear Oscillations, Dynamical Systems, and Bifurcations of Vector Fields*, Springer Verlag, 1983.

[28] P. Hartman, *Ordinary Differential Equations*, Second edition, Birkhauser, Boston, Basel, Stuttgart, 1982.

[29] C.Q. He, M.L. Lapidus, *Generalized Minkowski content, spectrum of fractal drums, fractal strings and the Riemann zeta-function*, Mem. Amer. Math. Soc. 127 (1997), no. 608.

[30] D. Henry, *Geometric theory of semilinear parabolic equations*, Lecture Notes in Mathematics, no 840, Springer-Verlag, Berlin, 1981.

[31] L. Horvat Dmitrović, Box dimension and bifurcation of one-dimensional discrete dynamical systems, Discrete and continuous dynamical systems - Series A, to appear

[32] L. Horvat Dmitrović, Fractal analysis of bifurcations of discrete dynamical systems and applications to continuous systems (in Croatian), PhD Thesis, University of Zagreb, 2011

[33] L. Horvat, D. Žubrinić, Maximally singular Sobolev functions, J. Math. Anal. Appl. 304 (2005), no. 2, 531–541.

[34] L. Korkut, M. Resman, Fractal oscillations of chirp-like functions, submitted

[35] L. Korkut, D. Vlah, D. Žubrinić, V. Županović, Generalized Fresnel integrals and fractal properties of related spirals, Appl. Mathematics and Computation, 206 (2008), 236–244.

[36] L. Korkut, D. Vlah, D. Žubrinić, V. Županović, work in progress.

[37] L. Korkut, D. Žubrinić, V. Županović, Box dimension and Minkowski content of the clothoid, Fractals 17 (2009), 485–492.

[38] V.P. Kostov, E.V. Degtiarova-Kostova, The planar motion with bounded derivative of the curvature and its suboptimal paths, Acta Math. Univ. Comenianae, Vol. LXIV, 2 (1995), 185-226.

[39] M.K. Kwong, M. Pašić, and J.S.W. Wong, Rectifiable oscillations in second order linear differential equations, J. Differential Equations, 245 (2008), 2333–2351.

[40] M.L. Lapidus, M. van Frankenhuysen, *Fractal Geometry, Complex Dimensions and Zeta Functions: Geometry and Spectra of Fractal Strings* , Springer–Verlag, 2006.

[41] W.T. Reid, *Sturmian Theory for Ordinary Differential Equations* , Springer-Verlag, New York, 1980.

[42] N.N. Lebedev, Special Functions and Their Applications, (Dover, 1972)

[43] A.M. Lyapunov, *The General Problem of the Stability of Motion* , Taylor and Francis, 1992.

[44] P. Mattila, *Geometry of Sets and Measures in Euclidean Spaces. Fractals and Rectifiability*, Cambridge, 1995.

[45] D.S. Meek, D.J. Walton, A controlled clothoid spline, Computers & Graphics 29, (2005), 353–363.

[46] J.-P. Milišić, D. Žubrinić, V. Županović, Fractal analysis of Hopf bifurcation for a class of completely integrable nonlinear Schrödinger Cauchy problems, Electronic J. of Qualitative Theory of Diff. Equations, (2010), No. 60, 1-32.

[47] M. Pašić, Minkowski - Bouligand dimension of solutions of the one-dimensional p-Laplacian, J. Differential Equations 190 (2003) 268-305.

[48] M. Pašić, Rectifiable and unrectifiable oscillations for a class of second-order linear differential equations of Euler type. J. Math. Anal. Appl., 335 (2007), 724–738.

[49] M. Pašić, Fractal oscillations for a class of second-order linear differential equations of Euler type, J. Math. Anal. Appl., 341 (2008), 211–223.

[50] M. Pašić, Rectifiable and unrectifiable oscillations for a generalization of the Riemann - Weber version of Euler differential equations, Georgian Math. J., 15 (2008), 759–774.

[51] M. Pašić, S. Tanaka, Fractal oscillations of Bessel type linear differential equations of second-order, submitted.

[52] M. Pašić, S. Tanaka, On fractal oscillations of chirp-like behaviour functions and differential equations, submitted.

[53] M. Pašić, J. S. W. Wong, Two-point oscillations oscillations in second-order linear differential equations, Diff. Eq. Appl., 1 (2009), 85–122.

[54] M. Pašić, J. S. W. Wong, Rectifiable oscillations in second-order half-linear differential equations, Annali di matematica pura ed applicata, Ann. Mat. Pura Appl., (4) 188, 3 (2009), 517–541.

[55] M. Pašić, D. Žubrinić, V. Županović, Oscillatory and phase dimensions of solutions of some second-order differential equations, Bulletin des sciences mathématiques 133 (2009), 859–874.

[56] M. Pašić, V. Županović, Some metric-singular properties of the graph of solutions of the one-dimensional p-Laplacian, Electronic J. of Differential Equations, 60, 2004(2004), 1–25.

[57] M. Pašić, D. Žubrinić, V. Županović, Oscillatory and phase dimensions of solutions of some second-order differential equations, Bulletin des Sciences Mathématiques, Vol 133(8) 2009, 859–874.

[58] L. Perko, *Differential Equations and Dynamical Systems*, Texts in Applied Mathematics, Springer–Verlag, 1991.

[59] P. Pucci, R. Servadei, Regularity of weak solutions of homogeneous or inhomogeneous quasilinear elliptic equations, Indiana Univ. Math. J., 57 (2008), 3329-3364.

[60] R. Roussarie, *Bifurcations of Planar Vector Fields and Hilbert's Sixteenth Problem*, Birkhäuser, 1998.

[61] J. Simon, *Sur des équations aux dérivées partielles non linéaires*. Thése. Paris (1977).

[62] E.M. Stein, *Singular Integrals and Differentiability Properties of Functions*, Princeton University Press, 1970.

[63] C.A. Swanson, *Comparison and Oscillation Theory of Linear Differential Equations*, Acad. Press, New York, 1968.

[64] F. Takens, Unfoldings of certain singularities of vector fields: Generalized Hopf bifurcations, J. Differential Equations, 14 (1973) 476–493.

[65] F. de Thélin, Local regularity properties for the solutions of a nonlinear partial differential equation, Nonlinear Analysis, Vol. 6, 8 (1982), 839–844.

[66] C. Tricot, *Curves and Fractal Dimension*, Springer–Verlag, 1995.

[67] H. Triebel, *Theory of Function Spaces*, Birkhäuser Verlag, 1983.

[68] J.S.W. Wong, On rectifiable oscillation of Euler type second-order linear differential equations, E. J. Qualitative Theory of Diff. Equ. 20 (2007), 1-12.

[69] J.S.W. Wong, On Rectifiable Oscillation of Emden-Fowler Equations Mem. Differential Equations Math. Phys. 42 (2007), 127-144.

[70] D. Žubrinić, Generating singularities of solutions of quasilinear elliptic equations, J. Math. Anal. Appl. 244 (2000), 10–16.

[71] D. Žubrinić, Singular sets of Sobolev functions, C. R. Acad. Sci., Analyse mathématique, Paris, Série I, 334 (2002), 539–544.

[72] D. Žubrinić, Singular sets of Lebesgue integrable functions, Chaos, Solitons and Fractals, 21 (2004), 1281–1287.

[73] D. Žubrinić, Extended singular set of potentials, Math. Inequal. Appl., 8 (2005), no. 2, 173–177.

[74] D. Žubrinić, Analysis of Minkowski contents of fractal sets and applications, Real Anal. Exchange, Vol 31(2), 2005/2006, 315–354.

[75] D. Žubrinić, V. Županović, Fractal analysis of spiral trajectories of some planar vector fields, Bulletin des Sciences Mathématiques, 129/6 (2005), 457–485.

[76] D. Žubrinić, V. Županović, Fractal analysis of spiral trajectories of some vector fields in $\mathbb{R}^3$, C. R. Acad. Sci. Paris, Série I 342, (2006), 959–963.

[77] D. Žubrinić, Maximally singular functions in Besov spaces, Arch. Math. 87 (2006), 154-162.

[78] D. Žubrinić, V. Županović, Box dimension of spiral trajectories of some vector fields in $\mathbb{R}^3$, Qualitative Theory of Dynamical Systems, Vol 6 (2005), 251–272.

[79] D. Žubrinić, V. Županović, Poincaré map in fractal analysis of spiral trajectories of planar vector fields, Bull. Belg. Math. Soc. Simon Stevin, Vol. 15 (2008), 947-960.

[80] D. Žubrinić, Hausdorff dimension of singular sets of Sobolev functions and applications, *More Progresses in Analysis*, Proceedings of the 5th International ISAAC Congress / Begher, H.G.W.; Nicolosi, F. (eds.). Singapore: World Scientific 2009, 793–802.

[81] D. Žubrinić, Generating singularities of solutions of p-Laplace equations on fractal sets, Rocky Mountain J. Math., 39 (2009), 359-366.

[82] D. Žubrinić, Singular dimension of solution set of a class of p-Laplace equations, Complex Var. Elliptic Equ. 55 (2010), no. 7, 669-676.

[83] D. Žubrinić, Loss of regularity of weak solutions of p-Laplace equations for $p \neq 2$, Differ. Equ. Appl. 2 (2010), no. 2, 217-226.

[84] V. Županović, D. Žubrinić, Fractal dimensions in dynamics, in *Encyclopedia of Mathematical Physics*, eds. J.-P. Françoise, G.L. Naber and Tsou S.T. Oxford: Elsevier, 2006, vol 2, 394–402.

[85] V. Županović, D. Žubrinić, Recent results on fractal analysis of trajectories of some dynamical systems, in *FUNCTIONAL ANALYSIS IX - Proceedings of the Postgraduate School and Conference held at the Inter-University Centre, Dubrovnik, Croatia*, 15-23 June, 2005. (eds. G. Muic, J. Hoffmann-Jørgensen), Aarhus, Denmark: University of Aarhus, Department of Mathematical Sciences, (2007), 126–140.

INDEX

#

20th century, xi

A

accounting, 59
adsorption, vii, 1, 3, 4, 18, 22, 23, 24, 26, 28, 44
affine group, 51
aggregation, vii, 2, 37, 38, 39, 40, 41, 45
algorithm, 37, 39, 66, 67, 74, 111, 112, 147
aqueous electrolyte solutions, vii, 1, 18
aqueous solutions, 3, 24, 25, 29, 35, 44
arithmetic, 87
Artificial Neural Networks, 145, 146, 148
assessment, 142, 144
atmosphere, 2, 22, 96, 97
atoms, 4, 5, 6, 7, 22
attachment, 37, 68
automata, 107
Avogadro number, 18

B

bandwidth, 104, 117, 119, 122, 123
Besicovitch-Taylor index, xi
biological behavior, 134
biological systems, 127, 129
biomarkers, 133, 134
blood vessels, 133
boils, 30
bonding, 47
bonds, 5, 10, 11, 12, 50
Borel logarithmic rarefaction, xi, 405
Bose–Einstein, ix
Bouligand dimension, xi, 405, 462
Brownian motion, 30, 34, 35, 37, 38, 44, 67

bubbles, stabilized by ions, vii, 1
bubston surface, vii, 1
Bubstons, 4
Butcher, 129, 135

C

calculus, 105
Cameroon, 149
cancer, vii, ix, 125, 126, 128, 131, 133, 134
Cantor set, 106
Cantor-Minkowski, xi
capacity dimension, xi
capillary, 134
capital expenditure, 65
carcinogenesis, 133
causality, 127, 129, 133, 134, 135
challenges, 65, 138, 141
chaos, 66, 80, 112
chaotic behavior, viii, 49, 50, 66, 132
charge density, 24, 25, 26, 27, 29, 30, 31
chemical, 130, 138, 142, 143
chemicals, 138, 139
classification, vii, 106, 130
cluster model, 37
clustering, 68, 69
clusters, vii, 1, 2, 4, 23, 37, 38, 40, 41, 42, 43, 44, 53
coding, 108
coefficient of variation, 143
compaction, 141
complement, 89
complex interactions, 59
complexity, viii, ix, 49, 50, 53, 80, 89, 112, 125, 126,
    127, 128, 129, 133, 134, 135, 136
compression, 12, 13, 16, 112
computation, 45, 95
computed tomography, ix, 137, 146, 147, 148
computer, 39, 40, 107, 136, 144

computing, 39
conception, 126
condensation, 16, 25, 26, 27
conductivity, ix, 23, 137, 138, 140, 144, 145, 146
conformity, 21, 38, 44
connective tissue, 136
consciousness, 53
conservation, 70, 139
constituents, 50
construction, 104, 127
contaminant, 138
contamination, 138, 143
controversial, 107
coordination, 5, 6, 7, 8, 10, 66
correlation, 33, 34, 51, 57, 63, 64, 73, 89, 94, 95, 140
Coulomb interaction, 25
Coulomb repulsion forces, vii, 1, 3, 21
counter-ion cloud, vii, 1
counter-ions, vii, 1, 18, 19, 23, 24, 25, 27, 28, 30, 44
cracking processes, ix, 137
cracks, 79, 143, 144, 145, 146
crises, 54, 59
crop, ix, 137, 139, 140, 141, 145
CRP, 139
crude oil, 52, 53, 59, 60, 62, 63, 64
crust, viii, 83, 84, 95, 96
crystal structure, 5
crystalline, 28
crystals, 130
CT scan, 142
cure, 133
cycles, x

D

data analysis, 85
data set, 142, 143
Debye – Huckel approach, vii, 1
deformation, 12, 16, 17, 100
degradation, 138
dependent variable, 131
depth, 86, 92, 96, 130, 140, 141, 145
derivatives, ix, 88, 89, 125
desiccation, 143, 145, 146
desorption, 23
destruction, 2, 17
detachment, 32
detection, vii, 95, 100
deviation, 55, 62
dielectric constant, 119
dielectric permittivity, 4
differential equations, 132
diffusion, ix, 3, 22, 23, 34, 38, 81, 134

diffusivity, 22, 23
dimensionality, 41
dipole moments, 4
discharges, 58, 95
discrimination, 103
disorder, vii, 1, 4, 7, 8, 12, 13, 15, 16, 17, 18, 21, 22, 44
distribution, viii, 2, 4, 5, 7, 8, 11, 12, 24, 25, 27, 29, 30, 31, 33, 38, 40, 41, 42, 43, 44, 45, 51, 63, 64, 66, 67, 68, 78, 81, 96, 131, 136, 139, 141, 142, 143, 147
distribution function, 4, 5, 8
divergence, 5
DNA, 53, 131
DOI, 45
DOL, 108, 109
drawing, 109, 112, 116
drug discovery, 127, 135
dynamic scaling, 59, 60, 63, 79
dynamical systems, x, 66, 106, 107

E

earthquake(s) (EQ), vii, viii, 83, 84, 85, 86, 90, 91, 92, 93, 94, 95, 96, 97, 98, 99, 100, 101, 393
ecological systems, 129, 131, 136
economic systems, 79
economics, viii, 49, 50
elasticity modulus, 8
electric charge, 24
electric field, 27, 31, 34, 44, 94
electrodes, 54
electrolyte, vii, 1, 18, 24, 31
electromagnetic, vii, viii, 41, 83, 84, 85, 92, 95, 96, 97, 98, 99, 100, 101, 103
electromagnetic fields, 100, 103
electron, 3
electrophoresis, 25, 31
emission, 92, 94, 95, 142
energy, x, 5, 7, 8, 12, 16, 19, 34, 44, 128
engineering, 50, 53, 80, 107
entropy, xi, 81, 128, 405
environment, 59, 107, 127, 128
environmental effects, 145
epilepsy, viii, 49, 53, 54, 58, 78
EQ prediction, viii, 83, 84, 97
equality, 32, 51
equilibrium, 2, 3, 4, 8, 13, 16, 21, 23, 24, 30, 31, 44, 79, 128
Euclidean space, 105
evolution, viii, ix, 12, 13, 14, 16, 18, 45, 83, 84, 87, 89, 90, 91, 97, 128, 135
extracellular matrix, 136

Index

F

fluctuations, 50, 54, 55, 57, 60, 66, 79, 91, 100, 107
fluid, 138, 140
force, 32, 53
forecasting, 59, 66, 80
formal language, 107
formation, vii, 1, 2, 3, 4, 5, 16, 18, 19, 22, 23, 25, 27, 37, 44, 50, 59, 134
formula, 2, 12, 15, 27, 33, 34, 35, 36
fractal analysis, vii, viii, ix, x, 49, 50, 53, 54, 78, 79, 83, 84, 85, 88, 91, 92, 93, 94, 96, 97, 107, 136, 137, 138, 139
fractal concepts, 107
fractal dimension, vii, ix, xi, 2, 40, 41, 45, 51, 86, 87, 88, 91, 98, 105, 115, 125, 130, 136, 137, 138, 139, 140, 141, 142, 143, 144, 145, 147, 148, 247, 248, 249, 250, 251, 256, 264, 281, 405, 406, 445
fractal entities, ix, 125
fractal geometries, vii, viii, 103, 104, 116, 122
Fractal Geometry of Nature, ix, 123, 125, 136, 241, 285
fractal objects, 41, 107, 130, 131
fractal properties, vii, viii, x, 1, 4, 37, 84, 139, 140, 141, 142, 146
fractal structure, x, 142, 144
fractal theory, viii, 49, 53, 99
fractality, x, 94, 95, 97
free energy, 12, 13, 18, 19
frequency responses, viii, 103
Frequency Selective Surfaces (FSSs), vii, viii, 103, 113, 115, 117, 118, 119, 122, 124
friction, 23

G

genes, ix, 133, 135
genetics, 126
genome, 68, 126
genomics, 129
geometry, 5, 50, 53, 104, 105, 107, 108, 109, 113, 115, 116, 118, 119, 123, 126, 135, 146
grain size, ix, 137, 145
graph, x, 10, 44, 55, 64, 66, 67, 68, 69, 74, 77, 78, 80
grass, ix, 137, 138, 139, 140, 141, 145, 146, 147, 148
Gross–Pitaevskii type nonlinear equations, ix
groundwater, 138, 143
growth, 4, 16, 17, 18, 22, 23, 35, 44, 45, 60, 62, 63, 79, 107, 108, 131, 132, 134, 138

H

heavy metals, 138
height, 60, 115, 118, 119
helium, 22
herbicide, 146
heterogeneity, 134
heterogeneous crust, viii, 83
hexagonal lattice, 44
human, 53, 125, 126, 129, 130, 133, 135, 136
hydraulic properties, ix, 137, 139, 140, 141, 147
hydrogen, 4
hydrogen bonds, 4
hypothesis, 3

I

image, 110, 111, 112, 136, 139, 143, 146, 147
imbalances, 59
impurities, 3
independent variable, 38
industry, vii, viii, 49, 53, 107, 127
inequality, 20, 21, 29, 30, 37, 89
initial state, 18, 109
integration, 130
interface, vii, 1, 3, 4, 16, 18, 21, 22, 24, 28, 30, 44, 60, 131, 136, 139
interference, 37, 41
interrelations, viii, 49, 50
intervention, 133
inversion, 25, 29, 30, 34, 35, 37, 44
ions, vii, 1, 3, 4, 18, 19, 21, 22, 23, 24, 25, 27, 28, 30, 44
Iterated Function Systems, viii, 103, 107, 110
iteration, 105, 106, 107, 111, 115, 116, 119, 120

K

kinetics, vii, 2, 3, 12, 13, 22, 134
Koch curves, 114, 124
Kolmogorov dimension, xi, 405
Kolmogorov multiplicative continuous cascades, x

L

laminar, 32
landscape, ix, 137, 141, 143, 144, 145
languages, 124
laws, 50, 78, 127, 134, 136
lead, 23, 33, 66, 70, 72, 73, 77, 78, 79, 80, 143
lens, 2

lesions, ix, 125, 136
lifetime, 3
light, vii, 2, 37, 38, 41, 45, 63, 78
limit capacity, xi, 405
linear fractals, 108
linear systems, 133
liquids, 3, 14, 18

M

macropores, ix, 137, 139, 140, 141, 142, 145, 146, 147
magnetic field, 98, 99, 100
magnetosphere, 91
magnitude, 3, 8, 18, 22, 34, 37, 38, 41, 85, 86, 92, 96, 133
management, ix, 137, 138, 139, 140, 141, 144, 145
mapping, 67, 68, 109
mass, ix, 32, 37, 38, 39, 137, 139, 143, 145, 146
materials, 53, 100
mathematical fractals, 104
mathematical methods, ix, 50, 125, 134
mathematics, 50, 104, 106, 107, 110, 147
matrix, vii, 2, 5, 37, 38, 41, 42, 43, 44, 45, 69, 136, 139
matter, 11, 128
measured porosity, 142, 145, 148
measurement(s), viii, x, 37, 49, 50, 98, 99, 100, 117, 129, 141, 146, 147
media, 3, 16, 37, 38, 99, 138, 141
medical, 79, 133, 134, 142
medicine, 50, 142
memory, 52, 89
messages, 53
metabolism, 53
microstructure, x
miniaturization, 104
Minkowski dimension, xi, 405, 442
mitochondria, 131
mitosis, 134
model system, ix
modelling, 129
models, vii, viii, 38, 41, 49, 50, 59, 67, 79, 80, 107, 108, 123, 127, 129, 132, 133, 141
modules, 107
modulus, 12, 16, 17, 93, 100
molecular biology, 126, 135
molecular dynamics, 16
molecular structure, 18
molecules, vii, 1, 2, 4, 5, 18, 22, 23, 28, 44
monomers, 41, 45
morphology, 50
multicellular organisms, 107

N

NaCl, vii, 2
nanometer, vii, 1, 26, 44
nanometer scale, vii, 1, 44
nanometer-sized voids, vii, 1
nanoscale structures, 4
network elements, 65
network theory, 74
neurological disease, 135
neurons, 53, 58
nitrogen, 22, 146
nodes, 67, 68, 69
nonlinear dynamics, viii, 66, 83, 84
normalization constant, 132
nucleation, viii, 2, 4, 18, 21, 22, 23, 83, 84, 97
nuclei, vii, 1
nucleus, 18, 21, 22, 23, 44
numerical analysis, 79
numerical computations, 38
nutrients, 137, 138

O

observed behavior, 64
obstacles, 132
oil, vii, viii, 49, 59, 78
ordinary differential equations, 132
organ, 129
organism, 107, 127, 129, 130
organize, 128
organs, 126, 128, 130
oscillation, ix, 66

P

parallel, x, 107
parenchyma, 135
particle-cluster, vii, 2, 37, 45
pasture(s), ix, 137, 139, 140, 145, 146
pathogenesis, 126
pathways, 129, 133
percolation, 95
permittivity, 19, 118, 119
phenotypes, 133
phosphorous, 146
photoelectron spectroscopy, 4
physical mechanisms, 126
physical phenomena, x, 132
physical properties, ix, 130, 137, 139, 145
physical sciences, 127
physics, viii, x, 49, 50, 53, 66, 78, 107, 135

Index 471

physiology, 53, 107, 127, 129
plant growth, 137, 138
point of origin, 8
Poisson equation, 27
polarity, vii, 2, 25
polarization, 4, 37, 119
population, 126, 133
porosity, ix, 137, 142, 145, 148
porous media, 138, 139, 141, 142, 144, 145, 146
positive correlation, 57, 127
potassium, 142, 144
principles, 127, 129, 133
probability, x, 5, 6, 7, 8, 9, 10, 11, 18, 19, 23, 51, 63, 64, 67, 68, 74, 111, 143
probability density function, 7
probability distribution, x, 64
programming, 109
propagation, 97
proposition, 127
prostate cancer, 135
protection, 138
proteins, 53, 133
proteomics, 129
prototypes, 117
public health, 125
pure water, 4

Q

quasi-equilibrium, 12

R

radial distribution, 11
radius, viii, 2, 5, 8, 9, 11, 18, 19, 21, 22, 23, 24, 25, 26, 29, 30, 31, 32, 33, 38, 39, 40
random ballistic-type clusters, viii, 2
random numbers, 39
random walk, 139
real fractals, 107
recommendations, iv
reconstruction, 89
recovery, 65, 66, 67, 68, 72, 77, 79
recovery process, 65, 66, 67, 68, 72, 77, 79
regulatory systems, 127
rejection, 104, 119, 122
relaxation, 12, 33
reliability, 93, 133
remote sensing, 99
René Descartes, 127
repair, 65, 70, 71
reproduction, 105

repulsion, vii, 1, 3, 21
researchers, 127, 142
resolution, 50, 139, 140, 146, 148
resources, 140
response, 104, 109, 113, 122, 126, 133
restoration, 147
rhythmicity, 53
room temperature, vii, 1, 2, 22, 24, 44
root(s), 20, 35, 36, 107, 138
rotations, 41, 42, 43, 111
roughness, 61
rules, viii, 49, 107, 108
runoff, 141, 146, 147

S

salts, 3, 24
samplings, 41
saturation, 111
scaling, ix, x, 51, 59, 60, 62, 63, 64, 66, 78, 79, 125, 130, 132, 139, 147
scaling approach, 59
scaling law, x
scaling relations, ix, 125
scanning electron microscopy, 4
scatter, 6
scattering, viii, 2, 37, 38, 41, 42, 43, 44, 45
schemata, 127
science, 50, 53, 107, 123, 135, 139, 146
scope, 65, 107
sediment, 141, 146, 147
seizure, 53, 54
self-affinity, 51, 105
self-organization, viii, 53, 83, 84, 91, 97, 101, 128
self-organized criticality (SOC), viii, 83, 84, 85, 91, 95, 97, 98
self-similarity, 50, 51, 104, 105, 108, 130
sensitivity, 86, 126
SES, 84
shape, ix, 35, 37, 105, 106, 111, 125, 130
shear, 12, 14, 17, 18
shear deformation, 14, 17
Sierpinski carpet, 106
Sierpinski gasket, 104
signals, 53, 54, 55, 84, 88, 90, 100
signs, 26, 29, 37
simulation, x, 4, 45, 133
simulations, 79, 117, 132, 134
social network, 68
sociology, 50
software, 54, 74, 117, 119
soil pores, ix, 137
solid matrix, 146

solute transport properties, ix, 137, 142
solution, vii, x, 2, 3, 14, 18, 19, 23, 24, 25, 26, 27, 31, 36, 41, 44, 142
space-time, 88, 89, 107
specialists, 2, 24
species, ix, 128
specifications, 113
spectroscopy, 4, 84, 85, 88, 90, 91, 98, 99, 100
stability, 66, 80
stable states, 35
stakeholders, 70
standard deviation, 41, 54, 55, 56, 62, 87
statistics, 11, 12, 72, 73, 136
stimulus, 126, 133
stochastic analysis, vii, 1
stochastic fractals, 106
stock markets, x
storage, 59
stress, 17, 18, 54, 107
stretching, 50
structural changes, 126, 138
structure, vii, ix, 4, 5, 6, 8, 11, 12, 13, 14, 15, 16, 21, 23, 25, 37, 41, 47, 50, 53, 69, 81, 84, 96, 97, 98, 100, 105, 107, 113, 116, 119, 125, 126, 127, 129, 130, 134, 138, 139, 140, 141, 145, 147
substitution, 26
substrate(s), 103, 113, 116, 118, 119, 122, 123
supply chain, vii, viii, 49, 53, 65, 66, 70, 71, 72, 73, 77, 79, 80
surface energy, 4
surface tension, 2, 3, 4, 18, 19, 21, 22
symmetry, 12, 16, 105

T

target, 66, 98
techniques, ix, 125, 139, 142
technology, ix, 125, 129, 134
TEM, 65, 70
temperature, 19, 22, 34, 35, 52
tension, 2, 3, 4, 19, 21
texture, 138, 144
therapeutic targets, 127
therapy, 126, 133
thermal energy, 34
thermodynamics, 127
time series, 51, 52, 54, 55, 56, 57, 59, 60, 66, 67, 68, 69, 71, 72, 73, 74, 78, 79, 80, 87, 90, 92, 96, 97, 99
tissue, 132, 134, 136
trade agreement, 59
transformation, 37, 51, 110, 111, 112

transformations, 12, 51, 106, 111, 112
translation, 133
transport, ix, 137, 138, 139, 140, 141, 142, 143, 144, 146, 147
transport processes, 138
transportation, 59
traveling waves, 134
treatment, 31, 55, 126, 133, 140, 141
tumor(s), 126, 132, 133, 134
turbulence, 79
turtle, 108, 109, 115

U

ULF (Ultra-low-frequency), viii, 83, 84, 85, 86, 87, 90, 93, 97, 98, 99, 100
universe, 131

V

variables, 59, 126, 132, 133
variations, 66, 86, 89, 99, 133
vascular system, 131
vector, 5, 8, 9, 10, 11, 39, 117
velocity, ix, 8, 12, 30, 32, 33, 34, 137, 139, 141, 142, 145, 146
viscosity, 13, 32, 34, 37
volatility, vii, viii, 49, 53, 59, 60, 61, 62, 63, 64, 65, 66, 67, 68, 73, 78, 80

W

water, vii, ix, 1, 2, 3, 4, 5, 18, 21, 22, 23, 24, 28, 30, 41, 47, 137, 138, 139, 141, 142, 143, 144, 145, 146, 147
wavelet, 100
weather patterns, 59
wetting, 3, 79
windows, 54, 57, 58

X

X-ray computed tomography (CT), ix, 137, 138, 139, 140, 141, 142, 143, 144, 145, 146, 147, 148

Y

Yaounde, 149